THE ALKALOIDS

Chemistry and Pharmacology

Volume XXIII

THE ALKALOIDS
Chemistry and Pharmacology

Edited by
Arnold Brossi
National Institutes of Health
Bethesda, Maryland

VOLUME XXIII

1984

ACADEMIC PRESS, INC.
(Harcourt Brace Jovanovich, Publishers)
Orlando • San Diego • New York • London
Toronto • Montreal • Sydney • Tokyo

ACADEMIC PRESS, INC.
Orlando, Florida 32887

United Kingdom Edition published by
ACADEMIC PRESS, INC. (LONDON) LTD.
24/28 Oval Road, London NW1 7DX

Library of Congress Catalog Card Number: 50-5522

ISBN 0-12-469523-X

PRINTED IN THE UNITED STATES OF AMERICA

84 85 86 87 9 8 7 6 5 4 3 2 1

CONTENTS

Chapter 4. Constituents of Red Pepper Species: Chemistry, Biochemistry, Pharmacology, and Food Science of the Pungent Principle of *Capsicum* Species

T. SUZUKI AND K. IWAI

Chapter 5. Azafluoranthene and Tropoloisoquinoline Alkaloids

KEITH T. BUCK

Chapter 6. Muscarine Alkaloids

PEN-CHUNG WANG AND MADELEINE M. JOULLIÉ

CONTRIBUTORS

Numbers in parentheses indicate the pages on which the authors' contributions begin.

ARNOLD BROSSI (1), National Institutes of Health, Bethesda, Maryland 20205

KEITH T. BUCK (301), Fries & Fries Division, Mallinckrodt, Inc., Cincinnati, Ohio 45216

HANS-GEORG CAPRARO (1), CIBA GEIGY Co., 4000 Basel, Switzerland

LIANG HUANG (157), Institute of Materia Medica Chinese, Academy of Medical Sciences, Beijing, People's Republic of China

K. IWAI (227), Department of Food Science and Technology, Faculty of Agriculture, Kyoto University, Kyoto 606, Japan

MADELEINE M. JOULLIÉ (327), Chemistry Department, University of Pennsylvania, Philadelphia, Pennsylvania 19104

PAUL J. REIDER (71), Merck Sharp & Dohme Research Laboratories, Rahway, New Jersey 07065

DENNIS M. ROLAND (71), Research Department, Pharmaceuticals Division, CIBA-GEIGY Corporation, Ardsley, New York 10502

T. SUZUKI (227), The Research Institute for Food Science, Kyoto University, Kyoto 611, Japan

PEN-CHUNG WANG* (327), Chemistry Department, University of Pennsylvania, Philadelphia, Pennsylvania 19104

ZHI XUE (157), Institute of Materia Medica Chinese, Academy of Medical Sciences, Beijing, People's Republic of China

* Present address: Chemical Products Laboratory, Dow Chemical USA, Midland, Michigan 48640.

PREFACE

Volume XXIII presents updated reviews of pharmacologically important classes of alkaloids. The articles on "Tropolonic *Colchicum* Alkaloids," last reviewed in Vol. XI (1968), and "*Cephalotaxus* Alkaloids," briefly discussed in a chapter on "*Erythrina* and Related Alkaloids" in Vol. XVIII (1981), with updated literature until 1983, offer pertinent information on biochemical, pharmacological, and clinical experience gained with representative compounds.

The additional four articles discuss alkaloids never reviewed in this text. "Azafluoranthene and Tropoloisoquinoline Alkaloids" are thus far a small and little explored class of alkaloids with unknown physiological potential. The article on "Muscarine Alkaloids," on the other hand, compounds related to muscarine (obtained from the toxic mushroom *Amanita muscaria*, which interacts with acetylcholine receptors), deals with alkaloids of great importance to experimental pharmacology. "Maytansinoids" belong to a relatively new class of antitumor agents of complex chemistry and challenging structural features. The article on "Constituents of Red Pepper Species" describes the pungent principles of pepper species widely used throughout the world as spices in food. A detailed discussion of these substances seems highly justified.

Articles on alkaloidal substances of relevance to health, medicine, and drug research, written by experts in these fields, will continue to be the primary objective for review in this text.

Arnold Brossi

CONTENTS OF PREVIOUS VOLUMES

Contents of Volume I (1950)

edited by R. H. F. Manske and H. L. Holmes

Contents of Volume II (1952)

edited by R. H. F. Manske and H. L. Holmes

Contents of Volume III (1953)

edited by R. H. F. Manske and H. L. Holmes

Contents of Volume IV (1954)

edited by R. H. F. Manske and H. L. Holmes

Contents of Volume V (1955)

edited by R. H. F. Manske

Contents of Volume VI (1960)

edited by R. H. F. Manske

Contents of Volume VII (1960)

edited by R. H. F. Manske and H. L. Holmes

Contents of Volume VIII (1965)

edited by R. H. F. Manske and H. L. Holmes

Contents of Volume IX (1967)

edited by R. H. F. Manske and H. L. Holmes

Contents of Volume X (1967)

edited by R. H. F. Manske and H. L. Holmes

Contents of Volume XI (1968)

edited by R. H. F. Manske and H. L. Holmes

Contents of Volume XII (1970)

edited by R. H. F. Manske and H. L. Holmes

Contents of Volume XIII (1971)

edited by R. H. F. Manske and H. L. Holmes

Contents of Volume XIV (1973)

edited by R. H. F. Manske and H. L. Holmes

Contents of Volume XV (1975)

edited by R. H. F. Manske and H. L. Holmes

Contents of Volume XVI (1977)

edited by R. H. F. Manske and H. L. Holmes

Contents of Volume XVII (1979)

edited by R. H. F. Manske and H. L. Holmes

Contents of Volume XVIII (1981)

edited by R. H. F. Manske and R. G. A. Rodrigo

Contents of Volume XIX (1981)

edited by R. H. F. Manske and R. G. A. Rodrigo

Contents of Volume XX (1981)

edited by R. H. F. Manske and R. G. A. Rodrigo

Contents of Volume XXI (1983)

edited by Arnold Brossi

Contents of Volume XXII (1983)

edited by Arnold Brossi

—CHAPTER 1—

TROPOLONIC *COLCHICUM* ALKALOIDS*

HANS-GEORG CAPRARO

CIBA-GEIGY Co., Basel, Switzerland

AND

ARNOLD BROSSI

Arthritis Institute, National Institutes of Health
Bethesda, Maryland

I. Introduction

Pelletier and Caventou (*1*) isolated amorphous colchicine (**1**), thought to be veratrine, as one of the active principles from the meadow saffron

* Dedicated to Professor Dr. F. Šantavý from the Institute of Chemistry of the Medical Faculty at the Palacky University in Olomouc, Czechoslovakia, who passed away on Palm Sunday, March 27, 1983.

THE ALKALOIDS, VOL. XXIII

ISBN 0-12-469523-X

Colchicum autumnale, a medicinal plant that had been described in the classical work *De Materia Medica* by the Greek physician Dioscorides. Further work by Geiger (*2*), Oberlin (*3*), Zeisel (*4*), and Houdé (*5*) established the correct molecular composition of crystalline colchicine as $C_{22}H_{25}NO_6$ and its relationship to colchiceine (**2**), obtained after mild acid hydrolysis.

Studies of the modern chemistry of colchicine started soon after the discovery of its powerful action on cell division in plants and animals by Dustin's school. Authoritative works by Eigsti and Dustin (*6*, and references therein) and by Dustin (*6a*) regarding these findings have been published. The progress made afterward in the chemistry of colchicine and its biological implications was discussed in Volumes II (*7*), VI (*8*), and XI (*9*) of this treatise as well as by Šantavý (*10*), Manitto (*11*), Wildman (*12*), Döpke (*13*), Yusupov and Sadykov (*14*), Šantavý (*15,15a*), Gaignault (*16*), and Kiselev (*17*).

Whereas the earlier reviews discussed primarily the structure elucidation of colchicine and some of its minor alkaloids as well as its total synthesis and biological properties, this review primarily considers novel congeners, the application of modern physical methods to their structure determination, the chemistry of colchicine itself, and its conversion into minor alkaloids. Emphasis is given to the biological effects of colchicine, its mode of action, and the exploration of structural features which affect potency, binding to tubulin, and toxicity.

To keep the discussion concise, certain compounds—such as the lumicolchicines, photoisomers of colchicine (*11,12*); the allocolchicines, colchicines with a benzenoic ring C; the homomorphinanes; homoaporphines; and 1-phenethyl-1,2,3,4-tetrahydroisoquinolines that occur beside colchicine and congeners (*15, 15a*)—are not included. The nomenclature used throughout the article is derived from the generic names colchicine (**1**), colchiceine (**2**), and isocolchicine (**3**), an unnatural isomer of colchicine obtained by O-methylation of **2**.

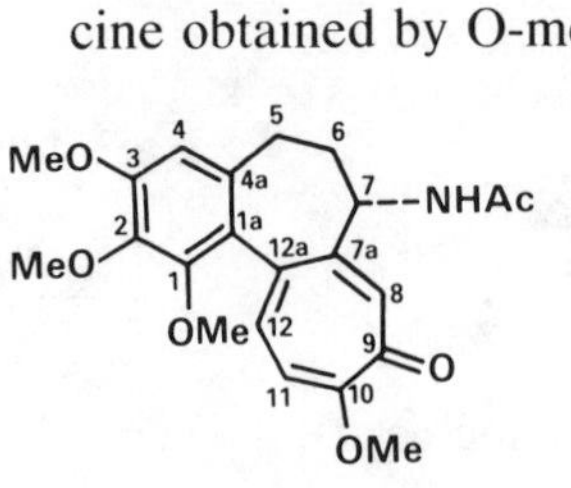

MeO 3 4 4a 5 6 7 NHAc 2 1a 12a 7a MeO 1 OMe 12 8 9 O 11 10 OMe

1

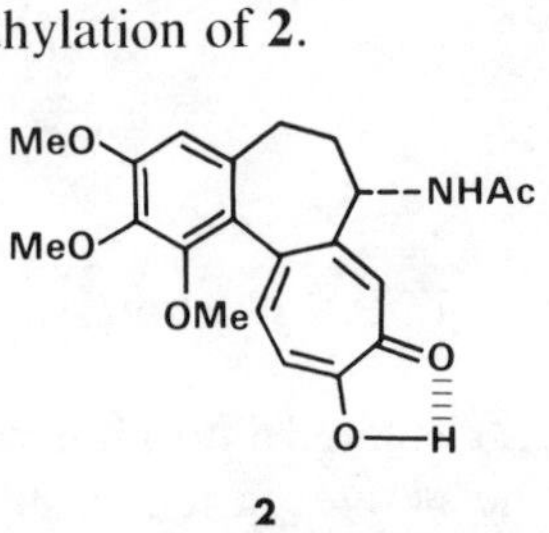

MeO MeO NHAc OMe O O—H

2

MeO MeO A B NHAc OMe C OMe O

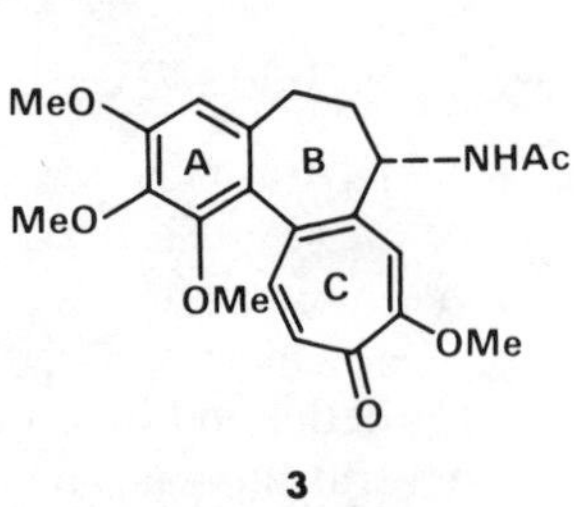

3

The complicated *Chemical Abstracts* nomenclature for colchicine as (−)-*N*-5,6,7,9-tetrahydro-1,2,3,10-tetramethoxy-9-oxobenzo(*a*)heptalene-7-yl(*S*)-acetamide can thus be avoided. Naturally occurring colchicine possesses one asymmetric center at C-7, and its absolute configuration shown in **1** was established by Corrodi and Hardegger (*18*) by ozonolysis to *N*-acetyl-L-glutamic acid. All the tropolonic alkaloids known so far have this absolute stereochemistry.

II. Occurrence in Nature

The presence of tropolonic *Colchicum* alkaloids is limited to plants. A search for chemotaxonomic purposes showed that only plants of the subfamily Wurmbaeiodeae and of the genus *Kreysigia* indigenous to Australia produce such alkaloids (*19, 19a*). The first part of this section deals with the isolation and analysis of tropolonic *Colchicum* alkaloids, whereas the second part summarizes the results obtained by microbial and metabolic degradation of colchicine and analog.

A. Plant Alkaloids

For more than 100 years colchicine was believed to be the only active principle of *Colchicum autumnale*, but a closer examination of this and other species resulted in the discovery of many new tropolonic alkaloids. Some of their structures are still unknown, but most of them are fully characterized and have been confirmed by partial synthesis. Most of the alkaloids were isolated before 1965 and are listed in Volume XI of this treatise (*9*). Plants investigated later (*19*) produced relatively few new tropolonic alkaloids. Different extraction, purification, and separation procedures used for their isolation and the advantages and disadvantages of the methods were critically discussed and simpler procedures elaborated (*20*). Methods of analysis for colchicine, described in several pharmacopeias, were evaluated, and appropriate modifications proposed (*21*). The presence of tropolonic alkaloids can be detected in the UV spectrum at 352 nm. For a qualitative analysis, paper and thin-layer chromatographic techniques were described, and separation of the alkaloids was easily accomplished by chromatography on alumina or silica gel (*20,22*). High performance liquid chromatography (HPLC) on normal or reversed phases was also used for the separation of *Colchicum* alkaloids (*23–27*) and was particularly useful in separating colchicine from its congeners, e.g., colchiceine (**3**) (*23a*). HPLC analysis of commercial samples of colchicine revealed the presence of minor alkaloids, one of which could be identified by mass spectrometry as the hitherto unknown alkaloid colchifoline (**16**) (*28*), prepared later by partial synthesis (*29*). These efforts resulted in the characterization of more than 30 naturally occurring *Colchicum* alkaloids that are listed in Table I.

TABLE I
TROPOLONIC *COLCHICUM* ALKALOIDS FROM PLANT MATERIAL

Structure	Formula (mol. wt.)	mp (°C) (solvent) (ref.)	$[\alpha]_D$ (conc.; solvent) (ref.)	Additional data
Colchicine 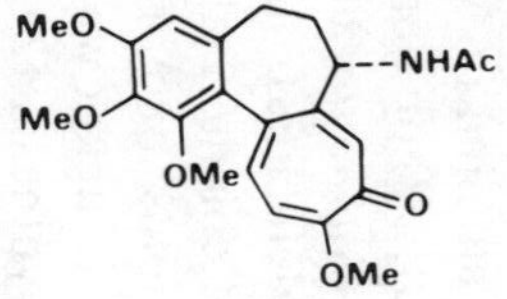 **1**	$C_{22}H_{25}NO_6$ (399.41)	155–157 (EtOAc)(*30,31*) 154–156 (EtOAc–Et_2O) (*32*) 152–155 (EtOH) (*33*)	−121° (0.879; $CHCl_3$) (*30–32*) −429° (1.771; H_2O) (*30*)	IR: *33* UV: *32,34* MS: *35,36* ^{1}H NMR: *34,35,37*[a] ^{13}C NMR: *38,39*
A ring different from colchicine				
Cornigerine 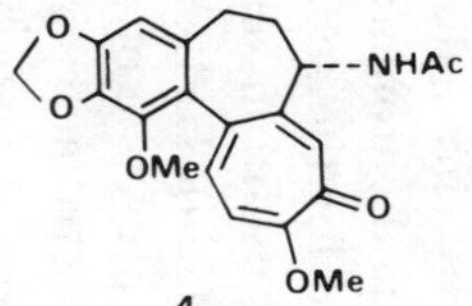 **4**	$C_{21}H_{21}NO_6$ (383.37)	268–270 (EtOAc–Et_2O) (*40*) 263–264 (EtOAc–Et_2O); dec.) (*41*)	−150° (0.631; $CHCl_3$) (*40,41*) −233° (0.728; MeOH) (*40*)	IR: *40,41* UV: *40* MS: *42* ^{1}H NMR: *41,42*
2-Demethylcolchicine[b] 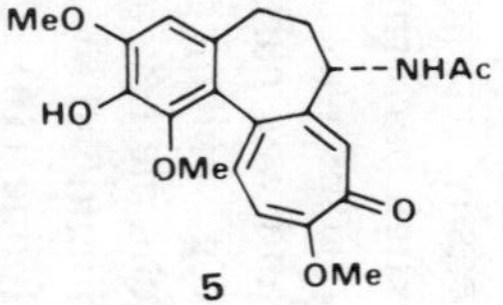 **5**	$C_{21}H_{23}NO_6$ (385.39)	110–140 and 178–180 ($CHCl_3$)[c] (*43,44*) 190–200 ($CHCl_3$; solvate) (*45*) 172–180 ($CHCl_3$; solvate) (*46*)	−133° (1.02; $CHCl_3$) (*43*) −137° (1.0; $CHCl_3$) (*45*)	IR: *45,47* UV: *32,34,35,43,45,48* MS: *35* ^{1}H NMR: *34,35,37*[a]*,45,47* ^{13}C NMR: *47*

3-Demethylcolchicine[b] 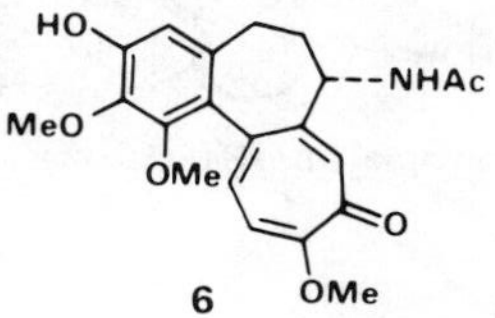 (6)	$C_{21}H_{23}NO_6$ (385.39)	176–182 ($CHCl_3$)(*32*) 176–179 (Acetone) (*49*) 178-180 (CH_2Cl_2–Et_2O) (*50*) 180–190 and 275–280 (Acetone; dec.)[c] (*51*) 171–177 and 270[c] (Acetone; dec.) (*46*) 276–278 (Acetone) (*52*)	−128° (1.0; $CHCl_3$) (*32,51*) −132° (0.85; $CHCl_3$) (*52*) −151° (1.0; $CHCl_3$) (*46*) −231° (1.0; MeOH) (*46,51*) −355° (1.0; H_2O) (*51*)	IR: *47* UV: *32,34,35,48* MS: *35* ^{1}H NMR: *34,35,37,*[a]*47,50* ^{13}C NMR: *47*
Colchicoside 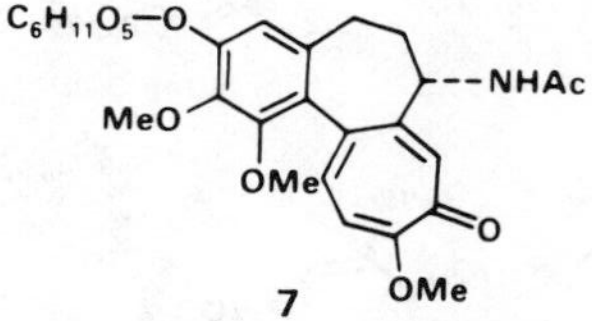 (7)	$C_{27}H_{33}NO_{11}$ (547.52)	216–218 (EtOH) (*53*) 193–195 (*54*)	−360° (1.0; H_2O) (*53*)	UV: *53*
2,3-Didemethylcolchicine[d] 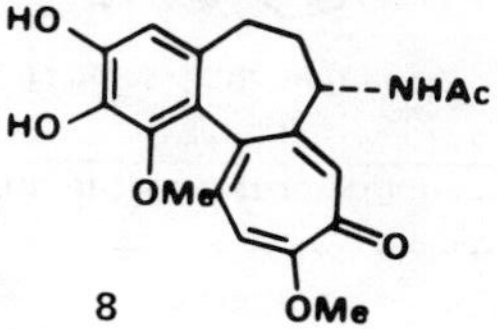 (8)	$C_{20}H_{21}NO_6$ (371.37)	Amorphous (*41,48*)		UV: *48* MS: *41* ^{1}H-NMR: *41,48*

(*Continued*)

TABLE I (*Continued*)

Structure	Formula (mol. wt.)	mp (°C) (solvent) (ref.)	$[\alpha]_D$ (conc.; solvent) (ref.)	Additional data
B ring different from colchicine				
N-Deacetylcolchicine 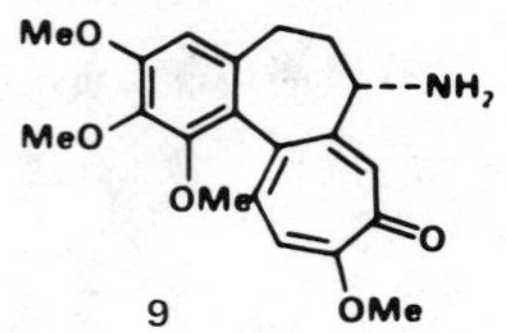 **9**	$C_{20}H_{23}NO_5$ (357.38)	125–127 (EtOAc) (*54a*)	−148° (1.1; $CHCl_3$) (*54a*) −152° (1.17; $CHCl_3$) (*56*)	IR: *56* UV: *56* MS: *37*[a],*56* ^{1}H NMR: *55,56* ^{13}C NMR: *55*
Demecolcine 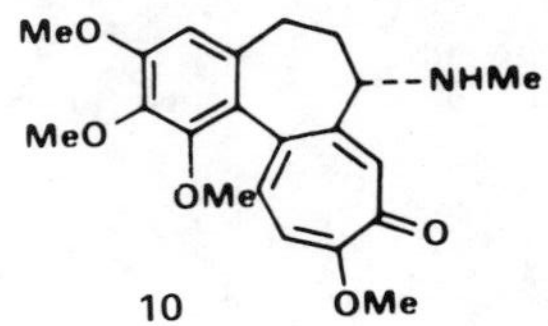 **10**	$C_{21}H_{25}NO_5$ (371.40)	184–186 (EtOAc or acetone) (*43,57,58*) 183–185 (EtOAc–Et_2O) (*56*)	−123° (0.99; $CHCl_3$) (*56*) −126° (1.03; $CHCl_3$) (*43*) −127° (0.99; $CHCl_3$) (*57,58*) −130° (0.978; $CHCl_3$) (*59*)	IR: *56* UV: *56,60* MS: *36,56* ^{1}H NMR: *55,56* ^{13}C NMR: *55*
N-Deacetyl-*N*-formylcolchicine 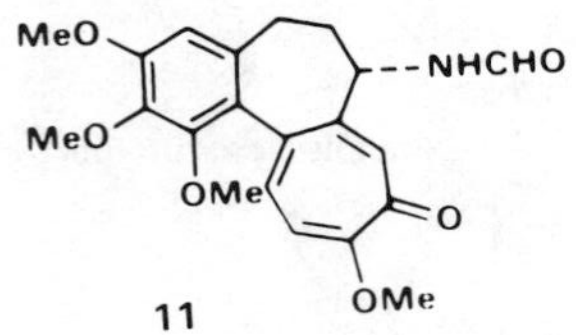 **11**	$C_{21}H_{23}NO_6$ (385.39)	264–267 (EtOAc–Et_2O; dec.) (*32,61*) 267 (MeOH) (*33*)	−171° (1.08; $CHCl_3$) (*32*) −173° (1.14; $CHCl_3$) (*61*)	IR: *33* UV: *32–34* MS: *36* ^{1}H NMR: *34,37*[a]

N-Methyldemecolcine MeO, MeO, OMe, --NMe$_2$, O, OMe **12**	$C_{22}H_{27}NO_5$ (385.45)	205–208 (EtOAc–Et_2O) (*56,60*)	−104° (1.08; $CHCl_3$) (*60*) −111° (1.78; $CHCl_3$) (*56*)	IR: *56,59* UV: *56,59,60* MS: *55,56* ^{1}H NMR: *55,56* ^{13}C NMR: *55*
N-Formyldemecolcine MeO, MeO, OMe, --NMeCHO, O, OMe **13**	$C_{22}H_{25}NO_6$ (399.41)	187–189 (MeOH–Et_2O) (*62–64*)	−202° (0.79; EtOH) (*63,64*) −189° (1.02; $CHCl_3$) (*62*)	IR: *64*
Alkaloid CC 12[e] OH, MeO, MeO, OMe, --NHAc, O, OMe **14**	$C_{22}H_{25}NO_7$ (415.44)	197–199 (Acetone) (*59*)	−83° (0.51; MeOH) (*59*) −45° (0.24; $CHCl_3$) (*65*)	IR: *59* UV: *59* ^{1}H NMR: *65*
Colchiciline[e] OH, MeO, MeO, OMe, --NHAc, O, OMe **15**	$C_{22}H_{25}NO_7$ (415.44)	170–171 (EtOAc; dec.) (*28,48,66*)	−121° (1.93; $CHCl_3$) (*48,66*)	IR: *66* UV: *66* MS: *66* ^{1}H NMR: *28,66* ^{13}C NMR: *66*

(*Continued*)

TABLE I (*Continued*)

Structure	Formula (mol. wt.)	mp (°C) (solvent) (ref.)	$[\alpha]_D$ (conc.; solvent) (ref.)	Additional data
Colchifoline 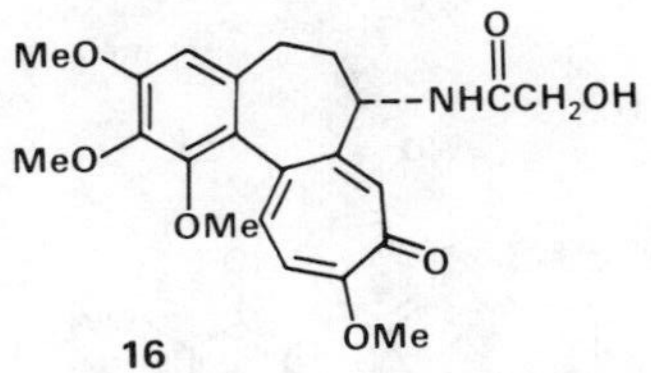 MeO, MeO, OMe, NHCCH$_2$OH, O, O, OMe **16**	$C_{22}H_{25}NO_7$ (415.44)	151–152 (EtOAc) (*29*)	−149° (0.49; $CHCl_3$) (*29*)	MS: *28* ^{1}H NMR: *29*
N-Methylcolchicine 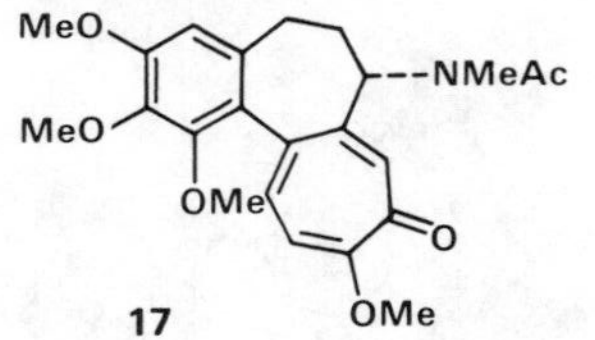 MeO, MeO, OMe, NMeAc, O, OMe **17**	$C_{23}H_{27}NO_6$ (413.44)	230–232 (EtOAc) (*59,62,67*) 199–200 and 233–235 (EtOAc–Et_2O)[c] (*55*)	−240° (1.91; $CHCl_3$) (*59,62,67*) −251° (1.17; $CHCl_3$) (*55*)	IR: *55* UV: *55* MS: *55* ^{1}H NMR: *55* ^{13}C NMR: *55*
Speciosine 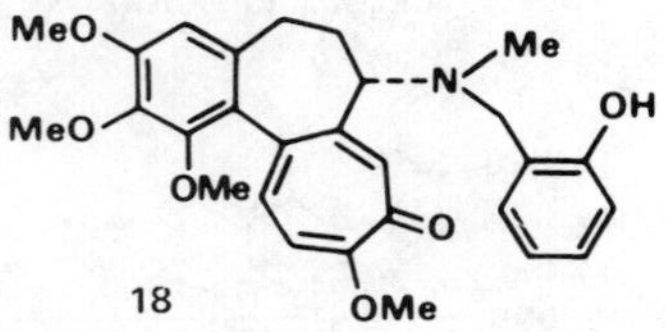 MeO, MeO, OMe, N, Me, OH, O, OMe 18	$C_{28}H_{31}NO_6$ (477.52)	211–214 (EtOAc–benzene) (*68*) 209–211 (Acetone) (*69*)	−22° ($CHCl_3$) (*68,69*)	IR: *68* UV: *68* MS: *68* ^{1}H NMR: *68*

C Ring different from colchicine				
Colchiceine 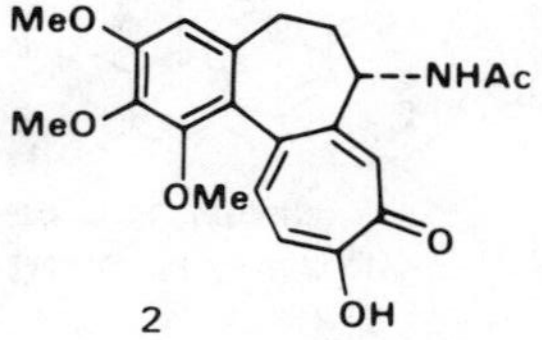 (2)	$C_{21}H_{23}NO_6$ (385.39)	175–177 (EtOAc) (*70*)	−256° (1.00; $CHCl_3$) (*70*)	UV: *35* MS: *35* ^{1}H NMR: *35,65* ^{13}C NMR: *39,47*
10,11-Oxy-10,12a-cyclo-10,11-secocolchicine 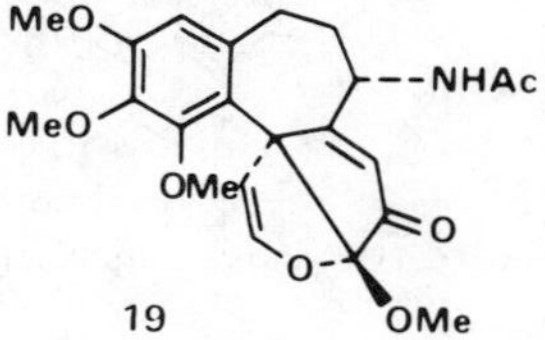 (19)	$C_{22}H_{25}NO_7$ (415.44)	251–253 (EtOAc) (*48,66*) 251 (*i*-Isopropanol; dec.) (*71*)	−237° (1.68; $CHCl_3$) (*48,66*) −211° (1.0; $CHCl_3$) (*71*)	IR: *66* UV: *66,71* MS: *66* ^{1}H NMR: *66,71* ^{13}C NMR: *71*
Dithiocolchicine[f] (structure unknown)	$C_{22}H_{25}NO_4S_2$ (431.54)	265–267 (MeOH; dec.) (*72*)	−366° (1.00; $CHCl_3$) (*72*)	UV: *72*
A and B Ring different from colchicine				
2-Demethyldemecolcine 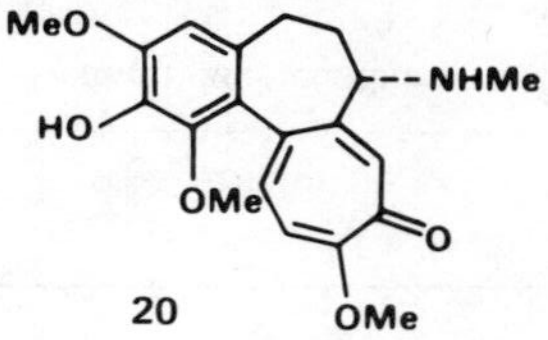 (20)	$C_{20}H_{23}NO_5$ (357.38)	136–138 (MeOH–Et_2O) (*73,74*) 137–139 (MeOH) (*44*) 131 (MeOH) (*75*)	−109° (1.00; $CHCl_3$) (*75*) −119° (1.00; $CHCl_3$) (*73,74*)	UV: *48,73* MS: *76* ^{1}H NMR: *76*

(*Continued*)

TABLE I (*Continued*)

Structure	Formula (mol. wt.)	mp (°C) (solvent) (ref.)	$[\alpha]_D$ (conc.; solvent) (ref.)	Additional data
3-Demethyldemecolcine 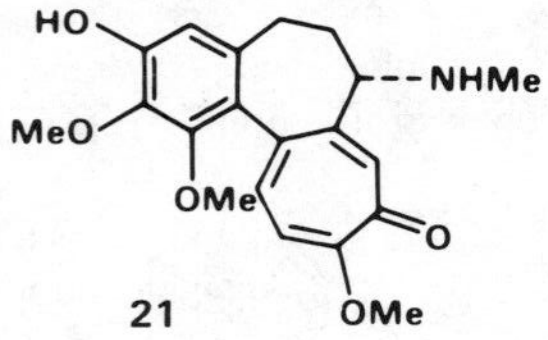 HO, MeO, OMe, OMe, O, NHMe **21**	$C_{20}H_{23}NO_5$ (357.38)	220–222 (*59*)	−128° (0.89; $CHCl_3$) (*59*)	IR: *59* UV: *59*
2-Demethyl-*N*-deacetyl-*N*-formylcolchicine[g] 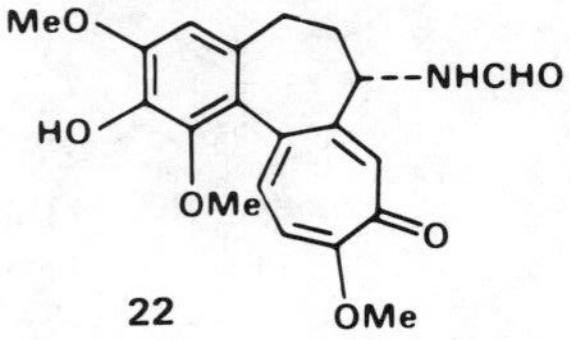 MeO, HO, OMe, OMe, O, NHCHO **22**	$C_{20}H_{21}NO_6$ (371.36)	Amorphous (*44,77*) Acetate: 237–239 (*44*)		UV: *48* ^1H NMR: *44*
3-Demethyl-*N*-deacetyl-*N*-formylcolchicine 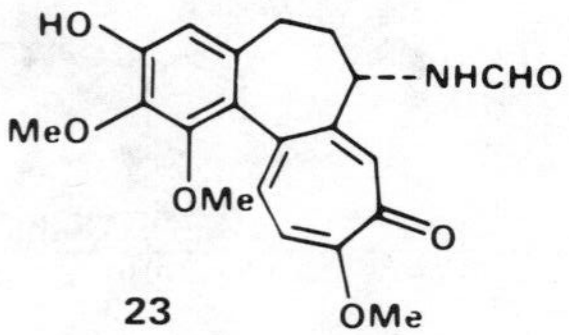 HO, MeO, OMe, OMe, O, NHCHO **23**	$C_{20}H_{21}NO_6$ (371.36)	263–267 (EtOAc) (*48,66*) 230–231 (Methyl ethyl-ketone) (*34*) 183–184 and 229–230 (MeOH)[c] (*34*)	−180° (0.30; $CHCl_3$) (*48,66*) −210° (0.41; $CHCl_3$) (*34*)	IR: *34* UV: *34,48* MS: *66* ^1H NMR: *34,37*[a]

2-Demethylcolchifoline 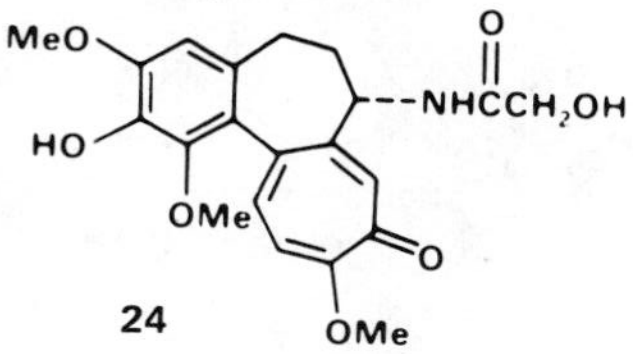 **24**	$C_{21}H_{23}NO_7$ (401.39)	Amorphous		
	Diacetate: $C_{25}H_{27}NO_9$ (485.46)	227–229 (EtOAc) (*44*) 229–232 (EtOAc) (*78*)	−106° (0.68; $CHCl_3$) (*44*) −109° (0.68; $CHCl_3$) (*78*)	IR: *78* MS: *78* ^1H NMR: *78* ^{13}C NMR: *78*
3-Demethyl-*N*-deacetyl-colchicine 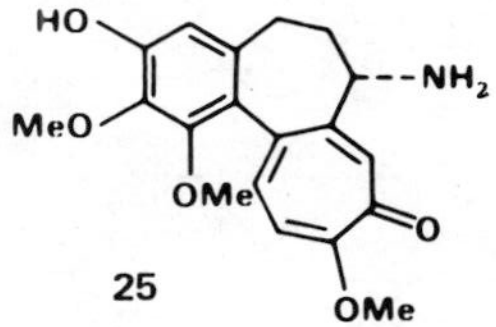 **25**	$C_{19}H_{21}NO_5$ (343.35)			
	N,O-Diacetate: $C_{23}H_{25}NO_7$ (427.42)	224–226 (EtOAc–Et_2O) (*73*)	−93° (0.82; $CHCl_3$) (*73*)	
2,3-Didemethyl-*N*-deacetylcolchicine 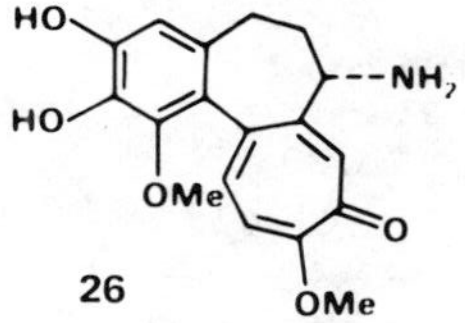 **26**	$C_{18}H_{19}NO_5$ (329.33)	amorphous (*77*)		UV: *48*

(*Continued*)

TABLE I (*Continued*)

Structure	Formula (mol. wt.)	mp (°C) (solvent) (ref.)	$[\alpha]_D$ (conc.; solvent) (ref.)	Additional data
2-Demethylcolchiciline[h] 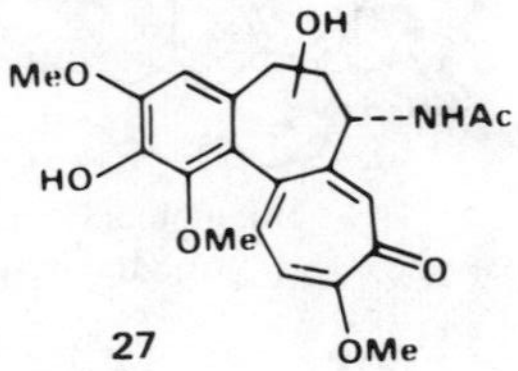 **27**	$C_{19}H_{22}NO_7$ (376.36)			
A and C Ring different from colchicine				
2-Demethylcolchiceine[g] 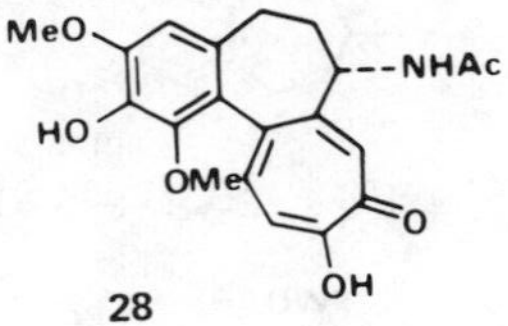 **28**	$C_{20}H_{21}NO_6$ (371.37)	(*77*)		UV: *48*
3-Demethylcolchiceine[g] 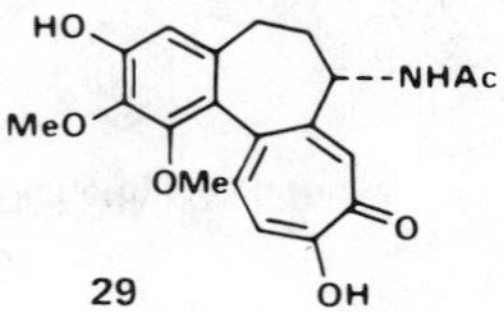 **29**	$C_{20}H_{21}NO_6$ (371.37)	179–183 (amorphous) (*52*)		IR: *52* UV: *48,52* ^{1}H NMR: *52*
	Diacetate: $C_{24}H_{25}NO_8$ (417.39)	119–122 (EtOAc–Et_2O) (*52*)	−170° (0.64; $CHCl_3$) (*52*)	IR: *52* UV: 52

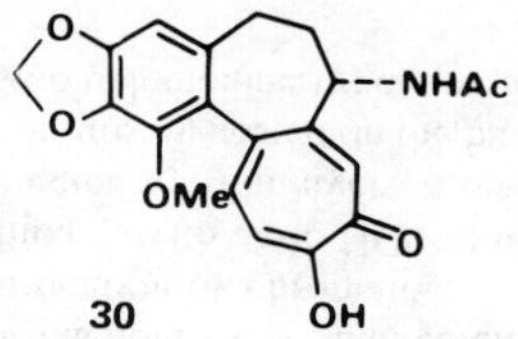 Cornigereine **30** (NHAc, OMe, OH)	$C_{20}H_{19}NO_6$ (369.35)	168–170 (EtOAc or acetone) (*42*)	−222° (0.76; $CHCl_3$) (*42*) −245° (0.70; MeOH) (*42*)	IR: *79* ^1H NMR: *79*
B and C Ring different from colchicine				
N-Deacetylcolchiceine 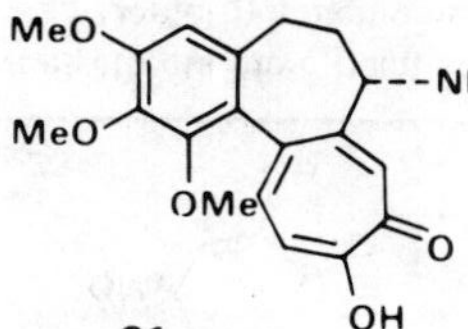 **31** (MeO, MeO, OMe, NH₂, OH)	$C_{19}H_{21}NO_5$ (343.35)	151–153 (*80*) 155–157 (*81*)	−152° (*80*) −184° (1.0; $CHCl_3$) (*81*)	UV: *80* ^{13}C NMR: *39,47*
Demecolceine 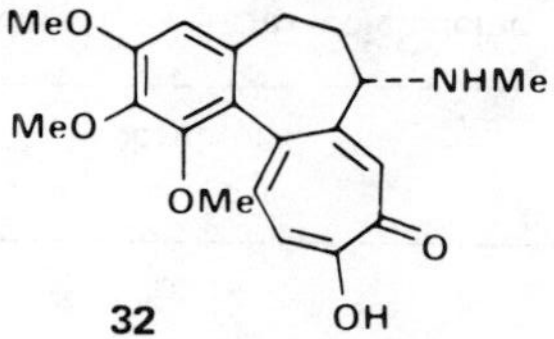 **32** (MeO, MeO, OMe, NHMe, OH)	$C_{20}H_{23}NO_5$ (357.38)	133–135 (MeOH) (*74,82*) 132–134 (MeOH–acetone) (*58*) 130–134 (MeOH) (*56*)	−220° (1.21; $CHCl_3$) (*74*) −211° (0.72; $CHCl_3$) (*82*) −206° (0.62; $CHCl_3$) (*58*) −99° (1.37; EtOH) (*74*) −228° (1.24; $CHCl_3$) (*56*)	IR: *56,74* UV: *56,82* MS: *56* ^1H NMR: *56* ^{13}C NMR: *55*

(Continued)

TABLE I (*Continued*)

Structure	Formula (mol. wt.)	mp (°C) (solvent) (ref.)	$[\alpha]_D$ (conc.; solvent) (ref.)	Additional data
N-Deacetyl-*N*-formylcolchiceine MeO MeO OMe NHCHO O OH **33**	$C_{20}H_{21}NO_6$ (371.38)	149–152 (EtOAc–Et_2O) (*32*) 150–152 (H_2O–MeOH) (*32*) 150–152 and 172–215 (H_2O)[c] (*32*)	−280° (1.15; $CHCl_3$) (*32*)	

[a] Solvent effects have been studied.
[b] 2- and 3-Demethylcolchicines do not occur in fresh plant extracts. They are present only in dried material, attached in an unknown manner to cell constituents and liberated enzymatically (*44*).
[c] Double mp are frequently seen in colchicines; the first mp usually indicates a loss of solvent.
[d] Compound only sparingly soluble in commonly used solvents (*54a*).
[e] For alkaloid CC-12 and colchiciline see discussion in section III,D,1.
[f] Purification by chromatography on alumina afforded a benzene–ether eluate which did not show a carbonyl group and gave no reaction by refluxing with conc HCl. Material was never isolated again and is possibly an artifact.
[g] Detected by thin layer chromatography (TLC) and isolated in amorphous form only (*77*).
[h] Structure suggested on the basis of the color which developed on TLC after spraying with different reagents and in the UV-light. Substances of the colchiciline type give intense violet spots after spraying with iodoplatinate (*48*).

B. Microbial and Metabolic Transformation Products

Few reports are available on the microbial and metabolic degradation of colchicine and its congeners. In attempting to biodegrade colchicine (**1**) and thiocolchicine (**34**) microbially, Velluz and Bellet in France (*83,83a*)

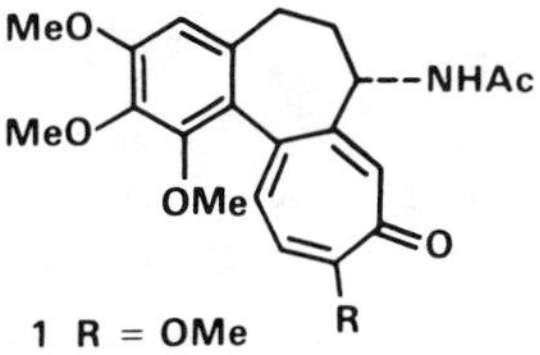

1 R = OMe
34 R = SMe

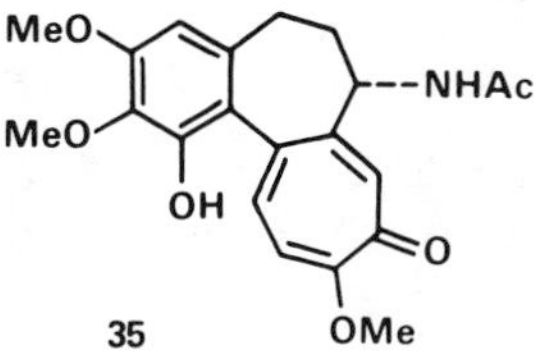

35

used a culture of *Streptomyces griseus*: 850 mg of colchicine was inoculated in 4 liters of a *S. griseus* culture and left for incubation for 48 hr at room temperature. Extraction with chloroform after chromatography on alumina afforded beside starting material 160 mg of a demethylated product of mp 185°C (dec.) and $[\alpha]_D$ −195° (MeOH) and −164° ($CHCl_3$). This material was not identical to either 2-demethylcolchicine (**5**) or the 3-demethyl analog (**6**), nor to 1-demethylcolchicine (**35**) prepared later (*45*). Its UV spectrum, however, indicated the presence of a tropolonic ring C, and O-methylation afforded colchicine. In a similar way 920 mg of thiocolchicine (**34**) afforded, after incubation with *S. griseus* and work-up, 175 mg of a demethylated thiocolchicine, mp 182–184°C and $[\alpha]_D$ −281° (MeOH) (*84*). Earlier attempts to repeat these experiments were unsuccessful (*85*), but similar investigations carried out with colchicine afforded 2-demethylcolchicine (**5**) in addition to 3-demethylcolchicine (**6**), both of which were characterized (*47*).

A total of 63 microorganisms screened for their ability to metabolize colchicine were investigated (*47*), but only *Streptomyces griseus* and *S. spectabilis* were able to biotransform colchicine. The overall yield with *S. griseus* was lower than that obtained with *S. spectabilis*. 2-Demethylcolchicine (**5**) was the major product using *S. griseus*, whereas 3-demethylcolchicine (**6**) was the major product when *S. spectabilis* was employed. *N*-Methylcolchiceinamide (**36**) was completely metabolized by *S. griseus* to three products (a total of 77 different microorganisms were examined) (*86*). On a preparative scale (200 mg of **36**), the predominant conversion was *N*-dealkylation of **36** to colchiceinamide (**37**, 130 mg, 65%), identified by spectral data and by comparison with an authentic sample. In addition to **37**, minor amounts of 2-demethyl-*N*-methylcolchiceinamide (**38**) and its 3-demethylated congener **39** were isolated and characterized. In light of these results it can not be ruled out that the microbial biodegradation of thiocolchicine (**34**) may have similarly afforded a mixture of 2- and 3-

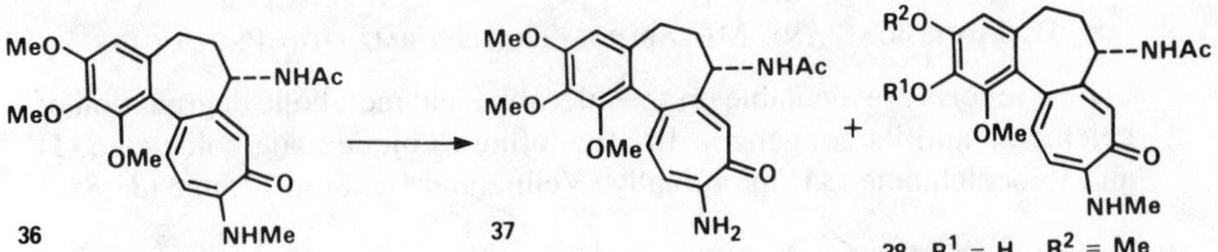

MeO MeO OMe NHAc O NHMe 36 → MeO MeO OMe NHAc O NH2 37 + R2O R1O OMe NHAc O NHMe
38 R^1 = H R^2 = Me
39 R^1 = Me R^2 = H

demethylthiocolchicine (*83,84*). Formation of homogenous crystals from mixtures of compounds is a well-known phenomenon among colchicine alkaloids (*87*).

Arthrobacter colchovorum, a soil bacterium that grows close to *Colchicum autumnale,* was found to biodegrade colchicine to *N*-deacetylcolchicine (**9**) and 7-oxodeacetamidocolchicine (**40**) depending on experimental

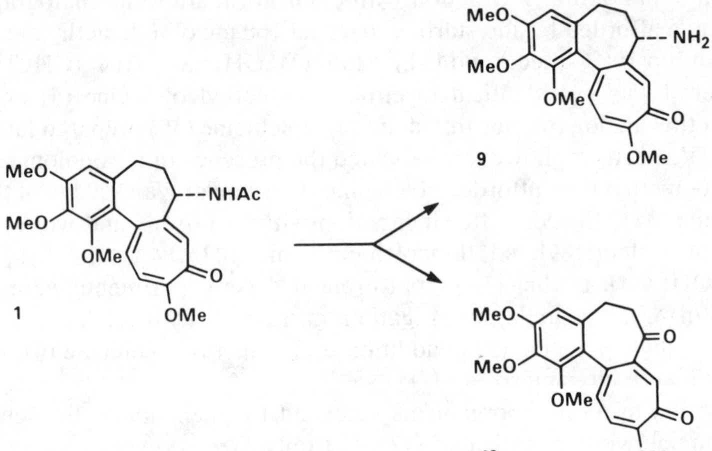

MeO MeO OMe NHAc O OMe 1 → MeO MeO OMe NH2 O OMe 9 ; MeO MeO OMe O O OMe 40

conditions (*85*). When grown in the presence of inorganic nitrogen, *A. colchovorum* utilized colchicine as the carbon source and afforded *N*-deacetylcolchicine (**9**) as the major reaction product and traces of **40**. When glucose was added to the culture medium with inorganic nitrogen absent, compound **40** was the major degradation product, and in the absence of both nutrients, **40** was the only metabolite. The diketone (**40**) was first synthesized at Roussel-UCLAF in France in 1963 (*82a*) and later obtained elsewhere by a quite different route (*88*) (see Section III,D,1).

To investigate the metabolic transformation of colchicine in mammals, a modified Udenfriend redox system, known to simulate microsomal reactions, was used (*35*). Under the conditions summarized in Table II, colchicine was converted to four major products (A–D) that were present

TABLE II
METABOLIC DEGRADATION OF COLCHICINE[a]

Udenfriend systems	Compound (%)			
	A = **5**	B = **6**	C = **2**	D = **41**
Conventional				
Addition of 56 m*M* ascorbate, 5.2 m*M* Fe^{3+}, 26 m*M* EDTA	7	5	12	4
Modified				
Addition of 40 m*M* and 8 m*M* Cu^{2+}	14	11	4	6

[a] Incubation system: 2 m*M* colchicine in 0.1 *M* phosphate buffer, pH 6.8, atmospheric O_2, 1.5 hr at 37°C.

beside starting material and minor amounts of secondary products. The products A–C were isolated by chromatographic techniques, and their structures were determined by spectral data as 2-demethylcolchicine (**5**), 3-demethylcolchicine (**6**), and colchiceine (**2**), respectively. Detection of radioactive formaldehyde after incubation with [^{14}C]colchicine (10-OMe) was sound evidence for an oxidative removal of the methyl group at 10-OMe in the formation of colchiceine (**2**). On the basis of spectral data and chemical behavior, product D, named colchinal, was believed to have the allocolchicine structure **41**, but the exact location of the CHO and OH groups in ring C remains undetermined.

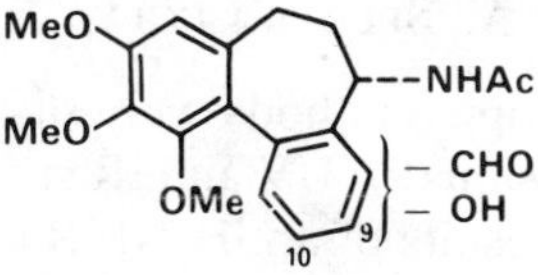

41

2- and 3-Demethylcolchicine and colchiceine, but not colchinal, were metabolites isolated through incubation of colchicine with liver microsomes from rats, mice, and Syrian gold hamsters (*89*). Both NADPH and atmospheric oxygen were required for the formation of the metabolites, indicating that oxidative O-demethylation had occurred. Hamsters were chosen since they are known to be highly resistant to colchicine. The turnover of colchicine with hamster liver microsomes, which converted about 40% of the colchicine into metabolites, was four times higher than that observed with microsomes from rats or mice. Furthermore, the hamster liver microsomes did not produce colchiceine, and 3-demethylcolchicine was obtained in glucuronated form; rat liver microsomes produced the glucuronide of 2-demethylcolchicine. The monodemethylated derivatives of colchicine were not further metabolized and seem to be the final products of microsomal metabolism *in vitro*. A possible participation of these metabolites in the biological effects observed after colchicine administration was discussed (*89*).

A number of colchiceinamides (**42**, R = alkyl, ω-OH-alkyl, and ω-NH_2-alkyl), were also subjected to microsomal degradation (*90*). It was found that the oxidative attack occurred at the side chain, as well as at the 3-OMe group. Hydroxylation at the ω position of the aliphatic side chain at C-10 was the preferred route of metabolism when R was larger than Me, but the amide **37** became the major metabolite when R = Me. A phenolic group in ring A of colchiceinamides often prevented oxidation of the side chain.

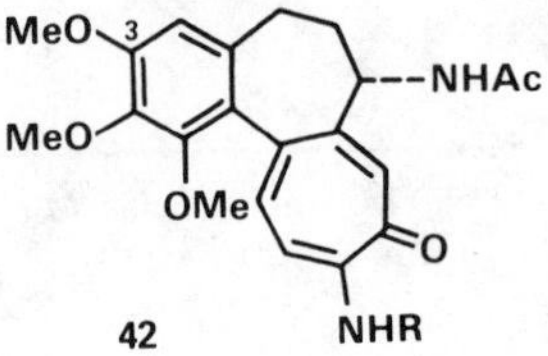

III. Chemistry of *Colchicum* Alkaloids

A. Spectroscopy

The following spectroscopic methods were discussed: optical rotation and optical rotatory dispersion (*9*), UV and IR spectroscopy (*79*), and MS spectrometry (*9,91*). The results of an ^{1}H-NMR analysis of more than 50 colchicine derivatives (*79,92*) were reviewed in Volume XI of this treatise (*9*). However, the proton assignment for H—C-11 and H—C-12 in isocolchicines remained an unresolved problem. In selective single-frequency off-resonance decoupling ^{13}C-NMR experiments, the carbon signal near 141 ppm in isocolchicine assigned to C-12 was recognized to hold the proton corresponding to the lowest field part of the AB quartet in the ^{1}H-NMR spectrum, suggesting that in the normal as well as in the iso series H-12 always appears further down field than H-11 (*55*).

^{13}C-NMR spectroscopy has become routine in the structural elucidation of natural products, and colchicine and many of its congeners have been analyzed (*38,39,47,55,93,94*). Modern spectroscopic techniques, such as ^{1}H noise decoupling, ^{1}H single-frequency selective decoupling (SFSD), and ^{1}H single-frequency off-resonance decoupling (SFORD), has allowed the assignment of all 22 carbon signals in colchicine. Interpretation of the ^{13}C-NMR chemical shifts of colchicine was first proposed by Singh *et al.* in 1977 (*93*). However, a more detailed analysis required reassignment of some signals (*38,47,94*).

Based upon a detailed analysis of colchicine, the carbon atoms of more than 20 colchicines and isocolchicines have now been correctly assigned. Table III lists the ^{13}C-NMR values of several colchicines and isocolchicines.

B. X-Ray Structure Determinations

The first crystallographic analyses of tropolonic colchicines were reported for a Cu salt of colchiceine (*95*) and for complexes of colchicine with methylene bromide and methylene iodide (*96*). They provided confirmation of the molecular structure of colchicine as proposed by Dewar and supported the reported chemistry. X-ray analyses of colchicine, colchiceine, and congeners, which allow predictions of structural features important for biological activity from solid state conformations, were also reported (*97*).

Colchiceine (**2**), prepared by standard procedures (*7*) and crystallized from ethyl acetate, showed an unusual ethyl acetate–water solvated structure (*98*) that formed a three-dimensional network of strongly hydrogen-bonded complexes including water molecules. Ring B essentially had a boat conformation with some of the torsion angles flattened because of the attached benzene and tropolone rings. The X-ray structure clearly indicated that the carbonyl function is at C-9. Because O-methylation of colchiceine (**2**) afforded a mixture of colchicine and isocolchicine, it was assumed that a chemical equilibrium exists between the two forms in solution. The three aromatic methoxy groups have different conformations, also seen with other colchicine congeners, and this, in fact, may play an important role in the biological activity noted with some of these compounds, but rotation in solution is possible. Brossi, with Sharma and Silverton at the National Institutes of Health (NIH), have investigated the biologically active 3-demethylcolchicine (**6**) and its rather inactive 1-demethyl isomer (**35**) by single crystal X-ray analysis. (Figs. 1 and 2). This study revealed that both phenolic isomers were intermolecularly hydrogen bonded and that this interaction may well influence the biological effects.

The space groups of both compounds are $P2_12_12_1$, but the cell dimensions are quite different, being 9.054(4), 11.910(3), and 17.787(5) Å for **35** and 8.0299(9), 9.7000(12), and 30.710(3) Å, for **6**. The crystals of **35** are

TABLE III

^{13}C-NMR DATA OF COLCHICINE AND ITS DERIVATIVES[a,b]

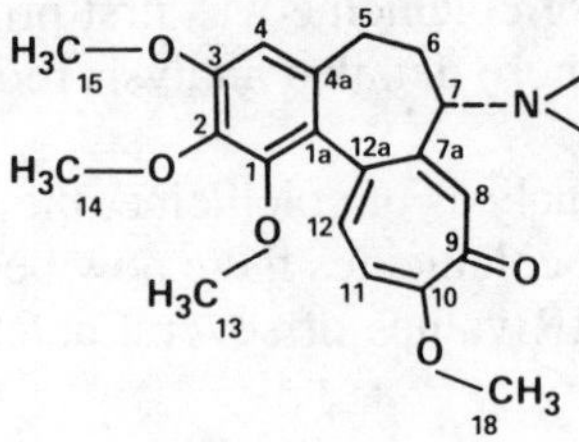

1

C Atom	Colchicine (1)[c]	Isocolchicine (3)[d]	Colchiceine (2)[c]	N-Deacetyl-colchiceine (31)[c]	2-Demethyl-colchicine (5)[c]
1a	126.0	125.9	126.1	126.0	125.5
1	151.4	150.8	153.9*	153.7*	147.8*
2	142.2	141.5	141.8	141.6	138.1
3	153.8	153.7	151.1*	153.3*	145.0*
4	107.9	107.6	107.7	107.3	107.0
4a	134.4	134.9	134.6	135.7†	129.9
5	30.1	30.1	29.9	30.6	29.8
6	36.6[f]	38.0	37.6	41.8	36.9
7	52.8[f]	52.9	52.9	53.7	52.8
7a	152.6	145.6	151.7	150.6	152.2
8	130.7	133.4	119.5	118.6	130.7
9	179.6	163.7	170.1†	172.2‡	179.7
10	164.3	179.3	170.2†	168.7‡	164.3
11	113.1	111.2	122.5	123.8	112.8
12	134.5	141.7	141.6	141.6	135.1
12a	137.2	134.9	136.5	135.6†	136.9
13	61.3*[g]	61.3	61.3‡	61.1	61.3
14	61.5*[g]	61.3	61.5‡	60.9	—
15	56.3†[g]	56.1	56.3	56.2	56.3
16	170.0	170.3	170.5†	—	170.0
17	22.7	22.6	22.8	—	22.9
18	56.5†	56.1	—	—	56.5

[a] For structures see Table I.

[b] Assignments bearing the same indices (*,†,‡) may be reversed.

[c] $CDCl_3$; values taken from (*47*), in agreement with those reported in (*38,39*). In (*47*), no made possible by comparing these signals with those of 1-demethyl-1-acetylcolchi-

[d] $CDCL_3$; values taken from (*38*).

[e] $CDCL_3$; values taken from (*55*).

[f] Assignments based on (i) deshielding β effect and shielding γ effects on the NH_2 colchicines, e.g., 4-formylcolchicine (*94*); and (iii) on a comparison with 7-oxo-7-

[g] Methoxy groups with diortho substituents, like C-13 and C-14, appear at lower field appear at higher field (55–57 ppm) (*47*).

[h] N—CH_3.

MeO
MeO
OMe
NHR
OMe
O

43 R = Me
44 R = H

3-Demethyl-colchicine (**6**)[c]	Demecolcine (**10**)[e]	*N*-Deacetyl-colchicine (**9**)[e]	Isodemecolcine (**43**)[e]	*N*-Deacetyliso-colchicine (**44**)[e]
125.1	126.0	125.9	126.0	125.9
150.3*	150.6	150.9	150.7	150.6
139.3	141.6	141.6	141.2	141.8
149.9*	153.5	153.6	153.6	153.7
110.3	107.5	107.4	107.5	107.4
134.2	135.3	134.5	135.7	135.9
29.7	30.4	30.7	30.4	30.7
36.6	38.7	40.6	40.1	42.4
52.8	62.8	53.8	63.0	53.0
152.5	150.9	154.5	145.4	147.2
130.7	132.3	132.0	111.2	111.1
179.7	179.8	179.8	164.1	163.9
164.2	164.1	164.0	179.6	179.6
112.9	111.9	111.9	133.8	133.6
135.3	134.6	135.3	141.4	141.3
137.1	137.2	136.5	135.2	134.2
61.3†	60.8*	61.0*	61.0*	60.9*
61.5†	61.2*	61.1*	61.2*	61.1*
—	56.2†	56.3†	56.1†	56.2†
170.2	34.5[h]	—	35.2	—
22.8	—	—	—	—
56.4	56.2†	56.3†	56.1†	56.2†

specific assignment for C-1 and C-3 in colchicine was made. These assignments were cine (*55*).

group like in 1-aminoindan (*47*); (ii) on a steric γ effect on C-5 observed in 4-substituted deacetamidocolchicine (*38*).

(60–63 ppm); methoxy groups with none or one ortho substituent, like C-15 and C-18,

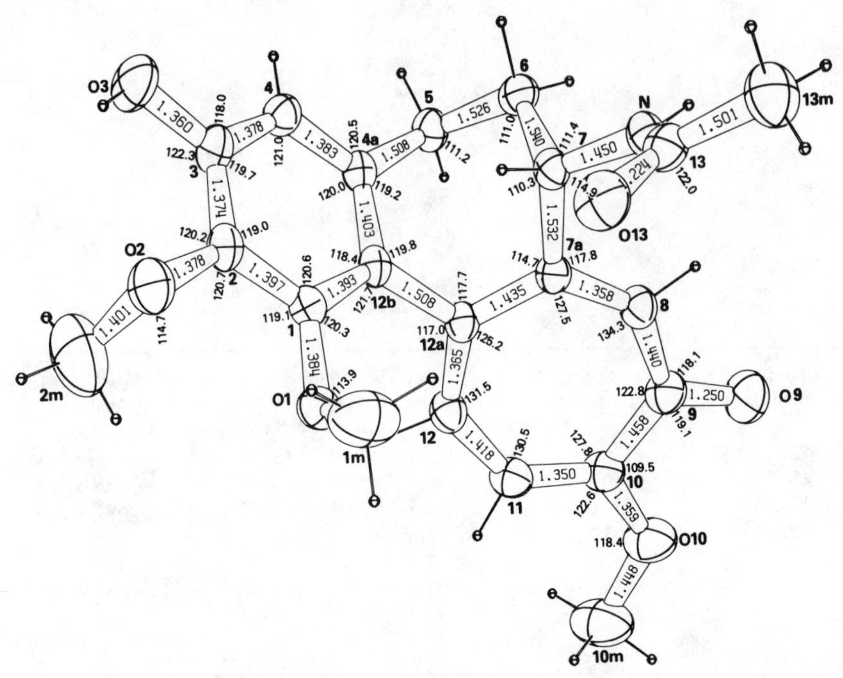

FIG. 1. Conformation of 3-demethylcolchicine (**6**) of mp 281°C and crystallized from acetone in the solid state. The ORTEP diagram showing the *S*-configuration was drawn by a PDP 10 computer using experimentally determined coordinates from X-ray diffraction analysis.

free of solvent but those of **6** incorporate one molecule of acetone for each molecule of **6**. The molecular conformations and dimensions of **6** and **35**, derived from least squares refinement including all H atoms, are shown in Figs. 1 and 2. The final R factors are 4.8% (2788 reflections, 1092 $<$ 1 sigma) and 8.5% (2214, 606 $<$ 1 sigma) for **35** and **6**, respectively. As can be seen from the figures, the molecular conformations and dimensions are very similar to those reported for other colchicinoids.

The packing and hydrogen bonding of the two compounds are very different, as might be deduced from the cell dimensions. Compound **6** has an N+H···O-9 intermolecular hydrogen bond similar to that observed in other colchicinoids (N···O-9 : 2.849 Å; H···O-9 : 2.065 Å), and the 3-hydroxyl H atom is also hydrogen bonded to the O-13 atom of the acetamido group of another molecule (O-3···O-13 : 2.674 Å; H···O-13 : 1.915 Å). The molecules of acetone are not hydrogen bonded and seem simply to fill voids in the structure. Solvated crystals have been reported frequently with other similar compounds. Compound **35** forms a reasonably strong

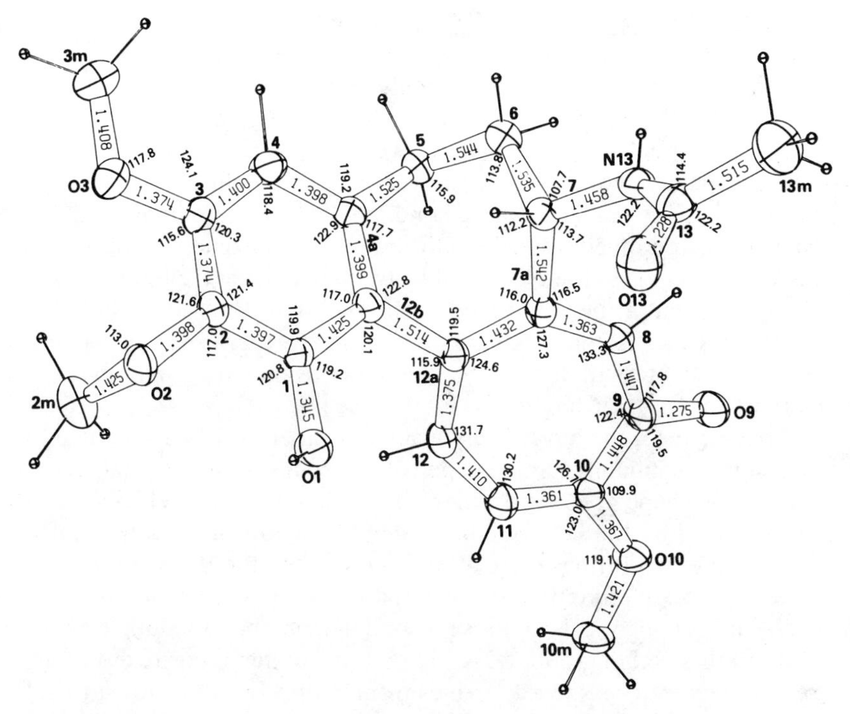

FIG. 2. Conformation of 1-demethylcolchicine (**35**) of mp 307°C (dec.) and crystallized from a mixture of chloroform and methanol in the solid state. The ORTEP diagram showing the *S*-configuration was drawn by a PDP 10 computer, using experimentally determined coordinates from X-ray diffraction analysis.

O-1—H···O-9 hydrogen bond (O-1···O-9 : 2.679 Å; H···O-9 : 2.093 Å) that seems to inhibit the usual NH···O-9 hydrogen bond, which is probably absent because, even though the NH···O-9 distance of 2.967 Å might indicate a weak hydrogen bond, the NH···O-9 distance is 2.317 Å, and this length is on the order of a van der Waals contact. As previously mentioned, the crystals of **35** are unsolvated.

In contrast to colchiceine, colchiceine acetate (**45**), existing in solution as a mixture of tautomers (*39*), was found by X-ray analysis to have

MeO
MeO
OMe
NHAc
OAc
O
45

crystallized as the isoform (*99*). The conformations and orientations of the three methoxy groups in ring A of **45** were similar to those in colchicine, but their accessibility to the molecule was decreased because of the almost perpendicular position of the acetoxy groups at C-9.

Crystals of colchicine from water were dihydrated, and the asymmetric unit consisted of conformationally very similar colchicine molecules. The tropolonic rings had alternating bond lengths and were nonplanar. The C rings formed a dihedral angle with the planar benzene ring of 53° in one molecule and 51° in the other. All water molecules were found to be near each other and to be held together through hydrogen bonds in a distinct "water region" of the crystal. The binding of water or solvent molecules is very characteristic for colchicine and congeners and may play an important role in their binding to tubulin (see Section V).

Isocolchicine, the biologically inactive isomer of colchicine, also forms two conformationally similar types of molecules in the crystal (*100*). Although the shape of the molecules differs from that of colchicine, a difference caused by the keto–methoxy interchange in ring C, the conformations of the two B rings were almost identical to the B ring of colchicine. The aromatic methoxy conformations in one molecule of isocolchicine were almost identical to those in colchicine, but the other molecule showed the methyl group at 2-OMe pointing in the opposite direction and restricting the access to the A rings from both sides. In contrast to colchicine, isocolchicine did not bind solvent or water when crystallized from ethanol–water, and the only hydrogen-bond donors were the H atoms of the amido groups. Beside the "basic" colchicines, the following natural and unnatural compounds were examined by X-ray analysis: demecolcine (**10**) (*101*), thiocolchicine (**34**) (*102*), *N*-deacetylthiocolchicine (**46**) (*102*), thiocolchicoside (**47**) (*103*), 3-demethylisothiocolchicine (**48**) (*104*), and *N*-acetylcolchinol (**49**) (*105*), an analog with a benzenoid ring C. These

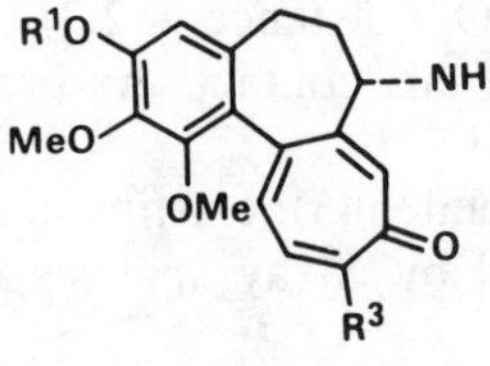

	R^1	R^2	R^3
10	Me	Me	OMe
34	Me	Ac	SMe
46	Me	H	SMe
47	$C_6H_{11}O_5$	Ac	SMe

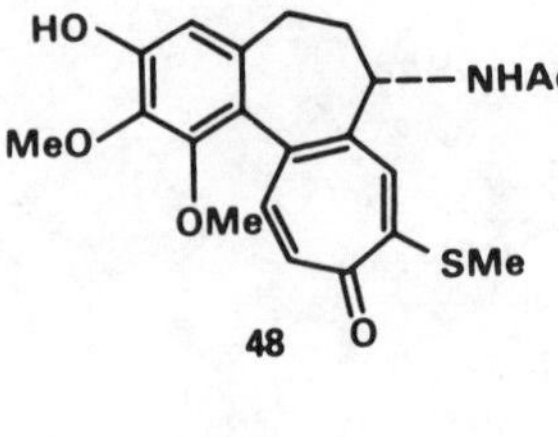

48 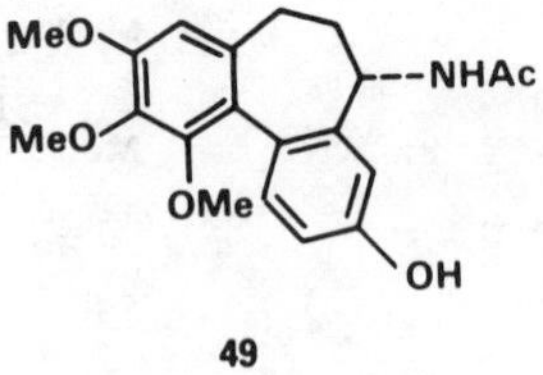 **49**

compounds were analyzed to gain a better understanding of the colchicine–tubulin interaction and its dependence on molecular conformations.

Demecolcine (**10**) showed stereochemical features similar to podophyllotoxin, a natural substance with antitumor properties almost identical to those of colchicine and demecolcine (*101*). Substitution of sulfur for oxygen in the thiocolchicines effects greater bulking of the tropolone rings to produce some changes in the interatomic distances between the atoms of rings A and C. Special attention was given to the conformation of the methoxy groups and, as already mentioned, to the relevance of these stereochemical features with regard to tubulin binding. It can not be ruled out that the stereochemistry associated with the aromatic substitution and hydrogen bonding is more important than the dihedral angle between rings A and C, which is remarkably similar in biologically active and inactive colchicines (*102*), as are the twin angles of the B ring.

The β-D-glucosides of 3-demethylcolchicine and 3-demethylthiocolchicine, having bulky substituents, do not bind to tubulin and have no antitumor activity. The overall shape of *N*-acetylcolchinol (**49**) was found to be similar to that of colchicine, and the H bonding, the hydrophobic regions, and the conformation of the aromatic methoxy groups were also quite similar.

A rather unusual, approximately 3:2 mixture (**50**) of two very similar isomers (**50a** and **50b**) was demonstrated by single crystal X-ray analysis of 7-oxodeacetamidocolchiceine (*82a,88;* Section III,D,1). The two possible tautomers are held together in the crystal by hydrogen bonds.

The structure of the enamide **51**, obtained in 73% yield from colchicine in refluxing acetic anhydride (*106*) (see Section III,F,1), was confirmed by X-ray analysis (*107*).

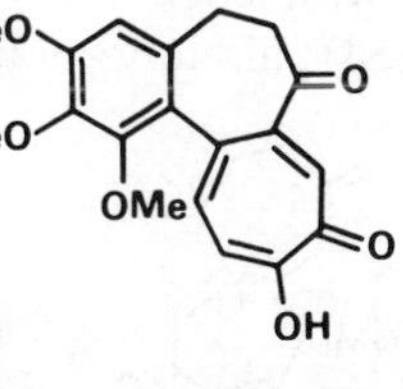

MeO MeO OMe O O OH **50a**

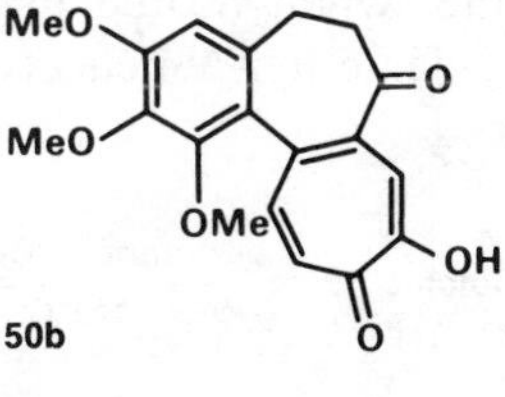

MeO MeO OMe O OH O **50b**

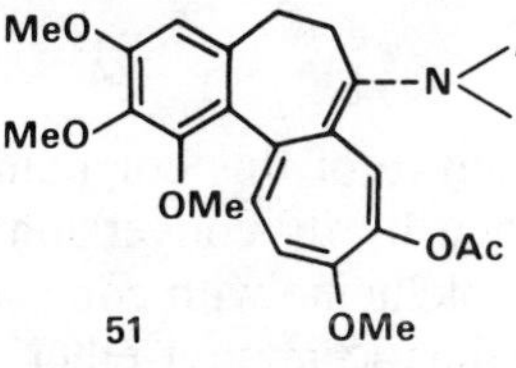

MeO MeO OMe N Ac Ac OAc OMe **51**

Puckering and junction angles of rings B and C in **51** were found to be quite different from those in colchicine. Moreover rings B and C were strongly deformed and out of the plane of the aromatic ring A. An unusual structure was assigned to 10,11-secocolchicine **19**, a natural congener of colchicine and isolated by Šantavý (*66*) from *Colchicum latifolium* S.S. The first proposed structure (**19a**) was based on ^{1}H-NMR and ^{13}C-NMR

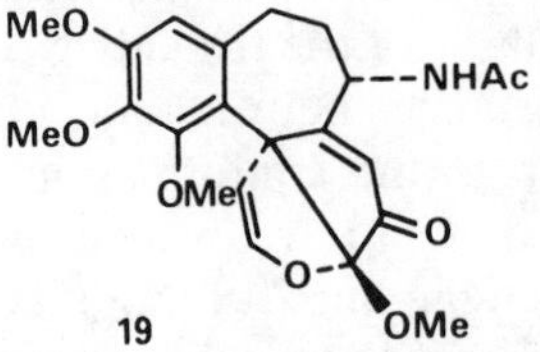

MeO MeO OMe NHAc O O OMe 19   MeO MeO OMe NHAc O O OMe 19a

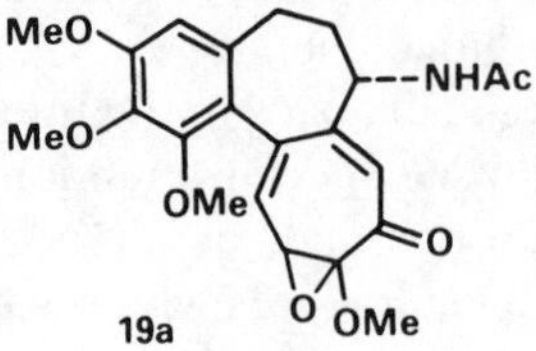

data, but had to be corrected in favor of **19** on the basis of a single crystal X-ray analysis of semisynthetic material (*71*). The 7-membered B ring shows a somewhat distorted twist–boat, rather than a boat conformation as observed with colchiceine and 7-oxodeacetamidocolchiceine (**50**) (*88,98*), while the two 5-membered rings are flattened half-chairs.

C. Chemistry of Ring A

1. Selective O-Demethylations

Whereas enzymic deglucosidation of colchicoside (**7**) afforded the naturally occurring 3-demethylcolchicine (**6**) (*83,108*), also obtained by microbial fermentation (*35,47*), other methods to selectively cleave the aromatic methyl ethers in colchicine have only more recently been developed. Procedures for preparing the three monophenols and two catechols of colchicine became available and proved important for the synthesis of methylenedioxy-substituted congeners of colchicine (*46*).

Colchicine, when heated in a sealed tube with conc HCl at 140°C, afforded 2-demethyl-*N*-deacetylcolchiceine (**54**) already obtained by

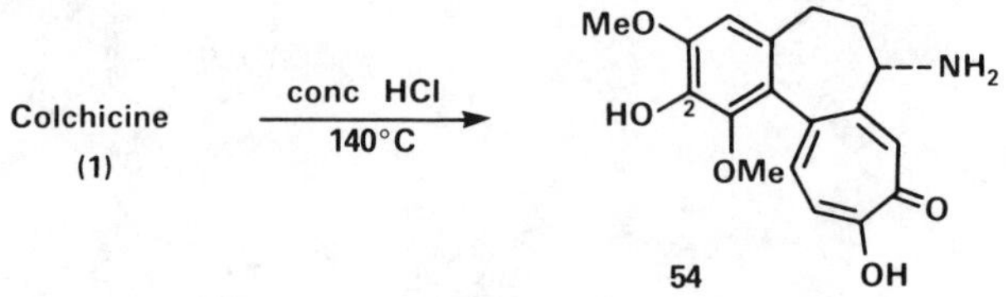

Colchicine (1) → conc HCl, 140°C → MeO HO 2 OMe NH_2 O OH 54

Zeisel in 1888 (*4*). The structure of this compound was, however, only relatively recently established by its conversion into 2-demethylcolchicine (**5**) (*45*). Treatment of colchicine with conc sulfuric acid cleaved the most sterically hindered aromatic methyl ether at C-2 and afforded **5**. Similar treatment of colchicoside (**7**) with 85% phosphoric acid gave 3-demethylcolchicine (**6**) (*46*). Thiocolchicine (**34**) and thiocolchicoside, the 10-SMe analog of **7**, behaved similarly (*84*). The 2-demethylthiocolchicine was found identical with the enzymic degradation product (*83a*). The unnatural 1-demethylcolchicine (**35**) was prepared by hydrolysis of the intermediate 1-O-acetate, obtained from colchicine with $SnCl_4$ in the presence of acetyl chloride (*45*).

Treatment of the monophenols **6** and **35** with conc sulfuric acid at 85–90°C afforded 2,3-didemethylcolchicine (**8**) and its 1,2-didemethyl isomer (**52**), respectively (*41*) (Scheme 1). The catechol **8** has properties similar to a compound isolated from *Colchicum latifolium* S. S. (*48*) and from *Gloriosa superba* of Indian and African origin (*77*). Methylenation of the catechol **52** afforded the methylendioxy-substituted compound **53**, previously claimed to represent the structure of cornigerine (**4**), an alkaloid isolated from the species *Colchicum cornigerum* (*40*). Comparison of the natural

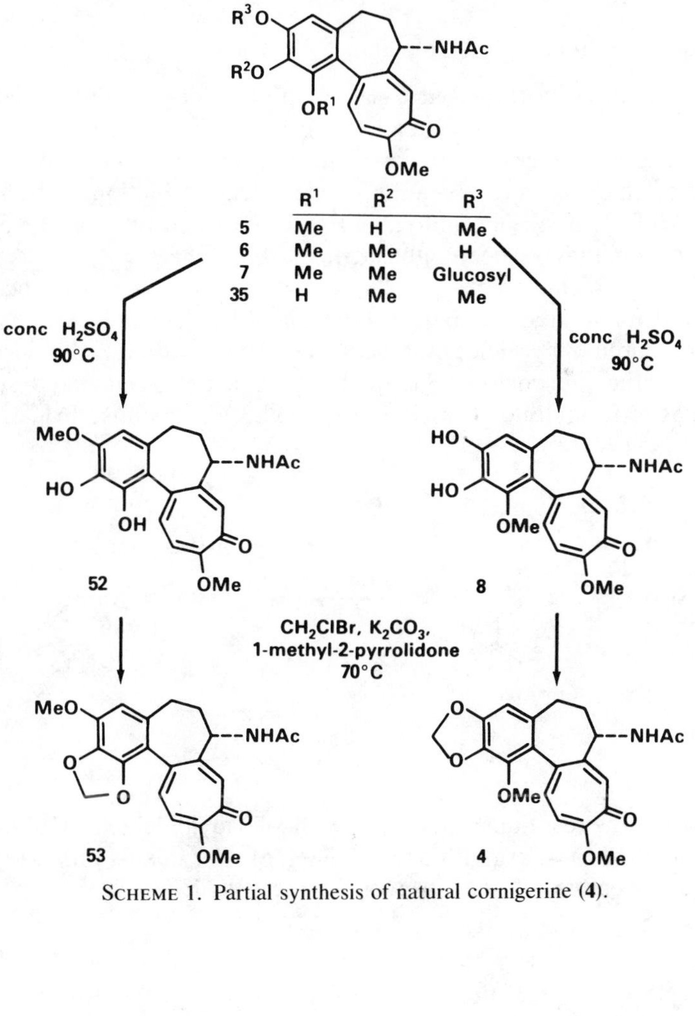

R3O
R2O
OR1
NHAc
O
OMe
R1 R2 R3
5 Me H Me
6 Me Me H
7 Me Me Glucosyl
35 H Me Me
conc H2SO4
90°C
conc H2SO4
90°C
MeO
HO
OH
NHAc
52
OMe
HO
HO
OMe
NHAc
8
OMe
CH2ClBr, K2CO3,
1-methyl-2-pyrrolidone
70°C
MeO
O
O
NHAc
53
OMe
O
O
OMe
NHAc
4
OMe

SCHEME 1. Partial synthesis of natural cornigerine (**4**).

alkaloid with the semisynthetic material similarly obtained from the catechol **8**, showed that **4** represents the structure of cornigerine (*41*).

Ethyl ether derivatives of 2-demethyl- and 3-demethylcolchicine and their corresponding demecolcines were isolated and characterized by paper and thin-layer chromatography. *O*-Acetates of these phenols were prepared *in situ* by treating plant extracts prior to chromatography with acetic anhydride (*109*). Extraction of the alkaloids was carried out with large volumes of chloroform contaminated with ethanol, and ethanol was frequently used during the chromatographic analysis as a cosolvent. It is thus most likely that these ethyl ethers are artifacts. The 3-ethyl ether derivative of naturally occurring 3-demethyl-*N*-deacetyl-*N*-formylcolchicine (**23**), isolated from *Gloriosa superba,* was prepared for characterization purposes (*34*).

2. Substitution at C-4 in Colchicine

Colchicines substituted at C-4 have not yet been encountered among natural products, and the chemistry thereof, mostly buried in the patent literature, has never been reviewed in this treatise. Almost all C-4 substituted colchicines were prepared for biological evaluation at the Roussel-UCLAF Research Laboratories in France between 1960 and 1965, but no biological data have been published.

4-Formylcolchicine (**55**) (mp 250°C; $[\alpha]_D$ +5° in $CHCl_3$) was the starting material for a large number of 4-substituted colchicines. Compound **55** was prepared (80% yield) by reacting colchicine with α,α-dichloromethyl ether in the presence of $SnCl_4$. Thiocolchicine (**34**) reacted similarly, yielding 4-formylthiocolchicine (**56**) (75%; amorphous, $[\alpha]_D$ +318° in $CHCl_3$) (*110–112*).

Cl_2HCOCH_3 / $SnCl_4$

	R	
1	OMe	55
34	SMe	56

Most of the reactions carried out with 4-formylcolchicine (**55**) have also been applied to 4-formylthiocolchicine (**56**). For details one should refer to the respective references given for the colchicine analogs.

Treating 4-formylcolchicine (**55**) with α,α'-dichloromethyl ether in the presence of $ZnCl_2$ afforded a moderate yield of the dichloro compound **57** (20%; mp 140°C and 170°C; $[\alpha]_D$ value reported oscillating between −25° and −80° in $CHCl_3$) (*111,113*). Dehydration in refluxing acetic anhydride of the oxime **58** (obtained from **55** with hydroxylamine) yielded 4-cyano-colchicine (**59**) (60%; mp 258°C; $[\alpha]_D$ −50° in $CHCl_3$) (*111,114*), while Beckmann rearrangement of **58** with polyphosphoric acid gave the *N*-formyl derivative **60** (mp 260°C; no yield or optical rotation are given) (*115*). The hydroxymethyl compound **61** (mp 260°C; $[\alpha]_D$ −104° in $CHCl_3$), a starting material for the synthesis of a series of 4-substituted ethers and esters (*111,116*), is accessible in different ways. Reduction of **55** with KBH_4 in methanol (60% yield) (*111*) seems to be the most straightforward method, while treatment of 4-chloromethylcolchicine (**62**) with aqueous $BaCO_3$ afforded a 40% yield of **61** (*117*). The chloromethyl derivative **62** was the main product of the reaction of colchicine with chloromethyl methyl ether in the presence of $SnCl_4$ (*117*). Dimeric colchicine, linked together by a methylene bridge at C-4, was a by-product of this reaction (see also *118*). Oxidation of **55** by chromic acid gave **63** (40%; mp 215°C; $[\alpha]_D$ −165° in EtOH) (*111,119*).

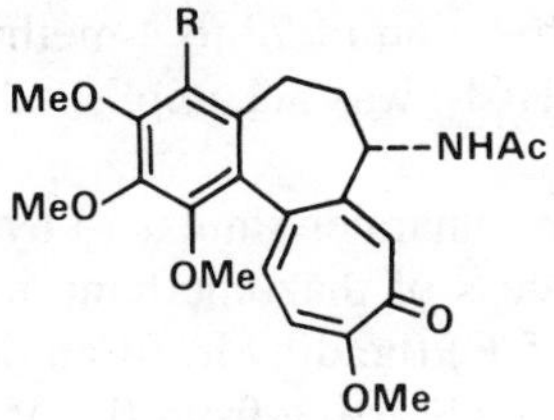

	R
57	$CHCl_2$
58	CH = NOH
59	CN
60	NHCHO
61	CH_2OH
62	CH_2Cl
63	COOH

The availability of 3-demethylcolchicine (**6**, see Section III,C,1) made substitution at C-4 also possible by a different methodology. The sodium salt of **6** on reaction with aqueous formaldehyde afforded a 60% yield of 4-hydroxymethyl-3-demethylcolchicine (**64**) (dec. above 210°C; $[\alpha]_D$ −156° in $CHCl_3$), and methylation with diazomethane yielded **61** (*120*). Similarly, diazo coupling of **6** with sulfanilic acid followed by reductive cleavage gave 3-demethyl-4-aminocolchicine (**65**) (33%; (mp 215°C; $[\alpha]_D$ −150° in $CHCl_3$), (*121*), converted subsequently with $K_3Fe(CN)_6$ and reductive work-up to 3-demethyl-4-hydroxycolchicine (**66**) (mp 200–208°C; $[\alpha]_D$ −268° in EtOH). O-Methylation with diazomethane gave 4-methoxycolchicine (**67**) (mp 235–236°C; $[\alpha]_D$ −93.5° in $CHCl_3$) (*122*).

The Mannich reaction of **6** with formaldehyde and dimethylamine afforded 4-dimethylaminomethyl-3-demethylcolchicine (**68**) (mp 198°C; $[\alpha]_D$ −92° in $CHCl_3$, and after quaternization and reduction of the methiodide

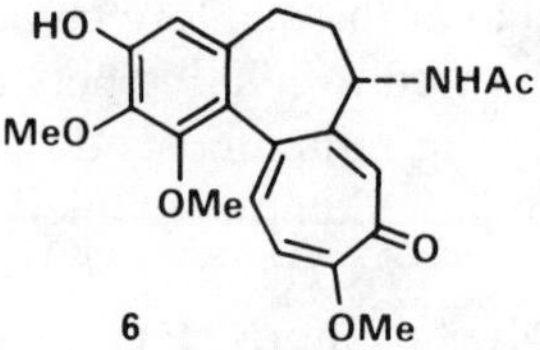

6

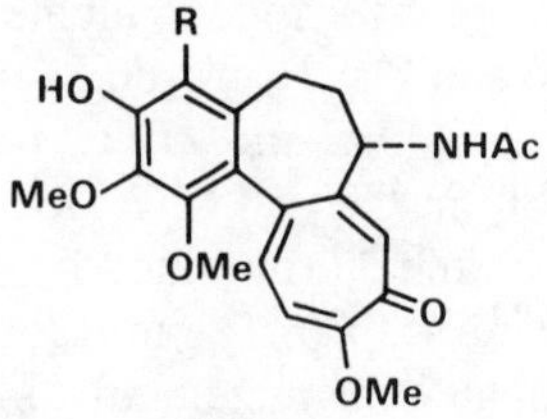

64 R = CH_2OH
65 R = NH_2
66 R = OH
68 R = CH_2NMe_2
69 R = $CH_2\overset{+}{N}Me_3\,I$
70 R = Me

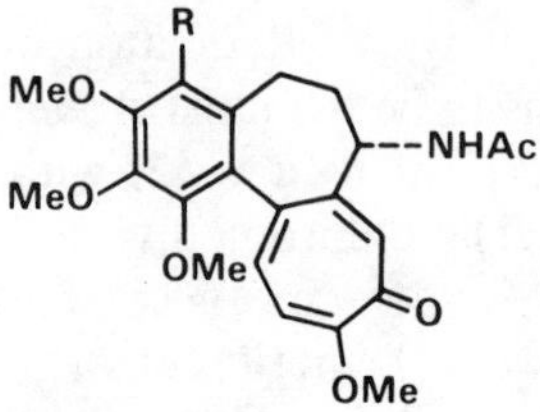

67 R = OMe
71 R = Me

69 with $NaBH_4$, gave 4-methyl-3-demethylcolchicine (**70**) (mp about 190°C; $[\alpha]_D$ −137° in $CHCl_3$). O-Methylation of **70** to 4-methylcolchicine (**71**) (mp 214–215°C; $[\alpha]_D$ −45.5° in $CHCl_3$) was accomplished with diazomethane (*123,123a*).

The thiomethiodide **72**, prepared in a manner similar to **69** (*123a,124*), afforded on reaction with a large excess of diazomethane in methylene chloride–methanol solution (48 hr at 5°C) the dihydrofuran derivative **73** (mp about 194°C; $[\alpha]_D$ −199° in $CHCl_3$). Starting from the Mannich base **74**, the dihydrofuran **73** was also obtained, but the reaction was sluggish. The methiodide **69** of 3-demethylcolchicine has been reported to react similarly with diazomethane, but the authors were unable to obtain the reaction product in pure form. A mechanism for this type of reaction was proposed (*124*).

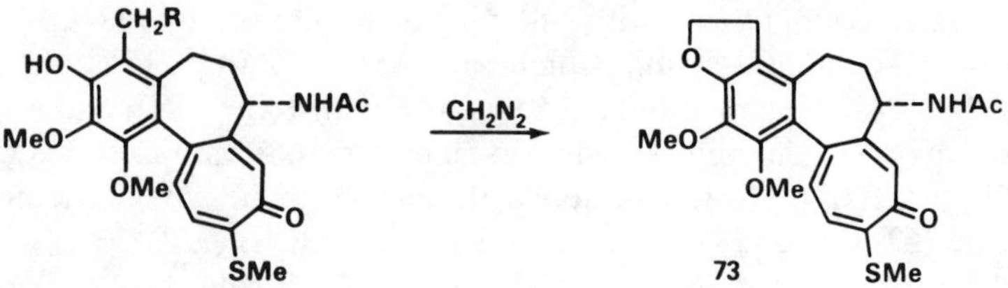

CH_2N_2

73

72 R = $\overset{+}{N}Me_3\ \overset{-}{I}$
74 R = NMe_2

D. Chemistry Affecting Ring B of Colchicine

1. Structural Changes in Ring B

Alkaloid CC-12 and colchiciline (**14** and **15** in Table I) are isomeric hydroxycolchicines of the molecular composition $C_{22}H_{25}NO_7$. Both alkaloids were isolated from the seeds of *Colchicum latifolium* S. S., which possibly also contain 2-demethylcolchiciline (*48*). It was found that commercial samples of colchicine may contain colchiciline as a minor impurity (*28*). Little information is available regarding the alkaloid CC-12 (*59,65*), except that the additional OH group, seen in its IR spectrum as a broad signal at 3300 cm^{-1}, is most likely hydrogen bonded. The similarity of the ^{1}H-NMR and UV spectra to those of colchicine suggest that alkaloid CC-12 is tropolonic in nature and probably belongs to the colchicine group of alkaloids. The extra OH group was for these reasons assigned to ring B; however, CC-12 is different from colchiciline.

Colchiciline afforded an *O*-acetate after acetylation (mp 150–151°C); after hydrolysis with 0.1 *N* HCl colchicileine (**75**) (mp 253–256°C) with a UV behavior identical to that of colchiceine (**2**) (*66*) was formed. The structure of colchiciline, as shown in **15**, was established by spectral data and by model inspections. The location of the additional OH group at C-6 was deduced from ^{13}C-NMR data, which showed that the triplet signal at 30.0 ppm assigned to C-6 in colchicine had largely disappeared. The IR spectrum of **15** suggested that the β-oriented OH group was hydrogen bonded with the tropolonic carbonyl, therefore making ring B conformationally different from that of colchicine. Two teams have more recently reported that the 30.0 ppm signal in the ^{13}C-NMR spectrum of colchicine has to be assigned to C-5 rather than to C-6 (see Section III,A), suggesting that the structures of colchiciline and colchicileine may have to be replaced by **15a** and **75a** or their epimers (*124a*).

An interesting structural variation of ring B in colchicine is illustrated by 7-oxodeacetamidocolchicine (**40**) (mp 229–230°C), a defined microbial

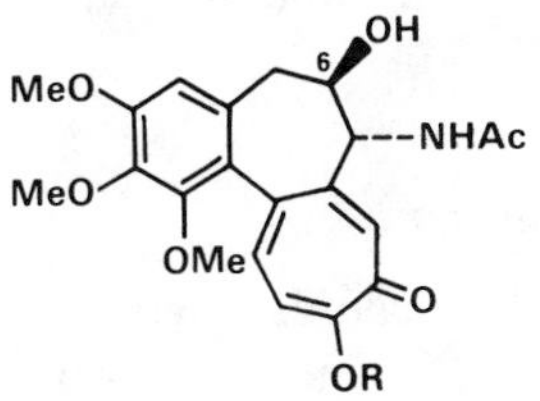

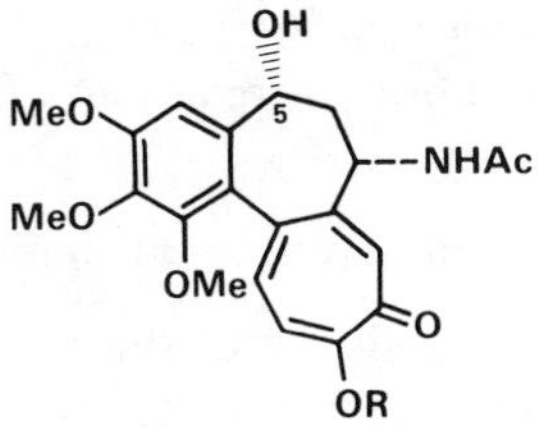

	R	
15	Me	15a
75	H	75a

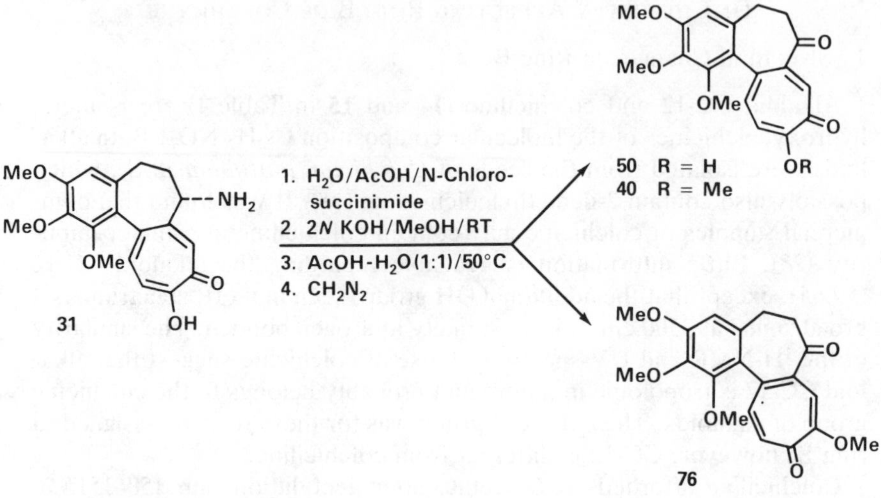

MeO MeO OMe O O OR
50 R = H
40 R = Me

MeO MeO OMe NH$_2$ O OH
31

1. H_2O/AcOH/N-Chloro-succinimide
2. 2*N* KOH/MeOH/RT
3. AcOH-H_2O(1:1)/50°C
4. CH_2N_2

MeO MeO OMe O OMe O
76

SCHEME 2. Roussel-UCLAF synthesis of 7-oxodeacetamidocolchicine.

degradation product of colchicine (see Section II,B). This diketone was first prepared at the Roussel-UCLAF Research Laboratories from deacetylcolchiceine (**31**) (*82a*) as shown in Scheme 2.

O-Methylation of the "oxocolchiceine" **50** (mp 154°C) afforded the methyl ether **40** of the natural series and **76** (mp 190–192°C) of the iso series. Compound **50** was later isolated during a base-catalyzed equilibration of the Schiff base prepared from **31** and benzaldehyde, a process known to result in racemization of colchicine (*125*) and now recognized to proceed through an aldimine–ketimine tautomerism (*88*). The correctness of structure **50** was supported by single crytal X-ray analysis, and the structures **40** and **76**, obtained after O-methylation, were determined by a detailed analysis of their ^1H-NMR and ^{13}C-NMR spectra in comparison with model compounds (*55,88*). Reaction of **40** with aqueous methylamine was reported to give a 10-demethoxymethylamide of mp 215°C (*82a*), whereas reaction of **50** with benzylamine, under the conditions applied in the equilibration reaction, afforded (±)-**31** (*88*).

2. Natural Variation at the Nitrogen Function

The abundance of the tropolonic *Colchicum* alkaloids originates from changes affecting the C-7 substitution (Table I). The four major variations encountered so far in nature are deacetylcolchicine (**A**), the *N*-acyldeacetylcolchicines (**B**), the demecolcines (**C**), and the colchifolines (**D**), all of which have the same absolute configuration at C-7. Similar changes in the

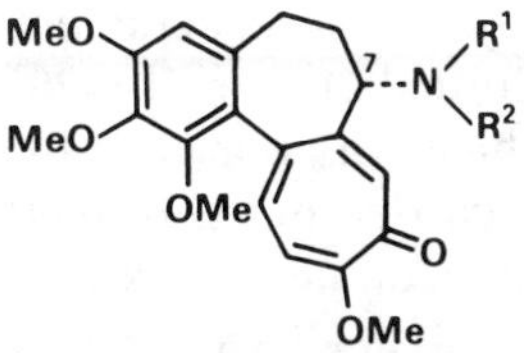

A R¹ = R² = H
B R¹ = H, R² = Acyl
C R¹ = Me, R² = H, Me or *o*-OH-benzyl
D R¹ = H, R² = COCH₂OH

side chain were also recognized to occur in the corresponding colchiceines, demecolceines, and demethylcolchicines, affording a large number of congeners.

Oxidation of the acetamido side chain to a glycolic acid amide is an oxidative conversion that, at least on chemical grounds, connects three important *N*-acyl modifications (-NHCHO,-NHCOCH$_3$, and -NHCO-CH$_2$OH) with each other.

A simple synthesis of deacetylcolchicine (**9**), prepared earlier from deacetylcolchiceine (**31**) by the Fernholz method (*126*), demecolcine (**10**), and demecolcine analogs was reported (*56*) as illustrated in Scheme 3. Trifluoroacetyldeacetylcolchiceine (**77**) (mp 171–172°C; $[\alpha]_D$ −216° in CHCl$_3$), a key compound in this synthesis, was obtained from deace-

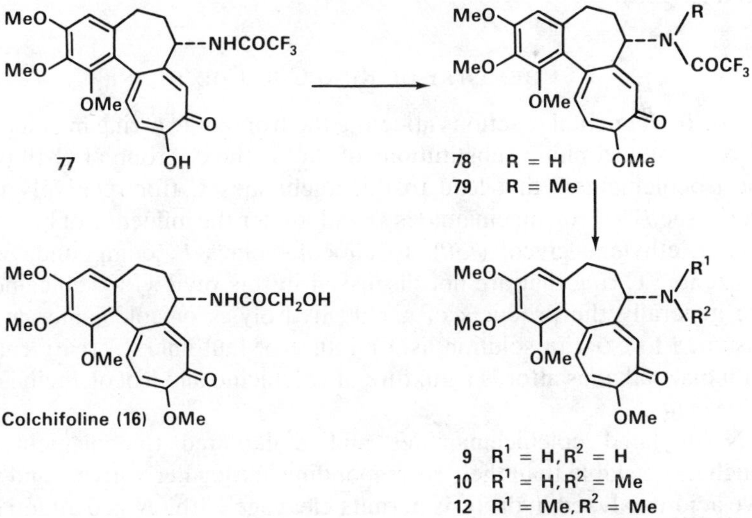

9 R¹ = H, R² = H
10 R¹ = H, R² = Me
12 R¹ = Me, R² = Me
18 R¹ = Me, R² = o-OH-benzyl

SCHEME 3. Partial synthesis of deacetylcolchicine (**9**) and demecolcine (**10**).

tylcolchiceine (**31**) with trifluoroacetic anhydride (*56*). O-Methylation with diazomethane afforded **78** (mp 203–205°C; $[\alpha]_D$ −79° in $CHCl_3$) beside the iso isomer which was removed by chromatography. When the methylation of **77** was carried out with methyl iodide in the presence of potassium carbonate in acetone, a mixture of the *N*-methyl analog (**79**) (mp 180–181°C; $[\alpha]_D$ −188° in $CHCl_3$) and its iso isomer was obtained and readily separated by chromatography. Hydrolysis of **79** with potassium carbonate in aqueous acetone at 60°C afforded demecolcine (**10**), whereas deacetylcolchicine (**9**) was similarly obtained from **78**. Methylation of demecolcine (**10**) afforded *N*-methyldemecolcine (**12**), and alkylation with *O*-acetoxybenzylbromide followed by mild hydrolysis afforded speciosine (**18**), which was reported to yield demecolcine again on heating for 15 min at 210°C (*127*).

A partial synthesis of colchifoline (**16**), detected in commercial samples of colchicine (*28*) and identical to the *O*-methyl ether of 2-demethylcolchifoline (**24**) isolated from *Colchicum autumnale* (*78*), was reported (*29*). For this synthesis a mixture of natural and isodeacetylcolchicine, obtained from **31** by O-methylation with diazomethane (*56*), was reacted with trifluoroacetylglycolyl chloride that, after work-up and chromatographic separation, directly afforded colchifoline (**16**) and its iso isomer. Both ethers after hydrolysis with 0.1 *N* HCC gave the identical colchifoleine (mp 227–228°C; $[\alpha]_D$ −254°, $CHCl_3$). Esterification of colchifoline with acetic anhydride in pyridine yielded after chromatography the *O*-acetate (*128*).

E. Chemistry of Ring C in Colchicine

The fundamental reactions affecting the tropolonic C ring in colchicines of **A** are nucleophilic substitutions of the methoxy group at C-10 (C-9 in the isocolchicines) that lead to thiocolchicines **C** (for relatively recent work, see *129*), colchiceinamides **D** and, under the influence of bases or in boiling ethylene glycol (*130*), to allocolchicines *E*, compounds with a benzenoic C ring that are not discussed in this review. Colchiceines (**B**) are generally the products of a mild hydrolysis of colchicines and are assumed to exist in solution as a mixture of tautomers. Treatment of **B** with diazoalkanes affords a mixture of colchicine and isocolchicine derivatives (*14*).

N-Alkylated colchiceinamides and S-alkylated thiocolchicines are much more stable than their corresponding alkoxy derivatives, and selective acid-catalyzed hydrolysis permits cleavage of the *N*-acetamido group in **C** (R = Me), leading to *N*-deacetylthiocolchicine (**46**) (80%; mp 198°C; $[\alpha]_D$ −201°) (*129,131*). Colchicide (**80**), which lacks the 10-methoxy group

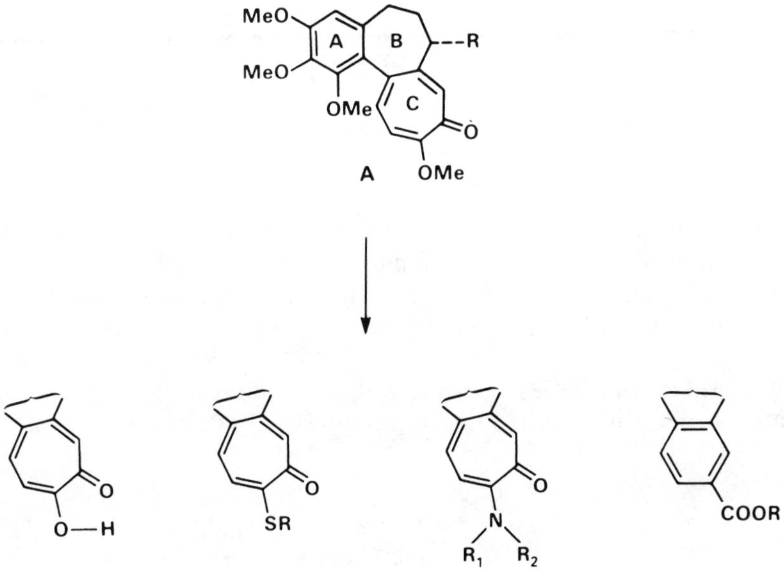

MeO MeO OMe OMe A B C --R O A

B C D E

O—H SR N R_1 R_2 COOR

of colchicine, obtained by Raney nickel desulfurization of thiocolchicine (**34**) (*132*), is one of the most biologically interesting compounds prepared by varying ring C (see Sections V and VI).

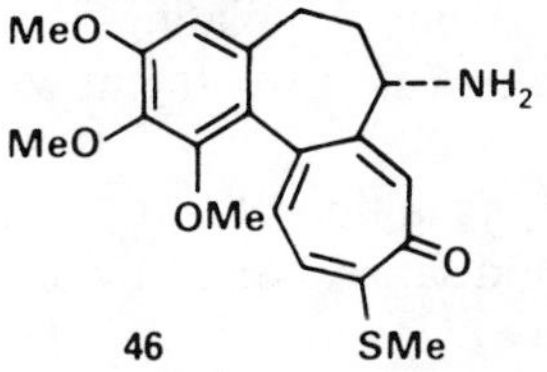

MeO MeO OMe --NH_2 O SMe **46**

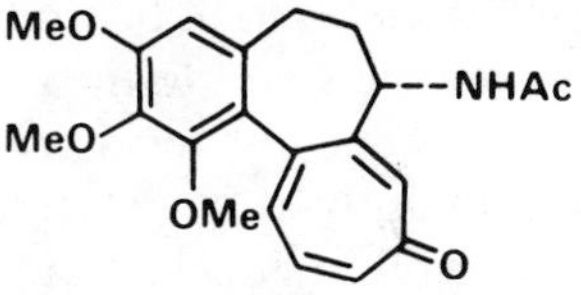

MeO MeO OMe --NHAc O **Colchicide (80)**

Details of these nucleophilic reactions have been reported in other reviews (e.g., *8,14*) and are not discussed here. The same holds for oxidation and reduction reactions affecting ring C of colchicine and carried out in connection with the structure elucidation of the alkaloid (*8,10,14*).

F. Chemical Conversions of Colchicine

1. Racemization

The classic racemization (*125*) of colchicine is based on alkali treatment of (−)-*N*-benzylidene-deacetylcolchiceine (**81**), which after treatment

81 $\xrightarrow{\text{1. KOH/MeOH; 2. HCl}}$ (±)-31

with KOH affords a 50% yield of (±)-deacetylcolchiceine (**31**) (mp 244°C). This reaction is the result of an aldimine–ketimine equilibration (*88*).

(±)-Deacetylcolchiceine (**31**) can be transformed into (±)-colchicine (mp 260°C) or converted into (+)-colchicine (see below). A more efficient method for preparing racemic colchicine has been developed by Bladé-Font (*106*): Treatment of colchicine with acetic anhydride at reflux temperatures for 24 hr afforded the triacetate **51** (73%; mp 170°C), and on

(−)-Colchicine $\xrightarrow{Ac_2O/\uparrow\downarrow}$ 51 $\xrightarrow{HOAc/\uparrow\downarrow}$ (±)-Colchicine

boiling **51** with acetic acid, (±)-colchicine was obtained in excellent yield. From a practical point of view, (±)-colchicine can be prepared in better than 70% yield directly from natural colchicine without isolating **51** by simply adding water to the reaction mixture obtained after treatment with refluxing acetic anhydride (*106*).

Acid hydrolysis of (±)-colchicine and optical resolution of (±)-colchiceine as well as N-acetylation, O-methylation, and separation of the (+)-colchicine from the (+)-isocolchicine has been achieved (*46,125*). (+)-Colchicine crystallized from water (mp 139–141°C; $[\alpha]_D$ +131° in $CHCl_3$) (*46*).

2. Transformations Affecting Ring B

Chromatography of triacetate **51** on acid-washed alumina led to **82** by selective N-monodeacylation (75%; mp 246°C), while aqueous base treatment of **51** yielded (±)-colchicine as the main product (61%) and the *N*-acetylcarbinolamine **83** (20%; mp 190°C) a new tropolonic compound, which was subsequently transformed into the *N*-acetylpyrrole **84** (mp 172°C) and the pyrrole **85** (mp 326°C). A similar reaction sequence was reported with propionic anhydride (*106*).

Brief treatment of the enol acetate **82** with base in methanol afforded the ketone **86** (60%; mp 242°C) as yellow crystals. The oxime **87** and the alcohol **88** were prepared by standard procedures. Racemic colchicine may be obtained by treating **86** with aqueous base at room temperature or with hot acetic acid, whereas acetic anhydride at reflux yielded a mixture of the triacetate **51** and the *N*-acetylpyrrole **84**. At room temperature and in the presence of *p*-TsOH, **86** gave **89** (60%; mp 202°C) in acetic anhydride. 6-Dehydrocolchicine (**90**) (90%; mp 254°C) was the main product when ketone **86** was reacted with trimethylphenylammonium perbromide. Finally, beside ketone **91** (18%; mp 230°C), (±)-isocolchiceine acetate (**45**) was obtained in 30% yield on reacting **82** with aqueous mineral acid (*133*) (Scheme 4).

Analogous pathways were reported with thiocolchicine, isocolchicine, and colchiceine (*134*).

3. Photochemical Transformations Not Leading to Lumicolchicines

Irradiation of **82** in acetone or benzene with a low-pressure Hg lamp gave **92** as the main product (25%; mp 166–167°C) in accordance with its spectral data (*135*). Rose Bengale sensitized the photooxidation of the same compound at low temperature in CH_2Cl_2 or MeOH and obtained a

	R	X
86	Me	O
87	Me	NOH
88	Me	<H, OH
91	Ac	O

SCHEME 4. Transformation of the B ring.

50–60% yield of compounds **93**, **94**, and **95** separated by column chromatography and fractional crystallization (Scheme 5a). The epoxide **94** was a minor product in the epoxidation of **82** with *m*-chloroperbenzoic acid.

In the autosensitized photooxidation of the pyrrole **85** in CH_2Cl_2 or MeOH at low temperature, the two unstable products **96** and **97** were formed and separated by preparative TLC (*136*). Compounds **96** and **97** yielded **85** again by reaction with $NaBH_4$. Reaction of **97** with SeO_2 in CH_2Cl_2 gave the aldehyde **98** as the only isolatable product (Scheme 5b). A mechanism for these transformations was proposed.

4. Transformation of the Acetamido Side Chain

In search for colchicine derivatives with therapeutic properties superior to those of colchicine, a series of compounds with modified acetamido side chains were prepared from colchicine (**1**) thiocolchicine (**34**), and 10-alkylamido-10-demethoxycolchicines (**42**).

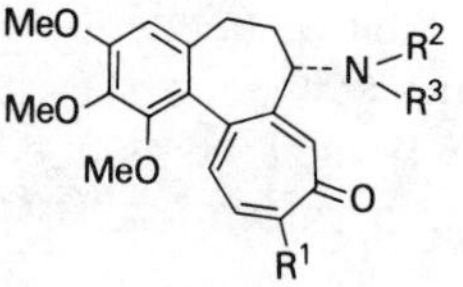

MeO MeO MeO N R² R³ O R¹

1 R^1 = OMe, R^2 = H, R^3 = Ac

34 R^1 = SMe, R^2 = H, R^3 = Ac

42 R^1 = NHAlkyl, R^2 = H, R^3 = Ac

Starting with *N*-deacetylcolchicine (**9**), Lettré (*137*) prepared a series of *N*-haloacetylcolchicines (**99**), *N*-haloacetyl-10-dimethylamino-10-demethoxycolchicines (**100**), and *N*-haloacetylthiocolchicines (**101**) by N-acylating the respective deacetyl compounds with the appropriate haloacetyl chlorides. The iodo compounds were obtained from their bromo derivatives by reaction with iodide in acetone. For biological data see Sections V and VI. A different approach to synthesize *N*-bromoacetylcolchicine (**99**, R^2 = Br) is shown below (*138*). This reaction sequence,

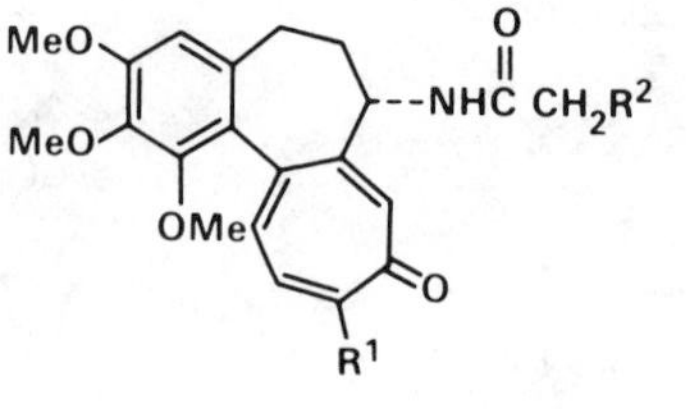

MeO MeO OMe NHC(O)CH₂R² O R¹

	R^1	R^2
99	OMe	F, Cl, Br, I
100	NMe_2	F, Cl, Br, I
101	SMe	F, Cl, Br, I

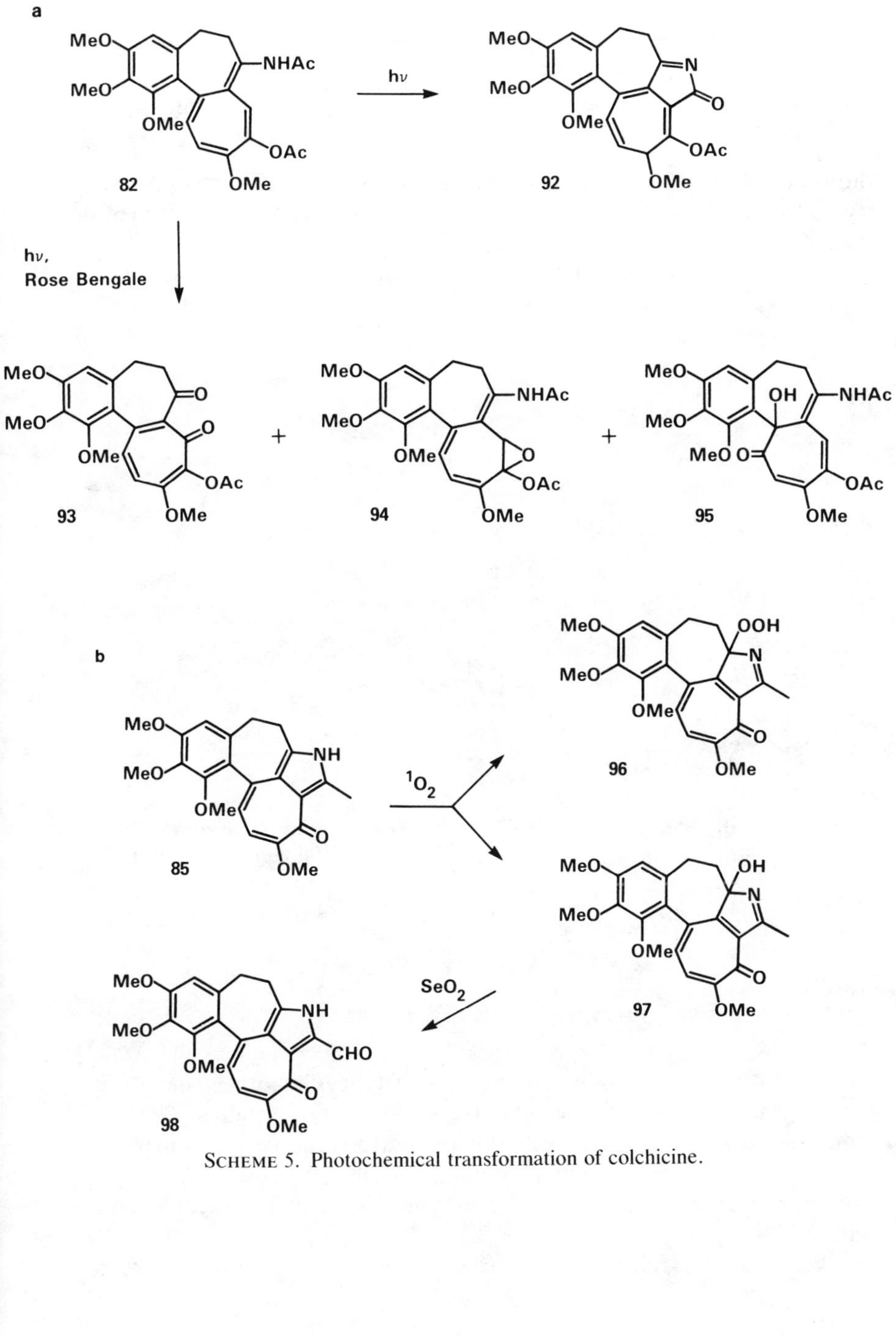

SCHEME 5. Photochemical transformation of colchicine.

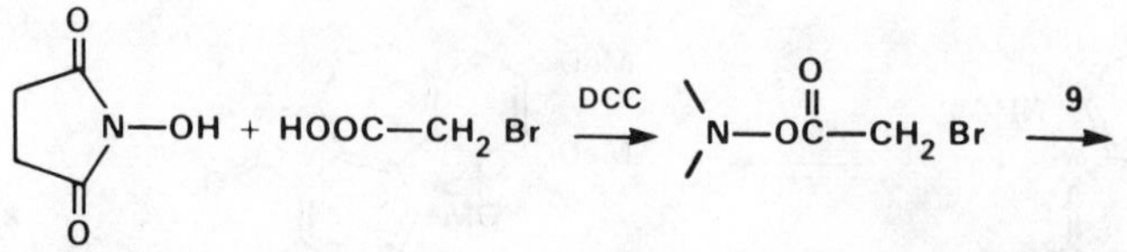

N—OH + HOOC—CH_2Br $\xrightarrow{DCC}$ N—OC(O)—CH_2Br $\xrightarrow{9}$ 99 (R^2 = Br)

however, was unfortunately carried out with a 25 : 75 mixture of *N*-deacetylcolchicine (**9**) and *N*-deacetylisocolchicine (**44**), and a separation of the isomers was not attempted (*138*).

By reacting *N*-deacetylalkylthiocolchicine (**102**) with isocyanates or isothiocyanates, a number of ureas and thioureas (**103–108**) were prepared

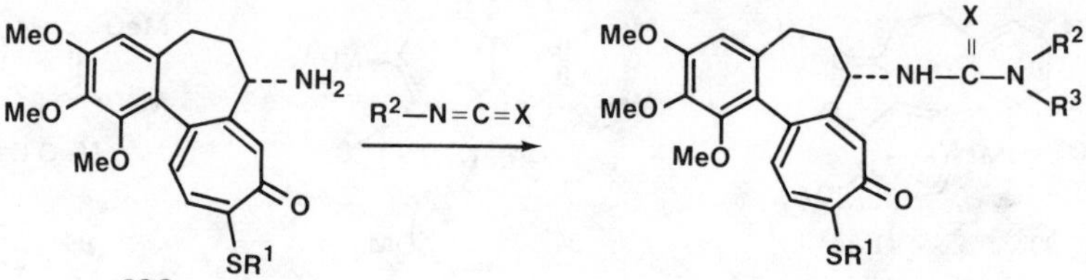

102 $\xrightarrow{R^2—N=C=X}$

	R_1	X	R_2	R_3
103 :	Me	S	allyl	H
104 :	Me	O	-CH_2-CH_2-Cl	H
105 :	Me	O	Me	H
106 :	Et	O	Me	H
107 :	Et	S	C_6H_5	H
108 :	Me	S	sugar[1]*	H

* R^2 = 2, 3, 4, 6 - Tetra-*O*-acetyl-β-D-glucopyranosyl
R^2 = 2, 3, 4 - Tri-*O*-acetyl-β-D-arabinopyranosyl
R^2 = 2, 3, 4 -Tri-*O*-acetyl-α-L-arabinopyranosyl

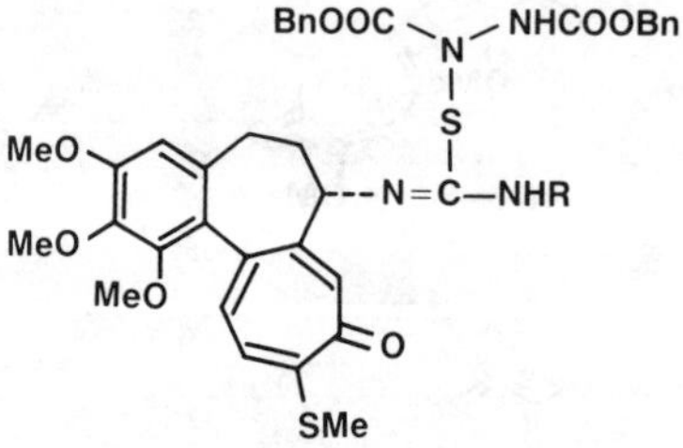

109 (Bn = benzyl)

(R = per-*O*-acetyl-β-D-glucopyranosyl and per-*O*-acetyl-α-L-arabinopyranosyl)

(*139,140,140a*). On reacting compounds of type **108** with dibenzylazodicarboxylate, the hydrazinoisothioureas (**109**) were obtained (*140a*). In addition, **104** and **105** were nitrosated to afford the desired nitrosourea analogs (R^3 = NO). No isomeric contamination was detected in the nitrosation, probably because of steric hindrance by the bulky colchicine molecule.

The same research group prepared a series of *N*-deacetyl-*N*-glycosylalkylthiocolchicines (**110**) by treating *N*-deacetylalkylthiocolchicines with the appropriate monosaccharides in MeOH (*141*).

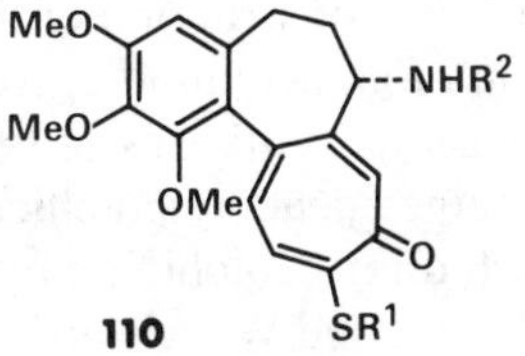

R^1 = Me, Et
R^2 = D-Glucosyl, D-Galactosyl, D-Mannosyl, D-Ribosyl, L-Arabinosyl

110

A few colchicine compounds without substitution at C-7 were reported to be biologically active (*142*). Hofmann degradation of the trimethyl ammonium salt of *N*-deacetylcolchiceine (**111**) led, through migration of the double bond, to the 5,6-dehydro compound **112** (*46,143*), and the products obtained after methylation with diazomethane, **114** and **115**, could be separated by chromatography and fractional crystallization (*55*) (Scheme 6). Hydrogenation of **112** gave **113** (*143*), and hydrogenation of **115** afforded deacetamidoisocolchicine (*143a*).

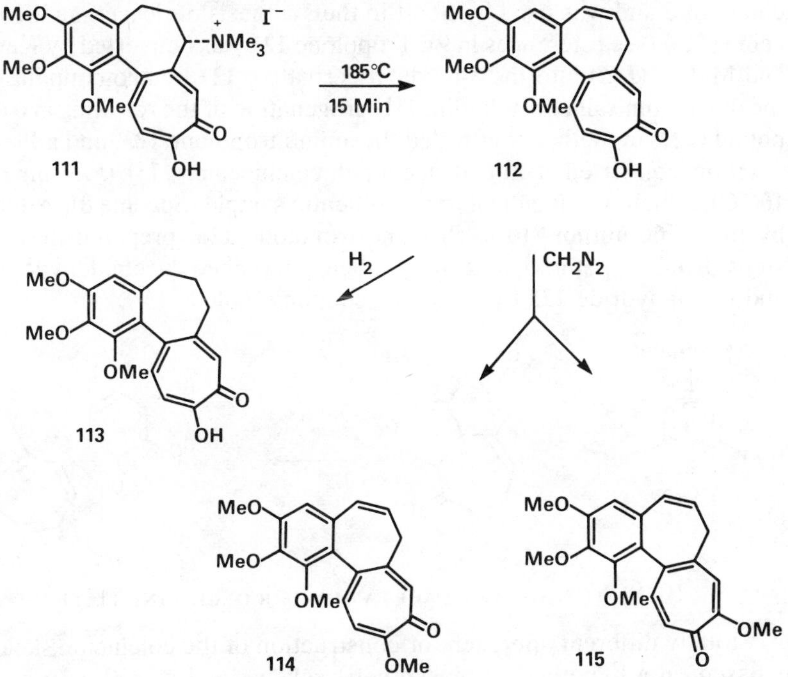

SCHEME 6. Deacetamidocolchicines.

IV. Novel Syntheses of Colchicine

Despite its relatively simple structure in comparison with other natural products, the synthesis of colchicine, requiring construction of the proper tropolonic moiety (ring C) and introduction of an acetamido group at C-7 (ring B), presents quite a challenge. The various syntheses of (±)-colchicine and its derivatives, including colchiceine, deacetylcolchiceine, and deacetamidocolchicine, previously converted into colchicine by Eschenmoser, vanTamelen, Nakamura, Scott, Martel, and Woodward, were discussed in detail in Volume XI of this treatise (*9*, and references therein). This chapter reviews the efforts made in the partial and total synthesis of colchicine subsequently reported.

A. Synthesis of Deacetamidocolchiceine (113)

Two syntheses of 4-[3-(3,4,5-trimethoxyphenyl)propyl]tropolone (**121**), an important intermediate in the synthesis of deacetamidocolchiceine (**113**), have been reported by Kato *et al.* (*144*) and by Matsui *et al.* (*145*). A key step in the Kato synthesis is the solvolytic fragmentation of the dichloroketene adduct **120** to **121** (Scheme 7). Tropolone **121** is a known compound and was used by Scott in the synthesis of deacetamidocolchiceine (**113**) (see references in *9*). Tropolone **121** was converted by Kaneko and Matsui (*146*) into the 5-tolylazo derivative **122** by azo coupling with the diazonium salt of *p*-toluidine. Hydrogenation of the resulting azo compound (**122**) in methanol afforded the amino tropolone **123**, and a Pschorr reaction converted it into deacetamidocolchiceine (**113**) (5%; mp 165–166°C), which was identical to an authentic sample (Scheme 8). Attempts by the same authors to cyclize the tropolone **126**, prepared in several steps from 3-methoxy-4-acetoxy-5-benzyloxyphenylacetaldehyde (**124**) and the anhydride **125** by oxidative coupling, failed.

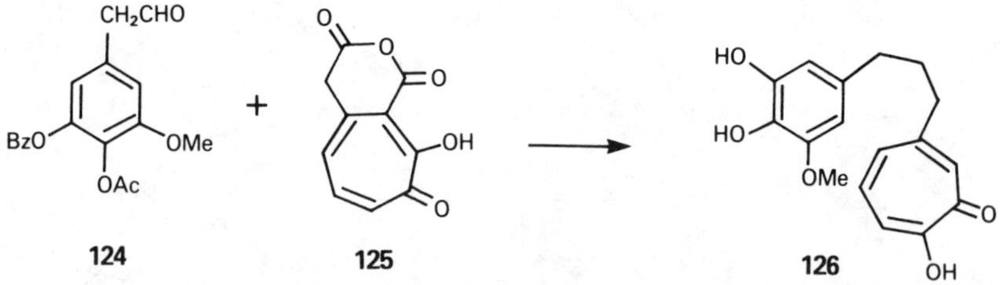

CH₂CHO BzO OMe OAc **124** + O O O OH O **125** → HO HO OMe O OH **126**

B. Synthesis of Deacetamidoisocolchicine (133)

A totally different approach for construction of the colchicine skeleton is based on a hypothetical biosynthetic scheme and was chosen by Tobinaga *et al.* (*147*), as shown in Scheme 9. The key step in this synthesis is

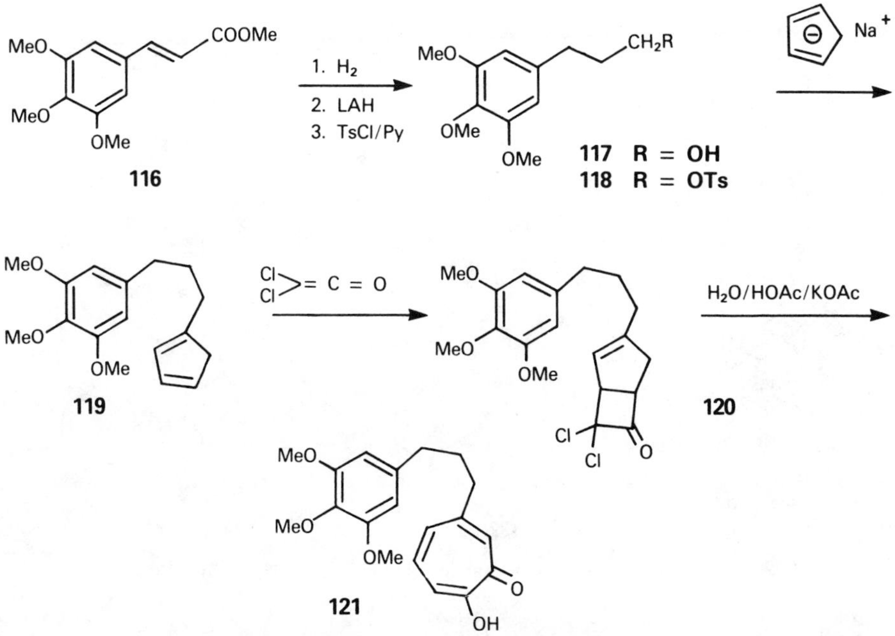

SCHEME 7. Synthesis of tropolone precursor **121** by Kato *et al.*

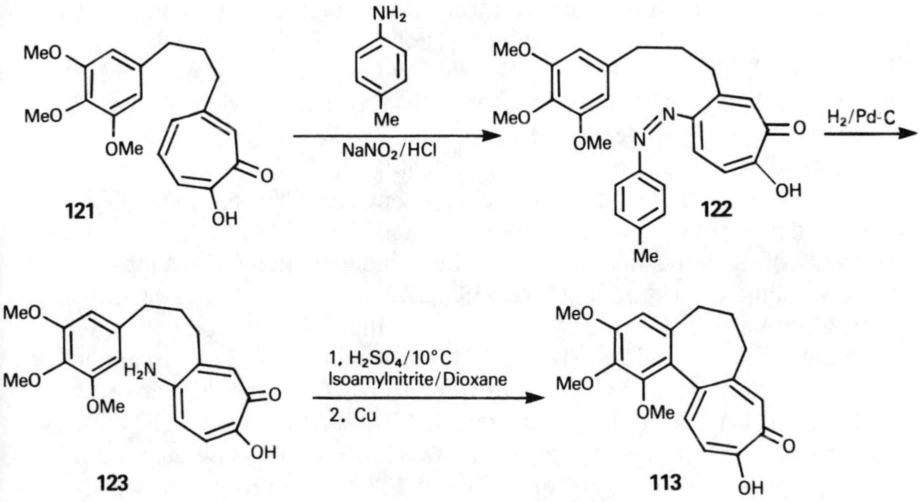

SCHEME 8. Synthesis of deacetamidocolchiceine (**113**).

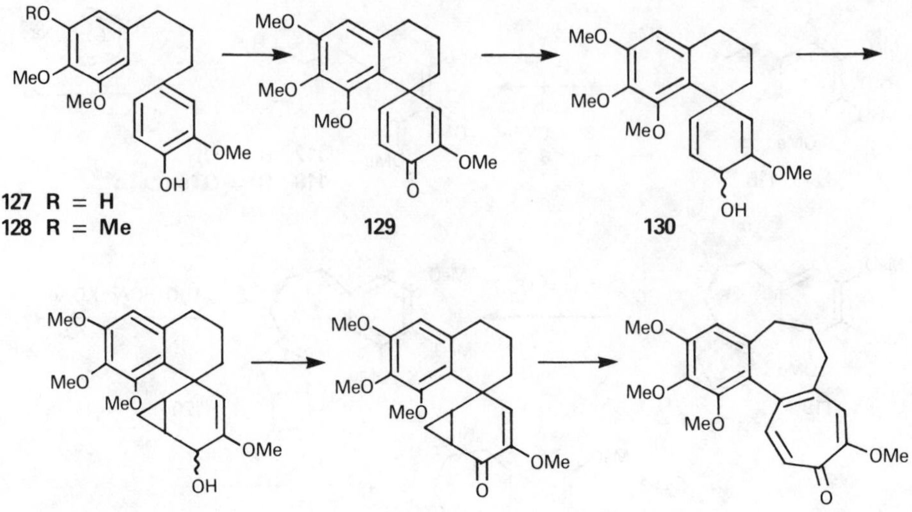

127 R = H
128 R = Me

SCHEME 9. Synthesis of deacetamidoisocolchicine (**133**) by Tobinaga.

a methylene transfer reaction followed by an acid-catalyzed rearrangement and dehydrogenation.

Hydride reduction of the spirodienone **129**, conveniently prepared in 80% yield by anodic oxidation of **128** or by oxidative phenolic coupling of **127** with an iron–DMF complex, afforded after treatment with diazomethane an epimeric mixture of the allyl alcohols **130**. Cyclopropanation by a Simmons–Smith reaction gave **131**, and oxidation with chromic acid afforded the ketone **132** (mp 163–165°C) in 42% overall yield from **129**. When treated with an acetic anhydride–sulfuric acid mixture, rearrangement and dehydrogenation occurred, possibly initiated by oxidation with SO_3, to give in a one-pot reaction the tropolone **133** (90%; mp 147–149°C).

A similar approach was chosen for the synthesis of **133** by Evans *et al.* (*148–149a*), as shown in Scheme 10. The quinone monoketal **134** afforded the crystalline cyclopropyl ketone **135** (93%) by treatment with dimethyloxosulfonium methylide, and reaction with the Grignard compound **136** gave the alcohol **137** in 70–90% yield. Rearrangement of **137** by treatment with trifluoroacetic acid afforded the dihydroisocolchicine derivative **138** (70%; mp 111–112°C) that converted to **133** by oxidation with DDQ (72%; mp 147–148°C). Shorter treatment of **137** with trifluoroacetic acid led to the conclusion that the rearrangement of **137** to **138** proceeded via the dienone **137a** and the diastereomeric spirans **137b** and **137c**. Spiran **137b**,

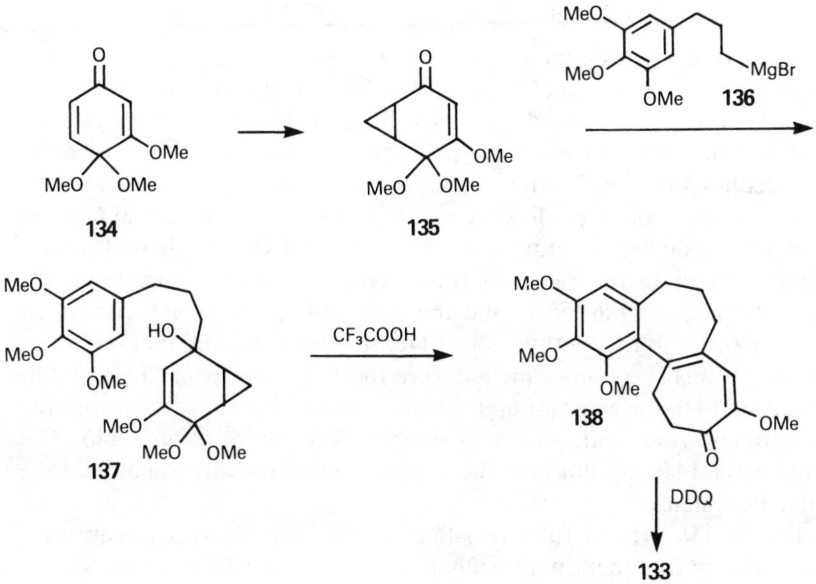

SCHEME 10. Synthesis of deacetamidoisocolchicine (**133**) by Evans.

identical to Tobinaga's intermediate **132**, was the only product with the stereochemical features required for the subsequent rearrangement to the isocolchicine structure.

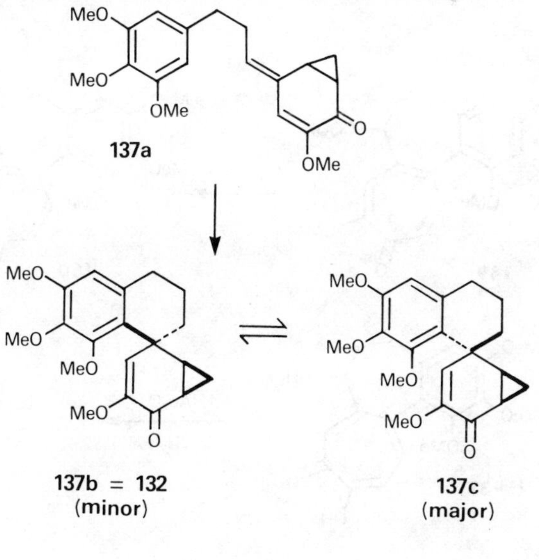

C. Synthesis of (±)-Colchicine by Evans

The tropolonic precursor **135** was utilized by Evans *et al.* in their total synthesis of racemic colchicine (*149*). The acetamido function was masked as a carboxylic acid ester (see Scheme 11). The *tert*-butyl ester **139** upon reaction with the ketone **135** in the presence of LDA yielded the labile alcohol **141**, which after immediate hydrolysis with oxalic acid afforded a mixture of three diastereomeric ketones with **143a** and **b** being the major products. Treatment of the mixture **143a** and **b** with boron trifluoride etherate for 5 min at room temperature in nitromethane afforded the lactone **146** (56%) and the acid **145** (23%), which converted advantageously to the methyl ester **144c**. It was postulated that the intermediates in this reaction sequence were the *tert*-butyl esters **144a** and **b**. For isomer **144b** the ester group is ideally situated for interaction with the cyclopropane ring, and, with loss of the alkyl group, affords **146**. The methyl ester **144c** did not give the corresponding dihydrotropolone by a similar treatment.

Whereas **146** proved fully resistant to the acid-catalyzed rearrangement, **144c** by treatment with trifluoroacetic acid for 75 min gave a mixture of the tricyclic esters **147**. Oxidation of **147** with DDQ gave a separable mixture of the tropolone ether **148** (mp 127–128°C) with a heptafulvene tautomer.

Base-catalyzed hydrolysis of **148** afforded the acid **149** (85%; mp 179–

148 → MeO, MeO, OMe, COOH, OMe, O **149** → MeO, MeO, OMe, NHCOOt-Bu, OMe, O **150**

H_3O^+ → MeO, MeO, OMe, NH_2, O, OH (±)-**31** → 1. Ac_2O 2. CH_2N_2 → (±) - Colchicine, (±) - Isocolchicine

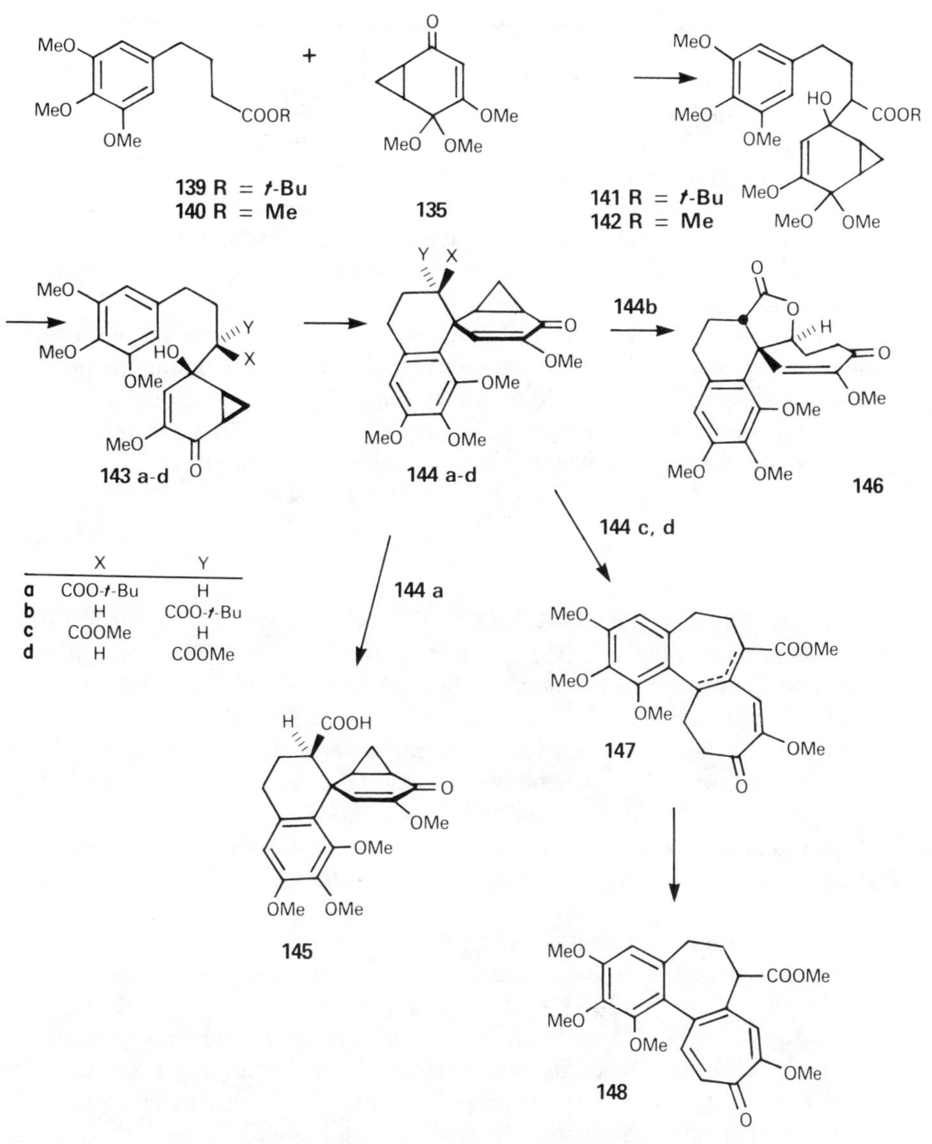

SCHEME 11. Total synthesis of (±)-colchicine by Evans.

180°C), which was further transformed into the carbamate **150** (54–62%) with diphenylphosphorylazide and triethylamine in *tert*-butanol. Removal of the *tert*-butoxycarbonyl group was accomplished by concomitant hydrolysis to afford (±)-deacetylcolchiceine (**31**; 72%). The low overall yield in the formation of **31** by this route is mainly due to loss of the *tert*-butyl group during the BF_3 treatment of **143a** and **b** affording the carboxylic acid **144** (X = H, Y = COOH) and the undesired lactone **146**, obtained during the acid-catalyzed opening of the cyclopropane ring. This could be avoided by beginning the synthesis with the methyl ester **140**. The intermediate methyl esters **143c** and **d** were more resistant in the cyclization reaction and afforded a 71% yield of the spiro compounds **144c** and **d**, and no formation of lactone **146** was observed. Treatment of **144c** and **d** with refluxing trifluoroacetic acid for 35 min afforded **147** in 78% yield after chromatography.

The keto ester **147** could be prepared in 92% yield from **142** with trifluoroacetic acid at different temperatures and without isolating the intermediates **144c** and **d**. With the conversion of **147** to deacetylcolchiceine [(±)-**31**] previously converted to (±)-colchicine by standard procedures, another total synthesis of this alkaloid was accomplished.

V. Biological Properties of Colchicine Alkaloids

The biological properties of colchicine, including toxicity, pharmacology, general cytological effects, action on tumor growth, effects on plant growth, the production of polyploidy in plants, and effects on lower animals were reviewed (*7,15,150,151*), and colchicine's interference with microtubule-dependent cell functions and its binding to tubulin, the subunit of microtubules, was authoritatively discussed (*6a,152,153*). Although colchicine, in arresting cell division, possesses unique biological features that cause it to be considered a potential antitumor agent in man, its high general toxicity precluded its clinical utilization (*154,155*). Colchicine remains, however, an important biochemical tool and is still being widely used for treatment of gout (*156*).

Colchicine, like Vinca alkaloids and podophyllotoxin, belongs to a group of natural products that bind to tubulin, the protein subunit of microtubules (*157–163*), and inhibits cell processes that depend on microtubule function by blocking polymerization. Microtubule protein has a molecular weight of ~110,000 (*164–167*), and has been isolated from diverse tissues such as HeLa cells (*168,169*), grasshopper embryos (*170*), mammalian brain (*167*), the cilia of *Tetrahymena* (*171*), and sea urchin sperm tails (*172,173*). Due to colchicine's inhibitory action, treated cells are arrested at metaphase because microtubules are essential for moving chromosomes during mitosis; the transport of vesicles is also affected (*163*). A colchicine-induced dissolution of microtubules present in polymorphonuclear leukocytes has been demonstrated (*173*). These leukocytes are intimately involved in the inflammatory process, and since colchicine inhibits leukocyte functions, such as ameboid migration and lysosomal degranulaton, it seems reasonable to assume that the leukocytes may mediate the antigout effect of colchicine (*173*).

In preventing microtubule formation, colchicine has been shown to inhibit catecholamine secretion from the adrenal medulla (*174,175*), iodine secretion from the thyroid gland (*176*), and prolactin secretion from pituitary tumor cells in culture (*177*). Moreover, as demonstrated *in vitro*, colchicine inhibits stimulated insulin secretion from the isolated perfused pancreas (*178*), from isolated perfused islets (*179*), and from slices of pancreas (*180*). Colchicine inhibits many processes involving the traffic in or out of cells; the marked inhibition of glucose-induced insulin secretion and the impairment of glucose tolerance in the intact rat (*181*) are such manifestations.

Although a relatively specific drug, colchicine acts directly on systems apparently unrelated to microtubules, including inhibition of isolated aldose reductases (*182*), blocking of ganglionic conduction due to direct postsynaptic antagonism to acetylcholine (*183*), lens epithelial elongation in chicks (*184*), and inhibition of nucleoside transport (*185*). Binding to membrane protein (*186*), which in many cases has been shown to be related to tubulin, and inhibition of collagen synthesis (*187,188*) may also represent effects of colchicine not necessarily related to microtubules.

The toxicity of colchicine and its analogs prevented their therapeutic application. Colchiceine, *N*-benzoyldeacetylcolchicine, and deacetylcolchicine, although considerably less toxic, showed only limited antimitotic effects (*189,190*). Deacetylcolchicine, lacking the *N*-acetyl side chain of colchicine, was selected for clinical trials (*191*) and tested at the National Cancer Institute (NCI) in the United States, under the code number NSC-36354, in form of its tartrate salt. It did show some activity in lymphoma and reasonable effects in chrome myelogenous leukemia, but the results were no better than those obtained with other antitumor agents, and the classic "colchicine toxicity" was still present (*191*).

Considerable efforts, particularly by Lettré in Germany (*192,193*) and by Velluz in France (*194*), to overcome the problem of toxicity by structural modification did not afford greatly improved compounds, although colchiceineamides (*192,193*), demecolcine (*195,196*), and thiocolchicines (*131,197*) had some advantages.

The most important biological effect of colchicine is the inhibition of processes that depend on microtubule function, which is caused by blockage of polymerization (*163*). The binding of colchicine to tubulin protein, the dimeric subunit composed of α- and β-units, has been extensively studied both *in vitro* and *in vivo* (*198–200*). The results of these investigations suggest that colchicine binds to the tubulin dimer, and binding to the intact microtubule has not been demonstrated (*163,164,172,201*). Each tubulin dimer holds one colchicine binding site (*163,202–204*), but the exact location of this high-affinity site is not known, although recent work suggests that it may be the β-subunit (*138*).

Tubulin isolated from widely diverse sources exhibits very similar colchicine binding properties (*163,164,166,169,172,205,206*). The binding of colchicine to rat brain tubulin is slow and virtually irreversible (*162,166, 167,207–210*). Reinvestigation of the kinetics of the tubulin-complex formation, using radioactive isotope labeling or fluorescent techniques, afforded data suggesting that the initial step is a colchicine-induced protein conformational change (*211*), which is essential to obtain nearly quantitative binding of one mole of colchicine and which is believed to be the cause for the shifting of assembled microtubules into their disassociated forms. The conformational change seems to be specific for the binding of the entire colchicine molecule, not just for one of its structural features. The altered conformation of tubulin may be involved in poisoning the microtubule assembly by the addition of colchicine–tubulin complex to the ends of the microtubules (*212*). The mechanism of tubulin assembly and disassembly has been extensively investigated (*213,214*).

The binding of colchicine to tubulin is both time and temperature dependent (*157,166*), and the binding sites, which are hidden in normal microtubules, are exposed in the cold or by Ca^{2+}, because both factors depolymerize microtubules (*215*). It has been postulated that the colchicine binding site on tubulin might also be a receptor site for endogenous cellular molecules, and it has been shown that the mammalian brain does in fact contain a number of molecules that interact with the colchicine binding site on tubulin (*216,217*). One of these molecules was found to be a protein and the other a heat-stable peptide. These endogenous ligands may play a significant role in the regulation of cellular microtubule function and assembly and possibly represent the natural ligands.

The binding of colchicine and the analogs podophyllotoxin and vinblastine to rat brain microtubule protein was investigated by measuring the ability to inhibit binding of [^{3}H]colchicine (*157,159,161,163*). Colchicine, podophyllotoxin, and vinblastine were noncovalently bound to tubulin, thus preventing its polymerization. Tubulin has two binding sites for vinblastine that are different from the colchicine binding site. Podophyllotoxin, on the other hand, competes with the colchicine binding site. The importance of substitution in the aromatic ring A of colchicine was demonstrated (*157,158,160*), and it was found that the tubulin binding potency paralleled the antigout effects (*157*). Although both demecolcine and deacetylcolchicine were good inhibitors, the presence of an *N*-acetyl or an *N*-formyl group appeared to enhance the binding (*218*), suggesting that *in vivo* N-acylation of primary or secondary colchicine amines to more potent N-acetylated analogs should also be considered (*157*). Good correlation was found between compounds active *in vivo* and their tubulin affinity *in vitro*, although certain exceptions were noted (*157,219*).

Relatively few studies explored the binding of colchicine to proteins other than tubulin, but antibodies of colchicine were prepared by coupling deacetylcolchicine to bovine serum albumin. The antibody binding site accommodated many changes in the tropolonic ring C and, in contrast to tubulin binding, the chemical and stereochemical features for binding were much less stringent (*220*). The effect of colchicine and derivatives on the induction of antibody responses in tissue cultures suggested that some may block transport through membranes (*221*), but no correlation was found between immunoinhibitory effect, toxicity, and antimitotic potency. The effects of inhibitors of tubulin polymerization on GTPase hydrolysis revealed stimulation by colchicine, but inhibition by vinblastine and podophyllotoxin suggested that colchicine's responses may depend on specific structural features (*222*). Investigation of the interaction of colchicine with DNA molecules suggested that intercalation occurred between the tropolonic moiety and the nitrogenous bases of DNA (*223*). Similar nonconclusive evidence was arrived at in a study with compounds containing sulfhydryl groups (*224*).

A spin probe for measuring colchicine binding to tubulin was prepared with allocolchicine, showing that this synthetic derivative did bind to the same site on tubulin as did colchicine (*225*). Other spin probes have recently been prepared from racemic and optically active 2,2,5,5-tetramethyl-1-pyrrolidinoxy-3-carboxylic acids with deacetylcolchicine and colchifoline (*226*). The latter derivatives showed good tubulin binding properties and will serve to further characterize the colchicine binding site on tubulin. The synthesis, purification, and biological activity of fluorescein colchicine (FC), a fluorescent affinity labeled colchicine with biological effects identical to those of colchicine, has been reported. FC was found to be of interest in studying colchicine-sensitive cellular pools (*227*).

VI. Structure–Activity Relationships

The antimitotic effects of colchicine and analogs have been assessed by measuring their influence on mitosis in rat and mouse liver (*228*), their inhibition of mitosis in sea urchin eggs (*229*), their antimitotic effect in the mouse ascites tumor (*230*), their effect in cell cultures using the L-1210 mouse leukemia assay (*231*), and their inhibition of human tumor growth (*232*). These methods are authoritatively discussed by Lettré (*233*).

Because chemical purity of the various compounds tested was not always secured and some of them had obviously been contaminated with the "inactive" iso isomer, quantitative assessments of the biological activity were difficult (*190*). Colchicine and congeners tend to crystallize with a considerable amount of water or solvents of crystallization (*234*), another factor which may have augmented experimental errors inherent in the biological evaluation. The structural features found essential for antitumor activity of colchinoids *in vivo* and *in vitro* lead to the following conclusions (*15,157,193,235–237*):

1. N-Acylation of deacetylcolchicine enhances both the *in vitro* activity and particularly the *in vivo* potency (*15*).
2. Substitution in the aromatic ring A of colchicine seems critical, as demonstrated by the inactivity of colchicoside, an *O*-glucoside of 3-demethylcolchicine (*15,157*).
3. Isocolchicines with the tautomeric arrangement of the tropolonic oxygen functionalities are inactive.
4. Colchiceines, keto enols with a 10-OH instead of a 10-OMe substituent, are inactive.

With the development of modern turbidometric and viscometric techniques for measuring the polymerization of tubulin (*157,158,162,163*), simpler methods became available to verify the above-mentioned conclusions and to extend the evaluation to new analogs of colchicine prepared since. Although colchiceine can exist in solution and probably also in biological media as the inactive isomer, it was shown that a colchiceine–ethyl acetate–water solvate used for single-crystal X-ray analysis did crystallize as the isomer having the natural configuration (*98*). Colchiceine, a metabolite of colchicine (*89*), does interact with gangliosides dependent on Ca^{2+} (*238*). It also has potent Mg^{2+}-, Zn^{2+}-, and Ca^{2+}-dependent inhibitory activity against alkaline phosphatases from *E. coli* and calf intestine, and was active against acid phosphatases from potato (*239*). The 10-OMe group in colchicine can, as already mentioned, be replaced without apparent loss of antimitotic potency by an SMe group (thiocolchicine), as shown by Velluz (*194*), or by a NHMe group (colchiceinamide), as demonstrated by Lettré (*192,193,240–242*).

Conversion of deacetylthiocolchicine and some of its *O*-alkyl homologs (*129*) into urea derivatives and evaluation in the rather unreliable L-1210 mouse leukemia screen *in vitro* (*191*) did not afford compounds superior to the starting material or to colchicine (*139–140a*).

The natural configuration of the *N*-acetamido side chain at C-7 seems essential for *in vivo* activity on the basis of data elaborated with (+)-colchicine (*125*), its unnatural antipode, which was much less active biologically in a number of *in vitro* and *in vivo* assays (*46,220,243*). The enhancement of the potency of N-acylated members of the colchicine alkaloids is paralleled by Lettré's findings of high potency in *N*-haloacetyldeacetylcolchicines (*137*), particularly the *N*-fluoroacetyldeacetylcolchicine and its 10-thiomethyl analog, both of which were several times more potent than colchicine in cultured chicken heart fibroblasts and mouse ascites tumors. Some activity was observed in patients with skin carcinoma (*244*).

The good correlation between the ability of a given colchicine analog to displace [^{3}H]colchicine from rat brain tubulin and its antimitotic and antigout properties (*157,245*), as illustrated in Table IV, marks the tubulin binding assay as a valuable first screening test for this group of compounds.

The disclosure of high *in vitro* potency of deacetamidocolchicine, which lacks the *N*-acetamido side chain (*246*), as supported by additional data (*142,247,248*), is surprising. Although excellent binding of deacetamidocolchicine and its 5,6-dehydro analog to tubulin could be demonstrated, the *in vivo* activity of these compounds is, however, much weaker than that of colchicine (*46,143a*). The Quinn–Beisler study (*237*) afforded additional data useful in the search for an improved colchicine. In this study colchicine and 26 analogs were compared with regard to *in vivo* activity in the leukemia lymphocytic P388 assay in relation to octanol–water partition coefficients. It was determined that N-acylated compounds were always more potent than their *N*-deacetyl congeners. Introduction of electron-withdrawing groups, such as fluorine or bromine, at the acetamido function often increased the potency, whereas replacing an *N*-acyl group with an *N*-aroyl group seemed to afford less potent compounds. With the availability of more specific and less time consuming assay methods, the need arose to reinvestigate systematically the structural features of colchicine.

These investigations, carried out at the NIH in Bethesda since 1977, included the comparison of well-characterized colchinoids with regard to tubulin binding affinity, acute toxicity, and *in vivo* activity in mice in the leukemia lymphocytic P388 screen (*46,249*). The tubulin binding was determined by measuring the ability of the compounds to displace [^{3}H]colchicine according to established procedures (*157*). The binding affinity of the compounds showing high affinity must be ascertained at much lower concentrations. A similar follow-up should to be carried out with compounds showing considerably different toxicity after single (LC) and multiple injections (NCI). Such a study will be carried out with compounds

TABLE IV

BIOLOGICAL ACTIVITY OF COLCHICINE AND ANALOGS

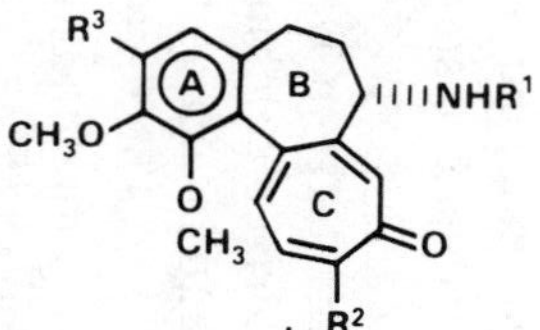

Compound	R^1	R^2	R^3	Inhibition of [^{3}H]colchicine binding to rat brain microtubule protein[a] (%)	MED[b] (mg/kg sc)	Inhibition of urate-induced rat hindpaw edema[c] (%)
Structure I						
Colchicine	$COCH_3$	OCH_3	OCH_3	90–95	1.5	87.7
Colchiceinamide	$COCH_3$	NH_2	OCH_3	82	2.0	78.7
N-deacetylthiocolchicine	H	SCH_3	OCH_3	67		50.4
Demecolcine	CH_3	OCH_3	OCH_3	60		87.9
TMCA methyl ether	H	OCH_3	OCH_3	52	2	78.9
TMCA ethyl ether	H	OC_2H_5	OCH_3	38	3	79
Colchicosamide	$COCH_3$	HN_2	$C_6H_{11}O_7$	2	>920	(18)
Colchicoside	$COCH_3$	OCH_3	$C_6H_{11}O_7$	3		(2)

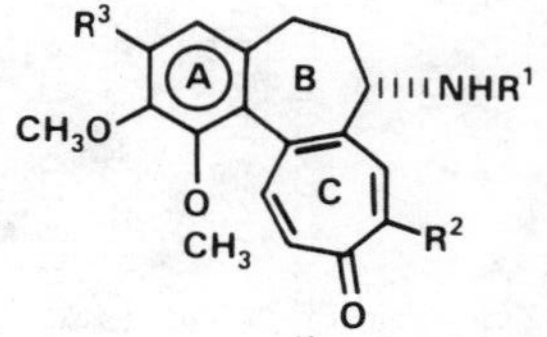

Structure II						
Isocolchicine	$COCH_3$	OCH_3	OCH_3	2	>200	(7)
Isocolchiceinamide	$COCH_3$	NH_2	OCH_3	1	>500	(0)
Iso-TMCA methyl ether	H	OCH_3	OCH_3	4	>1000	(16)
Colchiceine	$COCH_3$	OH	OCH_3	6	75	25
TMCA	H	OH	OCH_3	0	600	(26)
N-benzoyl TMCA	C_6H_5CO	OH	OCH_3	2	>200	(0)
Other antimitotic compounds						
Podophyllotoxin				94		
Vinblastine				25	2	64

[a] Incubation mixtures containing 2.5×10^{-6} *M* [^{3}H]colchicine, 2.5×10^{-5} *M* test compound and 100,000 *g* rat brain supernatant (1 to 1.5 mg/ml) were kept at 37° for 4 hr.

[b] MED is the minimum effective dose in mice required to produce hemorrhage and necrosis in sarcoma 37. Data taken from Leiter *et al.*

[c] Each value represents the mean percentage inhibition (eight rats) of mean control edema (eight rats) in one experiment. The dosage was 2 mg/kg (sc) for all drugs with the exception of TMCA where it was 10 mg/kg (sc). All values are significantly different from saline controls ($p < .05$), except those in parentheses. Data taken from Zweig *et al.* (*245*).

selected for further evaluation. The P388 assay of NCI employed 20- to 22-g mice with lymphocytic leukemia. The potency of the compounds listed in Table V is given in mg/kg and was calculated for an arbitrary value of *T*/*C* of 140 (40% increase in survival time in a treated versus a control set of mice in the P388 assay). The *T* represents the average number of days the group of treated mice survived at a particular dose level in the assay, and *C* stands for the survival time of a group of control mice (*46,249*).

Similar studies were carried out elsewhere using other plant constituents which bind to tubulin, such as podophyllotoxin (*161,250*), steganacin (*251*), mytanisine (*252*), taxol (*213*), the synthetic drug oncodazole (*253*), and other synthetic compounds (*254,255*). A direct relationship between tubulin binding and P388 potency was observed in the 3-demethyl-, 2-demethyl-, and 1-demethylcolchicine series, the 3-demethylcolchicine being the only monophenol that exhibited considerable potency and possibly lower toxicity than that of colchicine (Table V). Etherification or esterification of 2- and 3-demethylcolchicine is known to afford more potent derivatives in the ether series that approach the high potency of colchicine (*235*).

Acyl derivatives of demecolcine were more potent than the parent alkaloid whereas *N*-methyldemecolcine, a tertiary amine, showed only low *in vitro* and *in vivo* activity. The high *in vitro* but low *in vivo* potency of 5,6-dehydro-7-deacetamidocolchicine, a close relative of the deacetamidocolchicine investigated by Schindler (*142,246,247*), suggests that the presence of an *N*-acylamido substituent at C-7 is a prerequisite for high potency *in vivo*. The importance of the aromatic substitution in ring A in tropolonic colchinoids was confirmed not only by the high potency of cornigerine and its active 2,3-didemethyl congener, but also by the absence of activity of the 1,2-methylenedioxy isomer and its 1,2-didemethyl precursor (Table V). It is evident that aromatic substitution in ring A has a greater influence on the biological activity in colchinoids than hereto recognized, and it will be interesting to compare "active" and "inactive" molecules in the solid state by single crystal X-ray analysis in further searching for scientific explanations of this behavior.

In general, *in vivo* potency runs parallel with toxicity. Several *N*-aroyldeacetylcolchicines were prepared in pure form and evaluated. In contrast to earlier findings, the *N*-benzoyldeacetylcolchicine was found remarkably potent, demonstrating that QSAR studies will give useful results only if they are based on data elaborated with chemically pure compounds (*237*). In the aliphatic series of *N*-acyldeacetylcolchicines, the *N*-propionyl and *N*-butyryl analogs of colchicine were remarkably potent but also as toxic as the parent alkaloid. N-Methylated, O-ethylated, and

O-acylated derivatives of colchifoline were as potent and toxic as the parent alkaloid and as colchicine. The ethoxycarbamoyl derivative, containing a carbamate instead of an *N*-acylamido function, was highly potent and may well represent a novel and interesting structural variation in colchinoids.

Colchicide, first prepared by Rapoport from thiocolchicine and reported to be active *in vitro,* is another interesting molecule (*132*) that has also been found to be quite potent *in vivo* (*128,249*). This demonstrates that the 10-OMe group in colchicine, once considered vital for activity, is indeed not important. Studies to find less toxic colchicines useful as antigout and potentially for therapeutic antimitotic agents, possibly with selectivity toward tumor cells, was greatly stimulated by these developments. Derivatization of colchifoline afforded some novel biochemical tools which might be of interest for measuring and analyzing the colchicine binding site on tubulin (*226*). Recent X-ray crystallographic studies of inactive and highly potent colchinoids revealed that these compounds are differently hydrogen bonded in the solid state. Some active compounds retain solvents or water in "molecular holes," findings which might be of great importance for further exploration of structure–activity relationships (*254*).

Efforts to find a better and possibly less toxic colchinoid are being carried out by biochemists, who attempt to further characterize the colchicine binding site on tubulin and to arrive at an understanding as to how colchicine exerts its antimitotic effects at the molecular level. There is good reason to believe that interest in colchicine, a fascinating molecule which for sometime seemed forgotten, is revived and will hopefully stimulate renewed research in this important field.

VII. Biosynthesis of Colchicine

Scheme 12 delineates the biosynthesis of colchicine as obtained by the result of work accomplished mainly by Battersby *et al.* and to a minor extent by Leete *et al.* Since their results were reported (*256,257*), no details will be given here.

The basic building blocks in the biosynthesis of colchicine are cinnamic acid (from phenylalanine) and tyrosine (*258,259*). They form ring A, the carbon atoms C-5, C-6, and C-7 of ring B, and C-12 of ring C in colchicine. It has been proved that the tetrahydroisoquinoline alkaloid (*S*)-autumnaline (**151**) is an intermediate in the biosynthesis of colchicine (*260,261*). Therefore, colchicine can be considered to be a modified isoquinoline system rather than a "puzzling inconsistency" (*261*). Autumnaline itself

TABLE V
Biological Activity of Colchicine and Analogs

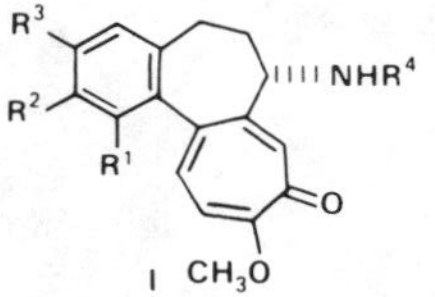

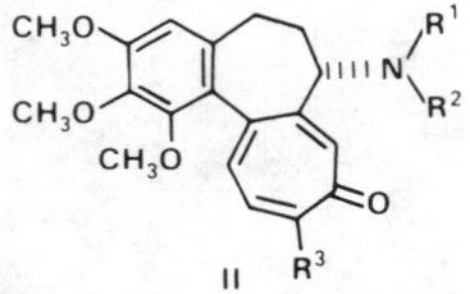

Compound	R^1	R^2	R^3	R^4	Potency[a] (μmol/kg)	Tubulin Binding[b] (%)	Toxicity[c] (NCI)	Toxicity[d] (LC)
Structure I								
Colchicine	OCH_3	OCH_3	OCH_3	$COCH_3$	0.4	90	2.5	3.0 (2.5–4.0)
N-Trifluoroacetyl-	OCH_3	OCH_3	OCH_3	$COCF_3$	0.4	100	0.9	1.1 (0.9–1.3)
Colchifoline	OCH_3	OCH_3	OCH_3	$COCH_2OH$	0.7	97	3.0	3.9 (2.5–4.0)
N-*n*-Butyryl-	OCH_3	OCH_3	OCH_3	$CO(CH_2)_2CH_3$	0.7	98	5.8	5.8 (4.0–8.9)
N-Pivaloyl-	OCH_3	OCH_3	OCH_3	$CO(CH_3)_3$	—	65	—	9.3 (6.8–12.9)
N-Propionyl-	OCH_3	OCH_3	OCH_3	$COCH_2CH_3$	—	98	—	8.2 (4.8–14.0)
Cornigerine	OCH_3	O—CH_2—O		$COCH_3$	<10.0	97	39.1	9.7 (7.0–13.1)
N-Benzoyl-	OCH_3	OCH_3	OCH_3	COC_6H_5	1.0	83	4.3	60.5 (43.2–85.0)
N-Phenoxyacetyl-	OCH_3	OCH_3	OCH_3	$COCH_2OC_6H_5$	1.3	81	5.1	61.3 (45.0–83.5)
N-Carbethoxy-	OCH_3	OCH_3	OCH_3	COOEt	0.7	86	11.7	114.0 (88.3–147.1)
O-Ethylcolchifoline	OCH_3	OCH_3	OCH_3	$COCH_2OEt$	0.4	74	2.8	9.0 (6.5–12.6)
O-Acetylcolchifoline	OCH_3	OCH_3	OCH_3	$COCH_2OAc$	—	86	—	8.5 (6.6–11.2)
3-Demethylcolchicine	OH	OCH_3	OCH_3	$COCH_3$	3.6	68	26.0	31.7 (22.6–44.4)
2,3-Didemethyl-	OCH_3	OH	OH	$COCH_3$	13.5	18	54	>450

2-Demethylcolchicine	OCH_3	OH	OCH_3	$COCH_3$	20.0	50	51.9	42.9 (28.6–63.9)
Trimethoxybenzoyl-	OCH_3	OCH_3	OCH_3	$COC_6H_2(OCH_3)_3$	14.0	46	72.6	>180
1,2-Methylenedioxy	O—CH_2—O		OCH_3	$COCH_3$	inactive	94	104.0	>167
1,2-Didemethyl-	OH	OH	OCH_3	$COCH_3$	inactive	0	>320	>450
Structure II								
N-Acetyldemecolcine	CH_3	$COCH_3$	OCH_3		0.4	98	3.0	3.6 (2.4–5.1)
N-Trifluoroacetyldem.	CH_3	$COCF_3$	OCH_3		1.5	83	5.4	8.8 (6.4–11.8)
N-Methylcolchifoline	CH_3	$COCH_2OH$	OCH_3		0.6	99	2.9	14.5 (11.2–18.9)
N-Formyldemecolcine	CH_3	CHO	OCH_3		1.7	76	6.3	179.7 (101.5–317.8)
Colchicide	H	$COCH_3$	H		9.2	87	108	168.9 (119.5–238.8)
Demecolcine	CH_3	H	OCH_3		10.8	93	21.6	236.7 (200.8–279.0)
Speciosine	CH_3	—[e]	OCH_3		22.0	62	83.9	296.4 (227.9–385.7)
N-Methyldemecolcine	CH_3	CH_3	OCH_3		68	38	51.9	519.5 (337.7–797.4)

[a] Obtained by graphical estimation of the potency which would give a $T/C = 140$.

[b] Percentage by which binding is reduced by the presence of the inhibitor, at 25 μM with [^3H] colchicine at 2.5 μM. Values are the average of triplicate assays.

[c] Toxicity in the P388 NCI screen. Defined as that dose, in μmol/kg, which caused an average weight loss >3 g in a set of mice or the death of one or more of the mice in the set. Deaths were noted on day 5 of a series of nine injections in 9 days (1 per day) of the compound.

[d] Toxicity found after a single im injection, in μmol/kg. The total number of deaths were counted after 7 days at various dose levels, and the LD_{50} was determined by probit analysis. Parenthesized numbers represent 95% confidence interval.

[e] CH_2— HO

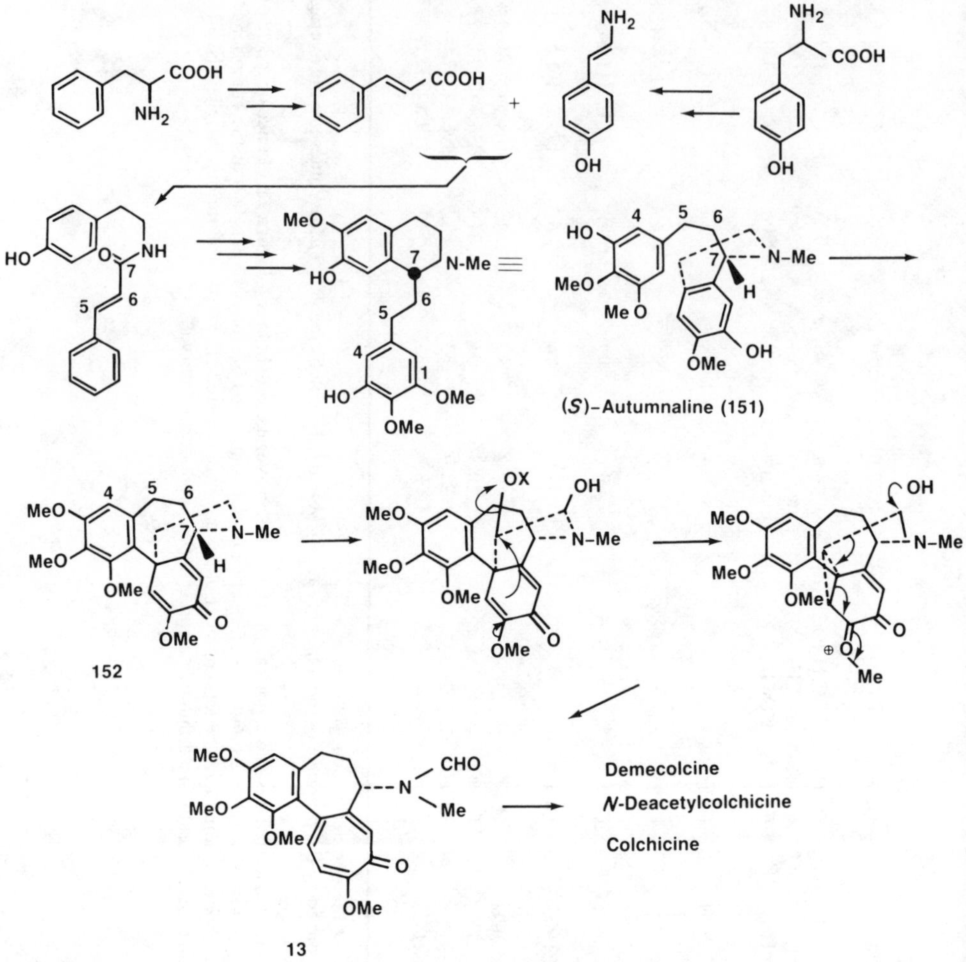

SCHEME 12. Biosynthesis of colchicine.

has been isolated from *Colchicum cornigerum,* confirming the biosynthetic pathway (*64*). [1-^{13}C]Autumnaline (**151**) has been incorporated into colchicine, and an approximately 2.5-fold enhancement of the ^{13}C-NMR signal for C-7 was found in colchicine (*262*). Oxidative coupling and O-methylation generated *O*-methylandrocymbine (**152**) as a second important biosynthetic intermediate. Labeled **152** was incorporated into colchicine to an extent (*261*). The demethylated compound androcymbine was found in *Androcymbium melanthioides* (*263–265*), a plant genus belonging to the same family, Colchiceae, as the genus *Colchicum*.

The bridge was attacked next, and ring expansion of the aromatic nucleus of tyrosine with inclusion of the first carbon atom of the side chain (C-12 in colchicine) generated the tropolone ring C. This has been proved by feeding *Colchicum* plants with [3-^{14}C]tyrosine and degrading the extracted colchicine (*266*). *N*-Formyldemecolcine (**13**) can be considered to be the common precursor for colchicine and its congeners.

Addendum

The slow kinetics of the colchicine-binding reaction (*211*) and a decrease in the dissociation constant of the tubulin α–β dimer when colchicine is bound (*267*) are consistent with the hypothesis that conformational change of colchicine itself may also occur in the reaction sequence leading to the formation of the colchicine–tubulin complex (*267,268*). Possible multiple conformations of colchicine were explained with a boat–boat interconversion of the tropinoid ring C of colchicine or by atropisomerism about the phenyl–tropolone axis (*267*). The *R* chirality of the biaryl, illustrated in the skewed conformation, appears to be the predominant molecular characteristic, permitting the recognition of colchicine by tubulin.

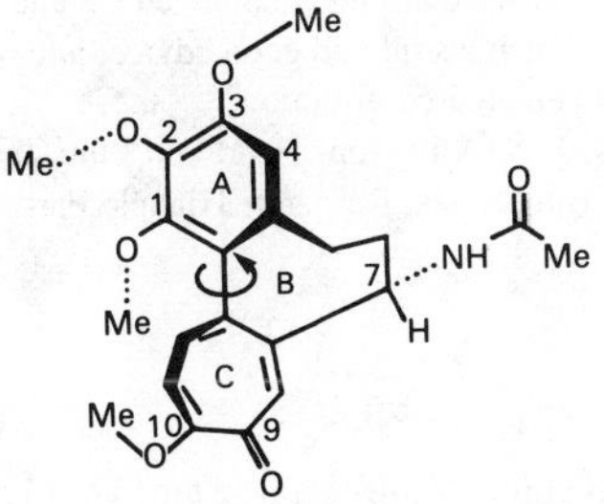

NORMAL

It has been postulated that the skewed conformation of colchicine first binds to tubulin with its primary recognition sites (C-1 OMe, NHAc, CO) by forming a tubulin–colchicine (TC) complex, changing later to a TC* complex of higher energy. A flattening of the biaryl system in the TC* complex is accompanied by a conformational change of tubulin itself (*268*). The solid state X-ray structures of tropolonic colchicines do not show skewed conformations (*254*). However, it can not be excluded that they occur in solution and that their formation is possibly favored by water molecules' occupying the molecular holes provided with the hydrogen-bonded network of primary recognition sites (*254*). A similar situation exists in the steganin family of compounds (*269*).

A novel affinity labeled analog of colchicine obtained by reacting deacetylcolchicine with *N*-succinimidyl-6-(4′-azido-2′-nitrophenylamino)hexanoate was incorporated into the α-tubulin monomer of renal tubulin in a 1 : 1 stoichiometry (*270*). Colchicine blocked the Na^+ currents of squid axons with little effect on K^+ currents, another effect of colchicine that is not directly dependent on microtubule function (*271*).

The structure of colchiciline (**15**), originally proposed by Santavy, was confirmed by an analysis of its 500-MHz proton spectrum (*272*). The spectrum showed that the NH proton (δ 8.34) is coupled to the hydrogen on C-7 (δ 4.47, *J*-6 Hz). The C-7 proton is also coupled to a proton on C-9 (δ 4.09, $J_{6,7}$-9.5 Hz). The value of $J_{6,7}$ requires these protons to have a trans diaxial relationship and therefore the substituents are trans diequatorial. The occurrence of atropisomers in a series of 1-acetyl-1-demethylcolchicines and their 1,2-diacetyl-1,2-didemethyl analogs has been reported to occur in solution (*54a*).

Acknowledgments

We thank Dr. Hanns H. Lehr, a dear friend, linguist, and scholar, for his help in finalizing this review, typed by Mrs. Kathleen Carter at our NIH office. We also would like to thank Dr. P. Bellet from Roussel-UCLAF in Paris for his constant encouragement and support, Drs. A. E. Jacobson and J. Wolff from our Institute for much useful and good advice, and particularly Dr. Padam N. Sharma for his enthusiasm, help, and personal contribution.

Thanks are also due to Drs. J. V. Silverton, Heart and Lung Institute, NIH, and C. F. Chignell, Laboratory of Environmental Biophysics, Research Triangle Park, North Carolina, for their critique and assistance.

References

1. P. J. Pelletier and J. Caventou, *Ann. Chim. Phys.* [2] **14,** 69 (1820).
2. P. L. Geiger, *Ann. Chem. Pharm.* **7,** 274 (1833).
3. M. L. Oberlin, *Ann. Chim. Phys.* [3] **50,** 108 (1857).
4. S. Zeisel, *Monatsh. Chem.* **4,** 162(1883); **7,** 557(1886); **9,** 1(1888).
5. A. Houdé, *C. R. Hebd. Seances Acad. Sci., Ser. A.* **98,** 1442 (1884).
6. O. J. Eigsti and P. Dustin, Jr., "Colchicine in Agriculture, Medicine, Biology and Chemistry." Iowa State Univ. Press, Iowa City, 1955.

6a. P. Dustin, "Microtubules," Springer-Verlag, 1978.

7. J. W. Cook and J. D. Loudon, *in* "The Alkaloids" (R. H. F. Manske, ed.), Vol. II, p. 261. Academic Press, New York, 1952.
8. W. C. Wildman, *in* "The Alkaloids" (R. H. F. Manske, ed.), Vol. VI, p. 247. Academic Press, New York, 1960.

9. W. C. Wildman and B. A. Pursey, *in* "The Alkaloids" (R. H. F. Manske, ed.), Vol. XI, p. 407. Academic Press, New York, 1968.
10. F. Šantavý, *Planta Med., Suppl.* 46 (1968).
11. P. Manitto, *Fitoterapia* **39,** 4 (1968).
12. W. C. Wildman, *in* "Chemistry of the Alkaloids" (S. W. Pelletier, ed.), p. 199. Van Nostrand-Reinhold, Princeton, New Jersey, 1970.
13. W. Döpke, *in* "Ergebnisse der Alkaloid Chemie," Vol. 1, Part 1, p. 68. Akademie Verlag, Berlin, 1976.
14. M. K. Yusupov and A. S. Sadykov, *Khim. Prir. Soedin.* 3 (1978).
15. F. Šantavý, *Acta Univ. Palacki. Olomuc., Fac. Med.* **90,** 15 (1979).
15a. F. Šantavý, *Heterocycles* **15,** 1505 (1981).
16. J.-C. Gaignault, *Actual. Chim.* **7,** 13 (1981).
17. V. V. Kiselev, *Khim. Prir. Soedin.* 3 (1977).
18. H. Corrodi and E. Hardegger, *Helv. Chim. Acta* **38,** 2030 (1955).
19. F. Šantavý, *Pharmazie* **37,** H1,56(1982).
19a. F. Šantavý, *Planta Med.* **36,** 231 (1979); *Herba Hung.* **18,** 111 (1979).
20. F. Šantavý, V. Preininger, V. Simánek, and H. Potesilová, *Planta Med.* **43,** 153(1981); F. Šantavý, V. Simánek, V. Preininger, and H. Potesilová, *Pharm. Acta Helv.* **57,** 243 (1982).
21. M. Carlassare, *Boll. Chim. Farm.* **118,** 343 (1979).
22. M Viková, H. Potesilová, M. Popović, I. Válka, and F. Šantavý, *Acta Univ. Palacki. Olomuc., Fac. Med.* **99,** 115 (1981).
23. A. E. Klein and P. J. Davis, *J. Chromatogr.* **207,** 247 (1981); P. J. Davis and A. E. Klein, *ibid.* **188,** 280 (1980).
23a. A. E. Klein and P. J. Davis, *Anal. Chem.* **52,** 2432 (1980).
24. D. Jarvie, J. Park, and M. J. Stewart, *Clin. Toxicol.* **14,** 375 (1979).
25. P. P. Petitjean, L. VanKerckhoven, M. Pesez, and P. Bellet, *Ann. Pharm. Fr.* **36,** 555 (1978).
26. G. Forni and G. Massarani, *J. Chromatogr.* **131,** 444 (1977).
27. E. Soczewinski and T. Dzido, *J. Liq. Chromatogr.* **2,** 511 (1979).
28. M. A. Iorio, A. Mazzeo-Farina, G. Cavina, L. Boniforti, and A. Brossi, *Heterocycles* **14,** 625 (1980).
29. M. A. Iorio, M. Molinari, and A. Brossi, *Can. J. Chem.* **59,** 283 (1981).
30. H. W. B. Clewer, S. J. Green, and F. Tutin, *J. Chem. Soc.* **107,** 835 (1915).
31. J. W. Ashley and J. O. Harris, *J. Chem. Soc.* 677 (1944).
32. F. Šantavý and T. Reichstein, *Helv. Chim. Acta.* **33,** 1606 (1950).
33. A. Ulubelen and M. Tanker, *Planta Med.* **34,** 216 (1978).
34. L. Canonica, B. Danieli, P. Manitto, G. Russo, and E. Bombardelli, *Chim. Ind.* (*Milan*) **49,** 1304 (1967).
35. M. Schönharting, P. Pfänder, A. Rieker, and G. Siebert, *Hoppe-Seyler's Z. Physiol. Chem.* **354,** 421 (1973).
36. J. M. Wilson, M. Ohashi, H. Budzikiewicz, F. Šantavý, and C. Djerassi, *Tetrahedron* **19,** 2225 (1963).
37. G. Severini-Ricca and B. Danieli, *Gazz. Chim. Ital.* **99,** 133 (1969).
38. B. Danieli, G. Palmisano, and G. Severini Ricca, *Gazz. Chim. Ital.* **110,** 351 (1980).
39. J. Elguero, R. N. Muller, A. Bladé-Font, R. Faure, and E. J. Vincent, *Bull. Soc. Chim. Belg.* **89,** 193 (1980).
40. A. El-Hamidi and F. Šantavý, *Collect. Czech. Chem. Commun.* **27,** 2111 (1962).

41. M. Rösner, Fu-Lian Hsu, and A. Brossi, *J. Org. Chem.* **46,** 3686 (1981).
42. A. D. Cross, A. El-Hamidi, J. Hrbek, and F. Šantavý, *Collect. Czech. Chem. Commun.* **29,** 1187 (1964).
43. F. Šantavý, *Collect. Czech. Chem. Commun.* **15,** 552 (1950).
44. V. Malichová, H. Potesilová, V. Preininger, and F. Šantavý, *Planta Med.* **36,** 119 (1979).
45. A. Bladé-Font, *Afinidad* **36,** 329 (1979).
46. M. Rösner, H.-G. Capraro, A. E. Jacobson, L. Atwell, A. Brossi, M. A. Iorio, T. H. Williams, R. H. Sik, and C. F. Chignell, *J. Med. Chem.* **24,** 257 (1981).
47. C. D. Hufford, C. C. Collins, and A. M. Clark, *J. Pharm. Sci.* **68,** 1239 (1979).
48. H. Potesilová, L. Hruban, and F. Šantavý, *Collect. Czech. Chem. Commun.* **41,** 3146 (1976).
49. F. Šantavý, P. Sedmera, G. Snatzke, and T. Reichstein, *Helv. Chim. Acta* **54,** 1084 (1971).
50. S. M. Kupchan, R. W. Britton, C. K. Chiang, N. Noyanalpan, and M. F. Ziegler, *Lloydia* **36,** 338 (1973).
51. P. Bellet, G. Amiard, M. Pesez, and A. Petit, *Ann. Pharm. Fr.* **10,** 241 (1952).
52. B. Chommadov, M. K. Yusupov, and A. S. Sadykov, *Khim. Prir. Soedin.* **6,** 82 (1970).
53. P. Bellet, *Ann. Pharm. Fr.* **10,** 81 (1952).
54. L. Szlavik, A. Zoltai, and J. Szabo, *Hung. Pat.* **156,224** (1967).
54a. P. Kerekes, A. Brossi, and J. L. Flippen-Anderson, *Can. J. Chem.*, in press.
55. C. D. Hufford, H.-G. Capraro, and A. Brossi, *Helv. Chim. Acta* **63,** 50 (1980).
56. H.-G. Capraro and A. Brossi, *Helv. Chim. Acta* **62,** 965 (1979).
57. F. Šantavý, *Pharm. Acta Helv.* **25,** 248 (1950).
58. K. M. Zuparova, B. Chommadov, M. K. Yusupov, and A. S. Sadykov, *Khim. Prir. Soedin.* **8,** 487 (1972).
59. M. Saleh, S. El-Gangihi, A. El-Hamidi, and F. Šantavý, *Collect. Czech. Chem. Commun.* **28,** 3413 (1963).
60. A. Uffer, O. Schindler, F. Šantavý, and T. Reichstein, *Helv. Chim. Acta* **37,** 18 (1954).
61. H. Potesilová, C. Alcaraz, and F. Šantavý, *Collect. Czech. Chem. Commun.* **34,** 2128 (1969).
62. F. Šantavý, R. Winkler, and T. Reichstein, *Helv. Chim. Acta* **36,** 1319 (1953).
63. H. Potesilová, J. Šantavý, E. El-Hamidi, and F. Šantavý, *Collect. Czech. Chem. Commun.* **34,** 3540 (1969).
64. A. R. Battersby, R. Ramage, A. F. Cameron, C. Hamaway, and F. Šantavý, *J. Chem. Soc. C* 3514 (1971).
65. A. D. Cross, A. El-Hamidi, L. Pijewska, and F. Šantavý, *Collect. Czech. Chem. Commun.* **31,** 374 (1966).
66. H. Potesilová, L. Dolejs, P. Sedmera, and F. Šantavý, *Collect. Czech. Chem. Commun.* **42,** 1571 (1977).
67. F. Šantavý, *Chem. Listy* **46,** 368 (1952).
68. R. Ramage, *Tetrahedron* **27,** 1499 (1971).
69. V. V. Kiselev, *Zh. Obshch. Khim.* **26,** 3218 (1956).
70. F. Šantavý and V. Macak, *Collect. Czech. Chem. Commun.* **19,** 805 (1954).
71. A. Brossi, M. Rösner, and J. V. Silverton, *Helv. Chim. Acta* **63,** 406 (1980).
72. P. Bellet and G. Muller, *Ann. Pharm. Fr.* **13,** 84 (1955).
73. F. Šantavý and M. Talas, *Collect. Czech. Chem. Commun.* **19,** 141 (1954).
74. F. Šantavý, *Collect. Czech. Chem. Commun.* **24,** 2237 (1959).
75. G. Muller and P. Bellet, *Ann. Pharm. Fr.* **13,** 81 (1955).

76. T. Baytop, N. Sutlupinar, and J. D. Phillipson, *Planta Med.* **38,** 273 (1980).
77. R. S. Thakur, H. Potesilová, and F. Šantavý, *Planta Med.* **28,** 201 (1975).
78. P. Sedmera, H. Potesilová, V. Malichova, V. Preininger, and F. Šantavý, *Heterocycles* **12,** 337 (1979).
79. A. D. Cross, J. Hrbek, Jr., J. L. Kaul, and F. Šantavý, *Beitr. Biochem. Physiol. Naturst.* p. 97 (1965).
80. K. Turdikulov, M. K. Yusupov, and A. S. Sadykov, *Khim. Prir. Soedin.* **8,** 247 (1972).
81. R. F. Raffauf, A. L. Farren, and G. E. Ullyot, *J. Am. Chem. Soc.* **75,** 5292 (1953).
82. F. Šantavý, Z. Hoscalkova, P. Podivinsky, and H. Potesilová, *Collect. Czech. Chem. Commun.* **19,** 1289 (1954).
82a. French Pat. 1,375,049 to Roussel-UCLAF (1963).
83. L. Velluz and P. Bellet, *C. R. Hebd. Seances Acad. Sci., Ser. A* **248,** 3453 (1959).
83a. Br. Pat. 923,421 to Roussel-UCLAF (1959).
84. P. N. Sharma and A. Brossi, *Heterocycles* **20,** 1587 (1983). The identity of this material with 2-demethylthiocolchicine has been established.
85. H.-J. Zeitler and H. Niemer, *Hoppe-Seyler's Z. Physiol. Chem.* **350,** 366 (1969).
86. P. J. Davis, *Antimicrob. Agents Chemother.* **19,** 465 (1981).
87. F. Šantavý, *Collect. Czech. Chem. Commun.* **35,** 2857 (1970).
88. M. A. Iorio, A. Brossi, and J. V. Silverton, *Helv. Chim. Acta* **61,** 1213 (1978).
89. M. Schönharting, G. Mende, and G. Siebert, *Hoppe-Seyler's Z. Physiol. Chem.* **355,** 1391 (1974).
90. B. Hasenmüller, M. Schönharting, and G. Siebert, *Hoppe-Seyler's Z. Physiol. Chem.* **359,** 725 (1978).
91. H. Budzikiewicz, C. Djerassi, and D. H. Williams, *in* "Structure Elucidation of Natural Products by Mass Spectrometry," Vol. I, Chapter 13. Holden-Day, San Francisco, California, 1964.
92. V. Delaroff and P. Rathle, *Bull. Soc. Chim. Fr.* 1621 (1965).
93. S. P. Singh, S. S. Parmar, V. I. Stenberg, and S. A. Farnu, *Spectrosc. Lett.* **10,** 1001 (1977).
94. A. Bladé-Font, R. Muller, J. Elguero, R. Faure, and E. J. Vincent, *Chem. Lett.* 233 (1979).
95. J. D. Morrison, *Acta Crystallogr.* **4,** 69 (1951).
96. M. V. King, J. L. DeVries, and R. Pepinsky, *Acta Crystallogr.* **5,** 437 (1952).
97. L. Lessinger and T. N. Margulis, *Acta Crystallogr., Sect. B* **B34,** 578 (1978).
98. J. V. Silverton, *Acta Crystallogr., Sect. B* **B35,** 2800 (1979).
99. C. Miravitlles, X. Solans, A. Bladé-Font, G. Germain, and J. Declercq, *Acta Crystallogr., Sect. B* **B38,** 1782 (1982).
100. L. Lessinger and T. N. Margulis, *Acta Crystallogr., Sect. B* **B34,** 1556 (1978).
101. T. N. Margulis, *J. Am. Chem. Soc.* **96,** 899 (1974).
102. C. Koerntgen and T. N. Margulis, *J. Pharm. Sci.* **66,** 1127 (1977).
103. J. I. Clark and T. N. Margulis, *Life Sci.* **26,** 833 (1980).
104. T. N. Margulis, *Biochem. Biophys. Res. Commun.* **76,** 1293 (1977).
105. T. N. Margulis and L. Lessinger, *Biochem. Biophys. Res. Commun.* **83,** 472 (1978).
106. A. Bladé-Font, *Tetrahedron Lett.* 2977 (1977).
107. B. Busetta, F. Leroy, M. Hospital, J. Elguero, and A. Bladé-Font, *Acta Crystallogr., Sect. B* **B35,** 1525 (1979).
108. Personal communication by Dr. P. Bellet, Roussel-UCLAF, Paris.
109. H. Potesilová, J. Hrbek, Jr., and F. Šantavý, *Collect. Czech. Chem. Commun.* **32,** 141 (1967).

110. French Pat. 1,330,908 to Roussel-UCLAF (1962); *Swiss Pat.* **407,102** to Roussel-UCLAF (1963).
111. G. Muller, A. Bladé-Font, and R. Bardoneschi, *Justus Liebigs Ann. Chem.* **662,** 105 (1963).
112. G. Muller and D. Branceni, *C. R. Hebd. Seances Acad. Sci., Ser. A* **255,** 2983 (1962).
113. French Pat. 1,344,446 to Roussel-UCLAF (1962).
114. French Pat. 1,359,639 to Roussel-UCLAF (1962).
115. French Pat. 1,359,652 to Roussel-UCLAF (1962).
116. French Pats. 1,359,632 and 1,359,642 to Roussel-UCLAF (1962); French Pat. 83,169 to Roussel-UCLAF (1962), first addition to Pat. 1,359,632.
117. French Pat. 1,359,637 to Roussel-UCLAF (1962).
118. French Pat. 1,359,638 to Roussel-UCLAF (1962).
119. French Pat. 1,347,136 to Roussel-UCLAF (1962).
120. French Pat. 1,369,483 to Roussel-UCLAF (1963).
121. French Pat. 1,359,648 to Roussel-UCLAF (1962).
122. French Pat. 1,512,320 to Roussel-UCLAF (1964).
123. French Pat. 1,347,137 to Roussel-UCLAF (1962).
123a. French Pat. 1,372,451 to Roussel-UCLAF (1963).
124. A. Bladé-Font, *Tetrahedron Lett.* 3607 (1969).
124a. We thank Dr. Herman Ziffer from the Laboratory of Chemical Physics (NIADDK, NIH) for having measured and interpreted the 500-MHz NMR spectrum.
125. H. Corrodi and E. Hardegger, *Helv. Chim. Acta* **40,** 193 (1957).
126. H. Fernholz, *Angew. Chem.* **65,** 319 (1953).
127. V. V. Kiselev, Y. V. Rashkes, and S. Y. Yunusov, *Khim. Prir. Soedin.* **4,** 536 (1974).
128. A. Brossi, P. N. Sharma, L. Atwell, A. E. Jacobson, M. A. Iorio, M. Molinari, and C. F. Chignell, *J. Med. Chem.* **26,** 1365 (1983).
129. G. T. Shiau, K. K. De, and R. E. Harmon, *J. Pharm. Sci.* **64,** 646 (1975).
130. V. V. Kiselev, M. E. Perel'son, B. S. Kikot, and O. S. Kostenko, *Zh. Org. Khim.* **13,** 2337 (1977).
131. L. Velluz and G. Muller, *Bull. Soc. Chim. Fr.* 1072 (1954).
132. H. Rapoport and J. B. Lavigne, *J. Am. Chem. Soc.* **77,** 667 (1955).
133. A. Bladé-Font, *Tetrahedron Lett.* 4097 (1977).
134. A. Bladé-Font, *Afinidad* **35,** 239 (1978).
135. R. Hunter, J. J. Bonet, and A. Bladé-Font, *Afinidad* **38,** 120 (1981).
136. R. Hunter, J. J. Bonet, and A. Bladé-Font, *Afinidad* **38,** 122 (1981).
137. H. Lettré, K. H. Dönges, K. Barthold, and T. J. Fitzgerald, *Justus Liebigs Ann. Chem.* **758,** 185 (1972).
138. H. Schmitt and D. Atlass, *J. Mol. Biol.* **102,** 743 (1976).
139. T.-S. Lin, G. T. Shiau, and W. H. Prusoff, *J. Med. Chem.* **23,** 1440 (1980); K. K. De, G. T. Shiau, and R. E. Harmon, *J. Prakt. Chem.* **318,** 523 (1976); for older reference, see also German Pat. 1,142,613 to Sandoz AG (1960).
140. K. K. De, G. T. Shiau, and R. E. Harmon, *J. Carbohydr., Nucleosides, Nucleotides* **2,** 171 (1975).
140a. K. K. De, G. T. Shiau, and R. E. Harmon, *J. Carbohydr., Nucleosides, Nucleotides* **2,** 259 (1975).
141. G. T. Shiau, K. K. De, and R. E. Harmon, *J. Pharm. Sci.* **67,** 394 (1978).
142. R. Schindler, *J. Pharmacol. Exp. Ther.* **149,** 409 (1965).
143. J. Schreiber, W. Leimgruber, M. Pesaro, P. Schudel, T. Threlfall, and E. Eschenmoser, *Helv. Chim. Acta* **44,** 540 (1961).
143a. A. Brossi, *Int. Conf. Chem. Biotechnol. Biol. Act. Nat. Prod.* [*Proc.*], *1st. 1981* Vol. 1, p. 114 (1981).

144. M. Kato, F. Kido, M. D. Wu, and A. Yoshikoshi, *Bull. Chem. Soc. Jpn.* **47,** 1516 (1974).
145. M. Matsui, K. Yamashita, K. Mori, and S. Kaneko, *Agric. Biol. Chem.* **31,** 675 (1967).
146. S. Kaneko and M. Matsui, *Agric. Biol. Chem.* **32,** 995 (1968).
147. E. Kotani, F. Miyazaki, and S. Tobinaga, *J. Chem. Soc., Chem. Commun.* 300 (1974); S. Tobinaga, *Bioorg. Chem.* **4,** 110 (1975).
148. D. A. Evans, D. J. Hart, and P. M. Koelsch, *J. Am. Chem. Soc.* **100,** 4593 (1978).
149. D. A. Evans, S. P. Tanis, and D. J. Hart, *J. Am. Chem. Soc.* **103,** 5813 (1981).
149a. D. A. Evans, D. J. Hart, P. M. Koelsch, and P. A. Cain, *Pure Appl. Chem.* **51,** 1285 (1979).
150. R. J. Ludford, *JNCI, J. Natl. Cancer Inst.* **6,** 89 (1945).
151. H. Lettré, *Angew. Chem.* **63,** 421 (1951).
152. K. Roberts and J. S. Hyams, "Microtubules," Academic Press, New York, 1979.
153. A. Zimmerman and A. Forer, "Mitosis/Cytokinesis." Academic Press, New York, 1981.
154. W. E. Dixon and W. Malden, *J. Physiol. (London)* **37,** 50 (1908).
155. R. M. Naidus, R. Rodvein, and H. Mielke, *Arch. Intern. Med.* **137,** 394 (1977).
156. A. Beck, *Naunyn-Schmiedebergs Arch. Exp. Pathol. Pharmakol.* **165,** 208 (1932).
157. M. H. Zweig and C. F. Chignell, *Biochem. Pharmacol.* **22,** 2142 (1973).
158. T. J. Fitzgerald, *Biochem. Pharmacol.* **25,** 1383 (1976).
159. B. Bhattacharyya and J. Wolff, *Proc. Natl. Acad. Sci. U.S.A.* **73,** 2375 (1976).
160. J. K. Kelleher, *Mol. Pharmacol.* **13,** 232 (1977).
161. F. Cortese, B. Bhattacharyya, and J. Wolff, *J. Biol. Chem.* **252,** 1134 (1977).
162. W. O. McClure and J. C. Paulson, *Mol. Pharmacol.* **13,** 560 (1977).
163. F. Wunderlich, *Biol. Unserer Zeit* **7,** 21 (1977).
164. R. C. Weisenberg, C. G. Borisy, and E. W. Taylor, *Biochemistry* **7,** 4466 (1968).
165. J. B. Olmsted, K. Carlson, R. Klebe, F. Ruddle, and J. Rosenbaum, *Proc. Natl. Acad. Sci. U.S.A.* **65,** 129 (1970).
166. L. Wilson, *Biochemistry* **6,** 3126 (1967).
167. C. G. Borisy and E. W. Taylor, *J. Cell Biol.* **34,** 525 (1967).
168. C. G. Borisy and E. W. Taylor, *J. Cell Biol.* **34,** 535 (1967).
169. L. Wilson and M. Friedkin, *Biochemistry* **6,** 3126 (1967).
170. F. L. Renand, A. J. Rowe, and I. R. Gibbons, *J. Cell Biol.* **36,** 79 (1968).
171. M. L. Shelanski and E. W. Taylor, *J. Cell Biol.* **34,** 549 (1967).
172. M. L. Shelanski and E. W. Taylor, *J. Cell Biol.* **38,** 304 (1968).
173. S. E. Malawista and K. G. Bensch, *Science* **156,** 521 (1967).
174. A. M. Poisner and J. Bernstein, *J. Pharmacol. Exp. Ther.* **177,** 102 (1971).
175. Ch. A. Strott and P. Ray, *Biochim. Biophys. Acta* **495,** 119 (1977).
176. J. A. Williams and J. Wolff, *Proc. Natl. Acad. Sci. U.S.A.* **67,** 1901 (1970).
177. K. W. Gautwik and A. H. Tashijian, Jr., *Endocrinology* **93,** 793 (1973).
178. G. Sommers, E. Van Obberghen, E. Decis, M. Ravazzola, F. Malaisse-Lagae, and W. J. Malaisse, *Eur. J. Clin. Invest.* **4,** 299 (1974).
179. P. E. Lacy, M. M. Walker, and C. J. Fink, *Diabetes* **21,** 987 (1971).
180. W. J. Malaisse, F. Malaisse-Lagae, M. O. Walker, and P. E. Lacy, *Diabetes* **20,** 257 (1971).
181. J. H. Shah and N. Wongsurawat, *Diabetes* **27,** 925 (1978).
182. K. H. Gabbay and W. J. Tze, *Proc. Natl. Acad. Sci. U.S.A.* **69,** 1435 (1972).
183. J. M. Trifaro, B. Collier, A. Lastoweka, and D. Stern, *Mol. Pharmacol.* **8,** 264 (1972).
184. D. C. Beebe, D. E. Faegans, E. J. Balanchette-Mackie, and M. E. Nau, *Science* **206,** 836 (1979).
185. S. B. Mizel and L. Wilson, *Biochemistry* **11,** 2573 (1972).

186. F. Wunderlich, R. Müller, and V. Speth, *Science* **182,** 1136 (1973).
187. M. Chvapil, E. C. Peacock, Jr., E. C. Carlson, S. Blau, K. Steinbronn, and D. Morton, *J. Surg. Res.* **1,** 49 (1980).
188. K. Trnsky, Z. Trnavska, and D. Mikulikova, *Cas. Lek. Cesk.* **18,** 564 (1977).
189. B. Goldberg, L. G. Ortega, A. Goldin, G. E. Ullyot, and E. B. Schoenbach, *Cancer* **3,** 124 (1950).
190. J. Leiter, V. Drowning, J. L. Hartwell, and M. J. Shear, *JNCI, J. Natl. Cancer Inst.* **13,** 379 (1952).
191. M. Suffness, National Cancer Institute, Washington, D.C. (personal communication).
192. H. Lettré, *Naturwissenschaften* **33,** 81 (1946).
193. H. Lettré and H. Fernholz, *Hoppe-Seyler's Z. Physiol. Chem.* **278,** 175 (1943).
194. L. Velluz, *Bull. Soc. Chim. Fr.* 591 (1962).
195. A. C. Banerjee and B. Bhattacharyya, *FEBS Lett.* **99,** 333 (1979).
196. A. S. Sherpinskaya, V. I. Gelfand, and B. P. Kopnin, *Biochim. Biophys. Acta* **673,** 86 (1981).
197. L. Velluz and G. Muller, *Bull. Soc. Chim. Fr.* 755 (1954).
198. S. Inoue, *Exp. Cell Res., Suppl.* **2,** 305 (1952).
199. E. Taylor, *J. Cell Biol.* **25,** 145 (1965).
200. J. B. Olmsted and G. C. Borisy, *Biochemistry* **12,** 4282 (1973).
201. J. R. Bamburg, E. M. Shoder, and L. Wilson, *Biochemistry* **12,** 1476 (1973).
202. L. Bryan, *Biochemistry* **11,** 2611 (1972).
203. R. J. Owellen, A. H. Owens, Jr., and D. W. Donigian, *Biochem. Biophys. Res. Commun.* **47,** 685 (1972).
204. L. Wilson and J. Bryan, *Adv. Cell Mol. Biol.* **3,** 21 (1974).
205. B. Bhattacharyya and J. Wolff, *Proc. Natl. Acad. Sci. U.S.A.* **71,** 2627 (1968).
206. J. Bryan, *Fed. Proc., Fed. Am. Soc. Exp. Biol.* **33,** 152 (1974).
207. D. L. Garland and D. C. Teller, *Ann. N. Y. Acad. Sci.* **253,** 232 (1975).
208. P. Sherline, J. T. Leung, and D. M. Kipnis, *J. Biol. Chem.* **250,** 5481 (1975).
209. L. Wilson, *Ann. N. Y. Acad. Sci.* **253,** 213 (1975).
210. A. Lambeir and Y. Engelborghs, *Eur. J. Biochem.* **109,** 619 (1980).
211. D. L. Garland, *Biochemistry* **17,** 4266 (1978).
212. H. W. Detrich, III, R. C. Williams, Jr., and L. Wilson, *Biochemistry* **21,** 2392 (1982).
213. P. B. Schiff and S. B. Horwitz, *Biochemistry* **20,** 3247 (1981).
214. D. Pantaloni, M. F. Carlier, C. Simon, and G. Batelier, *Biochemistry* **20,** 4709 (1981).
215. A. Banerjee, A. C. Banerjee, and B. Bhattacharyya, *FEBS Lett.* **124,** 285 (1981).
216. A. H. Lockwood, *Proc. Natl. Acad. Sci. U.S.A.* **76,** 1184 (1979).
217. P. Sherline, K. Schiavone, and S. Brocato, *Science* **205,** 593 (1979).
218. R. Jequier, D. Branceni, and M. Peterfalvi, *Arch. Int. Pharmacodyn. Ther.* **103,** 243 (1968).
219. F. Gaskin, C. R. Cantor, and M. Shelanski, *J. Mol. Biol.* **89,** 737 (1974).
220. J. Wolff, H.-G. Capraro, A. Brossi, and C. H. Cook, *J. Biol. Chem.* **255,** 7144 (1980).
221. J. Sterzl, F. Šantavý, P. Sedmera, and J. Cudlin, *Folia Microbiol.* (*Prague*) **27,** 256 (1982).
222. C. M. Lin and E. Hamel, *J. Biol. Chem.* **256,** 9242 (1981).
223. E. Buszman, T. Wilcock, B. Witman, and J. Siebert, *Hoppe-Seyler's Z. Physiol. Chem.* **358,** 819 (1977).
224. E. Schnell, M. Schönharting, and G. Siebert, *Hoppe-Seyler's Z. Physiol. Chem.* **357,** 567 (1976).
225. J. Deinum, F. Lincoln, T. Larsson, C. Lagercratz, and L. J. Erkell, *Acta Chem. Scand., Sect. B* **B35,** 677 (1981).
226. P. N. Sharma, C. F. Chignell, and A. Brossi, preliminary communication.

227. J. I. Clark and D. Garland, *J. Cell Biol.* **76,** 619 (1978).
228. E. C. Amoroso, *Nature (London)* **135,** 266 (1935).
229. H. Druckery, *Z. Krebsforsch.* **47,** 13 (1938).
230. H. Lettré, *Hoppe-Seyler's Z. Physiol. Chem.* **268,** 59 (1941).
231. Oughterson, Tennant, and Hirshfeld, *Proc. Soc. Biol. Med.* **36,** 661 (1937).
232. J. Stadler and W. W. Franke, *J. Cell Biol.* **60,** 297 (1974).
233. H. Lettré, *Ergeb. Physiol. Biol. Chem. Exp. Pharmacol.* **46,** 379 (1950).
234. M. Windhold, S. Budavari, L. Y. Stroumtsos, and M. N. Fertig, eds.), "The Merck Index," 9th ed., p. 2436. Merck & Co., Rahway, New Jersey, 1976.
235. M. Cernoch, J. Malinsky, O. Telupilova, and F. Šantavý, *Arch. Int. Pharmacodyn. Ther.* **99,** 141 (1954).
236. J. Leiter, J. L. Hartwell, G. E. Ullyot, and M. J. Shear, *JNCI, J. Natl. Cancer Inst.* **13,** 1201 (1952).
237. F. R. Quinn and J. A. Beisler, *J. Med. Chem.* **24,** 251 (1981).
238. H. Rösner and M. Schönharting, *Hoppe-Seyler's Z. Physiol. Chem.* **358,** 915 (1977).
239. G. Siebert, M. Schönharting, M. Ott, and S. Surjana, *Hoppe-Seyler's Z. Physiol. Chem.* **356,** 855 (1975).
240. J. L. Hartwell, M. V. Nadkarni, and J. Leiter, *J. Am. Chem. Soc.* **74,** 3180 (1952).
241. M. Peterfalvi and R. Jequier, *Arch. Int. Pharmacodyn. Ther.* **98,** 236 (1954).
242. J. Leiter, J. L. Hartwell, I. Kline, M. V. Nadkarni, and M. J. Shear, *JNCI, J. Natl. Cancer Inst.* **13,** 731 (1952).
243. H. Lettré, and R. Lettré, *Naturwissenschaften* **53,** 180 (1966).
244. H. Lettré and K. H. Donges, *in* "Aktuelle Probleme aus dem Gebiet der Cancerologie" (H. Lettré and G. Wagner, eds.), Vol. II, p. 200. Springer-Verlag, Berlin and New York, 1968.
245. M. H. Zweig, H. M. Maling, and M. E. Webster, *J. Pharmacol. Exp. Ther.* **182,** 344 (1972).
246. R. Schindler, *Nature* (*London*) **196,** 73 (1962).
247. G. Deysson, *Int. Rev. Cytol.* **24,** 99 (1968).
248. H. Lettré, T. J. Fitzgerald, and W. Siebs, *Naturwissenschaften* **53,** 132 (1966).
249. A. Brossi, *Trends Pharmacol. Sci.* **4,** 327 (1983).
250. J. K. Kelleher, *Cancer Treat. Rep.* **62,** 1443 (1978).
251. P. B. Schiff, A. S. Kende, and S. B. Horwitz, *Biochem. Biophys. Res. Commun.* **85,** 737 (1978).
252. B. Bhattacharyya and J. Wolff, *FEBS Lett.* **75,** 159 (1977).
253. J. Hoebeke, G. Van Nijen, and M. D. Brabander, *Biochem. Biophys. Res. Commun.* **69,** 319 (1976).
254. A. E. Brodie, J. Potter, and D. J. Reed, *Life Sci.* **24,** 1547 (1979).
255. M. K. Wolpert–DeFilippis, V. H. Bono, Jr., R. L. Dion, and D. G. Johns, *Biochem. Pharmacol.* **24,** 1735 (1975).
256. A. R. Battersby, *Ciba Found. Symp.* **53,** 25 (1978).
257. R. B. Herbert, *in* "The Alkaloids" (J. E. Saxton ed.), Vol. 4, p. 19. Chemical Society, Burlington House, London, 1974.
258. A. R. Battersby, R. Binks, J. J. Reynolds, and D. A. Yeowell, *J. Chem. Soc.* 4257 (1964).
259. E. Leete, *Tetrahedron Lett.* 333 (1965).
260. A. C. Barker, A. R. Battersby, E. McDonald, E. Ramage, and J. H. Clements, *J. Chem. Soc., Chem. Commun.* 390 (1967).
261. A. R. Battersby, R. B. Herbert, E. McDonald, R. Ramage, and J. H. Clements, *J. Chem. Soc., Perkin Trans. 1* 1741 (1972).
262. A. R. Battersby, P. W. Sheldrake, and J. A. Milner, *Tetrahedron Lett.* 3315 (1974).

263. J. Hrbek and F. Šantavý, *Collect. Czech. Chem. Commun.* **27,** 255 (1962).
264. L. Pijewska, J. L. Kaul, R. K. Joshi, and F. Šantavý, *Collect. Czech. Chem. Commun.* **32,** 158 (1967).
265. A. R. Battersby, R. B. Herbert, L. Pijewska, F. Šantavý, and P. Sedmera, *J. Chem. Soc., Perkin Trans. 1* 1736 (1972).
266. A. R. Battersby, T. A. Dobson, D. M. Foulkes, and R. B. Herbert, *J. Chem. Soc., Perkin Trans. 1* 1730 (1972).
267. H. W. Detrich, III, R. C. Williams, Jr., T. L. MacDonald, L. Wilson and D. Puett, *Biochemistry* **21,** 5999 (1981).
268. Personal communication by Dr. T. L. MacDonald, University of Virginia, Charlottesville, and presented at a lecture on May 13, 1983, at the National Cancer Institute in Silver Spring, Maryland.
269. F. Zavala, D. Guenard, J. P. Robin and E. Brown, *J. Med. Chem.* **23,** 546 (1980).
270. L. D. Barnes, The Robert A. Welch Foundation, Annual Report 1981–1982, p. 152.
271. D. Laudonne, J. B. Larsen and K. T. Taylor, *Science* **222,** 953 (1983).

—CHAPTER 2—

MAYTANSINOIDS

PAUL J. REIDER

Merck Sharp & Dohme Research Laboratories
Rahway, New Jersey

AND

DENNIS M. ROLAND

Research Department
Pharmaceuticals Division
CIBA-GEIGY Corporation
Ardsley, New York

I. Introduction

The maytansinoids are a relatively new and rare class of natural products that have generated a great deal of interest since their initial discovery in 1972. Professor Morris Kupchan (*1*) and co-workers at the University of Virginia first isolated maytansine (**1**), the key member of the class, from *Maytenus serrata* in their search for novel tumor inhibitors from plant sources (*2*). The potency, in the National Cancer Institute (NCI) antitumor screening protocol (*3,4*), of **1** and other maytansinoids subsequently isolated has led to a thorough investigation of their potential as chemotherapeutic agents.

THE ALKALOIDS, VOL. XXIII

ISBN 0-12-469523-X

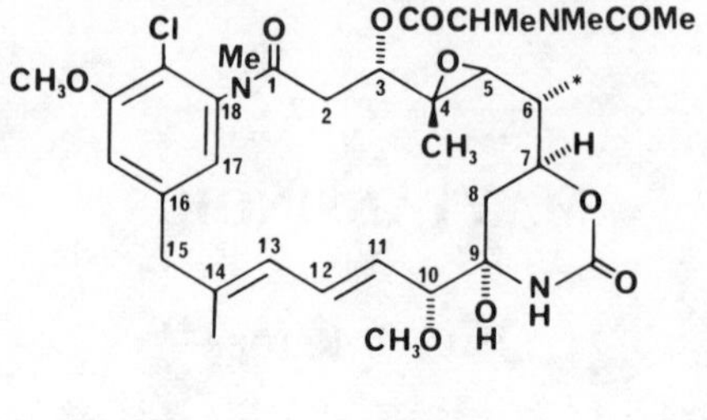

1

The novel structures of the maytansinoids are of interest also. Maytansinoids are ansa macrolides, that is, they are macrocycles containing an aromatic nucleus connected in nonadjacent positions by an aliphatic bridge, structures similar to the ansamycin antibiotics (*5*). All representatives are 19-membered lactam rings and are the first ansa compounds to exhibit tumor inhibiting properties. Most maytansinoids contain an aryl chloride and a C-4, C-5 epoxide. All contain a 4-hydro-2-oxo-2,3-dihydro[6*H*]-1,3-oxazine (usually referred to as the carbinolamide moiety) fused to C-7 and C-9. The biologically potent compounds contain an ester moiety at C-3 (*6*).

Maytansine (**1**) was in demand for clinical and biological studies after the initial screening results indicated its potential (*2*). Unfortunately the supply from plant materials was scarce, and it was exceedingly difficult to produce **1** in the large quantities needed (*7*). The chemical community was quick to respond with the first report of a synthetic approach to **1**, which was issued in 1974 by Professor Albert Meyers at Colorado State University (*8*). Professor E. J. Corey at Harvard University followed with his approach in 1975 (*9*). Subsequently, the two research groups mounted massive efforts (*10–19*) that ultimately lead to the total syntheses of several members of the class including (±)-*N*-methylmaysenine (**2**) (*20,21*), (−)-*N*-methylmaysenine [(−)-**2**] (*22*), (±)-maysine (**3**) (*23*), (−)-maysine

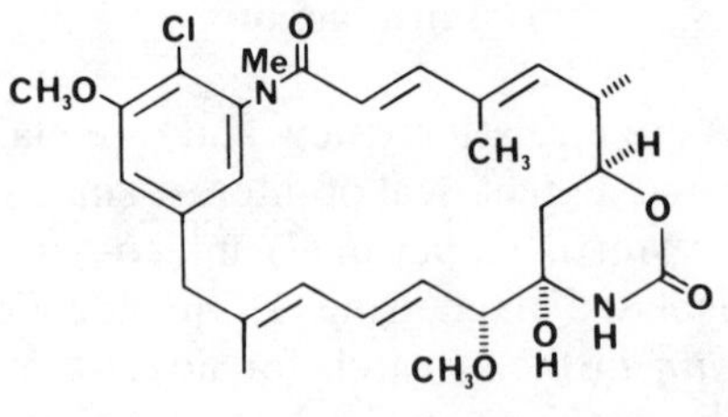

2

* This structural notation represents a methyl group.

[(−)-**3**] (*24*), (±)-maytansinol (**4**) (*25*), and maytansine (**1**) (*26*). More recently, a Japanese group led by Professor M. Isobe completed syntheses of (±)-maytansinol (**4**) (*27*), (±)-*N*-methylmaysenine (**2**), and (±)-maysine (**3**) (*28*) using an alternate approach (*29–31*). During the period 1974 to 1983 at least 11 other research groups reported their efforts toward various fragments of the maytansinoid framework (*32–52*).

In 1977 researchers from Takeda Chemical Industries (Osaka, Japan) reported their discovery of a microbial fermentation process for the production of several maytansinoid compounds which they named ansamitocins (*53*). Maytansinol (**4**), the penultimate synthon for maytansine (**1**) and other C-3 esters, could be obtained by reductive hydrolysis of the ansamitocins (*53,54*) and was later discovered in a fermentation broth (*55,56*). The supply problem has thus been greatly relieved. Not only can **1** be obtained but Takeda scientists also have a source of starting materials for exploring chemical and biological modifications to enhance activity and reduce toxicity.

U.S. Department of Agriculture scientists have isolated several new maytansinoids from *Trewia nudiflora* (*57–59*). Test results indicate these compounds have tumor-inhibiting properties and may act as agents for controlling several economically important insects (*60*). Research in this area may help answer questions of why plants produce maytansinoids. Kupchan suggested that maytansinoids were possibly produced by microorganisms living on or within the plants (*2*).

Maytansinoids have been reviewed on several occasions since 1974 (*61–69*). The 1980 review by Komoda and Kishi (Takeda Chemical Industries) (*63*) is particularly noteworthy for its thorough discussion of maytansinoid isolation and biological activity. Their review covers most of the synthetic chemistry reported up to the first total synthesis (*20*). The review by Issell and Crooke (*69*) is quite thorough in its coverage of early information relating to the clinical usefulness of maytansine (**1**). We have attempted to cover the synthetic work in its entirety up to mid-1983.

II. Isolation and Structural Variation

The plant-screening program operated by the National Cancer Institute has produced a number of biologically active compounds (*70–83*). Because the *Maytenus* species had been used for many years in herbal mixtures for the treatment of cancers by African people (*71,76*) samples were collected for examination. Professor Kupchan, working in the NCI program, isolated maytansine (**1**) as the most potent tumor-inhibiting component in *Maytenus serrata* (formerly *M. ovatus*) (*2*). Guided by bioassay against the KB cell culture and P388 tumor screens (*3,4*), he further fractionated the ethanolic extracts by solvent partitions and multiple chromatographies and obtained **1** in 2×10^{-5} % yield. The low yield represents only 0.2 mg per kg of dried plant material.

Requirements for larger quantities of **1** for clinical trials forced the search for other sources. *Maytenus buchananii* was found to contain a greater concentration of **1** (1.5 mg/kg) along with the new C-3 ester analogs maytanprine (**5**), maytanbutine (**6**), and maytanvaline (**7**) (*7,84,85*). Maysine (**3**), normaysine (**8**), and maysenine (**9**), all devoid of the C-3 ester group were also isolated. The demethyl compound normaytansine (**10**) was later found during a large scale extraction of *M. buchananii* (15,000 kg) to obtain **1** for clinical trials (*86*). *Putterlickia verrucosa* was found to be the richest plant source of **1** (12 mg/kg) (*87*). Further fractionation of the *P. verrucosa* extracts produced maytanprine (**5**), maytanbutine (**6**), and, for the first time, maytanacine (**11**), maytansinol (**4**), and normaytancyprine (**12**) (*87,88*). Several research groups have also isolated **1**, **5**, **6**, and **7** from other plant sources (see Table I).

In 1973 Wani *et al.* (*89*) reported the isolation of two new C-15 oxygenated maytansinoids, colubrinol (**13**) and colubrinol acetate (**14**), along with the known maytanbutine (**6**) from *Colubrina texensis*. A later report by Kupchan (*7*) disclosed the isolation of another C-15 oxygenated compound, maytanbutacine (**15**), from *M. serrata*. Powell *et al.* have reported the isolation of demethyltrewiasine (**16**), dehydrotrewiasine (**17**), trewiasine (**18**), 10-epitrewiasine (**19**), and nortrewiasine (**20**) from the seeds of *Trewia nudiflora*, all of which contain a C-15 methoxy group (*57,59*). An unusual new group of maytansinoids, trenudine (**21**), trefluorine (**22**), and *N*-methyltrenudone (**23**), containing an additional 12-membered macrocyclic ring fused at C-3 and the amide nitrogen at C-18, has also been isolated from *T. nudiflora* (*58*).

Higashide *et al.* (*53*) reported the isolation of a group of five ansamycin antibiotics from the fermentation broth of *Norcardia* C-15003. These compounds were found to be esters of maytansinol (**4**) and were named ansamitocins P-1, P-2, P-3, P-3′, and P-4. All of the ansamitocins yielded maytansinol (**4**) upon reductive hydrolysis with lithium aluminum hydride. Ansamitocin P-1 was shown to be identical with maytanacine (**11**) previously isolated from a plant source (*87*).

More recently a mutant strain of *Norcardia*, N-1231, has been reported to produce large amounts of ansamitocins P-3 and P-4 as major components as well as 19 other analogs (*55,56*) as minor components (see Table II). The group at Takeda has been very active in chemically and microbiologically modifying the ansamitocins (*90–119*).

The known maytansinoids of plant origin are listed in Table I along with

TABLE I

MAYTANSINOIDS OF PLANT ORIGIN

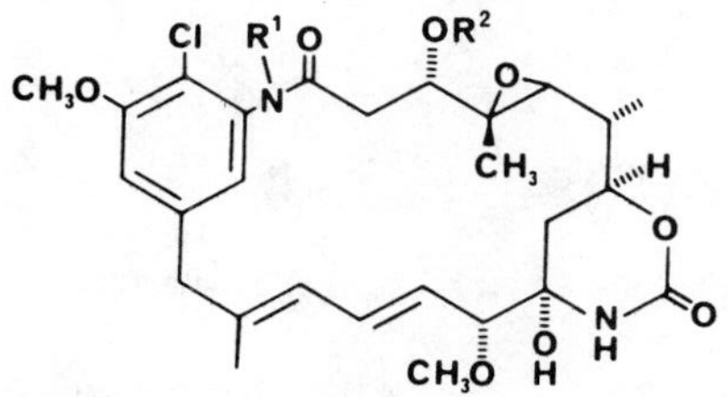

Name	R¹	R²	R³	R⁴	R⁵	Source	Reference
Maytansine (**1**)	CH_3	$COCHCH_3NCH_3COCH_3$				*Maytenus serrata*	*2,87*
						Putterlickia verrucossa	*7,87*
						Maytenus buchananii	*7,87*
						Maytenus rothina	*120*
						Maytenus phyllanthoides	*120*
						Maytenus ilicfloria	*121*
						Maytenus diversifolia	*122*
						Gymnosporia diversifolia	*123*
						Maytenus hookeri	*124*
						Maytenus graciliramula	*125,126*
						Maytenus ovatus	*127*
						Mihuameidenmu	*128*
						Maytenus confertiflorus	*129,130*
Maytanvaline (**7**)	CH_3	$COCHCH_3NCH_3COCH_2CH(CH_3)_2$				*Maytenus buchananii*	*85,87*
						Maytenus rothiana	*120*
Maytanprine (**5**)	CH_3	$COCHCH_3NCH_3COCH_2CH_3$				*Putterlickia verrucossa*	*7,84,87*
						Maytenus serrata	*7*
						Gymnosporia diversifolia	*123*

(*Continued*)

TABLE I (*Continued*)

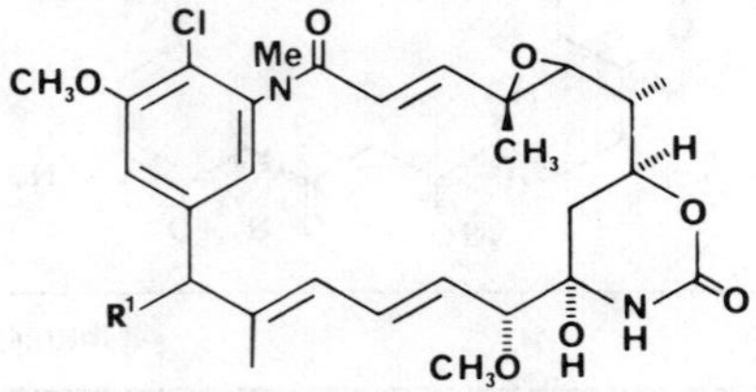

Name	R^1	R^2	R^3	R^4	R^5	Source	Reference
						Maytenus hookeri	*124*
						Maytenus confertiflorus	*129,130*
						Maytenus buchananii	*7*
						Maytenus ilicifloria	*121*
						Maytenus rothiana	*120*
						Maytenus graciliramula	*126*
Maytanbutine (**6**)	CH_3	$COCHCH_3NCH_3COCH(CH_3)_2$				*Putterlickia verrucossa*	*7,84,87*
						Maytenus serrata	*7*
						Maytenus buchananii	*7*
						Colubrina texensis	*89*
						Maytenus ilicifloria	*121*
						Maytenus rothiana	*120*
						Maytenus graciliramula	*126*
Maytansinol (**4**)	CH_3	H				*Putterlickia verrucossa*	*7,87*
Maytanacine (**11**)	CH_3	$COCH_3$				*Putterlickia verrucossa*	*7,87*
Normaytancyprine (**12**)	H	$COCHCH_3NCH_3CO$–(2,3-dimethylcyclopropyl: CH_3, CH_3)				*Putterlickia verrucossa*	*88*
Normaytansine (**10**)	H	$COCHCH_3NCH_3COCH_3$				*Maytenus buchananii*	*86*
Maysine (**3**)	H					*Maytenus buchananii*	*7,85*
						Maytenus buchananii	*7,85*

Trewsine	OCH$_3$					*Trewia nudifloria*	*57*
Normaysine (**8**)						*Maytenus buchananii*	*7,85*
Maysenine (**9**)							
Demethyltrewiasine (**16**)	CH$_3$	COCHCH$_3$NHCOCH(CH$_3$)$_2$	H	OCH$_3$	OCH$_3$	*Trewia nudiflora*	*57*
Dehydrotrewiasine (**17**)	CH$_3$	COCHCH$_3$NCH$_3$COCCH$_3$CH$_2$	H	OCH$_3$	OCH$_3$	*Trewia nudiflora*	*57*
Trewiasine (**18**)	CH$_3$	COCHCH$_3$NCH$_3$COCH(CH$_3$)$_2$	H	OCH$_3$	OCH$_3$	*Trewia nudiflora*	*57*

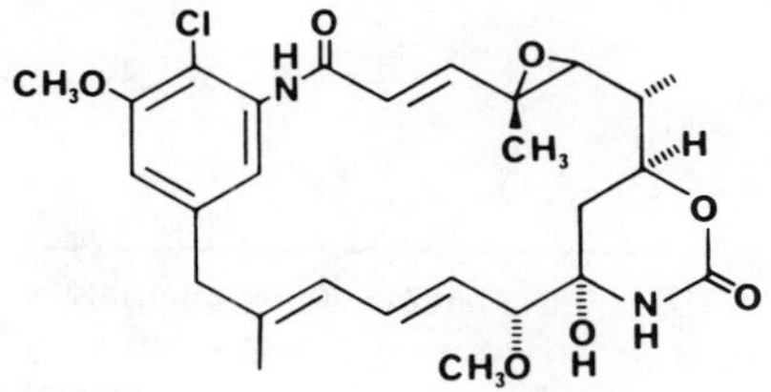

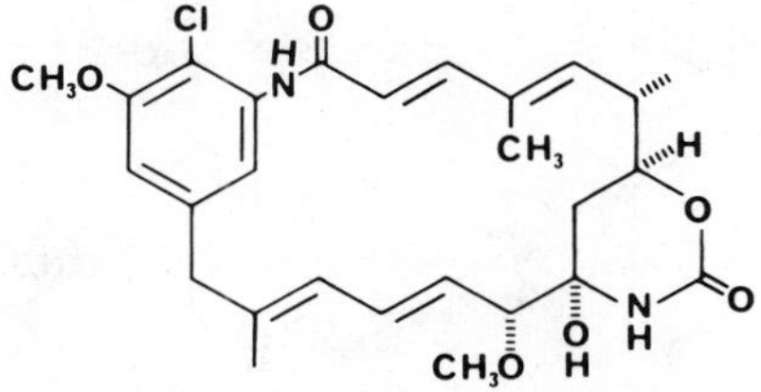

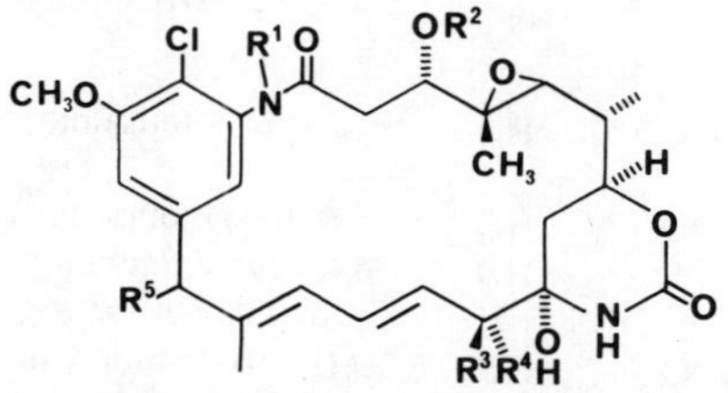

(Continued)

TABLE I (*Continued*)

Name	R^1	R^2	R^3	R^4	R^5	Source	Reference
10-Epitrewiasine (**19**)	CH_3	$COCHCH_3NCH_3COCH(CH_3)_2$	OCH_3	H	OCH_3	*Trewia nudiflora*	*59*
Nortrewiasine (**20**)	H	$COCHCH_3NCH_3COCH(CH_3)_2$	H	OCH_3	OCH_3	*Trewia nudiflora*	*59*
Maytanbutacine (**15**)	CH_3	$COCH(CH_3)_2$	H	OCH_3	$OCOCH_3$	*Maytenus serrata*	*7*
Colubrinol (**13**)	CH_3	$COCHCH_3NCH_3COCH(CH_3)_2$	H	OCH_3	OH	*Colubrina texensis*	*89*
						Trewia nudiflora	*59*
Colubrinol acetate (**14**)	CH_3	$COCHCH_3NCH_3COCH(CH_3)_2$	H	OCH_3	$OCOCH_3$	*Colubrina texensis*	*89*
Trenudine (**21**)	H, OH	H				*Trewia nudiflora*	*58*
Treflorine (**22**)	H_2	H				*Trewia nudiflora*	*58*
N-Methyltrenudone (**23**)	O	CH_3				*Trewia nudiflora*	*58*

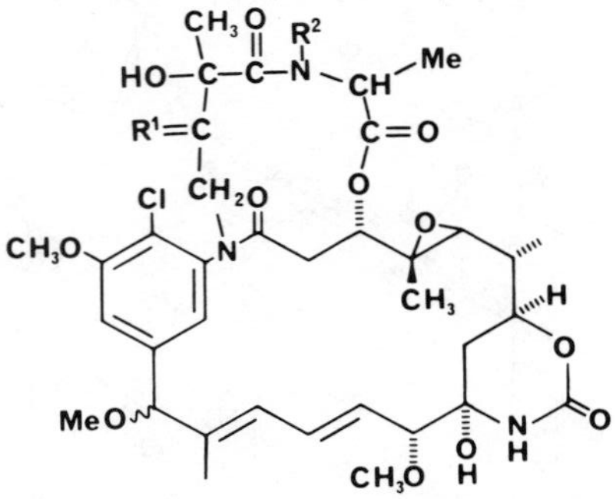

TABLE II

MAYTANSINOIDS ISOLATED FROM *NORCARDIA* FERMENTATION

Name	R^1	R^2	Reference
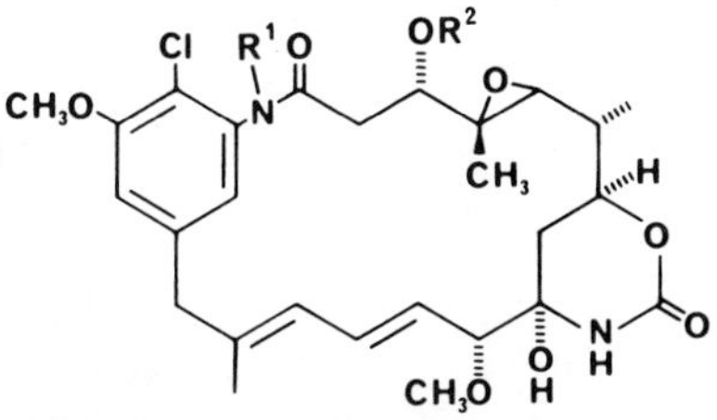			
Ansamitocin P-0 (**4**) (Maytansinol)	CH_3	H	*54–56*
Ansamitocin P-1 (**11**) (Maytanacine)	CH_3	$COCH_3$	*53–56,131*
Ansamitocin P-2 (**11a**)	CH_3	$COCH_2CH_3$	*53–56,131*
Ansamitocin P-3 (**24**)	CH_3	$COCH(CH_3)_2$	*53–56,131*
Ansamitocin P-3 (**25**)	CH_3	$COCH_2CH_2CH_3$	*53–56,131*
Ansamitocin P-4 (**24**)	CH_3	$COCH_2CH(CH_3)_2$	*53–56,131*
N-Demethylansamitocin P-4 (**27**) (PND-4)	H	$COCH_2CH(CH_3)_2$	*55,56*
N-Demethylansamitocin P-3 (**28**) (PND-3)	H	$COCH_2CH(CH_3)_2$	*55,56*
3-Propionyl-*N*-demethylmaytansinol (**29**) (PND-2)	H	$COCH_2CH_3$	*55,56*
3-Acetyl-*N*-demethylmaytansinol (**30**) (PND-1)	H	$COCH_3$	*55,56*
N-Demethylmaytansinol (**31**) (PND-0)	H	H	*55,56*
3-(β-Hydroxyisovaleryl)demethylmaytan-sinol (PND-4-βHY) (**32**)	H	$COCH_2C(OH)(CH_3)_2$	*55,56*
3-(γ-Hydroxyisovaleryl)maytansinol (**33**) (P-4-γHY)	CH_3	$COCH_2CH(CH_2OH)CH_3$	*55,56*
3-(β-Hydroxyisovaleryl)maytansinol (**34**) (P-4-βHY)	CH_3	$COCH_2C(OH)(CH_3)_2$	*55,56*
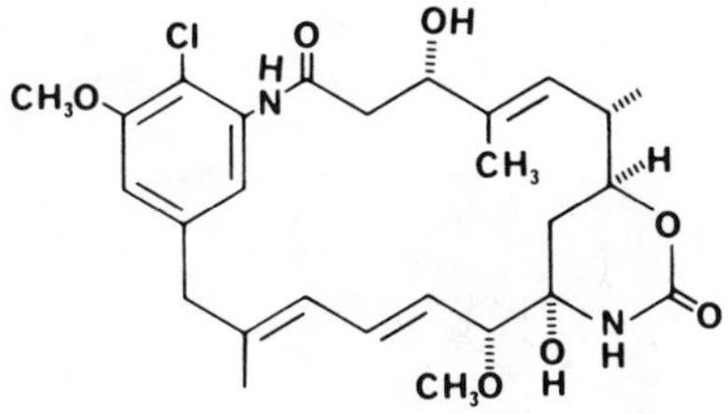			
N-Demethyl-4,5-deoxymaytansinol (**35**) (QND-0)			*55,56*

(*Continued*)

TABLE II (*Continued*)

Name	R^1	R^2	Reference
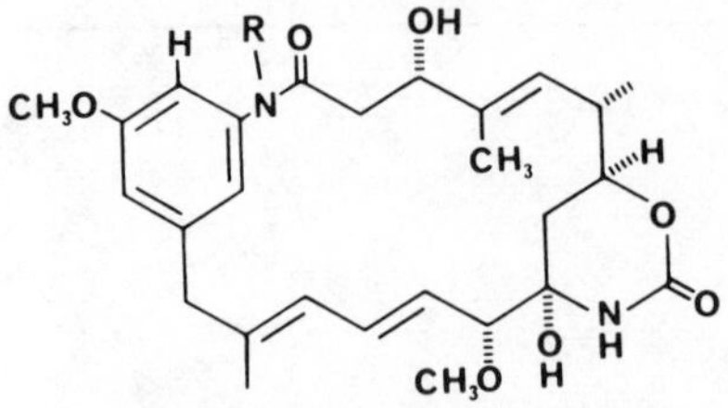			
19-Dechloro-*N*-demethyl-4,5-deoxymaytansinol (**36**) (deClQND-0) R = H			55,56
19-Dechloro-4,5-deoxymaytansinol (**37**) (deClQ-0) R = CH_3			55,56
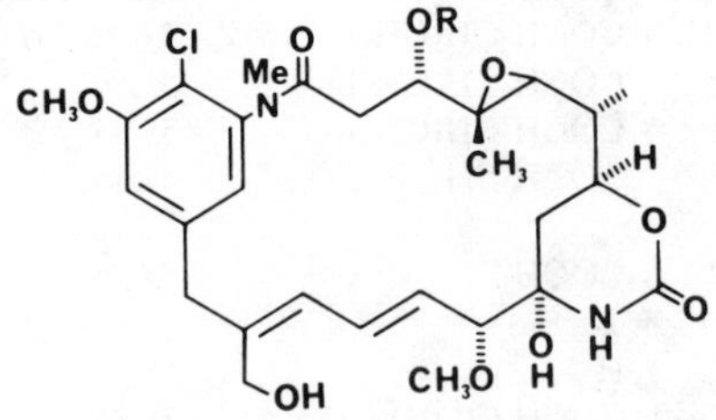			
3-Acetyl-26-hydroxymaytansinol (**38**) (PHM-1) R = $COCH_3$			55,56
26-Hydroxy-3-propionyl maytansinol (**39**) (PHM-2) R = $COCH_2CH_3$			55,56
26-Hydroxyansamitocin P-3 (**40**) (PHM-3) R = $COCH(CH_3)_2$			55,56
26-Hydroxyansamitocin P-4 (**41**) (PHM-4) R = $COCH_2CH(CH_3)_2$			55,56
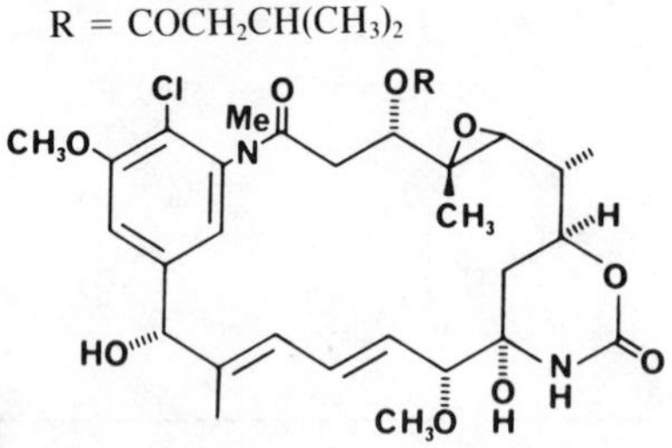			
Deacetylmaytanbutacine (**42**) (ansamitocin PHO-3) R = $COCH(CH_3)_2$			55,56

their respective plant sources. The ansamitocins from microbial culture fluids are listed in Table II. The various synthetic analogs prepared by modifying the natural products are not listed since they are not of natural origin.

III. Physical Properties

A. Nuclear Magnetic Resonance

1. Proton NMR

High-field ^{1}H-NMR data are available for nearly all of the known maytansinoids. This technique has been invaluable for the structural assignment of this complex class of macrocycles. Table III lists the proton assignments for maytansine (**1**) (*1*), maysine (**3**) (*7*), maytansinol (**4**) (*7,25,27*), maysenine (**9**) (*7*), maytanbutacine (**15**) (*7*), trewiasine (**18**) (*57*), trewsine (**43,** see p. 76) (*57*), and colubrinol (**13**) (*59,89*). These compounds reflect the structural variations within the macrocyclic ring. Variations of the ester side chain attached to C-3 will be found in Table IV.

When using maytansine (**1**) as the reference spectrum, certain differences should be highlighted as one examines the major substituent changes. Maysine (**3**) contains a C-2–C-3 double bond in which the C-2 H appears at δ5.65 ppm as a doublet (J = 16 Hz) and the C-3 H comes at 6.37, doublet (J = 16). In addition, the C-4 methyl and C-5 H have shifted. Relative to maytansine (**1**), maytansinol (**4**) shows the expected upfield shift of the C-3 H to 3.54; the C—H is now on an unacylated hydroxyl center. Maysenine (**9**), the deoxygenated analog of maysine (**3**), has a C-2–C-3, C-4–C-5 diene with three new vinyl hydrogens. Signals appear at 5.82 (C-2 H), 7.28 (C-3 H), and 5.50 (C-5 H). The methyl group attached to C-4 is now allylic and moved down to 1.56 ppm.

Maytanbutacine (**15**) contains a C-15 acetoxy group. Consequently, there is only one C-15 hydrogen, and it is shifted more than 3 ppm downfield to 6.21. The acetoxy methyl resonance is at 2.23, and the aromatic hydrogen at C-17 moves down to 7.09 ppm. Maytanbutacine (**15**) also differs with regard to the C-3 ester side chain. Whereas most of the plant-derived maytansinoids contain an *N*-acyl-*N*-methylalanine ester at C-3, this compound is the simple isobutyrate. As a result the C-2′ H is found as a multiplet between 2.0 and 2.6 ppm.

Trewiasine (**18**) is also oxygenated at C-15 in this case as a methyl ether. The lone C-15 H is seen as a singlet at 4.86 ppm with an accompanying methoxyl singlet at 3.37. Trewsine (**43**) is simply 15-methoxymay-

TABLE IIIA
PROTON NMR OF MAYTANSINE, MAYSINE, AND MAYSENINE

Proton assignment	Maytansine (**1**) (*7*)	Maysine (**3**) (*7*)	Maysenine (**9**) (*7*)
2A	2.21 dd	5.65 d	5.18 d
	$J_{2,2} = 15$; $J_2 = 3$	$J_{2,3} = 16$	$J_{2,3} = 16$
2B	2.65 dd	—	—
	$J_{2,2} = 15$; $J_{2,3} = 12$		
3	4.79 dd	6.37 d	7.28 d
	$J_{2,3} = 12,3$	$J_{2,3} = 16$	$J_{2,3} = 16$
4-Me	.87 s	1.06 s	1.56 br.s.
5	3.04 d	2.62 d	5.50 d
	$J_{5,6} = 9$	$J_{5,6} = 9$	$J_{5,6} = 10$
6-Me	1.34 d	1.30 d	1.25 d
	$J = 6$	$J = 6$	$J = 6$
7	4.28 m	4.24 m	4.11 m
10	3.50 d	3.39 d	3.45 d
	$J_{10,11} = 9$	$J_{10,11} = 9$	$J_{10,11} = 9$
11	5.66 dd	5.43 dd	5.42 dd
	$J_{10,11} = 9$; $J_{11,12} = 15$	$J_{10,11} = 9$; $J_{11,12} = 15$	$J_{10,11} = 9$; $J_{11} = 14$
12	6.42 dd	6.34 dd	6.37 dd
	$J_{11,12} = 15$; $J_{12,13} = 11$	$J_{12,13} = 11$; $J_{11,12} = 15$	$J_{12,13} = 10$; $J_{11,12} = 14$
13	6.70 br.d.	6.02 br.d.	6.00 br.d.
	$J_{12,13} = 11$	$J_{12,13} = 11$	$J_{12,13} = 10$
14-Me	1.69 br.s.	1.68 br.s.	1.65 br.s.
15A	3.13 d	3.02 d	3.07 d
	$J_{15,15} = 13$	$J_{15,15} = 12$	$J_{15,15} = 13$
15B	3.67 d	3.42 d	3.40 d
	$J_{15,15} = 13$	$J_{15,15} = 12$	$J_{15,15} = 13$
17	6.75 d	6.62 d	6.51 d
	$J_{17,21} = 1.5$	$J_{17,21} = 1.5$	$J = 1.0$
21	6.84 d	6.74 d	6.67 d
	$J_{17,21} = 1.5$	$J_{17,21} = 1.5$	$J = 1.0$
10-OMe	3.38 s	3.28 s	3.29 s
15-OMe	—	—	—
20-OMe	3.99 s	3.92 s	3.88 ss
18-NMe	3.22 s	3.22 s	(N—H) 7.15 s
2′	5.35 q	—	—
	$J = 7$		
2′-Me	1.37 d	—	—
	$J = 7$		
2′-NMe	2.89 s	—	—
4′	2.15 s	—	—
4′-Me	—	—	—

TABLE IIIB
PROTON NMR OF MAYTANBUTACINE, TREWIASINE, AND TREWSINE

Proton assignment	Maytanbutacine (**15**) (*7*)	Trewiasine (**18**) (*57*)	Trewsine (**43**) (*57*)
2A	2.43 dd	2.18 dd	5.67 d
	$J_{2,2}$ = 14; $J_{2,3}$ = 11	$J_{2,2}$ = 14.3; $J_{2,3}$ = 3.0	$J_{2,3}$ = 15.5
2B	4.79 dd	4.75 dd	6.42 d
	$J_{2,3}$ = 11.3	$J_{2,3}$ = 12.2, 3.0	$J_{2,3}$ = 15.5
4-Me	0.79 s	0.76 s	1.04 s
5	2.95 d	3.01 d	2.66 d
	$J_{5,6}$ = 8.5	$J_{5,6}$ = 9.6	$J_{5,6}$ = 9.8
6-Me	1.28 d	1.27 d	1.31 d
	J = 7	J = 6.2	J = 6.4
7	4.26 m	4.28 m	4.30 m
10	3.51 d	3.51	3.46 d
	$J_{10,11}$ = 9	$J_{10,11}$ = 9.1	$J_{10,11}$ = 9.4
11	5.60 dd	5.72 dd	5.58 dd
	$J_{10,11}$ = 9; $J_{11,12}$ = 14	$J_{10,11}$ = 9.1; $J_{11,12}$ = 15.3	$J_{11,12}$ = 15.3; $J_{10,11}$ = 9.4
12	6.30 dd	6.46 dd	6.44 dd
	$J_{12,13}$ = 9; $J_{11,12}$ = 14	$J_{12,13}$ = 11.1; $J_{11,12}$ = 15.3	$J_{11,12}$ = 15.3; $J_{12,13}$ = 10.9
13	6.44 d	6.98 d	6.24 d
	$J_{12,13}$ = 10	$J_{12,13}$ = 11.1	$J_{12,13}$ = 10.9
14-Me	1.67 s	1.52 s	1.56 s
15A	6.21 s	4.86 s	4.67 s
15B	—	—	—
17	7.09 d	6.54 d	6.60 d
	J = 1.5	J = 1.5	J = 1.6
21	6.85 d	7.22 d	7.20 d
	J = 1.5	J = 1.5	J = 1.6
10-OMe	3.35 s	3.35 s	3.34 s
15-OMe	2.23 s (OAc)	3.37 s	3.34 s
20-OMe	4.02 s	3.99 s	4.00 s
18-NMe	3.16 s	3.16 s	3.25 s
2′	2.0–2.6 m	5.37 m	—
2′-Me	1.28 d	1.28 d	—
	J = 7	J = 6.8	
	1.20 d		
	J = 7		
2′-NMe	—	2.88 s	—
4′	—	2.76 m	—
4′-Me	—	1.06 d	—
		J = 6.6	
		1.12 d	
		J = 6.8	

TABLE IIIC

PROTON NMR OF COLUBRINOL AND MAYTANSINOL[a]

Proton assignment	Colubrinol (**13**) (*59,89*)	Maytansinol (**4**) (*7,25,27*)
2A	2.18 dd	2.10 dd
	$J_{2,2}$ = 14.3; $J_{2,3}$ = 3.0	$J_{2,2}$ = 13.5; $J_{2,3}$ = 2
2B	2.56 dd	2.28 dd
	$J_{2,2}$ = 14.3; $J_{2,3}$ = 12.0	$J_{2,2}$ = 13.5; $J_{2,3}$ = 11
3	4.75 dd	3.54 dd
	$J_{2,3}$ = 12.0, 3.0	$J_{2,3}$ = 11.2
4-Me	0.77 s	0.84 s
5	3.02 d	2.57 d
	$J_{5,6}$ = 9.6	$J_{5,6}$ = 9.5
6-Me	1.27 d	1.29 d
	J = 6.1	J = 6.5
7	4.27 m	4.34 m
10	3.51 d	3.49
	$J_{10,11}$ = 9	$J_{10,11}$ = 9
11	3.51 dd	5.51 dd
	$J_{10,11}$ = 9; $J_{11,12}$ = 15.3	$J_{10,11}$ = 9; $J_{11,12}$ = 15
12	6.45 dd	6.43 dd
	$J_{11,12}$ = 15.3; $J_{12,13}$ = 11	$J_{11,12}$ = 15; $J_{12,13}$ = 11
13	6.95 d	6.14 d
	$J_{11,12}$ = 15.3; $J_{12,13}$ = 11	$J_{11,12}$ = 15; $J_{12,13}$ = 11
14-Me	1.59 s	1.69 s
15A	5.47 s	3.11 d
		$J_{15,15}$ = 12.5
15B	—	3.47 d
		$J_{15,15}$ = 12.5
17	6.55 d	6.80 d
	J = 1.5	J = 2
21	7.34 d	6.98 d
	J = 1.5	J = 2
10-OMe	3.34 s	3.20 s
15-OMe	—	—
20-OMe	4.00 s	3.98 s
18-NMe	3.16 s	3.35 s
2′	4.36 m	—
2′-Me	1.26 d	—
	J = 6.6	
2′-NMe	2.87 m	—
4′	2.76 m	—
4′-Me	1.04 d	1.10 d
	J = 6.5	J = 6.8

[a] Selected data for the C-3 ester side chains of maytansine (**1**) (*2*), maytanprine (**5**) (*84*), maytanbutine (**6**) (*84*), maytanvaline (**7**) (*85*), maytanacine (**11**) (*87*), colubrinol (**13**) (*59,89*), trewiasine **18** (*57*), dehydrotrewiasine *17*, (*57*), demethyltrewiasine (**16**) (*57*), treflorine (**22**) (*58*), trenudine *21* (*58*), *N*-methyltrenudone (**23**) (*58*), normaytancyprine (**12**) (*88*), maytanbutacine (**15**) (*7*), and nortrewiasine (**20**) (*59*) are presented in Table IV. In addition some semisynthetic esters (*6*) of maytansinol are reported.

TABLE IVA

PROTON NMR OF C-3 ESTER SIDE CHAINS: O—C(O)—CH(Me)—N(Me)—C(O)—R

Compound	R	2′	2′-Me	2′NMe	4′	4′Me	5′	5′Me	Reference
Maytansine (**1**)	CH_3	5.35 g (J = 7)	1.37 d (J = 7)	2.89 s	2.15 s				*7*
Maytanprine (**5**)	CH_2-CH_3	—[a]	—[a]	—[a]	2.37, 1H (m) 2.41, 1H (m)	1.18 t (J = 7)			*84*
Maytanbutine (**6**)[b] (exhibits 2 rotamers)	$CH(Me)_2$	—[a]	—[a]	2.87 (0.75H) 2.92 (2.25H)	2.80 (m)	1.12 d (J = 7) 1.19 *d* (J = 7)			*84*
Maytanvaline (**7**)[b] (rotamers)	$CH_2CH(Me)_2$	5.16 q (J = 7.6)	1.36 d (J = 7.6)	2.77 (0.5H) 2.89 (2.5H)	2.17 d (J = 2.0)	—	2.16 m	0.95 d (J = 6) 0.95 d (J = 6)	*85*
Normaytancyprine (**12**)	CHCH(Me)CH(Me)	5.29 q (J = 7)	1.28 d (J = 7)	2.90 s		—		0.64 d (J = 5)	*88*
Colubrinol (**13**)	$CH(Me)_2$	5.36 m	1.26 d (J = 6.6)	2.87	2.76 m	1.04 d (J = 6.5) 1.10 d (J = 6.8)			*59*
Demethyltrewiasine (**16**)	$CH(Me)_2$	4.90 m	1.35 d (J = 7)	6.87[d] d (J = 10.5)	2.30 m	1.08 d (J = 7) 1.16 d (J = 7)			*57*

(*Continued*)

TABLE IVA (*Continued*)

Compound	R	2′	2′-Me	2′NMe	4′	4′Me	5′	5′Me	Reference
Dehydrotrewiasine (**17**)	C(Me)=CH_2	5.29 m	1.33 d (J = 6.9)	2.88	—	1.92	5.02 5.22		*57*
Trewiasine (**18**)	$CH(Me)_2$	5.37 m	1.28 d (J = 6.8)	2.88	2.76 m	1.06 d (J = 6.6) 1.12 d (J = 6.8)			*57*
Nortrewiasine (**20**)	$CH(Me)_2$	5.34 m	1.29 d	2.89	2.76 m	0.91 d 1.08 d			*59*

[a] Not reported.
[b] Exhibits rotameric forms at ambient temperature.
[c] Unassigned envelope between 0.8 and 2.1 contains C-4′ H, C-5′ H, C-6′ H, and C-6′ Me.
[d] 2′-NH rather than 2′-NMe.

TABLE IVB

PROTON NMR OF C-3 ESTER SIDE CHAINS: O—C(O)—R

Compound	R	2′	2′-Me	3	3′	3′-Me	Reference
Maytanacine (**11**)	CH_3	2.08	—	—			*87*
Maytanbutacine (**15**)	$CH(Me)_2$	2.0–2.6 m	1.20 d (J = 7) 1.28d (J = 7)	—			*7*
Maytansinol 3-proprionate	$H_2C—CH_3$	2.44 g (J = 7)	1.19 t (J = 7)	4.87 dd (J = 3, 12)			*6*
Maytansinol 3-bromoacetate	$CH_2—Br$	3.83 s	—	5.01 dd (J = 3, 12)			*6*
Maytansinol 3-crotonate	$HC{=}CH(CH_3)$	5.20–5.90 m	—	4.83 m	6.1–6.6 m	1.65 d (J = 7)	*6*
Maytansinol 3-trifluoroacetate	CF_3	—	—	5.01 dd (J = 3, 12)			*6*

TABLE IVC
PROTON NMR OF C-3 ESTER SIDE CHAINS: CYCLIC ESTERS

Assignment	Trenudine *21*	Treflorine *22*	N-Methyltrenudone
2-H_a	2.22 dd (J = 13.6, 3.7)	2.21 dd (J = 14.9, 3.9)	2.09 dd (J = 14.7, 3.6)
2-H_b	2.53 dd (J = 13.6, 12.0)	2.50 dd (J = 14.9, 12.0)	2.62 dd (J = 14.7, 12.0)
3-H	4.63 dd (J = 12.0, 3.7)	4.51 dd (J = 12.0, 3.9)	4.51 dd (J = 12.0, 3.6)
2′-H	4.95 m	4.79 m	5.57 g
2′-NH	1.34 d (J = 7)	1.33 d (J = 6.9)	1.29 (J = 7)
2′-NMe	7.67 d (J = 9.9)	7.06 d (J = 10.7)	—
2′-NMe	—	—	2.75 s
4′-Me	1.48 s	1.40 s	1.53 s
5′-H_a	3.93 m (J = 2.5, 2.4)	2.78 m (J = 14.2, 3.0)	—
5′-H_b	—	1.45 m (J = 14.2, 3.0)	—
6′-H_a	3.55 dd (J = 15)	3.03 m (J = 14.2, 3.0)	4.17 d (J = 14.6)
6′-H_b	4.49 dd (J = 15)	4.46 m (J = 14.2, 3.0)	4.51 d (J = 14.6)

sine and shows the same OMe singlet as trewiasine (**18**) at 3.37 as well as a singlet at 4.67 for the C-15 H. In addition trewsine (**43**) contains the C-2–C-3 olefin found in maysine (**3**) with two extra vinyl hydrogen resonances at 5.67 (C-2 H) and 6.42 (C-3 H) plus noticeable shifts in the C-4 methyl and C-5 hydrogen.

Colubrinol (**13**) has a free hydroxyl function at C-15 showing only a single hydrogen at that position with a chemical shift of 5.47. Except for the aromatic hydrogens (C-17 H and C-21 H) the rest of the spectrum is normal. It should be pointed out that in many of the maytansinoids the shifts of these aromatic hydrogens are concentration dependent.

Finally, another paper (*59*) reports the C-10 isomer of trewiasine (**18**), 10-epitrewiasine (**19**). This unique compound showed a doublet (J = 3.5 Hz) at 3.72 that should be compared to the usual C-10 H doublet at 3.50 with a much larger coupling constant (J = 9.0 Hz).

2. Carbon NMR

Table V lists the ^{13}C chemical shifts, multiplicities, and assignments for maytansine (**1**) (*132*), maytansinol (**4**) (*132*), maysine (**3**) (*132*), maysenine

TABLE VA

^{13}C NMR OF MAYTANSINE, MAYTANSINOL, MAYSINE, MAYSENINE, AND TREWIASINE

Carbon	Maytansine (**1**) (*132*)	Maytansinol (**4**) (*132*)	Maysine (**3**) (*132*)	Maysenine (**9**) (*132*)	Trewiasine (**18**) (*57*)
2	32.5 t	35.6 t[a]	121.9 d	116.9 d	32.43 t
3	78.2 d	75.8 d	147.5 d	148.0 d	78.18 d
4	60.1 s	63.1 s	59.7 s	135.0 s[a]	59.99 s
5	67.2 d	66.6 d	66.9 d	140.9 d	67.73 d
6	39.1 d	37.9 d	38.7 d	39.2 d	38.86 d
7	74.2 d	75.4 d	75.0 d	75.9 d	74.15 d
8	36.5 t	35.8 t[a]	35.5 t	35.5 t	36.26 t
9	81.0 s	81.3 s	81.2 s	81.0 s	80.72 s
10	88.9 d	89.0 d	88. d	88.3 d	85.52 d
11	127.8 d	127.1 d	127.2 d	128.0 d	129.92 d
12	133.3 d	133.3 d	133.0 d	132.5 d	132.51 d
13	125.4 d	125.2 d	124.6 d	125.6 d	127.97 d
14	139.1 s[a]	138.9 s[b]	140.3 s[a]	139.3 s[a]	142.13 s
15	46.7 t	47.1 t	46.7 t	47.2 t	86.70 d
16	142.4 s[a]	142.5 s[b]	142.0 s[a]	138.9 s[a]	141.35 s
17	122.5 d	123.7 d	122.2 d	118.0 d	120.30 d
18	141.2 s[a]	140.2 s[b]	140.5 s[a]	136.4 s[a]	139.01 s
19	119.1 s	119.0 s	119.3 s	114.4 s	118.93 s
20	156.1 s[b]	155.8 s[c]	156.4 s[b]	155.8 s[b]	156.30 s
21	113.4 d	112.9 d	112.7 d	111.0 d	108.96 d
Carbamate					
C=0	152.2 s[b]	152.7 s[c]	152.1 s[b]	152.5 s[b]	152.40 s
Lactam					
C=0	168.8 s	171.8 s	164.3 s	166.3 s	168.84 s
C=0	170.2 s	—	—	—	170.71 s
C=0	170.8 s	—	—	—	170.92 s
4-Me	12.2 q	11.3 q	14.2 q	13.7 q[c]	11.96 q
6-Me	14.5 q	14.5 q	14.8 q	15.7 q[c]	14.62 q
14-Me	15.5 q	15.8 q	16.0 q	16.8 q[c]	10.01 q
10+20+			56.6 q		
15-OMe	56.7 2 q	56.6 2 q	56.7 q	56.6 2 q	56.3–56.7 3 q
18-NMe	35.4 q	36.0 q	36.0 q	—	35.22 q
2′	52.2 d	—	—	—	52.38 d
2′-Me	13.4 q	—	—	—	13.13 q
2′-NMe	31.7 q	—	—	—	30.42 q
4′	21.7 q	—	—	—	30.42 q
4′-Me	—	—	—	—	19.43 q
					18.85 q

[a] May be interchanged in any column.
[b] May be interchanged in any column.
[c] May be interchanged in any column.

TABLE VB
^{13}C NMR OF TREWSINE AND 10-EPITREWIASINE

Carbon	Trewsine (**43**) (*57*)	10-Epitrewiasine (**19**) (*59*)
2	119.71 d	32.62 t
3	147.59 d	78.18 d
4	59.53 s	60.24 s
5	66.68 d	67.26 d
6	38.54 d	38.79 d
7	74.80 d	74.28 d
8	35.16 t	31.84 t
9	80.91 s	82.73 s
10	129.01 d	128.81 d
12	132.06 d	130.24 d
13	126.08 d	126.21 d
14	141.42 s	142.52 s
15	87.22 d	87.08 d
16	141.09 s	141.41 s
17	121.66 d	120.23 d
18	140.64 s	136.86 s
19	Not observed	119.25 s
20	156.50 s	156.36 s
21	108.01 d	109.11 d
Carbamate		
C=0	152.21 s	154.80 s
Lactam		
C=0	164.23 s	169.03 s
C=0	—	170.92 s
C=0	—	177.09 s
4-Me	14.10 q	12.28 q
6-Me	14.82 q	14.75 q
14-Me	9.88 q	10.07 q
10+20+	56.61 3 q	56.3–56.7
15-Ome		58.4 3 q
18-NMe	35.87 q	35.41 q
2′	—	53.81 d
2′-Me	—	13.25 q
2′-NMe	—	30.54 q
4′	—	30.54 q
4′-Me	—	19.23 q
		18.84 q

(**9**) (*132*), trewiasine (**18**) (*57*), trewsine (**43**) (*57*), and 10-epitrewiasine (**19**) (*59*). Most of these data became available in recent years from the work of Sneden (*132*) and Powell (*57,59*). Again, using maytansine (**1**) as the base spectrum certain points should be emphasized.

One resonance of note is that due to C-3. Whereas in maytansine (**1**) it appears at 78.2, when the C-3 hydroxyl group is not acylated (maytansinol, **4**) the signal shifts to 75.8; when a C-2–C-3 double bond exists, the new olefinic carbon appears at 147.5 (maysine, **3**), 147.59 (trewsine, **43**), or 148 (maysenine, **9**). Other C-3 esters are quite similar to maytansine (**1**), showing C-3 at 78.18 (trewiasine, **18**, and 10-epitrewiasine, **19**).

The C-2–C-3 unsaturated compounds show dramatic shifts downfield for C-2. This carbon appears at 121.9 in maysine (**3**) and 119.71 in trewsine (**43**). Maysenine (**9**), a C-2–C-3, C-4–C-5 diene, shows C-2 at 116.9. In addition C-4 and C-5 move down to 135.0 and 140.9, respectively. The methyl groups attached to C-4 and C-6 are also affected by the presence of the double bonds.

Trewiasine (**18**), 10-epitrewiasine (**19**), and trewsine (**43**) all contain C-15 methoxyl groups, which result in a 40-ppm shift downfield of C-15 relative to Maytansine (**1**). In all three of these compounds the methyl group attached to C-14 moves upfield over 5 ppm. There are many other less dramatic differences between the various carbon resonances, and these can be found in Table V.

B. Mass Spectrometry

1. Electron Impact

Although no maytansinoids have shown a parent ion in electron impact mass spectrometry, these compounds do give very characteristic fragmentation patterns. Without exception each maytansinoid shows a peak at $M^+ - 61$ due to the loss of HNCO and H_2O from the cyclic carbamate. Kupchan (*84*) has termed this fragment the $M^+ - a$ ion (where $a = H_2O + HNCO$), and subsequently this has become custom.

For those compounds having an OR function at C-3 (R = H, amino acid-derived ester, acyl, etc.), a characteristic peak occurs at 485. This comes from the additional loss of the C-3 side chain as ROH and is therefore called the $M^+ - (a + b)$ ion (where b = ROH). This $m/e = 485$ ion can lose a methyl radical (470 amu) or a chlorine (450 amu). These peaks can be seen in Table VIA.

The nature of the ester side chain attached to C-3 can be inferred from some secondary fragmentations. After initial loss of the side chain as b in $M^+ - (a + b)$, two major ions are seen. One (b − OH) is due to loss of hydroxyl and the other (b − CO_2H) comes from decarboxylation. For these ions the homologous series of maytansine (**1**), maytanprine (**5**), maytanbutine (**6**), and maytanvaline (**7**) differs by a constant 14 amu (one methylene).

Variations in the substituents on the macrocycle result in expected changes in the mass spectrum. The C-15 methoxylated compounds trewiasine (**18**), 10-epitrewiasine (**19**), dehydrotrewiasine (**17**), and demethyl-

TABLE VI

MASS SPECTRAL CORRELATIONS

A. Compound	$M^+ - a$[a]	$M^+ - (a + b)$[b]	$M^+ - (a + b + CH_3)$	$M^+ - (a + b + Cl)$	$b - OH$	$b - CO_2H$	Reference
Maytansine (**1**)	630	485	470	450	128	100	*84*
Maytanprine (**5**)	644	485	470	450	142	114	*84*
Maytanbutine (**6**)	658	485	470	450	156	128	*84*
Maytanvaline (**7**)	672	485	470	450	170	142	*7*
Maytanacine (**11**)[c]	545	485	470	450	—	—	*7*
Maytansinol (**4**)[d]	503	485	470	450	—	—	*7*
Maytansinol 3-pruprimate	559	485	470	450	—	—	*6*
Maytansinol 3-bromoacetate	623	485	470	450	—	—	*6*
Maytansinol 3-crotoncete	571	485	470	450	—	—	*6*
Maytansinol 3-trifluoroacetate	599	485	470	450	—	—	*6*
P-1	545	485	470	450	—	—	*54*
P-2	559	485	470	450	—	—	*54*
P-3	573	485	470	450	—	—	*54*
P-3′	573	485	470	450	—	—	*54*
P-4	587	485	470	450	—	—	*54*
Trewiasine (**18**)[e]	688	515	500	None	156	128	*57*
10-Epitrewiasine (**19**)	688	515	500	480	156	128	*59*
Dehydrotrewiasine (**17**)[e]	686	515	500	—	154	126	*57*
Demethyltrewiasine (**16**)[e]	674	515	500	—	142	114	*57*
Normaytansine (**10**)[f]	616	471	456	436	128	100	*86*
Normaytancyprine (**12**)[f]	670	471	456	436	—	—	*88*
Nortrewiasine (**20**)[e,f]	674	501	—	—	—	—	*59*
Colubrinol (**13**)[g]	674	501	—	—	—	—	*57*

B. Compound	$M^+ - a$	$M^+ - (a + CH_3)$	$M^+ - (a + Cl)$	Reference
Maysine (**3**)	485	470	450	*7*
Normaysine (**8**)[f]	471	456	436	*7*
Maysenine (**9**)[h]	455	440	420	*7*
Trewsine (**43**)[e]	515	—	480	*57*

C. Compound	$M^+ - a$	$M^+ - (a + c)$[i]	$M^+ - (a + b + c)$	$483 - CH_3$	$483 - Cl$	Reference
Maytanbutacine (**15**)	631	571	483	468	448	*7*
Deacetylmaytanbutacine	589	571	483	468	448	*7*

D. Compound	$M^+ - a$	$M^+ - (a + OCH_3 - C1 + CO_2H)$	Reference
Treflorine (**22**)	688	577	*58*
Trenudine (**21**)	704	593	*58*
N-Methyltrenudone (**23**)	716	605	*57*

[a] a = H_2O + HNCO.
[b] b = ROH where RO = C-3 substituent.
[c] Here b = CH_3CO_2H.
[d] Here b = H_2O.
[e] C-15 OMe compounds show ions 30 amu higher.
[f] Compound contains N—H instead of N—CH_3, thus ions are 14 amu lower.
[g] Compound contains a C-15 hydroxyl, thus ions are 16 amu higher.
[h] Maysenine lacks the C-4–C-5 epoxide (−16 amu) and has N—H instead of N—CH_3 (−14 amu). Ions are 30 amu lower.
[i] In C-15 oxygenated compounds C = R′OH where R′O = C-15 substituent.

trewiasine (**16**) all show an $M^+ - (a + b)$ ion at 515. This 30-amu increase is due to the extra OCH_2. Compounds that have an N-H on the lactum nitrogen (normaytansine, **10**, normaytancyprine, **12**, nortrewiasine, **20**, and normaysine, **8**), instead of the more common *N*-methyl, show peaks 14 amu lower for $M^+ - a$, $M^+ - (a + b)$, $M^+ - (a + b + CH_3)$, and $M^+ - (a + b + Cl)$. Maysine (**3**), normaysine (**8**), maysenine (*9*), and trewsine (**43**) have no C-3 side chain and thus no $M^+ - (a + b)$ ion. Table VIB shows the $M^+ - a$, $M^+ - (a + CH_3)$, and $M^+ - (a + Cl)$ fragments for this series.

2. Chemical Ionization

Powell (*59*) has observed molecular ions using negative ion chemical ionization mass spectrometry. In addition to observing the parent anion M^-, one sees peaks for $(M - a)^-$, $[M - (a + b)]^-$, and $(b - H)^-$. Data for trewiasine (**18**), dehydrotrewiasine (**17**), demethyltrewiasine (**16**), treflorine (**22**), trenudine (**21**), *N*-methyltrenudone (**23**), 10-epitrewiasine (**19**), nortrewiasine (**20**), and colubrinol (**13**) are presented in Powell's paper (*59*).

C. Infrared and Optical Rotation

1. Infrared

Maytansine (**1**) (*2*) shows characteristic bands in the IR region at 1740, 1724, 1661, and 1577 cm^{-1} due to the ester, cyclic carbamate, lactam carbonyls, and the carbon–carbon double bonds, respectively. Maysine (**3**) (*85*) has carbonyl absorptions at 1709 (cyclic carbamate) and 1664 cm^{-1} (lactam) as well as C=C stretches at 1629 (C-2–C-3 double bond) and 1577 cm^{-1} (diene). In maytansinol (**4**) (*87*) the carbamate also appears at 1709 and the lactam carbonyl at 1650 cm^{-1}; the diene is at 1575. The additional unsaturation in maysenine (**9**) (*85*) is seen at 1610 cm^{-1}, the carbamate at 1704, and the lactam at 1664.

2. Optical Rotation

All of the known maytansinoids exhibit a negative rotation. The magnitude of the rotation, solvent used, temperature, and concentration are given in Table VII.

D. Crystal Structure and Absolute Configuration

Working with Kupchan's group, Bryan (*133*) performed a detailed single crystal X-ray analysis of maytansine (**1**) as its 3-bromopropyl ether.

TABLE VII
OPTICAL ROTATION

Compound	$[\alpha]_D$	(°) Temperature (°C)	Solvent	Concentration	Reference
Maytansine (**1**)	−145	26	$CHCl_3$	0.055	*2*
Maysine (**3**)	−173	30	EtOH	0.023	*85*
Maytansinol (**4**)	−309	23	$CHCl_3$	0.110	*87*
Maytanprine (**5**)	−125	30	$CHCl_3$	0.0559	*84*
Maytanbutine (**6**)	−122	30	$CHCl_3$	0.0492	*84*
Maytanvaline (**7**)	−135	26	$CHCl_3$	0.950	*85*
Normaysine (**8**)	−217	30	EtOH	0.051	*85*
Maysenine (**9**)	−57	30	EtOH	0.056	*85*
Maytanacine (**11**)	−119	23	$CHCl_3$	0.100	*87*
Colubrinol (**13**)	−94	22	—	—	*89*
Colubrinol (**13**)	−70	23	$CHCl_3$	0.555	*59*
Maytanbutacine (**15**)	−90	33	EtOH	0.055	*7*
Demethyltrewiasine (**16**)	−126	23	$CHCl_3$	0.049	*57*
Dehydrotrewiasine (**17**)	−90	23	$CHCl_3$	0.120	*57*
Trewiasine (**18**)	−94	23	$CHCl_3$	0.159	*57*
10-Epitrewiasine (**19**)	−48	23	$CHCl_3$	0.103	*59*
Nortrewiasine (**20**)	−58	23	$CHCl_3$	0.040	*59*
Trenudine (**21**)	−114	23	$CHCl_3$	0.240	*58*
Treflorine (**22**)	−138	23	$CHCl_3$	0.045	*58*
N-Methyltrenudone (**23**)	−110	23	$CHCl_3$	0.183	*58*
Trewsine (**43**)	−114	23	$CHCl_3$	0.055	*57*

This study was critical to the initial structural assignment and defined the absolute configuration. The asymmetric centers of maytansine (**1**) are 3*S*, 4*S*, 5*S*, 6*R*, 7*S*, 9*S*, 10*R*, 2′*S*. This structure and stereochemistry has been borne out by total synthesis as will be seen in the following sections.

IV. Synthetic Approaches

A. Introduction

In attempting a total synthesis of any of the maytansinoids one undertakes a formidable task. A brief look at the general structure (Fig. 1) shows seven chiral centers, a conjugated diene, a trisubstituted epoxide, a tetrasubstituted aromatic ring, and a cyclic carbamate. Putting all these pieces together has provided a challenge that many organic chemists have answered.

Early in the synthetic work the topological comparison of maytansine (**1**) to the United States was made (*13*). Since that time many of the partial

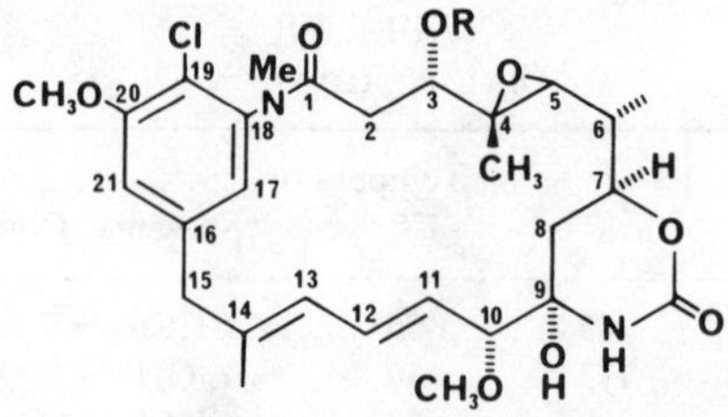

FIG. 1. General structure of maytansinoids.

syntheses have used geographic terminology to describe various fragments. In the sections that follow we will adhere to the same convention. For example, the aromatic moiety is referred to as the west or aromatic west. The literary potential of this nomenclature has not been overlooked; one issue of *Tetrahedron Letters* containing three different syntheses of the aromatic portion is commonly called "How the West Was Won." The southwestern zone stretches from C-10 to the aromatic ring without including the amide linkage. The north begins with C-1 and moves across the top of the molecule to C-6, while the northeast generally consists of C-3 through C-10. Although the heterocycle connecting C-7 and C-9 has been likened to Florida we will forego that analogy and refer to it as either the cyclic carbamate or the carbinolamide.

To review the synthetic work on the maytansinoids we have chosen to present first approaches to the various zones (west, southwest, north, northeast, carbinolamide) and then see how these pieces were used in successful total syntheses. Finally, work on asymmetric synthetic approaches will be discussed.

B. THE SYNTHESIS OF THE MAJOR ZONES

1. Aromatic West

The synthetic problem of the aromatic west of the maytansinoids was solved by four groups almost simultaneously. This was not a trivial problem; what was required was the introduction of three contiguous heteroatoms plus a handle for connection to the southern zone. Traditional electrophilic aromatic substitution chemistry did not offer an obvious solution.

a. Meyers. Meyers (*14*) started with commercially available methyl vanillate (**44**) (Scheme 1) and after nitration (HNO_3–HOAc, 83%), replaced the phenolic OH with chlorine. Originally thionyl chloride–dimethylformamide (DMF) was used in the next step, but later (*24*) it was found that oxalyl chloride–DMF converted **45** to **46** in 90% yield. The aromatic west could conceivably be connected to the south as either a

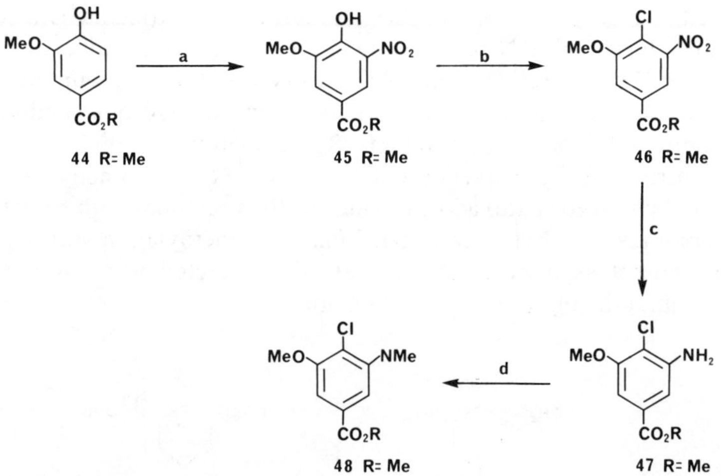

SCHEME 1. Meyers electrophilic west synthesis. (a) HNO_3, HOAc. (b) $C_2O_2Cl_2$, DMF. (c) $SnCl_2$, HOAc. (d) NaH, CH_3I.

nucleophile or an electrophile. Both synthons were made from **46**. Reduction ($SnCl_2$–HOAc) to the aniline **47** followed by methylation (NaH, MeI) gave the *N*-methylaniline **48**, which could be reacted with an organometallic southern piece at the carbomethoxy group.

Alternatively, the ester in **46** (Scheme 2) was hydrolyzed to the carboxylic acid **49** and then converted to the arylbromide **50**. This was done by a photoassisted Cristol–Firth–Hunsdieker reaction (*134*) in which visible light is thought to mediate the decomposition of the intermediate acylhypobromide. Reduction of the nitro group ($SnCl_2$–HOAc) was again fol-

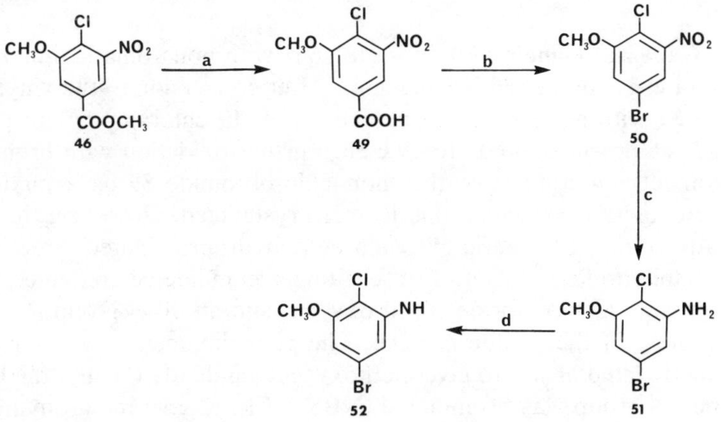

SCHEME 2. Meyers bromo-west synthesis. (a) aq. KOH. (b) HgO, Br_2, CCl_4, $h\nu$. (c) $SnCl_2$, HOAc. (d) NaH, CH_3I.

lowed by methylation to give the "bromo-west" **52** which would be metallated, after suitable N-protection, to produce a nucleophilic west.

Later (*21*), a fluoride-removable nitrogen protecting group was chosen that would come off under essentially neutral, anhydrous conditions and generate only volatile by-products. The trimethylsilylethylurethane (*135*) was introduced by conversion of the aniline **51** to the phenylurethane **53** (phenyl chloroformate) and exchange of the urethanes with trimethylsilylethanol and *tert*-butoxide to **54**. Finally, N-methylation with methyl iodide and potassium *tert*-butoxide gave the protected bromo-west **55** ready for lithium–halogen exchange (Scheme 3).

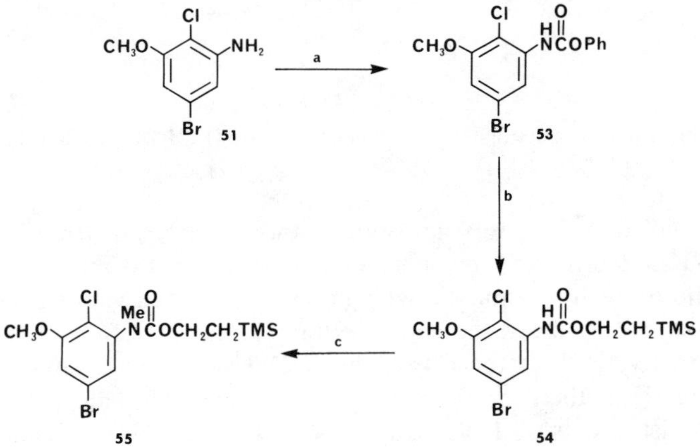

SCHEME 3. Meyers protected bromo-west. (a) PhOCOCl. (b) $TMSCH_2CH_2OH$, KO*t*-Bu. (c) CH_3I, KO*t*-Bu.

b. Ganem. Ganem (*40*) chose to start with nonaromatic 5-methylcyclohexane-1,3-dione (**56**) (Scheme 4). After conversion to the vinylogous amide **57** with aqueous methylamine, the C-19 chlorine was introduced with *N*-chlorosuccinimide to give **58**. Further oxidation with bromine in carbon tetrachloride gave the monochlorobromide **59** as a mixture of diastereomers from which one isomer crystallized. However, the crude mixture of diastereomeric **59** could be dehydrobrominated directly with acetic anhydride–*p*-TsOH. The resulting acetanilide **60** contained the C-18 nitrogen, C-19 chlorine, C-20 oxygen, and an alkyl group (C-15) for attachment of the southern piece. The phenolic acetate was saponified and methylated *in situ* to give methoxyacetanilide **61**. Finally, the benzylic methyl group was brominated (NBS–CCl_4) to give the aromatic west **62** with both N-protection and a site for C-14–C-15 bond formation.

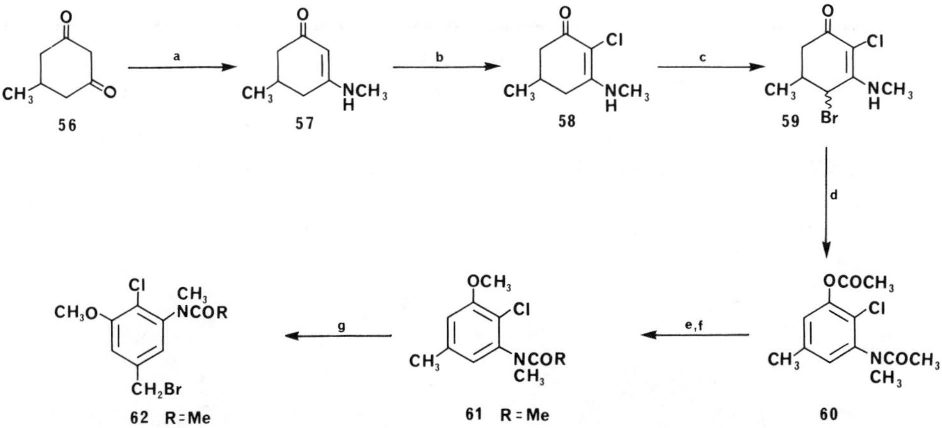

SCHEME 4. Ganem west synthesis. (a) aq. CH_3NH_2. (b) NCS, CH_2Cl_2. (c) Br_2, CCl_4, 0°C. (d) Ac_2O, *p*-TsOH, 110°C. (e) K_2CO_3; MeOH. (f) CH_3I, reflux. (g) NBS, CCl_4, *hν*.

c. Corey. As in the previous synthesis Corey (*10*) also began with nonaromatic material (Scheme 5). This cyclohexanedione (as its enol ether) **63** contained a carboxyl group for further elaboration of C-15. The enone ester **63** was treated with *N*-methylbenzylamine to give the vinylogous amide **64** which was chlorinated with *tert*-butyl hypochlorite at −50°C. The resulting chloro compound **65** was further oxidized by generation of the anion (presumably next to the ester) and quenching with phenylselenium bromide. Silica gel chromatography gave the aromatic system **66**. Methylation of phenolic **66** and debenzylation of the amine

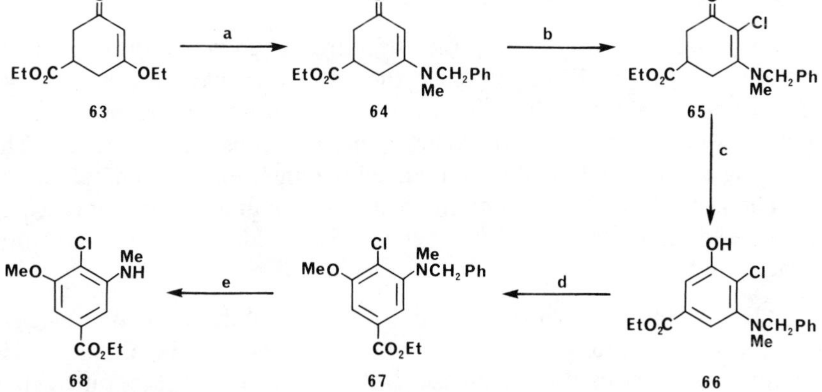

SCHEME 5. Corey west synthesis. (a) HNMeBzl. (b) *t*-BuOCl, $CHCl_3$. (c) $LiNEt_2$, PhSeBr, −78°C. (d) CH_3I, K_2CO_3. (e) H_2, Pd–C, EtOH.

gave the desired amino ester **68**. In later work (*11*) this ester was reduced to the corresponding benzylic alcohol and then converted to the benzylic iodide analogous to Ganem's bromide **62**.

d. Roche. The approach of the Roche group (*49,50*) was quite similar to the previously described Meyers route. Ethyl vanillate (**44**) (Scheme 6)

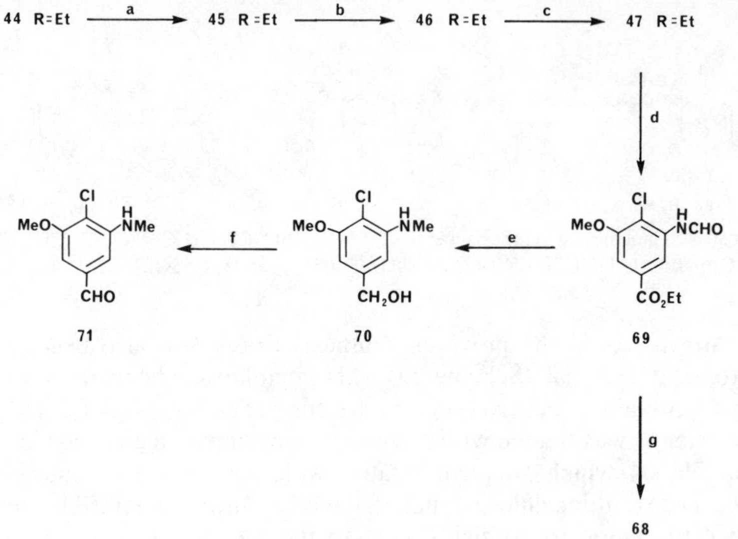

44 R=Et —a→ 45 R=Et —b→ 46 R=Et —c→ 47 R=Et

d

Cl MeO H NMe CHO 71 ←f— Cl MeO H NMe CH_2OH 70 ←e— Cl MeO H NCHO CO_2Et 69

g

68

SCHEME 6. Roche west synthesis. (a) HNO_3, HOAc. (b) LiCl, $POCl_3$, DMF. (c) Zn, HOAc, or Pd–C, HCl, EtOH. (d) PhOH, 80°C, HCO_2Ph. (e) LAH. (f) CrO_3, pyridine. (g) aq H_2CO,CH_3OH, H_2, Ra–Ni, 100 psi.

was nitrated and chlorinated, and the nitro group was then reduced to yield amino ester **47**. In this case they chose to have the west as an electrophilic synthon. Thus the nitrogen was formylated to produce **69**, and then both the formyl and carboethoxy groups were reduced. The resulting amino alcohol **70** was oxidized (Collins) to the amino aldehyde **71**. The versatility of the Roche chemistry is seen in the ready synthesis of Corey's key intermediate **68** (see Scheme 5) by reduction of the *N*-formyl group of **69**.

e. Ho. Three years after the initial synthesis of the aromatic west, Ho (*45*) reported an interesting but lengthy synthesis of Ganem's piece. He opted for protection of nitrogen as a carbamate **62** (R = OMe) rather than the original *N*-acetyl. In addition, ortho metallation of the aromatic ring was used as a way of introducing the C-18 substitutents (Scheme 7).

Commercially available 6-amino-*m*-cresol (**72**) was first N-acylated

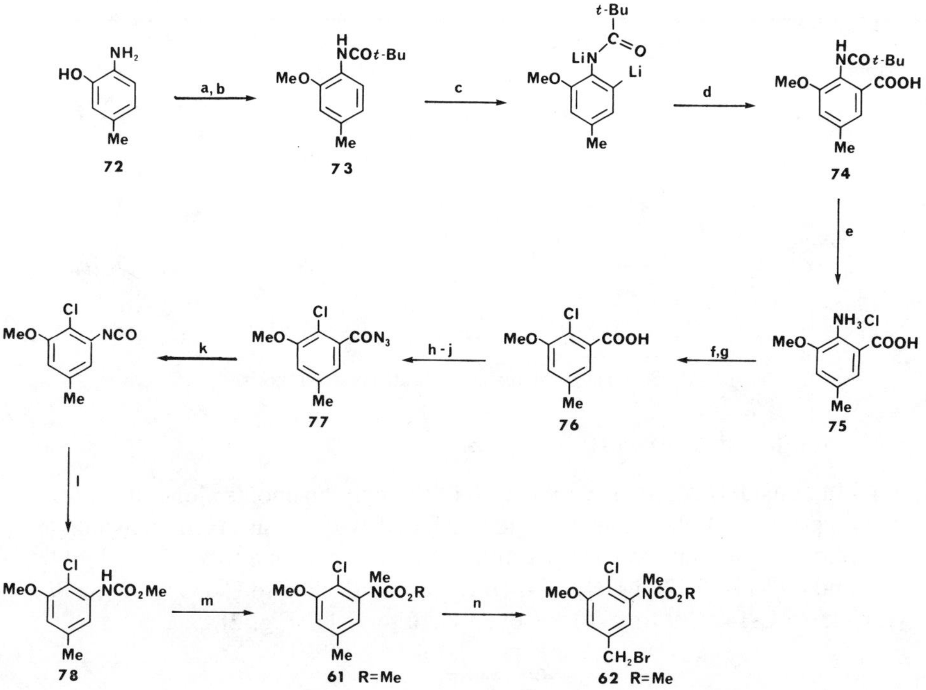

SCHEME 7. Ho west synthesis. (a) *t*-BuCOCl. (b) K_2CO_3, CH_3I. (c) 2.2 equiv BuLi, THF, RT, 6 hr. (d) CO_2, −78°C. (e) aq HCl, reflux, 16 hr. (f) $NaNO_2$, aq HCl, 0°. (g) CuCl, 60°C, 2 days. (h) $ClCO_2Et$. (i) Et_3N. (j) NaN_3. (k) PhH, 60°C, 30 min. (l) MeOH, 80°C. (m) NaH, CH_3I. (n) NBS, CCl_4, *hν*.

then O-methylated to give the methoxyamide **73**. This compound was bis metallated with *n*-butyllithium at nitrogen and the ortho position of the ring. Quenching with CO_2 gave the benzoic acid **74**. The amino group was deprotected and replaced by chlorine in a Sandmeyer reaction, and the resulting chlorobenzoic acid **76** was converted to its acyl azide **77**. Curtius rearrangement gave the isocyanate that was treated with methanol to form the urethane **78**. Finally, N-methylation and bromination with NBS resulted in the benzylic bromide **62**.

f. Zhou. A group of workers at the Shanghai Institute (*37*) converted 2-methoxy-6-nitroaniline (**79**) to the Roche (*49,50*) amino aldehyde **71**. Their approach (Scheme 8) entailed chlorination and cyanide displacement in order to put in C-16, Sandmeyer reaction to introduce the C-19 chlorine, and reduction of the nitro group to give the C-18 amine. After N-methylation (**83** to **84**) the cyano group was reduced to the desired aldehyde **71**.

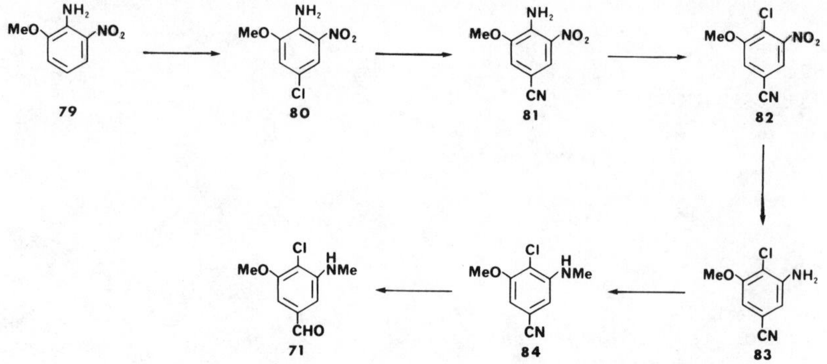

SCHEME 8. Shanghai Institute west synthesis.

2. South and Southwest

In considering any disconnection of the maytansinoids along the southern portion of the molecule, one is faced with a variety of reasonable options. You may imagine a synthesis that connects either of the double bonds (C-11–C-12 or C-13–C-14), the aromatic ring to C-15, the benzylic C-15 to C-14, C-9 to C-10, or even C-10 to C-11 (Fig. 2).

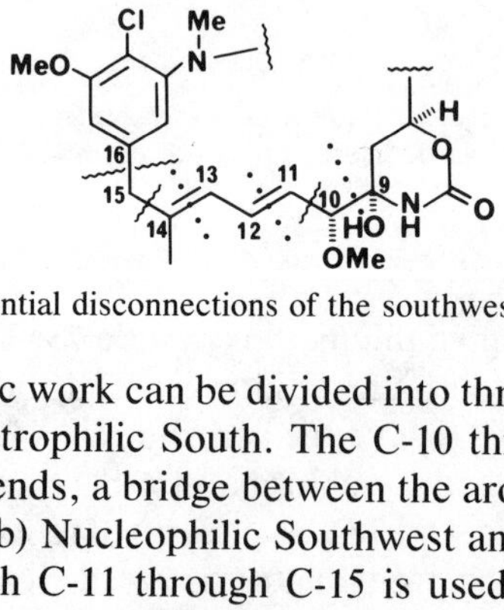

FIG. 2. Potential disconnections of the southwestern zone.

The actual synthetic work can be divided into three classes. (a) Nucleophilic West and Electrophilic South. The C-10 through C-15 synthon is electrophilic at both ends, a bridge between the aromatic west (C-16) and the northeast (C-9). (b) Nucleophilic Southwest and Electrophilic Northeast. In this approach C-11 through C-15 is used as a bridge, but it is electrophilic at one end (C-15) and nucleophilic at the other (C-11). The key is the connection of C-10 to C-11. (c) Electrophilic Southwest and Nucleophilic Northeast. This is the most common approach, and in it an acyl anion equivalent (C-9) is condensed with an electrophilic C-10. In any of these routes the geometry of the *E,E*-diene must be taken into account (Fig. 3).

a. West (Nuc^-) and South (E^+). Meyers' initial approach (*13,136*) used lithiodithiane **86A** as an acyl anion equivalent. In a model study (Scheme 9) 1,2-epoxybutane was converted to the substituted dithiane

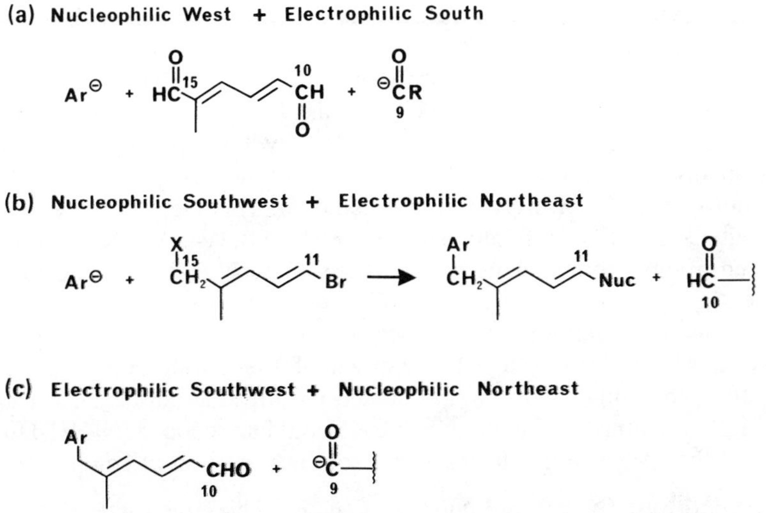

(a) Nucleophilic West + Electrophilic South

Ar⊖ + HC(=O) 15 ... 10 CH(=O) + ⊖C(=O)R 9

(b) Nucleophilic Southwest + Electrophilic Northeast

Ar⊖ + X 15 CH_2 ... 11 Br → Ar CH_2 ... 11 Nuc + HC(=O) 10

(c) Electrophilic Southwest + Nucleophilic Northeast

Ar ... CHO 10 + ⊖C(=O) 9

FIG. 3. Retrosynthetic analysis of the maytansinoid skeleton.

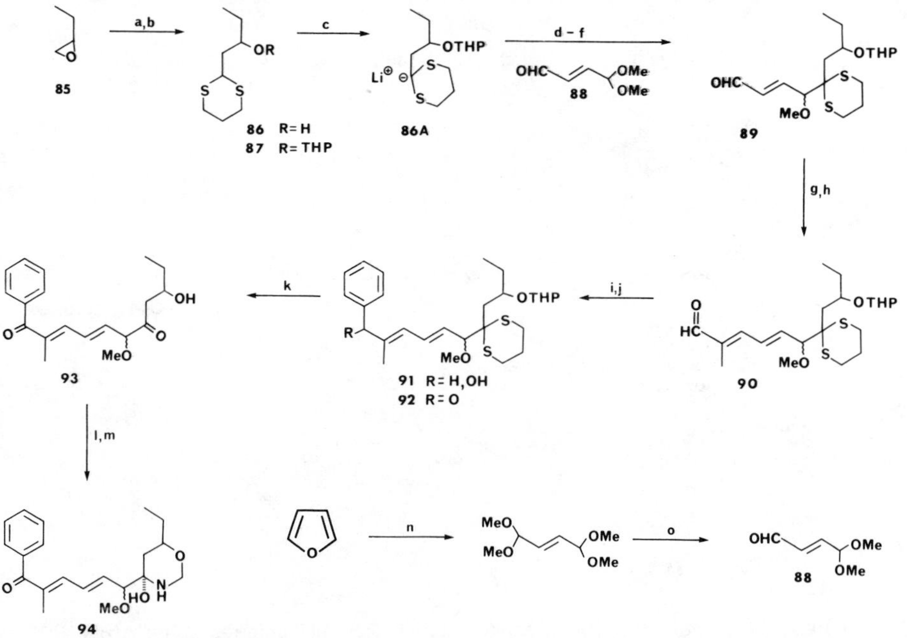

85 → a,b → 86 R=H, 87 R=THP (OR) → c → 86A ($Li^{\oplus}$ ⊖, OTHP) → d – f (OHC, OMe, OMe, 88) → 89 (OHC, MeO, OTHP)

89 → g,h → 90 (HC=O, MeO, OTHP)

90 → i,j → 91 R=H,OH; 92 R=O (R, MeO, OTHP) → k → 93 (O, MeO, O, OH)

93 → l,m → 94 (O, MeO, HO, N H, O)

furan → n → MeO, MeO, OMe, OMe → o → 88 (OHC, OMe, OMe)

SCHEME 9. Meyers nucleophilic west and electrophilic south. (a) lithiodithiane, $-25°C$. (b) H^+, DHP. (c) n-BuLi, $-30°C$. (d) **88**. (e) CH_3I, HMPA. (f) oxalic acid, H_2O, THF. (g) $LiCHCH_3CH{=}NC_6H_{11}$. (h) aq oxalic acid. (i) PhMgBr. (j) DDQ. (k) $HgCl_2$, H_2O, CH_3CN. (l) $COCl_2$. (m) NH_3 (l). (n) Br_2, MeOH. (o) $HClO_4$, THF, H_2O.

(**87**) and then deprotonated with *n*-butyllithium. Reaction with the monoacetal of fumaraldehyde **88** gave, after *in situ* O-methylation and hydrolysis of the acetal, the α,β-unsaturated aldehyde **89** that has the necessary functionality at C-7 through C-13. This aldehyde was chain extended by Wittig's directed aldol condensation (*N*-cyclohexylpropylimine, lithium diisopropylamide, −78°C), and the resulting hydroxy intermediate was dehydrated to afford the (*E,E*)-dienal **90**. This piece represents C-7 through C-15—the south (already connected to the east) is thus ready for condensation with a nucleophilic west. Instead of the functionalized aromatic ring, phenyl magnesium bromide was added to dienal **90** to give allylic alcohol **91** as a mixture of diastereomers. This mixture was oxidized to the C-15 ketone **92**. Removal of the tetrahydropyranyl and dithiane protecting groups was followed by introduction of the carbinolamide, producing **94**. The ketone at C-15 could be reduced ($NaBH_4$) to an alcohol that corresponds to the exact southern portion of colubrinol.

b. Southwest (Nuc⁻) and Northeast (E⁺). The construction of a nucleophilic southwest that would add to a C-10 aldehyde was first reported by Meyers in the synthesis of *N*-methylmaysenine (*21*). (Scheme 10) The C-11–C-15 unit as a bromomesylate **101** or dibromide **102** (*24*) was con-

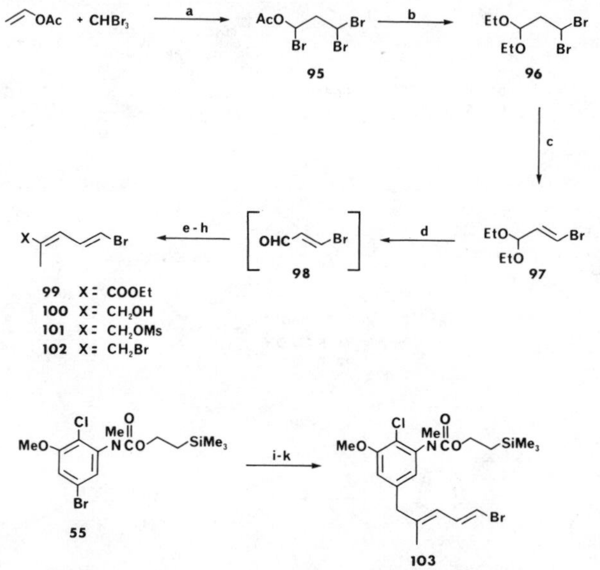

OAc + $CHBr_3$ —a→ AcO(Br)(Br)Br **95** —b→ EtO(EtO)(Br)Br **96** —c→ EtO(EtO)Br **97** —d→ [OHC—Br **98**] —e-h→ X—Br

99 X = COOEt
100 X = CH_2OH
101 X = CH_2OMs
102 X = CH_2Br

55 (MeO, Cl, Br, Me NCO SiMe₃) —i-k→ **103**

SCHEME 10. Meyers nucleophilic southwest and electrophilic northeast. (a) AIBN. (b) EtOH. (c) KO*t*-Bu. (d) SiO_2, oxalic acid. (e) $(EtO)_2POCH_2(CH_3)CO_2Et$, KO*t*-Bu (f) DIBAL. (g) MSCl, NEt_3, CH_2Cl_2. (h) NBS, CH_3SCH_3. (i) *n*-BuLi. (j) $C_3H_7C{\equiv}CCu[(Me_2N)_3P]_2$. (k) **101** or **102**.

densed with a mixed cuprate of the protected bromo-west **55** to give the crystalline southwest **103**.

Starting with the radical addition of bromoform to vinyl acetate to give tribromoacetate **95**, the route is based on the *in situ* preparation of β-bromoacrolein **98**. Solvolysis of **95** (ethanol) gave the dibromoacetal **96** which was dehydrohalogenated to the stable acetal of β-bromoacrolein **97**. An ethereal solution of **98**, generated by cleavage of the acetal, was added directly to the potassium salt of triethyl-2-phosphonopropionate, thus furnishing the (*E,E*)-diene ester **99** in 62% yield. After reduction to the bromoallylic alcohol **100** with diisobutylaluminum hydride (DIBAL), either the allylic mesylate **101** (mesyl chloride, Et_3N) (*21*) or the dibromide **102** (NBS, dimethyl sulfide) (*24*) could be formed.

Treatment of the protected bromo-west **55** with *n*-butyllithium followed by the addition of pentynyl copper–hexamethyl phosphorous triamide gave the mixed cuprate. Addition of either **101** or **102** resulted in the crystalline southwest **103**. After metal–halogen exchange **103** could be reacted with a variety of northeastern aldehydes (C-3–C-9).

c. Southwest (E^+) and Northeast (Nuc^-). At least six different research groups have worked on the synthesis of electrophilic southwestern fragments; most synthons contain an aldehyde at C-10. These may be condensed with a nucleophilic northeast, generally a C-9 dithiane, to form nearly the entire framework of the maytansinoids.

The Roche group (*49,50*), led by Bernauer and Götschi, was the first to report dienal **108** (Scheme 11). Their amino aldehyde **71** (Scheme 6) underwent a classic "Claisen–Schmidt" condensation with propionaldehyde to give the unsaturated aldehyde **104**. A 3-carbon elongation (C-10–C-12) was effected by reaction with the Grignard reagent prepared from 2-(2′-bromoethyl)dioxolane, yielding the alcohol **105**. Dehydration, catalyzed by acid, gave the diene **106** which was N-protected as the trichloroethyl carbamate **107**. This route was based on the prediction that when the acetal in **107** was hydrolyzed, the double bonds would migrate into conjugation with the resulting aldehyde, hopefully forming the more stable (*E,E*)-dienal. Indeed, when **107** was treated with aqueous HCl the desired migration occurred. Unfortunately the thermodynamic ratio of (*E,E*)- to (*E,Z*)-dienes (**108**:**109**) was only about 2:1. These could be separated by medium-pressure liquid chromatography on silica gel and the wrong isomer **109** recycled to the 2:1 equilibrium mixture. The right isomer **108** was elaborated using lithiodithiane as a model for the northeast. After methylation of the resulting alcohol, the nitrogen protecting group was removed to give the aminodithiane **110**. This piece contains C-9 through the C-18 nitrogen and foretells one route to maytansine.

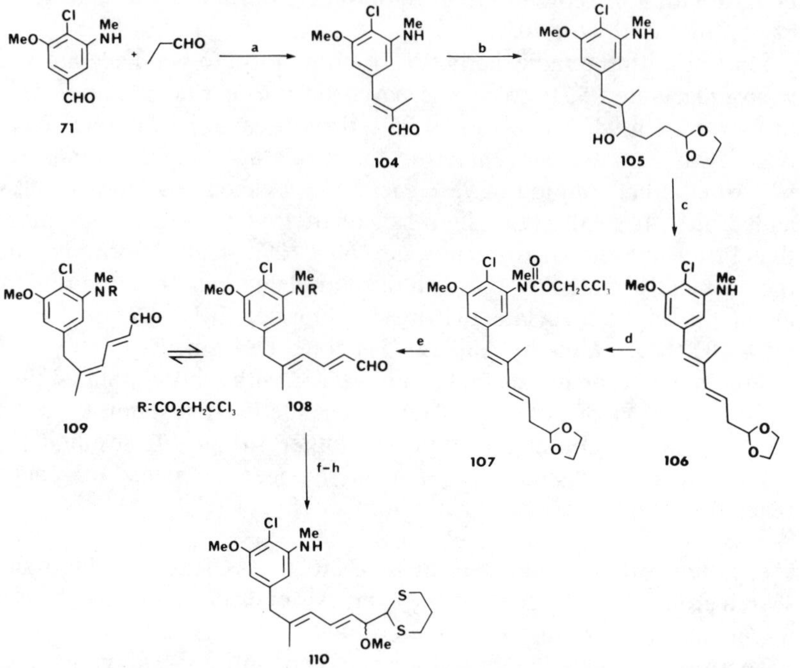

SCHEME 11. Roche electrophilic southwest and nucleophilic northeast. (a) KOH, EtOH. (b) 2-(2′-Magnesiumbromoethyl)dioxolane. (c) *p*-TsOH, $PhCH_3$, reflux. (d) $ClCO_2CH_2CCl_3$, pyridine. (e) 1 *N* HCl, acetone. (f) lithiodithiane. (g) CH_3I. (h) Zn, MeOH, HOAc.

Corey (*11*) then published a stereochemically controlled synthesis of the key dienal **108** (Scheme 12). The previously described amino ester **68** (Scheme 5) was reduced to the benzylic alcohol **111** and the nitrogen was protected as its methyl carbamate **112**. This was achieved by reaction of methyl chloroformate at both the nitrogen and the benzylic oxygen followed by selective solvolysis of the benzylic carbonate. The hydroxyl group of **112** was activated as its mesylate and displaced with iodide. The resulting iodide **113** was used in a cross-coupling reaction with a mixed Gilman reagent to give the (*E*)-olefin **119** selectively. The cuprate was made from *trans*-crotyl alcohol **114** as shown in Scheme 12. Bromination of **114** gave *erythro*-2,3-dibromobutan-1-ol (**115**), which was converted to the allylic alcohol **116** and then protected as the tetrahydropyranyl ether **117**. After metallation with *n*-butyllithium the resulting vinyl lithium species was treated with a cuprous acetylide derived from 3-methoxy-2-methyl-1-butyne. This gave the mixed Gilman reagent **118** which was reacted with a solution of the benzylic iodide **113** in THF to produce the (*E*)-olefin **119**. After cleavage of the THP group the allylic alcohol was

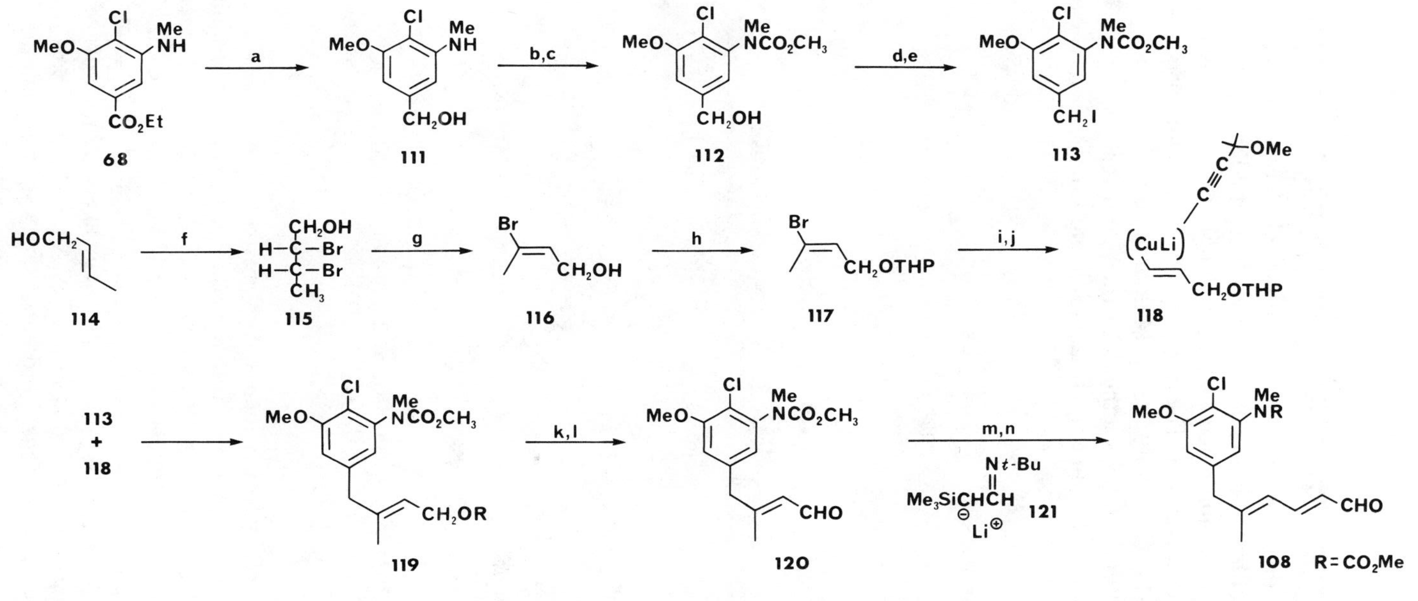

SCHEME 12. Corey route to southwest dienal **108**. (a) LAH. (b) MsCl. (c) NaOH, MeOH. (d) MsCl, NEt_3. (e) NaI. (f) Br_2, CCl_4. (g) LDA, HMPA. (h) DHP, H^+. (i) *n*-BuLi. (j) $(CH_3)_2MeOC—C\equiv CCu$. (k) *p*-TsOH, MeOH. (l) MnO_2, CH_2Cl_2. (m) **121**. (n) H^+.

oxidized with manganese dioxide to the corresponding aldehyde **120**. The chain extension of **120** to the target southwest fragment **108** was done with a reagent that combines Wittig's directed aldol with a Peterson olefination. The lithiated imine of trimethylsilylacetaldehyde (**121**) was added to aldehyde **120**, eliminated to form a double bond, and hydrolyzed to the aldehyde of dienal **108**. The resulting stereochemistry was *E,E* as required for the C-11–C-12 and C-13–C-14 double bonds.

Soon after Corey's synthesis of the southwest appeared, Meyers (*15*) published a very similar approach (Scheme 13). The benzylic bromide **122**

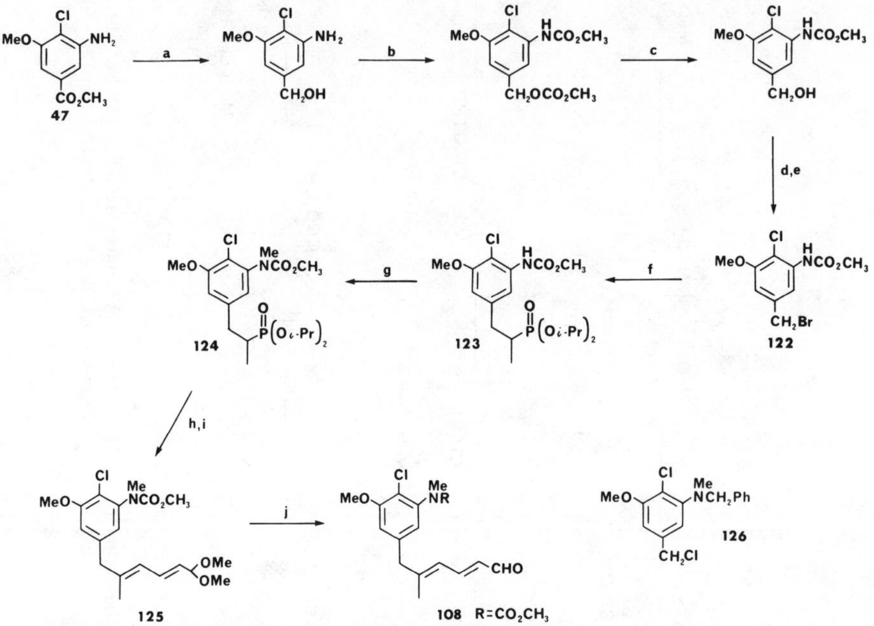

SCHEME 13. Meyers route to southwest dienal **108**. (a) LAH. (b) $ClCO_2CH_3$. (c) K_2CO_3, MeOH. (d) MsCl, NEt_3, CH_2Cl_2. (e) LiBr, DMF. (f) $[CH_3CH^-PO(O\text{-}i\text{Pr})_2]\ Li^+$. (g) KO*t*-Bu, CH_3I. (h) *n*-BuLi. (i) **88**. (j) H^+.

was prepared by nearly identical chemistry in the demethyl series. Here, however, the similarity stops. Reaction of the benzyl bromide **122** with lithioethyldiisopropyl phosphate gave **123** which was N-methylated to **124**. The overall yield of phosphonate **124** from amino ester **47** was 71%. This phosphonate (**124**) was lithiated and treated with the monoacetal of fumaraldehyde (**88**) to produce the diene acetal **125**. Hydrolysis gave the (*E,E*)-dienal **108** (50% yield from **124**). An analogous set of reactions was carried out in the *N*-benzyl-protected series starting with the benzyl chloride **126** and producing **108** in comparable yield.

Wollenberg (*47*), in developing a method for the conversion of carbonyl compounds to conjugated dienals, proposed a conversion of the ketone **127** (*49*) to dienal **108** (Scheme 14). Reaction with the anion of 1-ethoxy-

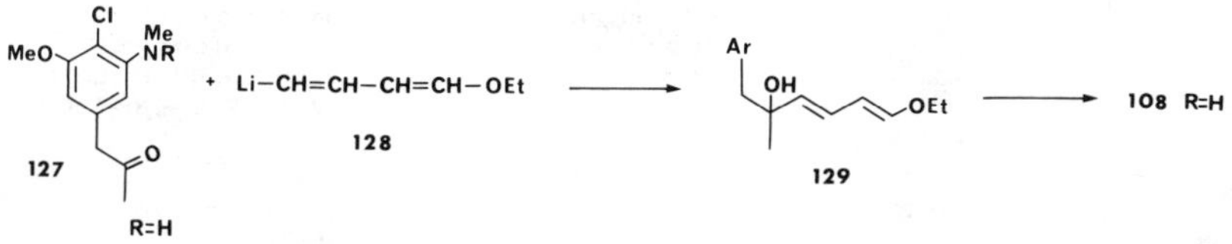

SCHEME 14. Wollenberg's proposed route to dienal **108**.

1,3-butadiene (**128**) should give the intermediate dienol ether **129** which would undergo elimination to form the dienal **108** (R = H) upon treatment with acid.

Another synthesis of dienal **108** came from Ganem (*42*) who converted the benzyl bromide **62** (Scheme 4) to the phosphonium salt **130** (Scheme 15). Deprotonation and reaction with ethyl fumaraldehydate (**131**) af-

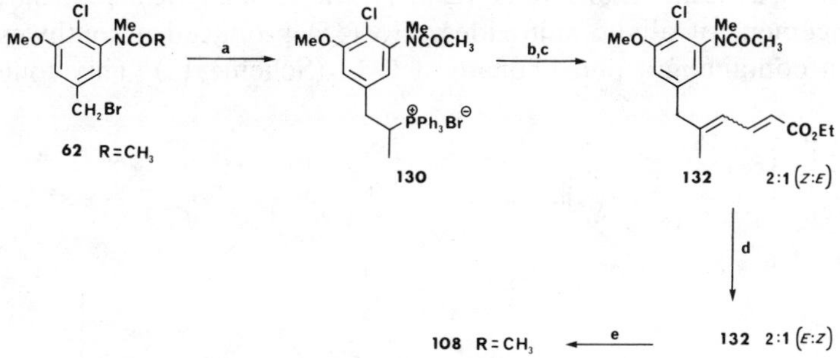

SCHEME 15. Ganem route to southwest dienal **108**. (a) $CH_3CH{=}PPh_3$. (b) *n*-BuLi. (c) (*E*)-$OHCCH{=}CHCO_2Et$ (**131**). (d) I_2, PhH. (e) reduction.

forded diene ester **132** as a 2 : 1 mixture of *Z* and *E* isomers. Exposure of the mixture to iodine in benzene shifted the ratio in favor of the desired *E,E* isomer (2 : 1). Conversion of the mixture of isomeric **132** to the corresponding dienals was followed by a chromatographic separation to give the pure (*E,E*)-dienal **108** (R = Me) as the acetanilide.

A combination of most of the earlier work came from a group at the Shanghai Institute (*37*) (Scheme 16). The Roche ketone **127** (*49*) was chain extended in a repetitive series of double bond formation (**127** to **133**), reduction of ester to alcohol (**133** to **134**), and reoxidation to an aldehyde (**134** to **135**) until the dienal **108** was reached (N-protected as the trifluoroacetanilide). This was reacted with lithiodithiane as in Scheme 11 to give the aminodithiane **110**.

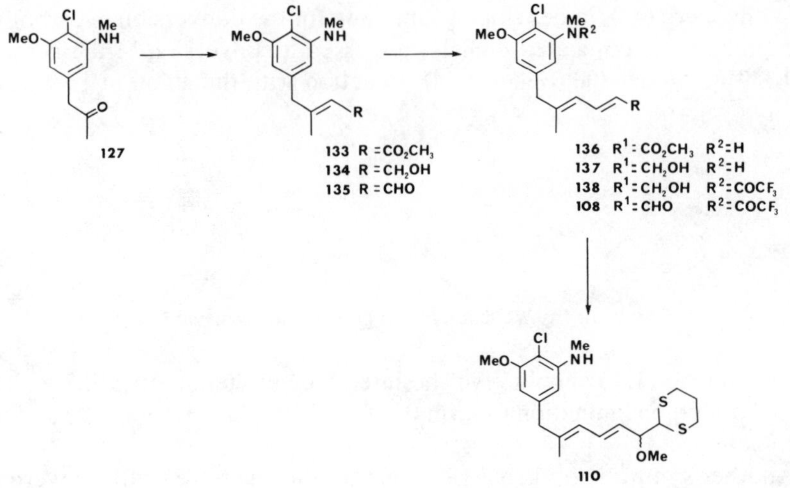

SCHEME 16. Shanghai Institute approach to southwest dienal **108**.

Applying a variation of Evans' (*137*) synthesis of allylic alcohols by the rearrangement of allylic sulfoxides, Ho (*45*) produced a southwestern synthon containing a phosphonate at C-12 (Scheme 17). This could be

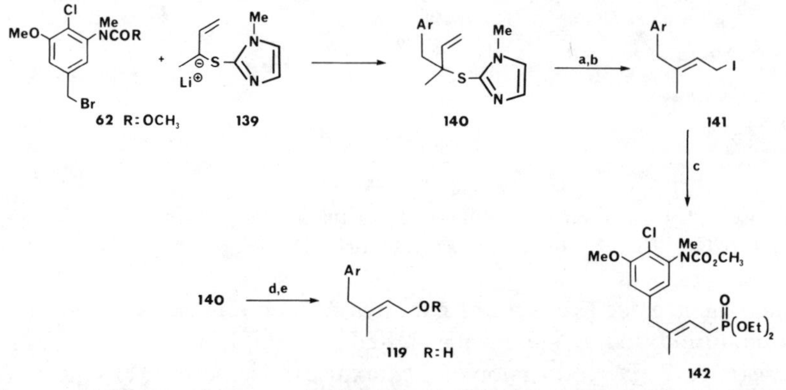

SCHEME 17. Ho's southwestern phosphonate. (a) CH_3I, LiI. (b) DMF, NaI. (c) $P(OEt)_3$, $PhCH_3$, reflux. (d) *m*-CPBA. (e) Et_2NH, MeOH.

reacted with a C-11 aldehyde to connect the northeast. In addition, the unprotected version of Corey's alcohol **119** was formed with Evans' standard conditions.

Thus, benzyl bromide **62** (Scheme 7) was treated with the lithium salt of the allylic imidazole sulfide **139** to give the fully substituted allylic sulfide

140. Treatment with methyl iodide in the presence of iodide salts produced the allylic iodide **141** which was immediately subjected to Arbuzov conditions to give the allylic phosphonate **142**. Alternatively, oxidation of sulfide **140** with *m*-chloroperbenzoic acid followed by treatment with diethylamine initiated the [2,3]sigmatropic rearrangement to form preferentially the (*E*)-allylic alcohol **119**. **119** is formally two steps away from the key dienal **108**.

Another route to the southwestern zone of the maytansinoids appeared in 1982. Isobe (*27*) synthesized the phosphonium salt virtually identical to Ho's (*45*) allylic phosphonate **142** (Scheme 18). What is interesting is how

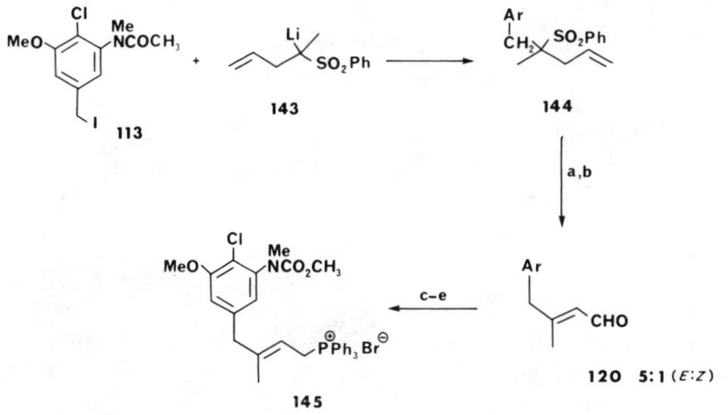

SCHEME 18. Isobe's southwest phosphonium salt. (a) O_3. (b) Et_3N. (c) $NaBH_4$. (d) PBr_3, LiBr, collidine. (e) PPh_3.

it was formed. The now familiar benzyl halide **113** was condensed with the sulfone stabilized carbanion **143** to give the terminal olefin **144**. Ozonolysis followed by elimination yielded the α,β-unsaturated aldehyde **120**. This is the same compound made by Corey (*11*) in 1978, but in this case it is a 5 : 1 mixture of the *E* and *Z* isomers. Reduction of the aldehyde, conversion to the allylic bromide, and finally treatment with triphenylphosphine gave the desired phosphonium salt **145**. This piece is ready for attachment to a northeastern section containing a C-11 aldehyde. The stereochemistry of the new C-11–C-12 double bond can present new isomer problems, however, and the wisdom of such a bond formation late in the synthesis will be discussed below.

3. North

The northern zone of the maytansinoids (C-1–C-6) has been approached by only two groups. As will be discussed later, most routes entail a synthesis of the northeast (C-3–C-9) as a discrete unit with the introduction of C-1 and C-2 as part of the ring closure.

In 1975 Meyers (*12*) reported the synthesis of a stereochemically correct C-1–C-6 moiety (Scheme 19). Aldehyde **146**, made from methallyl

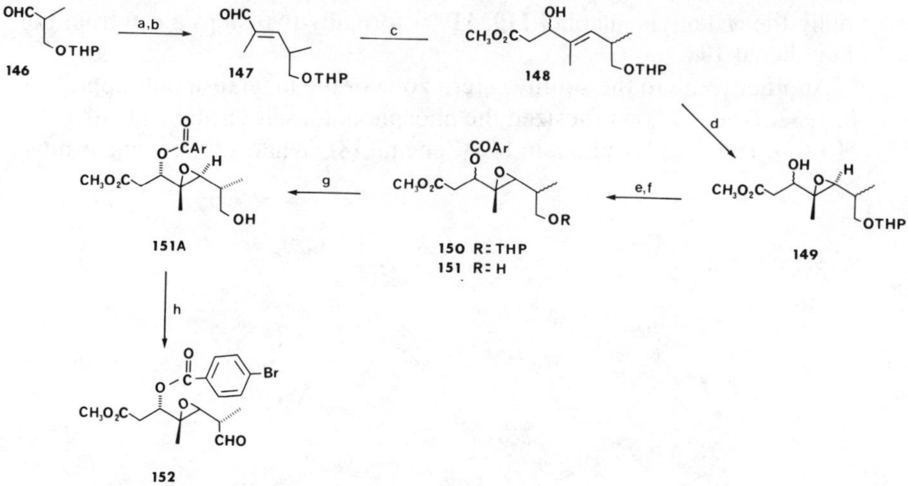

SCHEME 19. Meyers northeast zone synthesis. (a) Li^+ $[CH_3CH^-CH{=}N{-}C_6H_{11}]$. (b) H^+. (c) $LiCH_2CO_2CH_3$. (d) *t*-BuOOH, VO $(acac)_2$. (e) *p*-BrPhCOCl, pyridine. (f) HCl, MeOH. (g) PTLC. (h) CrO_3, pyridine.

alcohol (*8*), was treated with the lithiated cyclohexylimine of propionaldehyde to form, after dehydration, the unsaturated aldehyde **147**. Condensation of **147** with lithio methylacetate gave a mixture of diastereomeric β-hydroxy esters **148**. This mixture was not separated; instead the vanadium acetylacetonate-mediated epoxidation of Sharpless (*138*) was used to converted **148** to **149**. This mixture of epoxides was acylated with *p*-bromobenzoyl chloride, and then the tetrahydropyranyl ether of **150** was cleaved to give the epoxy alcohol **151** as a mixture of four diastereomers. Purification on silica gel gave a major product **151A** which accounted for 42% of the total epoxide mixture. The stereochemistry was assigned on the basis of single crystal X-ray analysis. Assuming that further elaboration to incorporate an eastern zone would require an aldehyde at C-7, Meyers oxidized the diastereomerically pure alcohol **151A** à la Collins (CrO_3–pyridine) to **152**. To ensure that epimerization of the C-6 methyl had not occurred, Meyers reduced **152** back to alcohol **151A** with sodium borohydride. Analysis of this reduction product indicated only the single diastereomer **151A**. Thus, a stereochemically correct, racemic C-1–C-6 synthon had been made.

Barton (*33*), in a projected partial synthesis of a bis-*n*-4,6-maytansinoid with the correct absolute stereochemistry, has arrived at chiral intermediate **153** (Scheme 20). This would be used to link an eastern epoxide

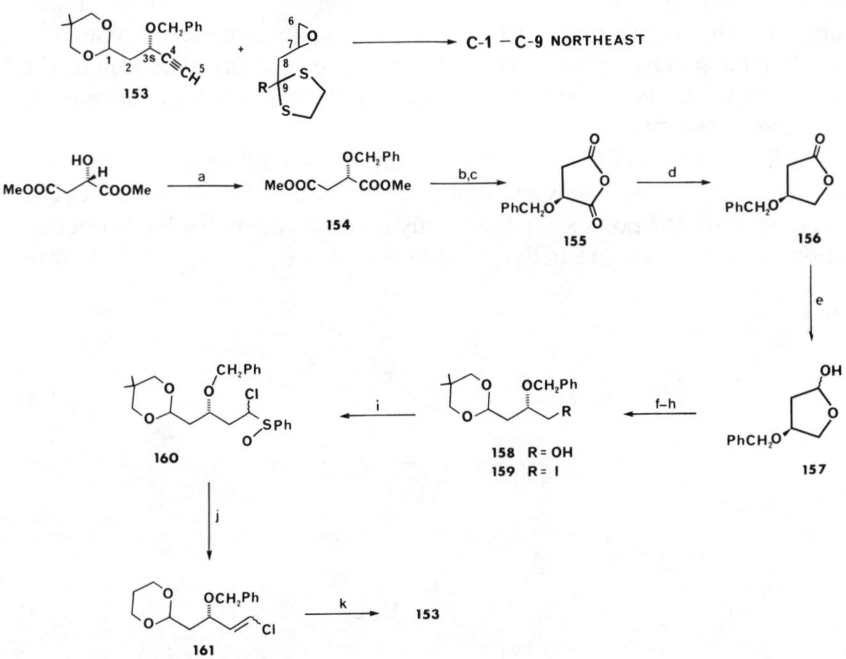

SCHEME 20. Barton's approach to northern zone. (a) $PhCH_2Br$, Ag_2O. (b) OH^-. (c) AcCl. (d) $NaBH_4$. (e) DIBAL, −40°C. (f) 2,2-dimethylpropane-1,3-diol, H^+. (g) TsCl. (h) NaI. (i) Li^+ [ClCH $^-$SOPh]. (j) xylene, 140°C. (k) $NaCH_2SOCH_3$.

containing C-6–C-9 in order to create a C-1–C-9 northeast that lacks methyl groups at C-4 and C-6. Dimethyl (*S*)-(+)-malate was protected as its benzyl ether **154**. The anhydride **155** was formed by saponification of the esters followed by dehydration with acetyl chloride. Reduction to lactone **156**, then further reduction to the epimeric lactols **157**, preceeded opening of the ring. Reaction of **157** with 2,2-dimethylpropane-1,3-diol gave the hydroxyl acetal **158**. Tosylation of the hydroxyl group and displacement with NaI led to **159**, which was further elaborated with the lithium derivative of chloromethylphenylsulfoxide. The resulting sulfoxide **160** afforded the isomeric vinyl chlorides **161** on heating. This mixture when treated with dimsylsodium gave the desired acetylenic compound **153**.

4. Northeast

The northeastern region (C-3–C-9) of the maytansinoids is probably the most challenging synthetic target. For maytansine (**1**) this portion contains five of the eight asymmetric centers, and two more are generated when the northeast is connected to the remaining portions of the molecule. In addition, the unusual hydroxylated cyclic carbamate linking C-7 to C-9 must also be formed. All of this must be considered as well as the need to have the versatility to reach the unsaturated analogs maysine (**3**) and maysenine (**9**).

The first work directed toward the synthesis of maytansine (**1**) appeared in 1974. Meyers and Shaw (*8*) described an efficient route to the cyclic carbinolamide **162** possessing the requisite functionality for further elaboration (Scheme 21). Methallyl alcohol was protected as its tetrahydro-

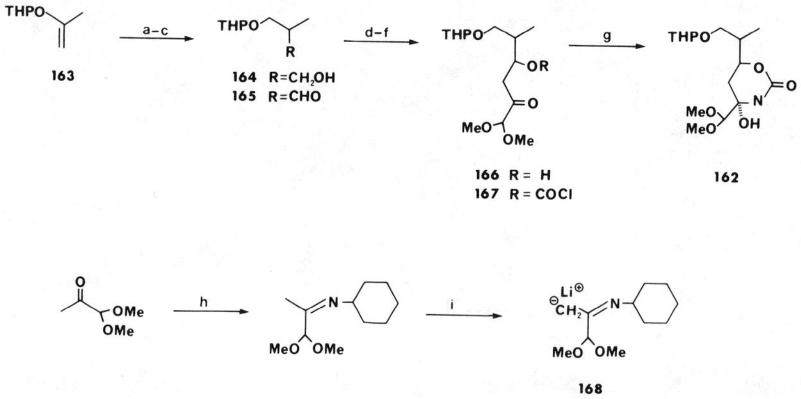

SCHEME 21. Meyers carbinolamide synthesis. (a) B_2H_6, THF. (b) H_2O_2. (c) DCC, DMSO, TFA, pyridine. (d) **168**. (e) H^+. (f) $COCl_2$, pyridine. (g) NH_3, MeOH, −50°C. (h) $C_6H_{11}NH_2$, Na_2SO_4. (i) LDA, −78°C.

pyranyl ether **163** and converted to primary alcohol **164** by hydroboration–oxidation. Further oxidation under Moffat conditions gave the aldehyde **165**. The β-hydroxy ketone **166** was generated by the reaction of aldehyde **165** and the lithioimine of pyruvaldehyde dimethyl acetal (**168**). This hydroxy ketone contains C-5–C-10 and is ready for formation of the heterocycle. Addition of 1.0 equiv of phosgene gave chloroformate **167** which was treated directly with an excess of ammonia to yield the target **162**. It was discovered that the center at C-9 readily epimerizes to the more stable axial alcohol—the correct stereochemistry for maytansine.

Corey and Bock (*9*) chose a very different route to the northeast (Scheme 22). They dealt with the relative stereochemistry at C-6 and C-7

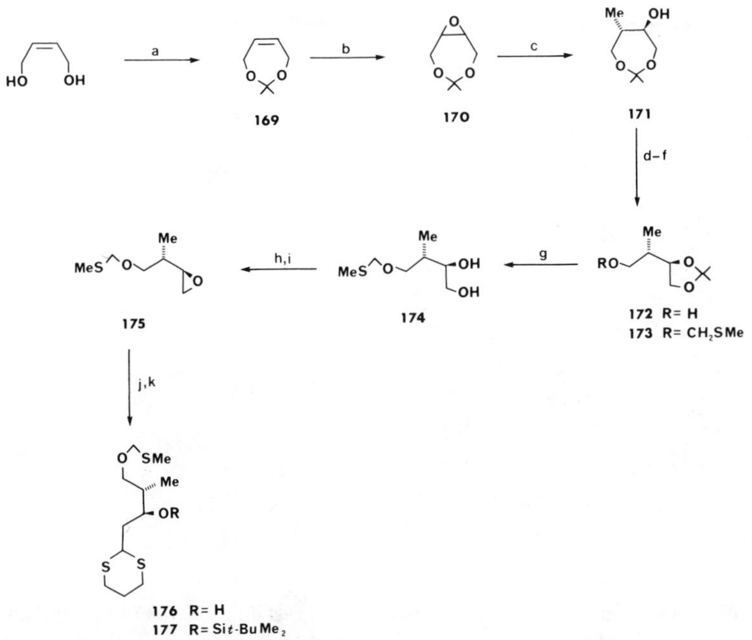

SCHEME 22. Corey northeast synthesis, (a) dimethoxy propane, (b) *m*-CPBA. (c) Me_2CuLi. (d) Cat $BF_3 \cdot Et_2O$. (e) NaH. (f) $ClCH_2SCH_3$. (g) $AcOH–H_2O$ (4 : 1), 45°C. (h) TsCl, pyridine. (i) NaH. (j) Lithodithiane. (k) t-$BuMe_2SiCl$, imidazole, DMF.

from the start. *cis*-2-Buten-1,4-diol was reacted with dimethoxypropane to generate acetonide **169**. Epoxidation was carried out under standard conditions to give **170** which was stereospecifically opened with lithium dimethylcuprate to hydroxyacetal **171**. Acid-catalyzed transacetalization led to the more stable 1,3-dioxolane **172**. The primary hydroxyl was protected as its thiomethyl ether **173**, and then the acetonide was cleaved to diol **174**. This diol was converted to an epoxide by monotosylation at the primary site followed by generation of the alkoxide at the secondary hydroxyl. The resulting epoxide **175** was opened with 2-lithio-1,3-dithiane to afford the hydroxydithiane **176** which was protected as its *tert*-butyldimethylsilyl ether **177**. This fragment contained C-5–C-9 with the correct erythro relationship at C-6–C-7. In addition the hydroxyl protecting groups could be removed selectively.

In the same paper (*9*), model studies were done to (1) form the dithiane anion and condense it with a dienal, thus generating the fully substituted C-9 in **178**, and (2) develop methodology for the carbinolamide (Scheme 23). The model hydroxydithiane **179** was transformed into the corre-

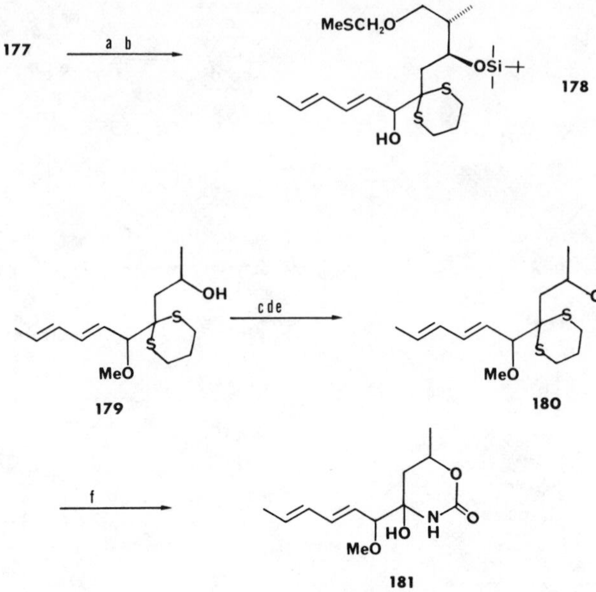

SCHEME 23. Model study for elaboration of northeast dithiane. (a) *n*-BuLi. (b) Sorbaldehyde. (c) NaH. (d) $COCl_2$. (e) NH_3. (f) $HgCl_2$, $CaCO_3$.

sponding urethane **180** upon sequential treatment with sodium hydride, phosgene, and ammonia. Under conditions that hydrolyzed the dithiane group, the urethane nitrogen attacked to give the cyclic carbamate **181**.

An independent route producing the benzyl ether **185** directly analogous to Corey's diol **174** came from Fried (*38*) (Scheme 24). In addition to the racemic hydroxyacetal **171**, this approach involved a resolution of diastereomeric urethanes **182** and **183**. The correct (5*R*,6*S*) enantiomer (−)-**171** was obtained in 22.5% yield by fractional crystallization followed by lithium aluminum hydride (LAH) reduction of the urethane. As in Corey's work, the acid-catalyzed transacetalization gave the dioxolane **172** which was O-benzylated to **184**, and then the acetonide opened to chiral diol **185**. After monoacetylation of the primary hydroxyl, the secondary hydroxyl was treated with sodium cyanate and trifluoroacetic acid (TFA) to yield the urethane **187**. Urethane **187** was elaborated using styrylglyoxal as a model for the southwest. The resulting hemiaminal **188** contained C-5 through C-9 but lacked a bond between C-8 and C-9. Fried's hope was to eventually generate an anion at C-9 and displace the C-8 acetoxy group to from the cyclic carbinolamide.

As an alternative to using a dithiane at C-9 to attach C-6–C-8 (*13*) or a southwest dienal (*9*), Ganem (*41*) (Scheme 25) turned to the dianion of

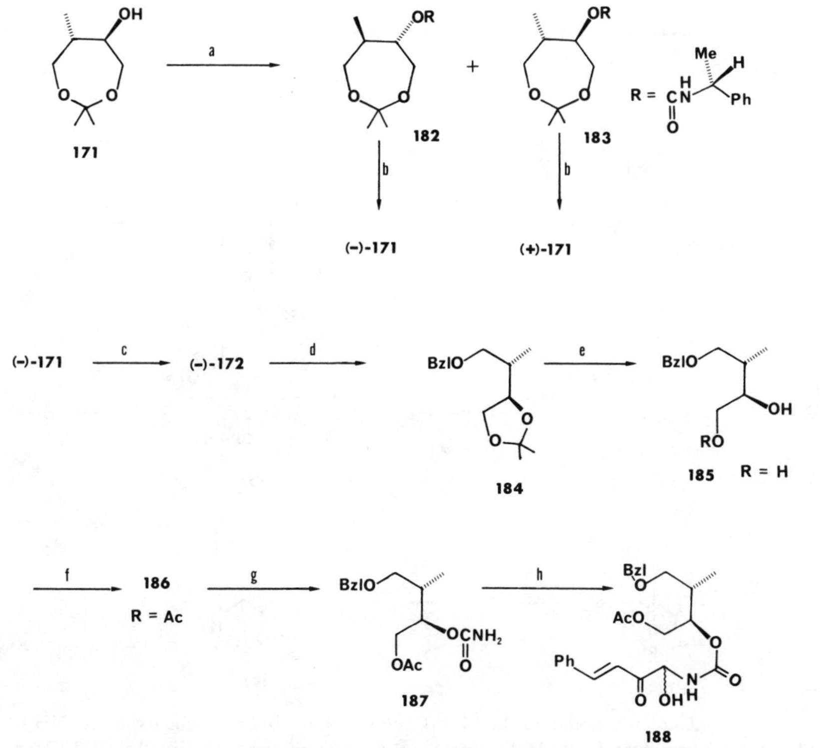

SCHEME 24. Fried's route to the northeast. (a) (+)-(R)-α-Phenethylamine isocyanate. (b) $LiAlH_4$. (c) H^+. (d) NaH, benzyl chloride. (e) H_3O^+. (f) Ac_2O, pyridine. (g) NaOCN, TFA, benzene. (h) Styrylglyoxal.

phenoxy acetic acid (**189**). In a model study, condensation with propylene oxide followed by an acidic work-up gave a 1 : 1 mixture of the epimeric γ-lactones **190**. The lactone enolate was generated and quenched using sorbaldehyde as a model dienal; an *in situ* O-methylation allowed isolation of the alkylated lactone **191**. To synthesize the cyclic carbamate Ganem relied on the Curtius rearrangement. The lactone **191** was opened with hydrazine to the hydroxyhydrazide **192**. On immediate treatment with N_2O_4 followed by chlorotrimethylsilane–triethylamine, the hydrazide was oxidized to the silyloxyazide **193**. After Curtius rearrangement to isocyanate **194**, the trimethylsilyl ether was cleaved with tetra-*n*-butylammonium fluoride leaving the intermediate hydroxy isocyanate to close spontaneously to cyclic carbamate **195**. The phenoxy substituent was readily exchanged, presumably through an acidic elimination–hydration, to yield the desired carbinolamide **196**.

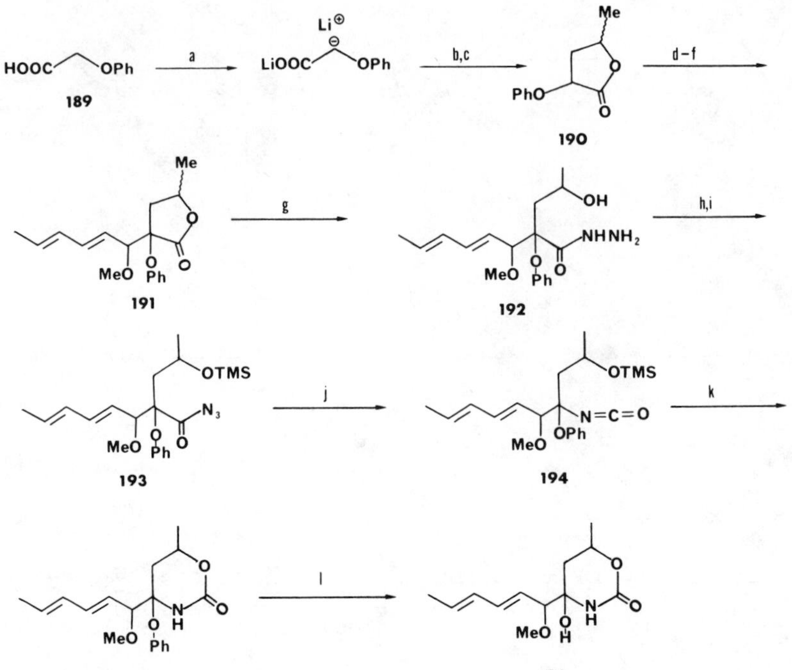

SCHEME 25. Ganem's model northeast. (a) 2 equiv LDA. (b) Propylene oxide. (c) NH_4Cl. (d) LDA. (e) Sorbaldehyde. (f) MeI, HMPA. (g) Hydrazine. (h) N_2O_4. (i) TMSCl, Et_3N. (j) Benzene, sodium acetate, reflux. (k) tetra-*n*-butyl ammonium fluoride. (l) H_3O^+.

In later work Ganem (*42*) moved to a dithioacetal as an acyl anion equivalent at C-9 (Scheme 26). The correct northeast of maytansine (**1**) was actually used as the diepoxide **203** in a condensation with 2-lithio-1,3-dithiane. The plan was eventually to use the elaborated dithiane **204** as a nucleophile in reaction with the electrophilic southwest dienal **108** (cf. Scheme 11).

The acetaldehyde **197** was made from *cis*-2-butene-1,4-diol presumably by the method of Corey and Bock (*9*). Aldol condensation with the dianion of propanoic acid gave β-hydroxy acid **198** as a mixture of isomers which was cyclized to the unsaturated γ lactone **199**. After tosylation of the primary hydroxyl, the double bond was converted into the cis diol **200** which was first acetylated at the secondary alcohol and then mesylated at the tertiary site. The resulting lactone **201** contains a primary, secondary, and tertiary hydroxyl—each uniquely functionalized. Treatment with methoxide resulted in the opening of the lactone, cleavage of the acetate, and displacement of the mesylate and tosylate to form the bisepoxide **202**.

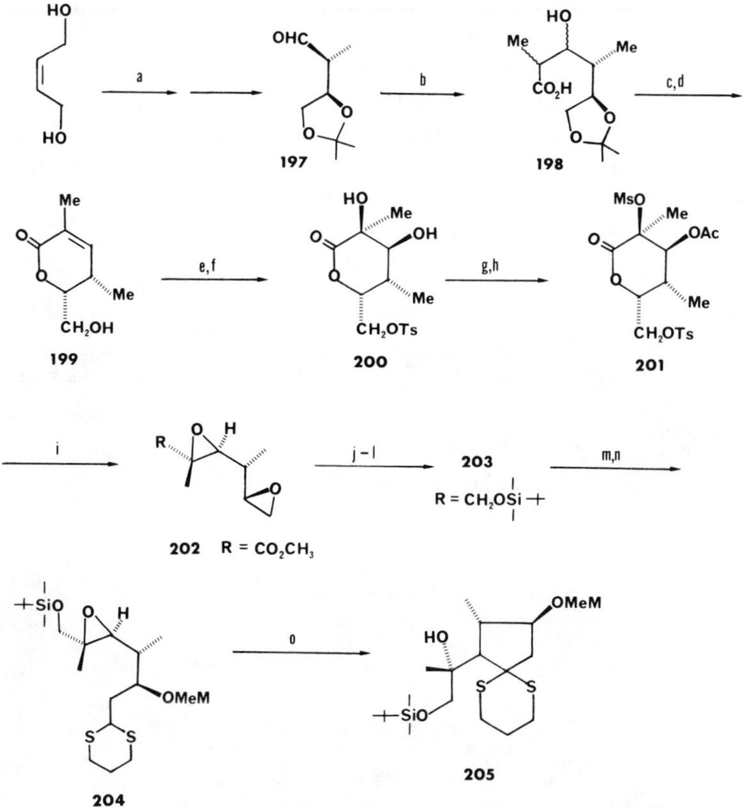

SCHEME 26. Ganem's C-9 dithiane. (a) Five steps. (b) $CH_3CH(Li)CO_2Li$. (c) Acetic acid, H_2O. (d) Triflic acid, Amberlyst 15, heat. (e) Tosyl chloride. (f) OsO_4. (g) Acetylation. (h) Mesyl chloride. (i) Potassium methoxide. (j) Saponification. (k) Mixed anyhydride reduction. (l) *t*-Butyldimethylsilyl chloride. (m) 2-Lithio-1,3-dithiane. (n) MEM-Cl. (o) *n*-BuLi, TMEDA.

Reduction of the methyl ester and protection of the resulting alcohol with *tert*-butyl dimethylsilylchloride yielded **203**, a compound that contained the correct relative stereochemistry at C-4, C-5, C-6, and C-7 for maytansine (**1**). Reaction with lithiodithiane proceeded regioselectively, attacking at the less-hindered monosubstituted epoxide to yield, after protection with β-methoxyethoxymethyl chloride (MEM chloride), the elaborated northeast dithiane **204**.

It should be noted that the dianion of phenoxyacetic acid (**189**), which was used effectively in the model study (*41*) (cf. Scheme 25), could not be used with the real piece. In addition, when dithiane **204** was treated (*139*) with *n*-butyllithium an intramolecular attack at C-5 occurred to generate cyclopentane **205**.

The stereochemical complexity of the northeast has attracted the interest of numerous chemists around the world. A Belgian team under Vandewalle (*51*) (Scheme 27) turned 4-cumyloxy-2-cyclopentenone (**206**) into

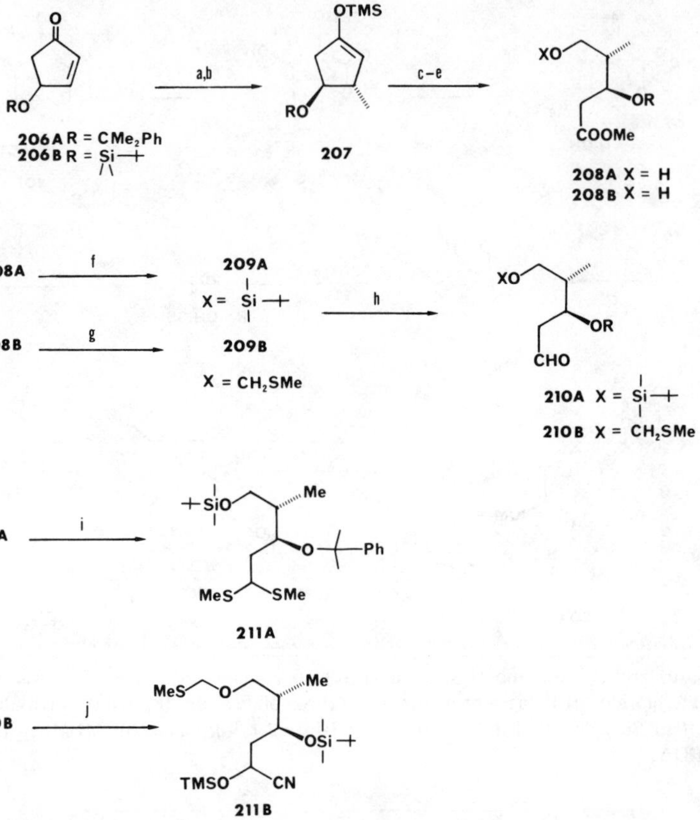

SCHEME 27. Vandewalle's northeast. (a) Me_2CuLi. (b) TMSCL. (c) O_3. (d) $NaBH_4$. (e) CH_2N_2. (f) *t*-Butyl dimethylsilylchloride, imidazole. (g) DMSO, HOAc, NaOAc, 25°C. (h) DIBAL. (i) Trimethyl orthothioborate. (j) TMS-CN.

the potential acyl anion equivalents **211A** and **211B**. If the anions of these masked aldehydes could be generated they could be used with the southwest dienal **108**. A chiral synthesis from (+)-tartaric acid was also alluded to.

Isobe (*140*) used the heteroconjugate addition of methyllithium to vinyl sulfone **217** to selectively establish the C-6–C-7 stereochemistry (Scheme

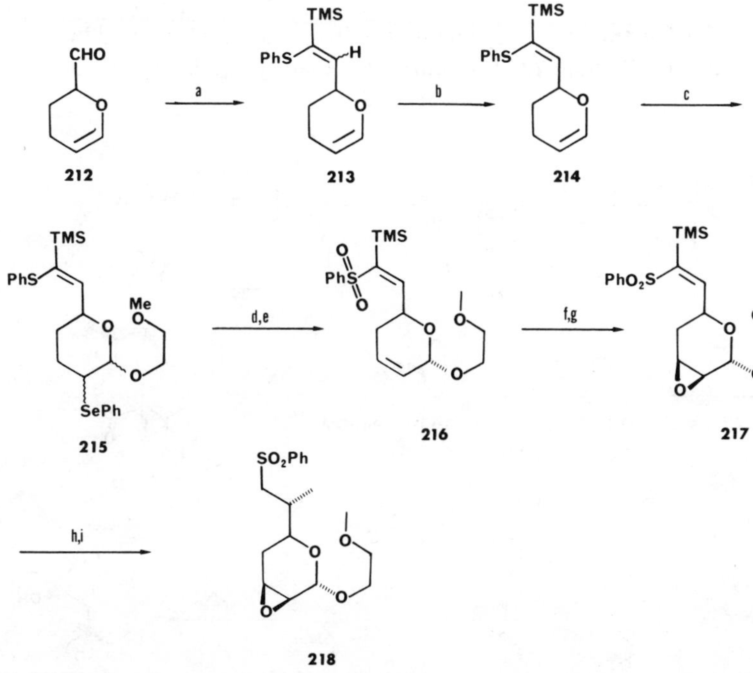

SCHEME 28. Isobe's northeast. (a) $PhS(TMS)_2CLi$. (b) Silica gel isomer separation. (c) PhSeCl, $HOCH_2CH_2OMe$. (d) 3.5 equiv MCPBA. (e) CH_2Cl_2, reflux. (f) MCPBA. (g) Isomer separation. (h) Methyllithium. (i) KF.

28). After the elaboration of the acrolein dimer **212** to the mixture of α and β epoxides **217**, the predominant β isomer was treated with methyllithium. The metal is presumably complexed by the three oxygens as depicted in Figure 4, and the methyl group is delivered selectively to generate the threo product **218**.

5. Cyclic Carbamate

In addition to the studies directed at the various carbons in the macrocyclic ring of the maytansinoids, work on the cyclic carbamate has also

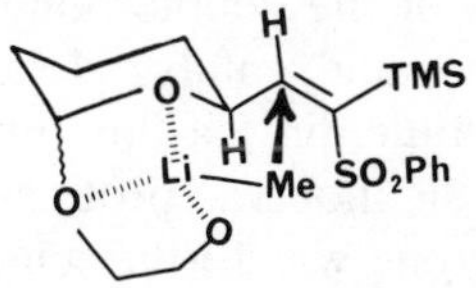

FIG. 4. Isobe's heteroconjugate addition proposal.

been published (*19,39,43*). Ho and Edwards (*43*) predated Ganem (*41*) in the use of a Curtius rearrangement to convert an acyl azide **226** to the model cyclic carbamate **227** (Scheme 29). Readily available 3,4-epoxycy-

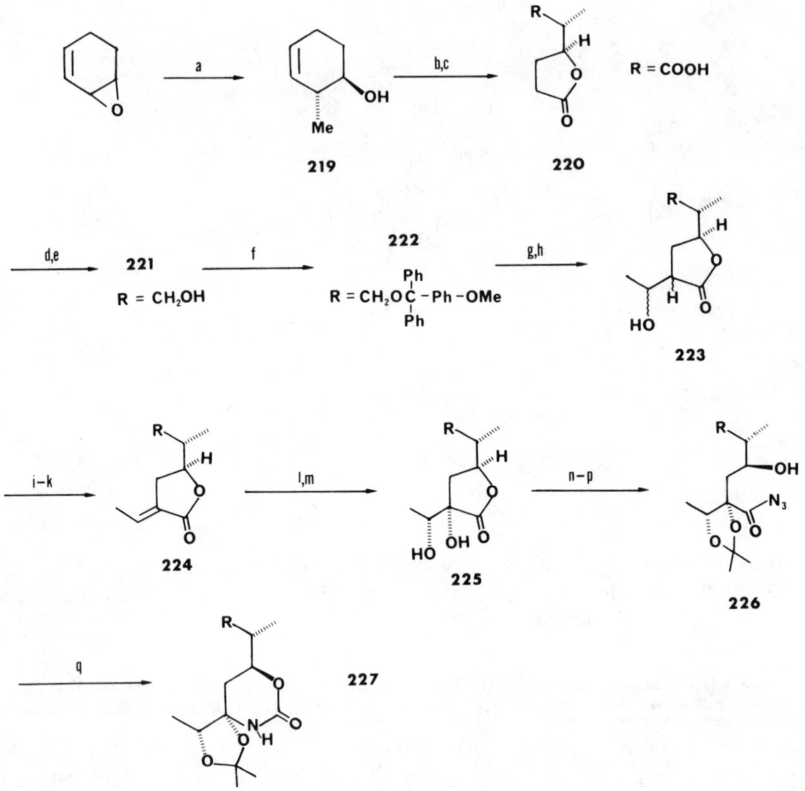

SCHEME 29. Ho's model cyclic carbamate. (a) MeLi. (b) IO_4^-–MnO_4^-. (c) AcOH. (d) Ethyl chloroformate, triethylamine. (e) $NaBH_4$. (f) $ClC(Ph)_2C_6H_4OMe$, benzene, pyridine. (g) LDA. (h) acetaldehyde. (i) MsCl. (j) DBN. (k) SiO_2 isomer separation. (l) OsO_4, pyridine. (m) Reductive work-up ($NaHSO_3$). (n) 2,2-dimethoxypropane, H^+. (o) Hydrazine. (p) N_2O_4, NaOAc. (q) Xylene, reflux a.

clohexene was opened to the *trans*-methylcyclohexenol **219** with methyllithium. Oxidative cleavage of the double bond with periodate–permanganate was followed by an acid catalyzed lactonization to give **220**. Reduction of the carboxyl function via the anhydride gave the hydroxylactone **221**. The primary alcohol was protected as its *p*-anisyl diphenylmethyl ether **222**. The lactone was further elaborated by reaction of its lithium enolate with acetaldehyde to produce an epimeric mixture of alcohols **223** which was dehydrated by elimination of the corresponding mesylates with 1,5-diazobicyclo[4.3.0]non-5-ene (DBN). The resulting olefins **224** were separated on silica gel and the predominant *E*-isomer was treated with osmium tetroxide to form the *cis*-diol **225**. After protection of the diol as an acetonide, the lactone was opened with hydrazine to give an unstable hydroxyhydrazide; this was immediately oxidized to the acyl azide **226** with N_2O_4 and then heated in xylene to effect the Curtius rearrangement. By doing this chemistry in the presence of a free hydroxyl group, the intermediate isocyanate was trapped as the desired cyclic carbamate **227**.

We have already seen that Fried (Scheme 24) planned on building the cyclic carbamate by forming the C-8–C-9 bond last. As a model (*39*) (Scheme 30) for the northeastern piece **188**, the simpler open chain ure-

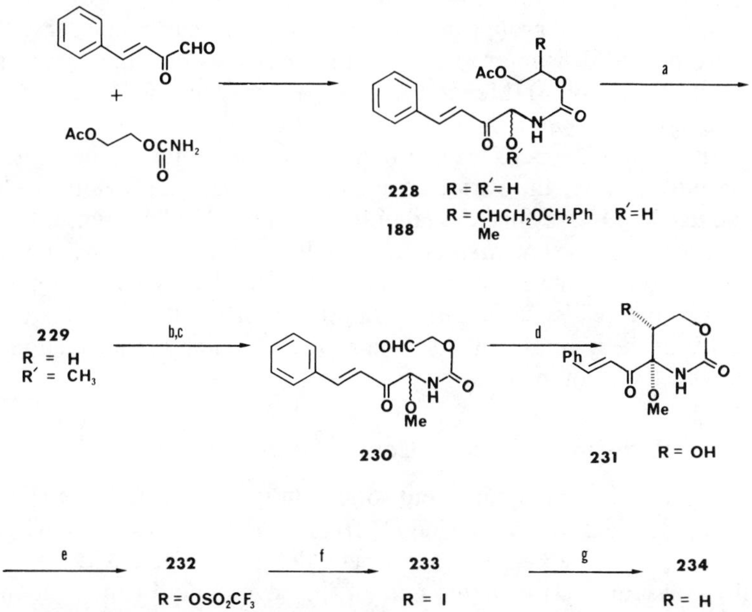

CHO
+
AcO OCNH$_2$
AcO R
a
228 R = R′ = H
188 R = CHCH$_2$OCH$_2$Ph (Me) R′ = H
229 R = H, R′ = CH$_3$
b,c
OHC
230
d
Ph
231 R = OH
e
232 R = OSO$_2$CF$_3$
f
233 R = I
g
234 R = H

SCHEME 30. Fried's model cyclic carbamates. (a) HCl, methanol. (b) KOH, MeOH. (c) PCC, NaOAc. (d) K_2CO_3, MeOH. (e) Triflic anyhydride. (f) NaI, acetone. (g) Tri-*n*-butyltin hydride.

thane **228** was synthesized by condensation of 2-acetoxyethyl carbamate with styrylglyoxal. Unfortunately all attempts at alkylating C-9 of **229** with an activated C-8 (acetate, tosylate, iodide) were unsuccessful. Eventually cyclization was achieved by an intramolecular aldol reaction using aldehyde **230** to give rise to the hydroxylated cyclic carbamate **231**. Deoxygenation by conversion to the corresponding iodide **223** and treatment with tri-*n*-butyltin hydride yielded the model carbinolamide (methyl ether) **234**.

C. Total Synthesis

The road to the maytansinoids was long and hard. After Kupchan reported the isolation and structure determination (1972), it took over 6 years until a maytansinoid [(±)-*N*-methylmaysenine (**2**) (*20,21*), 1978] was synthesized—and this was still not a naturally occurring member of the family. After 7 years (±)-maysine (**3**) (*23*) (1979), a naturally occurring maytansinoid, yielded, and the following year (1980) (±)-maytansinol (**4**) (*25*) and (−)-maytansine (**1**) (*26*) were synthesized. These milestones were followed by alternate racemic syntheses (*27,28*) (1982) and an asymmetric synthesis of (−)-maysine (**3**) (*24*) in 1983.

Of all the research groups involved over this 11-year period, only three reached the end. It is interesting that after 7 years work, the manuscripts on (±)-maytansinol (**4**) [Meyers (*25*)] and (−)-maytansine (**1**) [Corey (*26*)] were received for publication only 31 days apart.

In this section the steps leading to the successful total syntheses will be discussed. At times the various fragments seen in the previous sections will be used and their further elaboration will be shown; often, however, new pieces will arise and *their* coupling will appear. Finally, the closure of the maytansinoid ring and the appropriate methodology will be reported. Hopefully the reader will gain an appreciation for the complexity and difficulty of these total syntheses and for the feeling of satisfaction that came with their completion.

1. (±)-*N*-Methylmaysenine (2), Corey (*20*) (Scheme 31)

The first racemic maytansinoid total synthesis appeared in 1978. The target is identical to maytansine (**1**) from C-6 all the way to C-1, but it lacks the C-4–C-5 epoxide and has been dehydrated across C-2–C-3. *N*-Methylmaysenine (**2**) contains four of the seven chiral centers found in maytansinol (**4**).

Starting with the hydroxydithiane **235**, which was presumably made by deprotection of the previously reported methylthiomethyl ether **177** (Scheme 22), the primary hydroxyl group was reprotected as the 2-methoxy-2-propyl ether **236**. Reaction of **236** with *n*-butyllithium and TMEDA in THF (−78° to −24°C) gave the lithiated dithiane, which was concentrated to dryness and then suspended in toluene. Addition of the aforementioned southwest dienal **108** gave the coupling product **237** as a

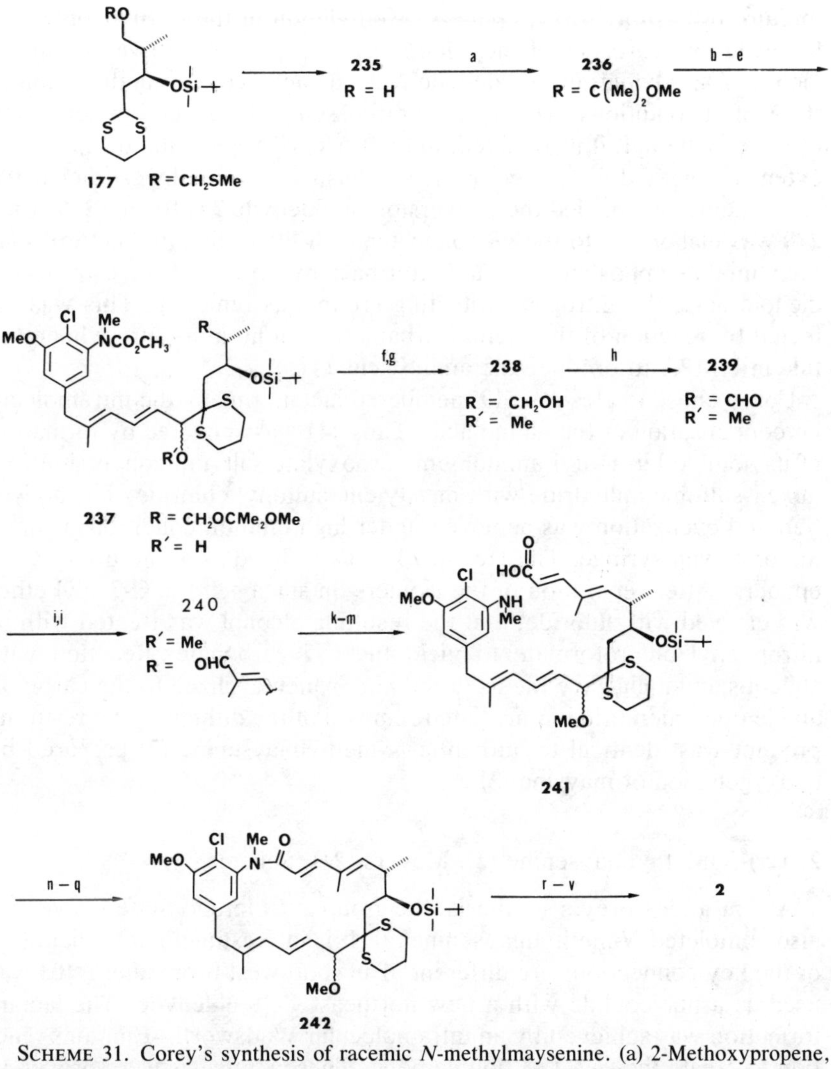

SCHEME 31. Corey's synthesis of racemic *N*-methylmaysenine. (a) 2-Methoxypropene, $POCl_3$. (b) *n*-BuLi, TMEDA, THF, −78° to −24°C. (c) *In vacuo* concentration. (d) Toluene. (e) Dienal **108**. (f) NaH, MeI. (g) THF:Et_2O:1% HCl (1:1:1). (h) Diethylcarbodiimide DMSO, pyridine, TFA. (i) TMS—CMe(Li)—C (H) = *N*-*tert*-butyl. (j) H_3O^+. (k) $(MeO)_2P(O)C(Li)CO_2Me$. (l) aq NaOH. (m) LiS–Pr. (n) $(n\text{-}Bu)_4NOH$, toluene. (o) Azeotropic drying. (p) Mesitylene sulfonyl chloride, diisopropyl ethylamine, benzene. (q) 35°C, 3 hr. (r) Chromatographic separation of C-10 epimers. (s) $(n\text{-}Bu)_4NF$. (t) *p*-Nitrophenyl chloroformate, pyridine. (u) NH_4OH, *tert*-butanol. (v) $HgCl_2$, $CaCO_3$.

mixture of C-10 hydroxyl epimers. Methylation of the C-10 alcohol was followed by hydrolysis of the 2-methoxy-2-propyl ether to yield primary alcohol **238**. Oxidation to aldehyde **239** was achieved by modification of the Moffat conditions. Although not discussed in the paper, epimerization of the C-6 methyl, it is assumed, did not occur. Application of the chain-extension methodology used in the synthesis of the southwest dienal **108** (cf. Scheme 12) enabled the conversion of aldehyde **239** to enal **240**. Enal **240** was elaborated to the dienoic ester with lithio dimethylmethoxycarbonylmethane phosphonate, and after base hydrolysis of the ester to the dienoic acid, the nitrogen protecting group was removed. This was effected by reaction of the methylcarbamate with lithium *n*-propylmercaptide in HMPA to produce the amino acid **241**.

Corey chose to close the 19-membered lactam ring by the intramolecular condensation of the amino acid. Thus **241** was activated by formation of its soluble tetrabutyl ammonium carboxylate salt and conversion to a mixed sulfonic anhydride with mesitylene sulfonyl chloride. The activation and cyclization was achieved under high dilution conditions using a motor-driven syringe. The lactam **242** was isolated as a mixture of C-10 epimers. After separation of the isomers on silica gel, the C-7 silyl ether was cleaved with fluoride, and the resulting alcohol was treated with *p*-nitrophenyl chloroformate to yield the C-7 carbonate. Reaction with aqueous ammonia gave the C-7 urethane, which cyclized to the carbinolamide upon mercuric ion-mediated removal of the dithiane. The resulting product was identical to authentic *N*-methylmaysenine (**2**) prepared by deoxygenation of maysine (**3**).

2. (±)-*N*-Methylmaysenine (**2**), Meyers (*21*) (Scheme 32)

A year after Corey's synthesis, the group at Colorado State University also completed *N*-methylmaysenine (**2**). It is interesting to note that most of the key connections are different. The southwest bromodiene **103** was used as a nucleophile with a new northeast C-10 aldehyde. The lactam formation was achieved by an intramolecular Wadsworth–Emmons reaction to create the C-2–C-3 double bond. These strategies were proven to be quite versatile in the application to (±)-maysine (**3**) (*23*) later.

The already discussed (Scheme 19) enal **147** was deprotected under acidic conditions to hydroxyaldehyde **243**. After acetylation the aldehyde was protected as an acetal and the acetate removed to yield the hydroxyacetal **245**. In this way the monoprotected dialdehyde **246** could be produced. Collins oxidation of **245** gave aldehyde **246**. The northeast was further elaborated by the addition of lithioethyl dithioacetate to **246** to generate the β-hydroxy dithioester **247** as a 4.7 : 1 mixture of the erythro

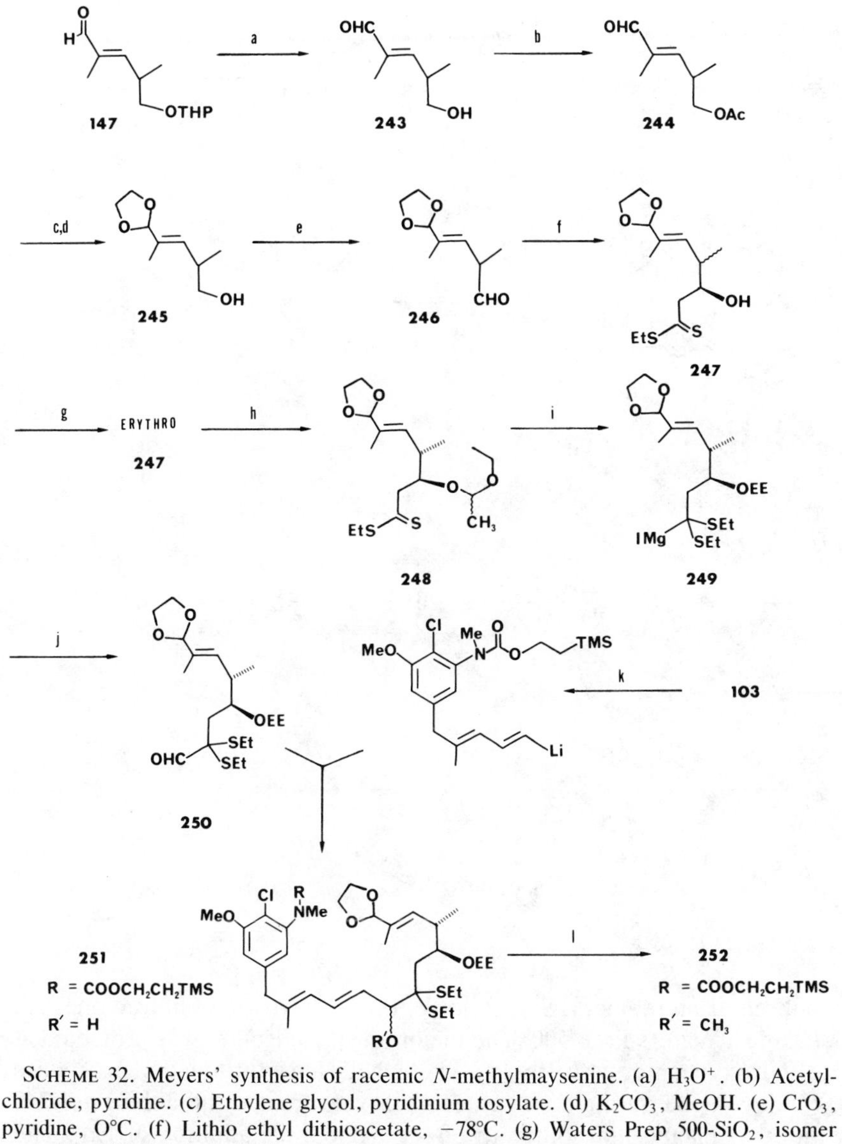

147 243 244 245 246 247 ERYTHRO 247 248 249 250 103 251 252

R = COOCH$_2$CH$_2$TMS R' = H

R = COOCH$_2$CH$_2$TMS R' = CH$_3$

SCHEME 32. Meyers' synthesis of racemic *N*-methylmaysenine. (a) H_3O^+. (b) Acetyl-chloride, pyridine. (c) Ethylene glycol, pyridinium tosylate. (d) K_2CO_3, MeOH. (e) CrO_3, pyridine, O°C. (f) Lithio ethyl dithioacetate, −78°C. (g) Waters Prep 500-SiO_2, isomer separation. (h) Ethyl vinyl ether, *p*-TsOH. (i) 3 equiv EtMgI. (j) MAPF. (k) 2.0 equiv *tert*-butyl lithium. (l) K-*t*-butoxide, MeI. (m) Tetrabutyl ammonium fluoride, CH_3CN, 40°C. (n) Pyridine. (o) Oxalic acid, H_2O, THF. (p) K-*t*-butoxide. (q) $HgCl_2$, $CaCO_3$. (r) 0.1 *N* HCl, THF, O°C. (s) Phenyl chloroformate. (t) NH_3. (u) Chromatographic isomer separation.

(*Continued*)

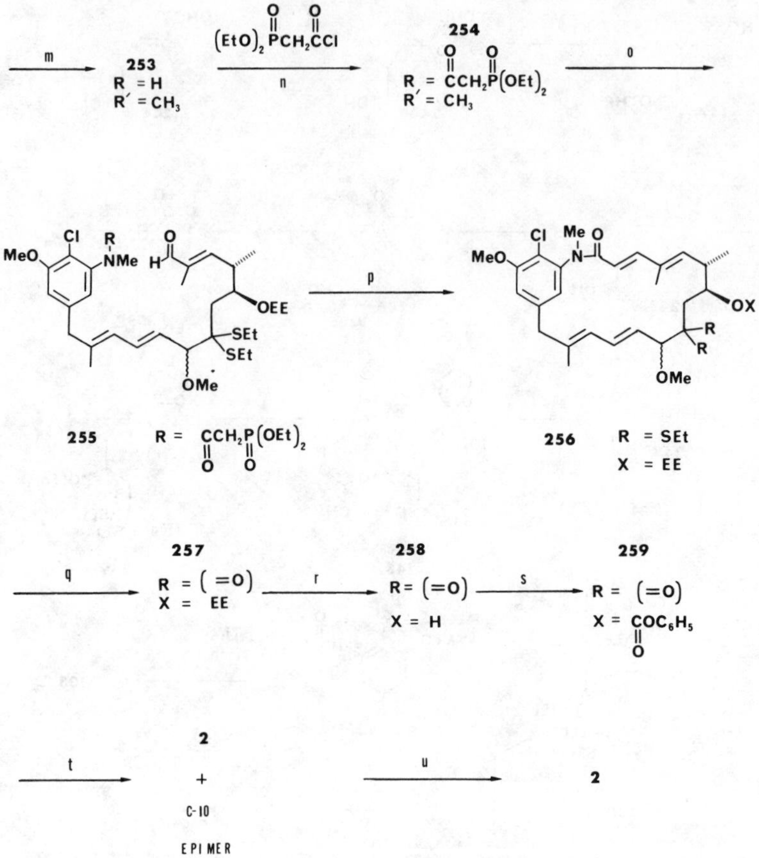

—m→ **253** R = H, R′ = CH_3 —$(EtO)_2P(O)CH_2C(O)Cl$, n→ **254** R = $C(O)CH_2P(O)(OEt)_2$, R′ = CH_3 —o→

255 R = $C(O)CH_2P(O)(OEt)_2$ (MeO, Cl, NMe, H, O, OEE, SEt, SEt, OMe) —p→ **256** R = SEt, X = EE (MeO, Cl, Me, N, O, OX, R, R, OMe)

—q→ **257** R = (=O), X = EE —r→ **258** R = (=O), X = H —s→ **259** R = (=O), X = $C(O)C_6H_5$

—t→ **2** + C-10 EPIMER —u→ **2**

SCHEME 32. (*Continued*)

and threo diastereomers. Separation of the isomers was achieved on silica gel using a Waters Prep 500. The major (erythro) isomer was protected at the hydroxyl center with ethyl vinyl ether. Using a formylation procedure (*141*) developed during the synthesis, Meyers treated the dithioester **248** with ethyl magnesium iodide to effect a thiophilic addition of Et^-. The resulting acyl anion equivalent **249** was formylated with 2-(*N*-methyl-*N*-formyl)aminopyridine (MAPF) to yield the northeast aldehyde **250**. The southwest diene bromide **103** (Scheme 10) was then added as a nucleophile. Thus **103** was treated with *tert*-butyl lithium to get lithium–halogen exchange, and the newly formed vinyllithium was reacted with aldehyde **250**. The product **251**, a 1 : 1 mixture of C-10 alcohols, was methylated with potassium *tert*-butoxide–methyliodide to give **252**. Removal of the silyl carbamate with fluoride gave the free amine **253**. This piece contained the complete maytansinoid skeleton except for the two carbon unit which bridges C-3 to the aniline nitrogen. These two carbons were introduced by acylation with $(EtO)_2P(O)CH_2C(O)Cl$, and the phosphono amide **254** was converted to aldehyde **255** by removal of the dioxolane.

The cyclization was done via a Wadsworth–Emmons reaction, and even though a 19-membered ring was being formed it proceeded quite smoothly. The macrocycle **256** was still a mixture of epimeric C-10 methyl ethers. Introduction of the carbinolamide was effected by removing the dithioacetal, hydrolyzing the ethoxyethyl ether, and then functionalizing the resulting hydroxy ketone **258**. The mixed carbonate **259** was made with phenyl chloroformate and on treatment with ammonia yielded a 1 : 1 mixture of (±)-*N*-methylmaysenine (**2**) and its C-10 epimer. Chromatographic separation gave the pure maytansinoid **2** that was identical with an authentic sample derived from maysine (**3**).

3. (±)-Maysine (3), Meyers (*23*) (Scheme 33)

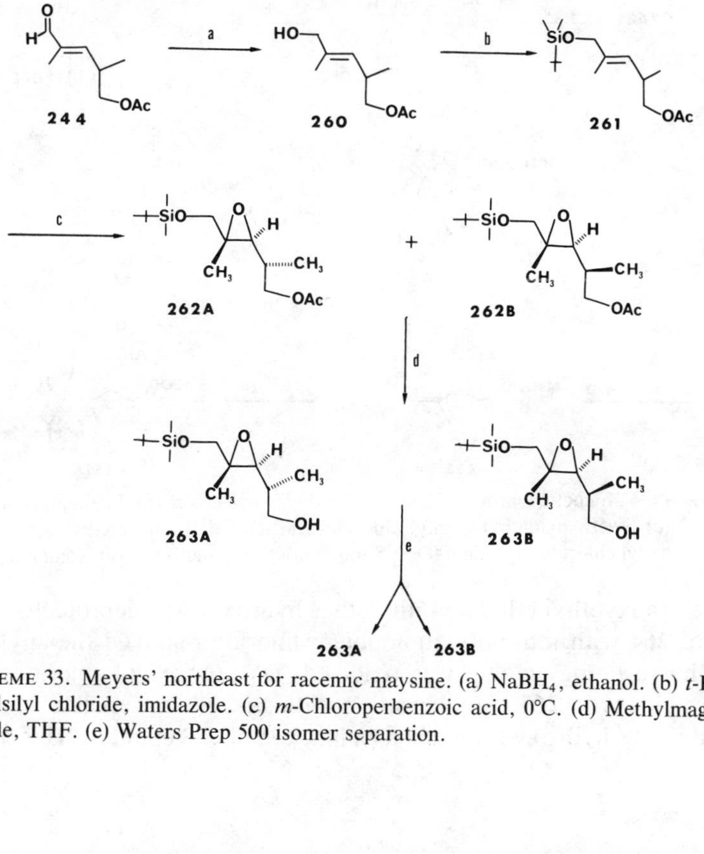

SCHEME 33. Meyers' northeast for racemic maysine. (a) $NaBH_4$, ethanol. (b) *t*-Butyldimethylsilyl chloride, imidazole. (c) *m*-Chloroperbenzoic acid, 0°C. (d) Methylmagnesium chloride, THF. (e) Waters Prep 500 isomer separation.

In 1979 the total synthesis of (±)-maysine (**3**) was reported. This was the first natural maytansinoid ever synthesized. Maysine (**3**) has a C-4–C-5 epoxide containing two additional chiral centers, bringing the total to six. To form a northeast synthon containing the necessary trisubstituted epoxide, the α,β-unsaturated aldehyde **244** from the above scheme was elaborated. Reduction with sodium borohydride gave the allylic alcohol **260**, which was protected as its *tert*-butyldimethylsilyl ether **261**. This compound contained two primary alcohols specifically protected so that each could be freed selectively. Although stereoselective epoxidation could be achieved, in all cases the predominant product was the undesired epimer **262B**. Meyers settled for a nonselective procedure (MCPBA) that gave a 1 : 1 mixture of **262A** and **262B**. The acetate protecting group was cleaved by reaction with methyl magnesium chloride to yield a mixture of epimers **263A** and **263B** from which the correct isomer **263A** was isolated by normal phase chromatography on a Waters Prep 500 system. The overall yield of **263A** from **244** was 25%.

In order to verify that **263A** was indeed the correct isomer, a structural proof was carried out (Scheme 34). After protection of the free hydroxyl

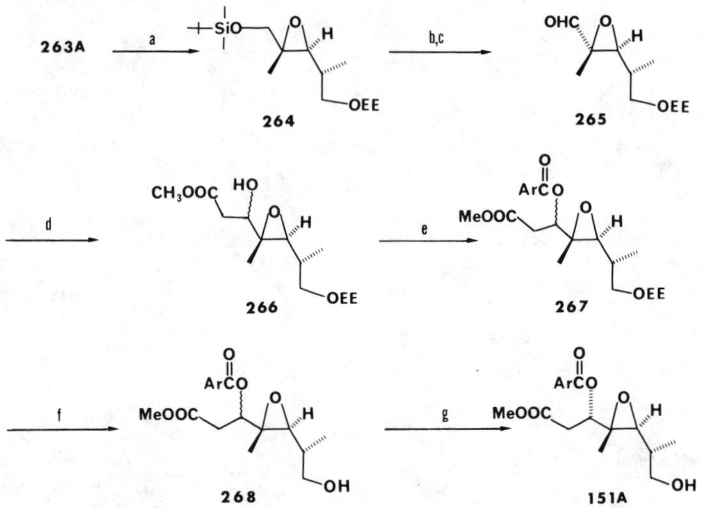

263A → (a) 264 → (b,c) 265 → (d) 266 → (e) 267 → (f) 268 → (g) 151A

SCHEME 34. Structure proof of **263A**. (a) Ethyl vinyl ether. (b) Tetrabutyl ammonium fluoride. (c) Swern oxidation (oxalyl chloride–DMSO). (d) Lithio methyl acetate. (e) *p*-Bromobenzoyl chloride, pyridine. (f) Pyridinium tosylate, methanol. (g) Isomer separation.

as the ethoxyethyl ether **264**, the other hydroxyl was deprotected. Reaction of **264** with tetrabutyl ammonium fluoride removed the silyl ether, and the resulting alcohol was oxidized (DMSO, oxalyl chloride) to the epoxy aldehyde **265**. Condensation with the lithium enolate of methyl acetate gave hydroxy ester **266** as a mixture of C-3 epimers. This mixture was acylated with *p*-bromobenzoyl chloride to the isomeric benzoates **267**. Hydrolysis of the ethoxyethyl ether gave **268** as a mixture of C-3 epimers from which isomer **151A** was separated. This compound was identical to the material synthesized earlier (Scheme 19) on which an X-ray analysis had been carried out.

With the "correct" stereochemistry for C-4–C-6 in hand, the synthesis continued (Scheme 35). The hydroxy epoxide **263A** was oxidized (CrO_3–

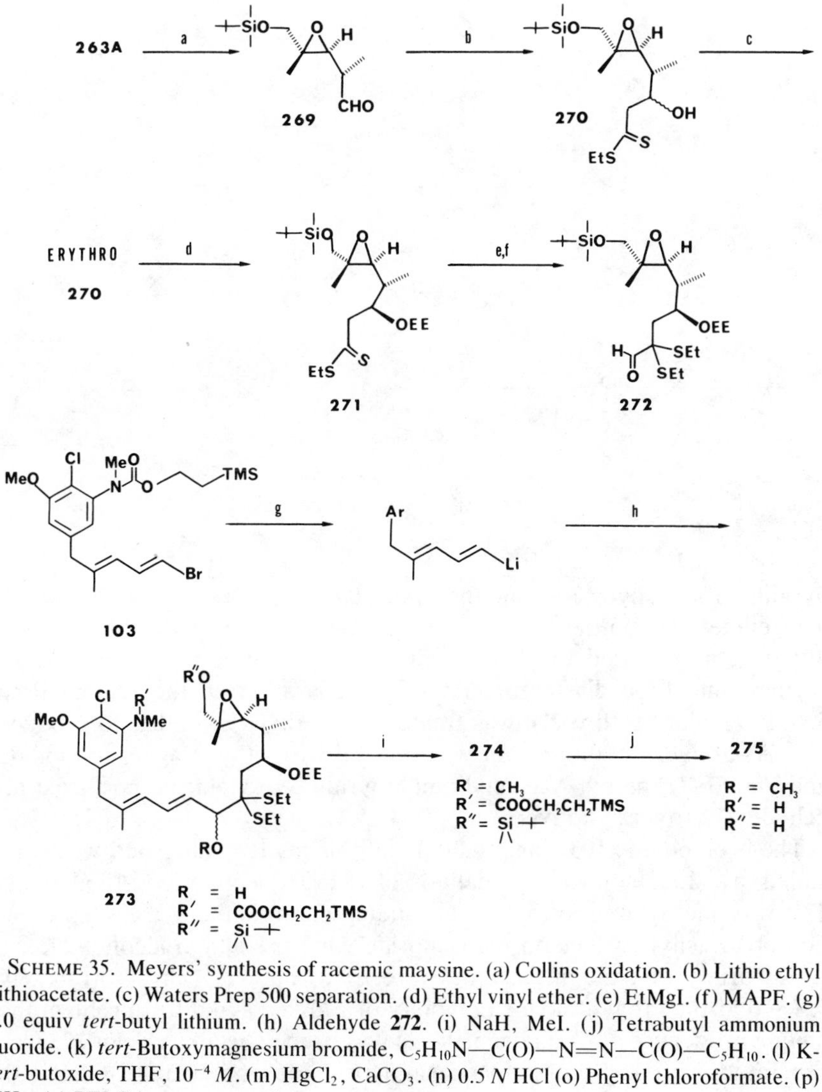

SCHEME 35. Meyers' synthesis of racemic maysine. (a) Collins oxidation. (b) Lithio ethyl dithioacetate. (c) Waters Prep 500 separation. (d) Ethyl vinyl ether. (e) EtMgI. (f) MAPF. (g) 2.0 equiv *tert*-butyl lithium. (h) Aldehyde **272**. (i) NaH, MeI. (j) Tetrabutyl ammonium fluoride. (k) *tert*-Butoxymagnesium bromide, $C_5H_{10}N—C(O)—N{=}N—C(O)—C_5H_{10}$. (l) K-*tert*-butoxide, THF, 10^{-4} *M*. (m) $HgCl_2$, $CaCO_3$. (n) 0.5 *N* HCl (o) Phenyl chloroformate. (p) NH_3. (q) PTLC isomer separation.

(*Continued*)

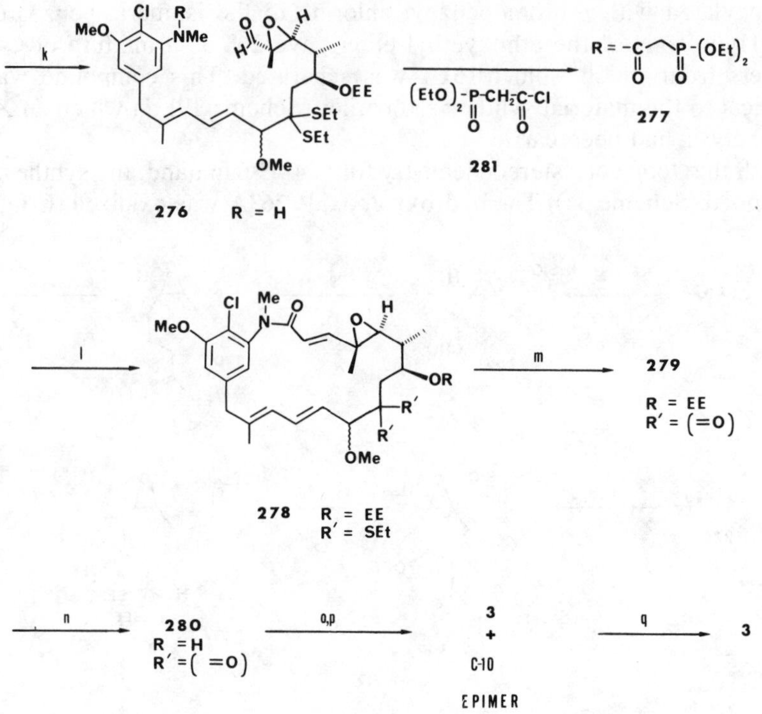

SCHEME 35. (*Continued*)

pyridine) to aldehyde **269**, and the same elaboration seen in the *N*-methylmaysenine (**2**) synthesis was used. Condensation with lithio ethyl dithioacetate yielded the hydroxydithioesters **270** as a 3 : 1 mixture of erythro and threo diastereomers. After separation on the Waters Prep 500, the major erythro **270** was treated with ethyl vinyl ether to produce **271**. As above, the dithioester was reacted with ethyl magnesium iodide and then 2-(*N*-methyl-*N*-formyl)aminopyridine to yield the northeast aldehyde **272** (overall conversion of **244** to **272** was 6.9%).

The bromodiene **103** was lithiated and this nucleophilic southwest was added in 1,2 fashion to the aldehyde of **272**. The epimeric C-10 alcohols **273** were methylated to the corresponding methyl ethers **274**. Deprotection of the silyl carbamate with fluoride gave the amino alcohol **275**. A wide variety of oxidizing agents were tried, but they all unfortunately reacted at the ethyldithioacetal moiety of **275**. To avoid this problem the unique procedure of Mukaiyama (*142*) was used. The alkoxide of **275** was generated with *tert*-butoxymagnesium bromide and then reacted with

1,1′-(azodicarbonyl)dipiperidine. Only the alkoxide can add to the nitrogen–nitrogen double bond and be oxidized so that the sulfurs at C-9 remain intact; the desired amino aldehyde **276** was formed in 71% yield. Acylation of the amine with phosphonoacetyl chloride **281** gave the phosphono amide **277**. Base-induced cyclization to the macrocyclic olefin **278** gave the pure *E*-stereochemistry. Once again a 19-membered ring had smoothly closed under Wadsworth–Emmons reaction conditions. The final steps in the route to (±)-maysine (**3**) involved formation of the carbinolamide and separation of the C-10 epimer. Thus **278** was converted, through ketone **279**, to hydroxy ketone **280** which was treated with phenyl chloroformate followed by ammonia to give (±)-maysine (**3**) plus its C-10 isomer. Separation by preparative thick-layer chromatography on silica gel gave pure (±)-maysine (**3**).

4. (−)-*N*-Methylmaysenine (**2**), Corey (*22*) (Scheme 36)

This asymmetric synthesis involved the use of a chiral northeast dithiane **292** derived from tri-*O*-acetyl-*d*-glucal **282**. As in many other cases where sugars are used as asymmetric building blocks, there were a large number of protection–deprotection steps required to convert glucose to a useful synthon. The concept, however, is quite exciting. It should be remembered that this work was done before the advent of the Sharpless–Katsuki asymmetric epoxidation, which, in retrospect, makes much of this sugar chemistry look cumbersome.

The acetyl groups of **282** were removed with methoxide, and then the double bond of **283** was subjected to oxymercuration. The resulting organomercurial **284** was reduced with borohydride to triol **285**. After protection of the primary hydroxyl as its trityl ether **286**, both secondary alcohols were converted to the sodium alkoxides **287**. Treatment with triisopropylbenzene sulfonyl imidazole gave the chiral epoxide **288**. Opening of the epoxide with methyl lithium set up the desired stereochemistry for C-6 and C-7. The methylated glucose derivative **289** was transketalized with 1,3-propanedithiol and then detritylated with acid to yield triol **290**. Selective reprotection was achieved by reaction with 1-ethoxycyclopentene–BF_3–Et_2O. The cyclic acetal dithiane **291** was further protected at the C-7 hydroxl with MEM chloride. Thus, the desired chiral northeast dithiane **292** was in hand.

As in the earlier racemic synthesis of *N*-methylmaysenine (**2**) (*20*), the dithiane anion was formed and condensed with the southwest dienal **108** to form the "three-quarter" piece **293**. The mixture of C-10 epimers was separated on silica gel and the correct isomer **293A** carried on. In order to recycle precious material the wrong alcohol **293B** was oxidized with man-

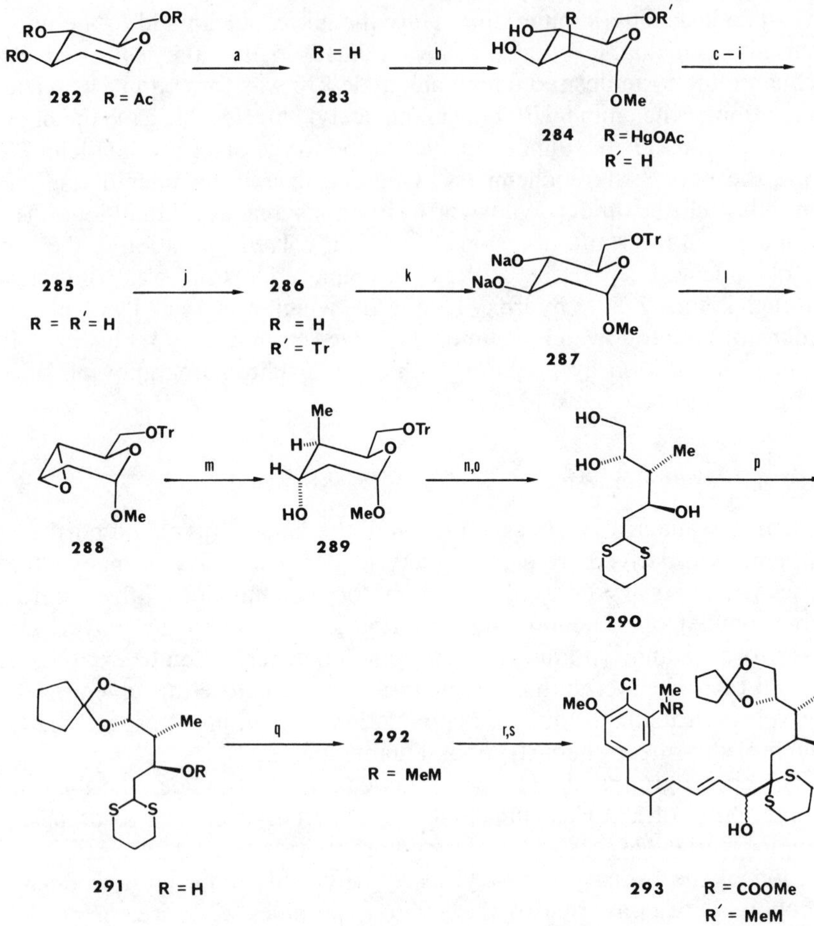

SCHEME 36. Corey's synthesis of *N*-methylmaysenine. (a) Sodium methoxide, methanol. (b) $Hg(OAc)_2$, methanol. (c) NaCl, MeOH. (d) $NaBH_4$, isopropanol. (e) Concentrate *in vacuo*. (f) 12 *N* HCl–ethyl acetate. (g) $NaHCO_3$. (h) 4 Å molecular sieves. (i) Concentrate, dry with P_2O_5. (j) Trityl chloride, pyridine. (k) Sodium hydride, HMPT. (l) Triisopropyl benzene sulfonyl imidazole. (m) MeLi, CuI. (n) 1,3-Propandithiol. (o) 12 *N* HCl. (p) 1-Ethoxycyclopentene, BF_3–Et_2O. (q) MEM-chloride, Hunig's base. (r) *n*-BuLi, TMEDA. (s) dienal **108**. (t) C-10 epimer separation. (u) NaH, MeI. (v) LiSMe, DMF. (w) $HClO_4$. (x) $Pb(OAc)_4$. (y) TMS—C(Me)(Li)—CH=N—*t*-Bu. (z) H_3O^+. (aa) MeO_2C—CH(Li)—$P(O)(OMe)_2$. (bb) Bu_4NOH. (cc) Azeotropic drying. (dd) Mesitylene sulfonyl chloride, DIEA. (ee) 10% H_2SO_4, 40°C, (ff) *p*-Nitrophenyl chloroformate, pyridine. (gg) NH_4OH, *tert*-butanol. (hh) $HgCl_2$, $CaCO_3$.

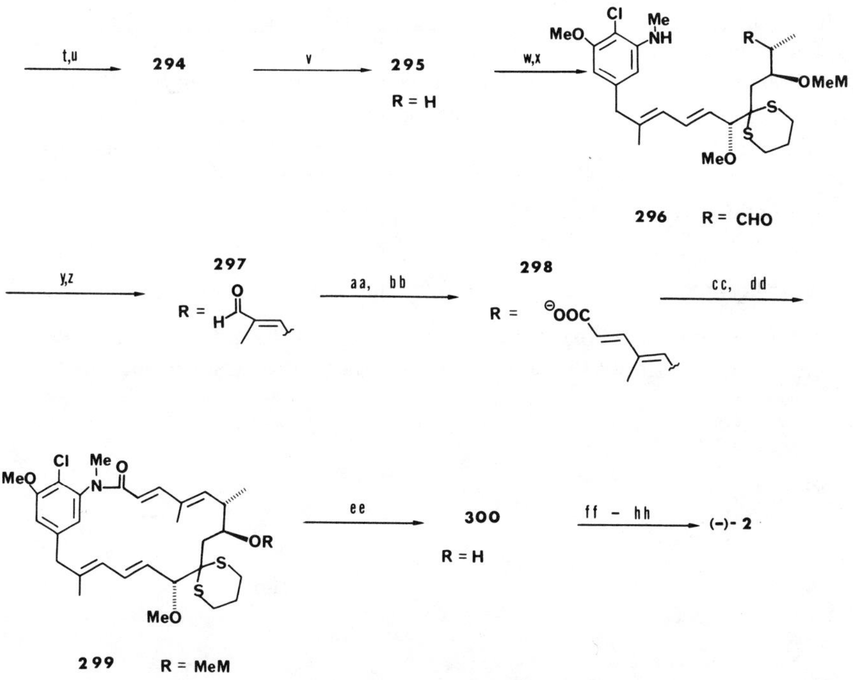

t,u → 294 —v→ 295 R = H —w,x→ 296 R = CHO

—y,z→ 297 R = —aa, bb→ 298 R = —cc, dd→

299 R = MeM —ee→ 300 R = H —ff – hh→ (−)-2

SCHEME 36. (*Continued*)

ganese dioxide and reduced with lithium borohydride to a mixture of **293A** and **293B**. These could again be separated, and the result was good recovery of the desired isomer. Alcohol **293A** was then methylated with sodium hydride–methyl iodide to methyl ether **294**. At this point a somewhat delicate cleavage of the methyl carbamate was needed, and finally lithium thiomethoxide was used. This reagent freed the C-18 nitrogen but also resulted in some demethylation of the aromatic (C-20) methyl ether. The product could be remethylated to give pure amino acetal **295**. Opening of the cyclic acetal with perchloric acid and oxidative cleavage of the intermediate diol with lead tetraacetate gave the somewhat unstable aminoaldehyde **296**. The aldehyde was elaborated to the enal **297** as in Corey's earlier synthesis and then to the corresponding dienoic acid. This amino acid, as its tetra-*n*-butyl ammonium salt **298**, was cyclized and carried on to (−)-*N*-methylmaysenine (**2**) by essentially the same reactions used in the racemic synthesis (cf. Scheme 31). The major differences in this part of the route were the point at which the C-10 epimers were separated, the time when the methyl carbamate was cleaved, and the use of the MEM ether at C-7 instead of the *tert*-butyldimethyl silyl ether. It is quite surprising that in the deprotection of the C-7 alcohol of **299** the entire molecule held up to 10% aqueous sulfuric acid–toluene at 40°C for 5 hr. One would have expected formation of the allylic carbonium ion at C-10 followed by attack of the C-7 hydroxyl to give a tetrahydrofuran. It would be interesting to see if the corresponding C-4–C-5 epoxy macrocycle needed for maysine (**3**) would hold up to these conditions. In any case, the final product was indistinguishable from authentic (−)-*N*-methylmaysenine (**2**) derived from natural (−)-maysine.

5. (±)-Maytansinol (**4**), Meyers (*25*) (Scheme 37)

In the spring of 1980 the major goal of maytansinoid synthesis was reached. Maytansinol (**4**), the common synthetic precursor to the biologi-

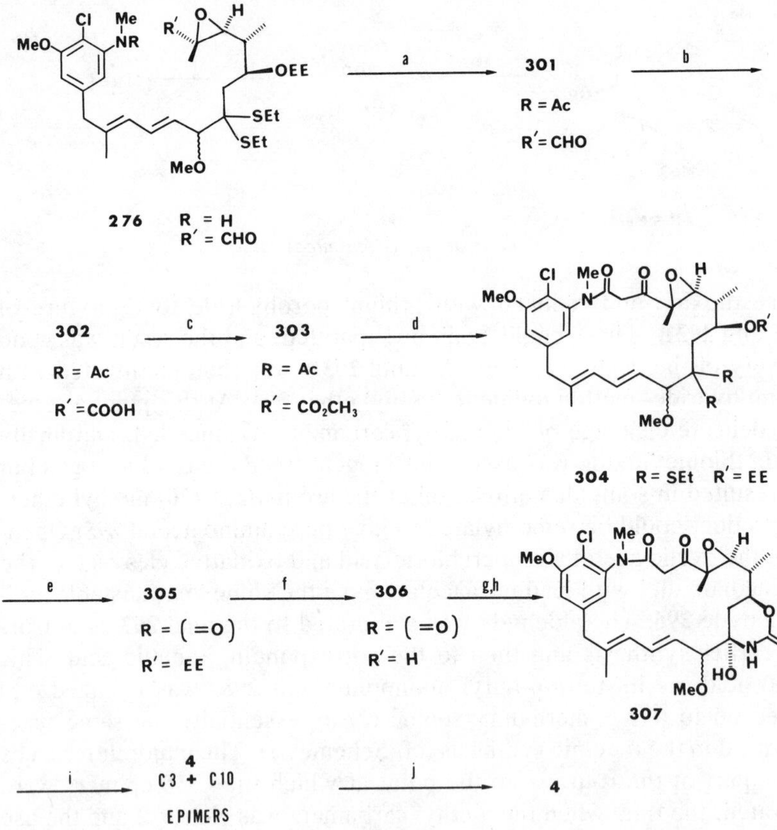

276 R = H, R′ = CHO —a→ 301 R = Ac, R′ = CHO —b→

302 R = Ac, R′ = COOH —c→ 303 R = Ac, R′ = CO_2CH_3 —d→ 304 R = SEt R′ = EE

—e→ 305 R = (=O), R′ = EE —f→ 306 R = (=O), R′ = H —g,h→ 307

—i→ 4 C3 + C10 EPIMERS —j→ 4

SCHEME 37. Meyers' synthesis of racemic maytansinol. (a) Acetyl chloride, pyridine. (b) $AgNO_3$, NaOH. (c) Diazomethane. (d) LiHMDSA. (e) $HgCl_2$, $CaCO_3$. (f) 1 *N* HCl. (g) Phenylchloroformate. (h) NH_3. (i) $NaBH_4$, −40°C. (j) Isomer separation.

cally active maytansinoids, was made in racemic form. The amino aldehyde **276** (cf. Scheme 35) used in the synthesis of (±)-maysine **3** (*23*) differed by only two carbons and a change in oxidation state from maytansinol (**4**).

The two carbons were introduced by acylation of the C-18 nitrogen with acetyl chloride to give acetanilide **301**. A vast number of aldol condensations were attempted to close the 19-membered ring and generate the necessary C-3 hydroxyl. Eventually it was decided that the reversibility of the aldol product was too much of a barrier to overcome. The Yamamoto (*143*) aldol conditions [ArN(Me)C(O)CH_2Br + RCHO + Zn + R_2AlCl], which generate a stable aluminum complexed aldol product, had worked nicely in model 19-membered ring cyclizations (*19*). In the real case the C-4–C-5 epoxide rapidly decomposed. Finally, the acetamido aldehyde **301** was oxidized to the corresponding acid **302** with Ag_2O. This oxidation was chosen to avoid complications with the C-9 thioacetal. Esterification with diazomethane gave **303** and set the stage for an anionic cyclization to ketoamide **304**. This was accomplished by treatment of **303** with the lithium salt of hexamethyldisilazane. The resulting macrocycle **304** was reacted with $HgCl_2$–$CaCO_3$ to cleave the dithioacetal at C-9, and the product **305** was further deprotected with acid to the hydroxy ketone **306**. The order of this sequence was critical; if the C-7 alcohol of **304** was freed under acidic conditions *without* a ketone at C-9, two major byproducts were formed. These proved to be tetrahydrofurans formed by attack at a C-10 allylic carbonium ion or by opening of the epoxide at C-4.

The hydroxy ketone **306** was converted to its carbinolamide **307** on stepwise treatment with phenyl chloroformate and ammonia. Reduction of the C-3 ketone with sodium borohydride gave maytansinol (**4**) along with three minor C-3 and C-10 diastereomers. Chromatographic separation yielded pure racemic maytansinol (**4**) identical to the natural product except for rotation.

6. (−)-Maytansine (1), Corey (*26*) (Scheme 38)

One month after Meyers completed (±)-maytansinol, the Harvard group reached the same target in optically active form. Corey chose to introduce the C-4–C-5 epoxide after the 19-membered ring had been formed—virtually at the end of the synthesis. In this work the same amino aldehyde **297** used for (−) *N*-methylmaysenine (**2**) (cf. Scheme 36) was chosen as starting material. One of the major differences between Corey's and Meyers' approaches is that here the two carbon piece (C-1–C-2) needed to complete the ring was first attached to C-3 and then bridged to the C-18 nitrogen. It is entirely possible that Corey's methodology could be applied to Meyers' amino aldehyde **276** (Scheme 37).

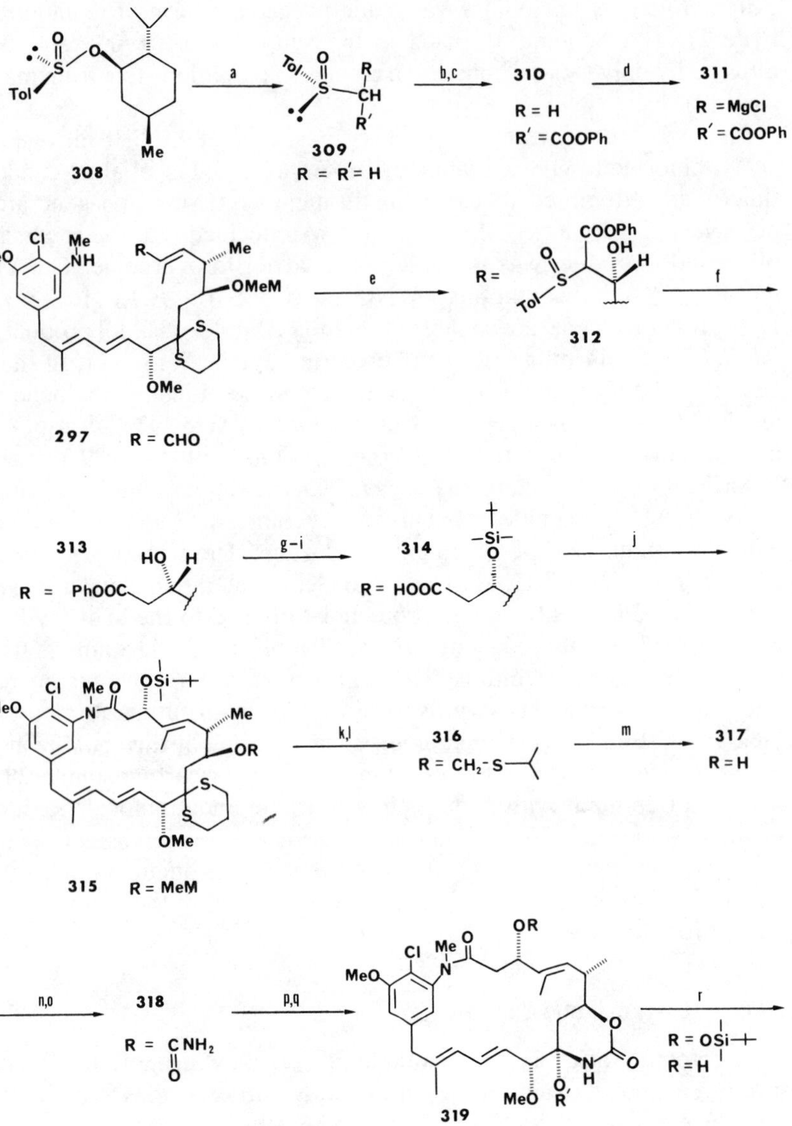

SCHEME 38. Corey's synthesis of maytansine. (a) MeMgBr. (b) LDA. (c) Phenyl chloroformate. (d) *tert*-Butyl magnesium chloride. (e) **311**. (f) Al(Hg). (g) TBDMS-Cl, DMF, imidazole. (h) SiO_2. (i) LiOH. (j) Cyclization. (k) Isopropyl mercaptan, $BF_3 \cdot Et_2O$. (l) n-Bu_4NOH. (m) $AgNO_3$, 2,6-lutidine. (n) p-Nitrobenzoyl chloride, pyridine. (o) NH_4OH. (p) $HgCl_2$, $CaCO_3$. (q) DIEA. (r) CH_3CN, HF, H_2O. (s) p-TsOH, MeOH. (t) *tert*-Butyl hydroperoxide, VaOAcAc. (u) 6-step work-up. (v) H_3O^+. (w) Takeda route. (x) CH_3—CH-(NAc—Me)C(O) Im. (y) 1% pyridinium hydrochloride.

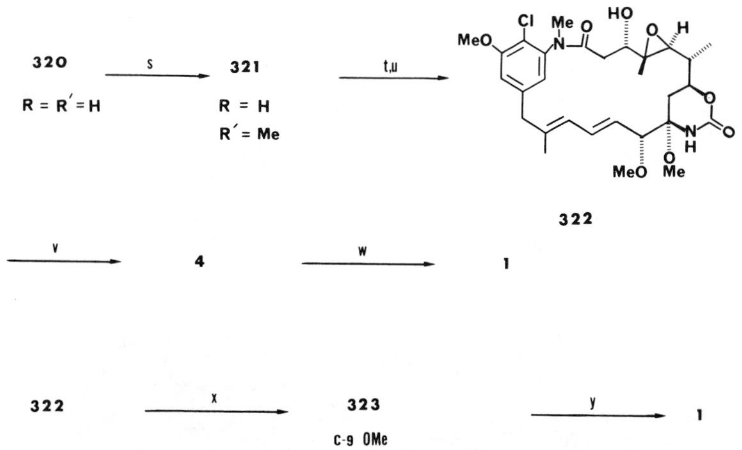

320 —s→ 321 —t,u→ 322

R = R′ = H; R = H, R′ = Me

—v→ 4 —w→ 1

322 —x→ 323 (C-9 OMe) —y→ 1

SCHEME 38. (*Continued*)

The two-carbon homologation of amino aldehyde **297** began with *S*-(−)-*p*-toluene sulfinic acid (−)-menthyl ester **308**. Reaction with methyl magnesium bromide gave the chiral sulfoxide **309** which was converted to its carbanion and acylated with phenyl chloroformate. The resulting α-carboxy sulfoxide **310** was then treated with *tert*-butyl magnesium chloride to produce the chiral Grignard **311**, the synthon used to elaborate **297**. When the amino aldehyde was condensed with the organomagnesium species, the C-3 alcohol **312** was formed with the correct α stereochemistry. Having served its purpose the chiral sulfoxide was reductively cleaved to the corresponding β-hydroxy ester **313** which was O-silylated. Hydrolysis of the phenyl ester resulted in the amino acid **314** which was cyclized by the mixed sulfonic anhydride method to **315**. In this case, with a protected hydroxyl function at C-3, the sulfuric acid removal of the C-7 MEM protecting group was not used. Instead the acetal of the MEM ether was transformed to monothio acetal **316** with 2-propanethiol. This more labile protecting group was readily cleaved with silver ion to the free C-7 alcohol **317**. The carbinolamide **319** was introduced in the usual fashion, and after fluoride removal of the C-3 silyl ether the allylic alcohol **320**, lacking only the C-4–C-5 epoxide, was isolated.

In order to insert the correct epoxide at this stage, the carbinolamide had to be protected as its C-9 methyl ether **321**. Once this was done the vanadium-catalyzed oxidation with *tert*-butyl hydroperoxide proceeded to give maytansinol-C-9-methyl ether **322**. Hydrolysis gave (−)-maytansinol (**4**). Alternatively, the *N*-acetyl-*N*-methyl-L-alanine side chain of maytansine (**1**) could be attached to **322** to yield, after hydrolysis, (−)-maytansine (**1**) that was identical to the natural product.

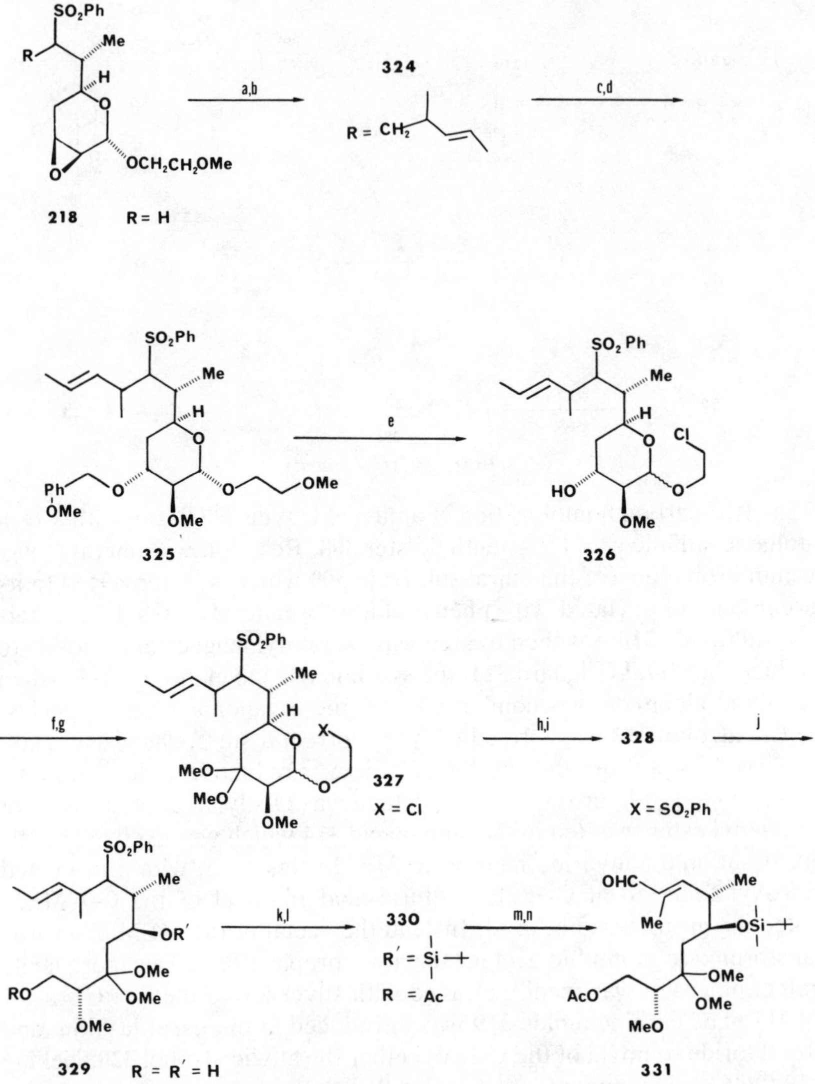

SCHEME 39. Isobe's synthesis of racemic maytansinol. (a) *n*-BuLi, (b) 5-Bromo-2-hexane. (c) Sodium *p*-anisyloxide. (d) MeI. (e) 2-Chloroethanol, CSA, trimethyl orthoformate, 80°C. (f) Collins oxidation. (g) $(MeO)_3CH$, CSA, MeOH. (h) NaSPH. (i) MCPBA. (j) $NaBH_4$. (k) AcCl, pyridine. (l) *t*-butyl dimethyl silyl chloride, imidazole. (m) O_3, −78°C. (n) triethylamine. (o) pyridinium tosylate, $(MeO)_3CH$. (p) $NaOCH_3$. (q) Collins oxidation. (r) **145**. (s) H_3O^+. (t) HPLC separation. (u) $NaBH_4$. (v) 12 *N* KOH–EtOH, reflux 22 hr. (w) Titanium tetraisopropoxide, *t*-BuOOH. (x) oxidation. (y) Lithio ethyl acetate. (z) TBDMS-Cl. (aa) 3 *N* KOH. (bb) Mesitylene sulfonyl chloride, DIPA, n-Bu_4NOH. (cc) n-Bu_4NF. (dd) H_3O^+. (ee) Excess *p*-nitrophenyl chloroformate, pyridine. (ff) NH_3–MeOH.

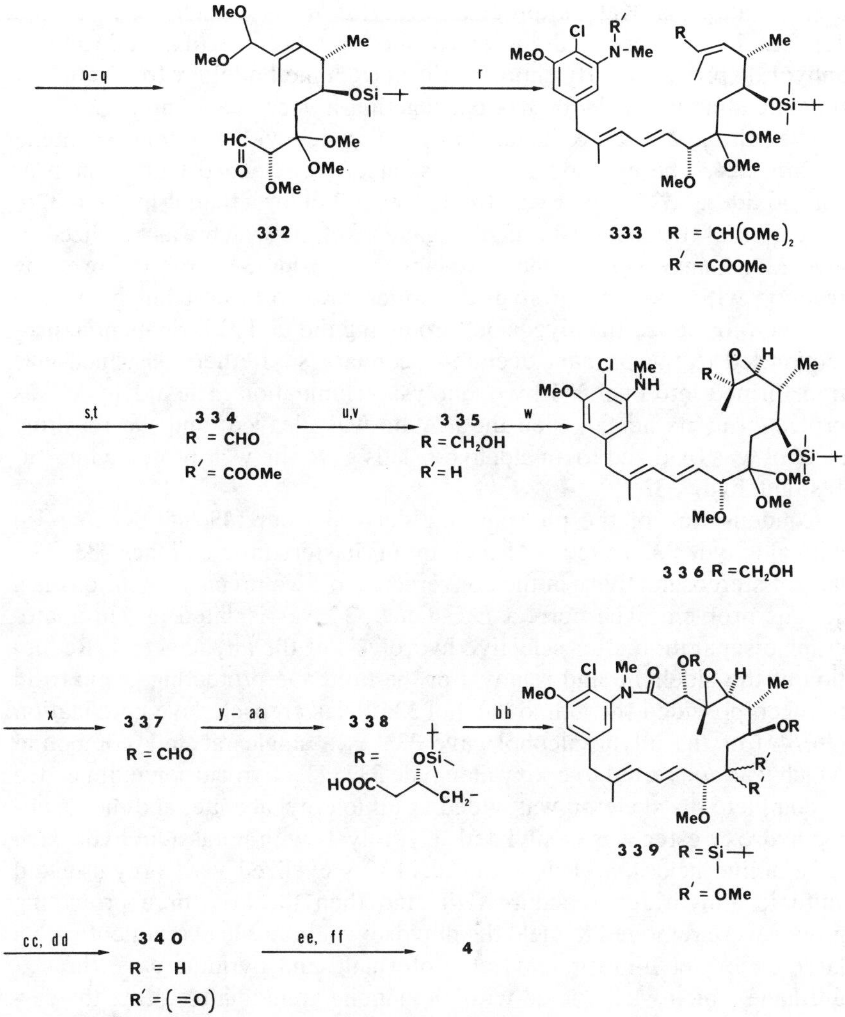

o–q → 332 → r → 333 R = $CH(OMe)_2$, R′ = COOMe

s,t → 334 R = CHO, R′ = COOMe → u,v → 335 R = CH_2OH, R′ = H → w → 336 R = CH_2OH

x → 337 R = CHO → y – aa → 338 R = HOOC–CH(OSi)–CH_2– → bb → 339 R = Si, R′ = OMe

cc, dd → 340 R = H, R′ = (=O) → ee, ff → 4

SCHEME 39. (*Continued*)

7. (±)-Maytansinol (**4**), Isobe (*27*) (Scheme 39)

The latest synthesis of (±)-maytansinol was a combination of Isobe's northeast (cf. Scheme 28) with the best features of many of the earlier approaches. The aromatic west **113** came from the Roche group (*50*) and was converted to a phosphonium salt **145** (Scheme 18) almost identical to Ho's phosphonate **142** (Scheme 17); the C-4–C-5 epoxide was synthesized by the Sharpless–Katsuki procedure (*144*) *à la* Meyers (*16*) (cf. Scheme 41); and the 19-membered ring was formed by Corey's (*20*) mixed sulfonic anhydride procedure. By choosing the correct methodology from much of the preceding work, Isobe has put together a very neat synthesis.

The sulfone **218** (cf. Scheme 28) was alkylated with 4-bromo-2-pentene to form **324**. The epoxide group was selectively opened with sodium *p*-anisyloxide to **325** which was treated with 2-chloroethanol to yield **326**. Oxidation of the secondary alcohol gave a ketone which was ketalized to give **327**. Conversion to the base-labile glycoside **328** was followed by reaction with sodium borohydride, which proved basic enough to open and then to reduce the glycoside, producing the diol **329**. Sequential protection led to the primary acetate–secondary silyl ether **330** which was transformed into enal **331** by ozonolysis–elimination. The aldehyde was protected as its acetal, then the acetate was cleaved, and the resulting alcohol was oxidized to an aldehyde. This gave the well-protected northeast aldehyde **332**.

Condensation of the phosphorane derived from **145** (cf. Scheme 18) with aldehyde **332** gave a 1 : 1 mixture of diastereomeric dienes **333**. The lack of stereoselectivity in the convergence of two precious synthons is a serious problem. The correct *E,E*-diene **334** was isolated by chromatographic separation after selective hydrolysis of the allylic acetal. Reduction of the aldehyde and removal of the urethane protecting group from nitrogen provided the amino alcohol **335**. Diastereoselective epoxidation (*16,144*) of the allylic alcohol gave **336** as a single racemic compound which was converted to epoxy aldehyde **337**. The two-carbon unit needed to complete the skeleton was added as lithio ethyl acetate, and the resulting hydroxy ester was O-silylated. Hydrolysis with potassium hydroxide gave amino acid **338**. This amino acid was cyclized via Corey's mixed sulfonic anhydride procedure (*20*), and then the last three protecting groups were removed to yield the dihydroxyketone **340**. Treatment with a large excess of *p*-nitrophenyl chloroformate and pyridine gave the C-7 urethane which was reacted with methanolic ammonia to create the carbinolamide and thus (±)-maytansinol (**4**). No explanation is given, nor is it readily understood, why the C-3 hydroxyl function, so easily esterified by Kupchan (*6*), is inert to chloroformate while the C-7 alcohol reacts so easily.

8. (±)-Maysine (**3**) and (±)-*N*-Methylmaysenine (**2**), Isobe (*28*)

Isobe used his synthetic intermediates **334** and **337** to make (±)-*N*-methylmaysenine (**2**) and (±)-maysine (**3**), respectively. The chemistry is outlined in Scheme 40.

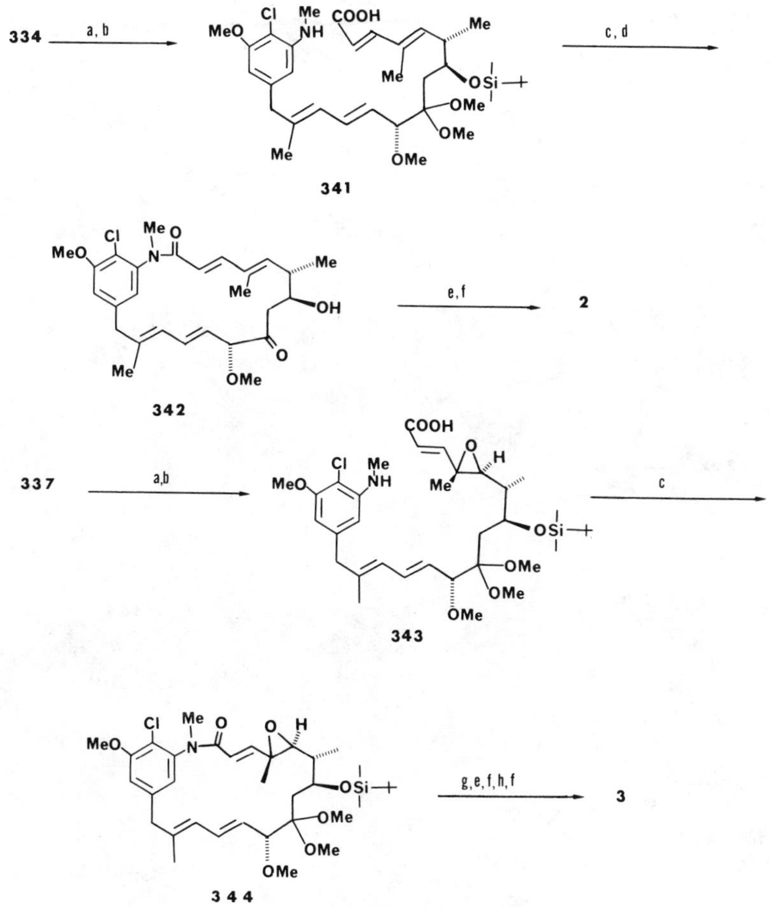

SCHEME 40. Isobe's synthesis of racemic *N*-methylmaysenine and maysine. (a) $(EtO)_2P(O)CH_2CO_2Me$. (b) 12 *N* KOH, 90°C. (c) Corey lactamization. (d) CSA, MeOH. (e) *p*-Nitrophenyl chloroformate. (f) NH_3. (g) *n*-Bu_4NF. (h) AcOH, THF, H_2O.

9. (−)-Maysine (**3**), Meyers (*24*) (Scheme 41)

In his total synthesis of (±)-maysine **3** (*23*) Meyers used the racemic northeast aldehyde **272** (cf. Scheme 35). Later (*16*) this synthon was prepared in the optically active form. More recently (*24*), this chemistry has been incorporated into a convergent synthesis of natural (−)-maysine that involves only a single separation of epimers (C-10). Unlike most of the previous maytansinoid syntheses, this work was reported with full experimental details.

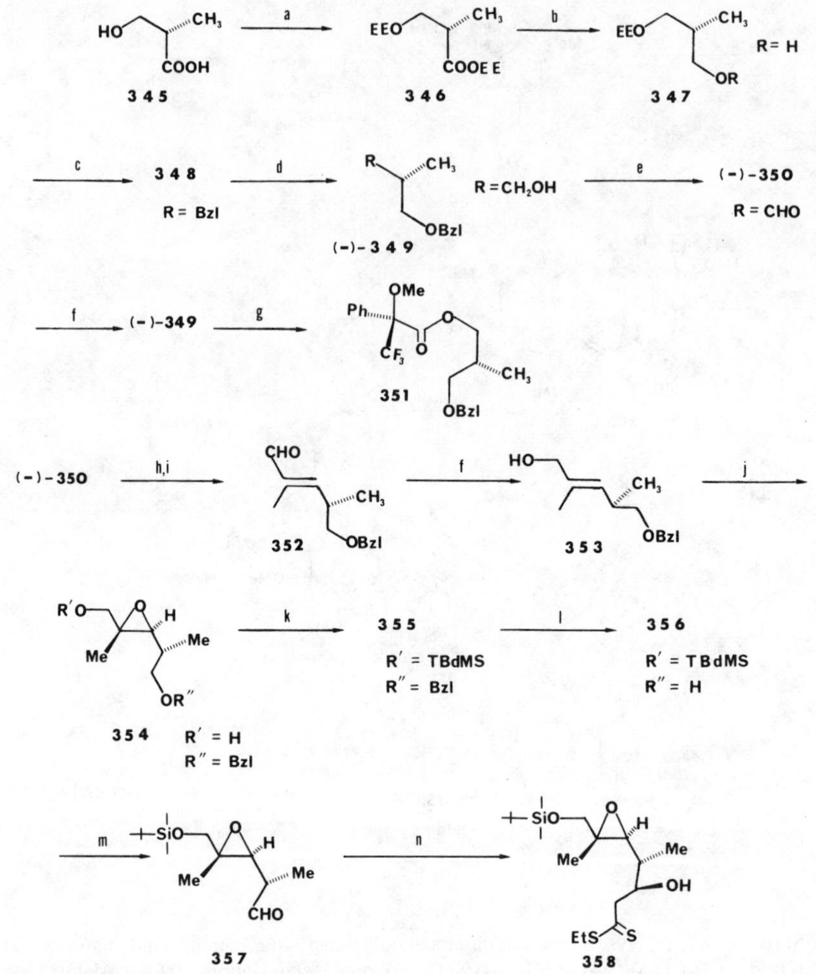

SCHEME 41. Meyers' synthesis of maysine. (a) Ethyl vinyl ether, H^+. (b) $LiAlH_4$. (c) K-*t*-butoxide, benzylbromide. (d) H_3O^+. (e) Swern oxidation. (f) $NaBH_4$. (g) Mosher's acid. (h) $CH_3CH{=}CH{-}N(Li){-}$cyclohexyl. (i) H^+. (j) Sharpless epoxidation. (k) TBDMS—Cl, imidazole. (l) Na, NH_3. (m) Collins oxidation. (n) Lithio ethyl dithioacetate. (o) EtMgI. (p) MAPF. (q) **103**. (r) Separation C-10 epimers, recyclization of wrong isomer. (s) NaH, MeI. (t) Final steps as in racemic synthesis.

Starting with *S*-(+)-3-hydroxy-2-methylpropanoic acid (**354**) available from fermentation (*145,146*), the chirality about the methyl group had to be inverted. This was done by a sequence of protection as **346**, reduction to **347**, benzyl ether formation to **348**, selective hydrolysis to **349**, and Swern oxidation to the aldehyde **350**, which now had the desired *R* configuration at C-6. The enantiomeric purity was checked by reducing **350**

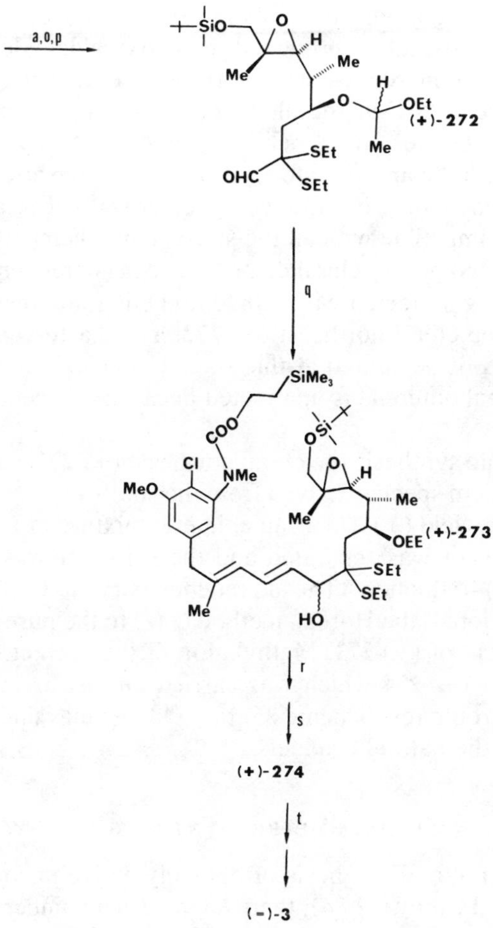

SCHEME 41. (*Continued*)

back to **349** with sodium borohydride and then forming the diastereomeric Mosher esters **351**. The product was >99% enantiomerically pure. When the oxidation of **349** to **350** was carried out with pyridinium chlorochromate (PCC) and the product checked for racemization, only an 80% ee was achieved.

Elaboration of the chiral northeast continued with the conversion of **350** to enal **352** by the previously discussed (Scheme 19) directed aldol condensation. Reduction with sodium borohydride gave the chiral allylic alcohol **353** which was subjected to the Sharpless–Katsuki epoxidation conditions (*144*) (titanium tetraisopropoxide, diethyl tartrate, *tert*-butyl hydroperoxide). The resulting chiral epoxide **354** was shown to be >99%

pure. After protecting-group manipulation of the two primary alcohols (**354** to **355** to **356**), oxidation produced the epoxy aldehyde **357** without racemization. This unit represents C-3 through C-7 of maysine (**3**). In order to introduce C-8–C-9, the aldehyde of **357** was condensed with lithio ethyldithioacetate to generate stereoselectively (10:1 erythro:threo) **358** with the appropriate erythro relationship at C-6–C-7. Protection of the C-7 alcohol as its ethoxyethyl ether was necessary; only this group was fully compatible with all the subsequent chemistry. The introduction of a new, temporary chiral center was considered a minor annoyance. Thus **358** was protected with ethyl vinyl ether and then formylated (*21,141*) to give the chiral northeast (+)-**272**. The diastereomeric ethoxy ethyl ethers could be separated at this stage or carried on as a mixture. The stereochemical outcome is unaffected because this center is eventually removed.

As in the racemic synthesis the chiral northeast (+)-**272** was condensed with the vinyl lithium species derived from the southwest bromodiene **103** (cf. Scheme 35) to yield (+)-**273** as an epimeric mixture of C-10 alcohols. The correct α isomer was separated and the β isomer was recycled by oxidation (MnO_2)–reduction. The stereochemistry at C-10 was determined by application of the Horeau method (*147*) to the pure ethoxy ethyl ether diastereomers of (+)-**273**. Methylation of the correct C-10 alcohol gave methyl ether (+)-**274** which was carried on just as in the original racemic maysine route (cf. Scheme 35) to yield (−)-maysine (**3**) identical in all respects to the natural product.

D. Asymmetric Synthesis

In addition to the total synthesis of optically active maytansinoids by Meyers (*16,24*) and Corey (*22,26*), there have been a number of asymmetric partial syntheses. The work of Ho (*44*), Barton (*32,34*), and Isobe (*148*) will be discussed in this section.

Ho and Edwards (*43*) had already demonstrated the synthesis of an appropriately functionalized carbinolamide **272** from a 5-membered lactone **225** (cf. Scheme 29). Ho (*44*) then proceeded to convert glucose derivative **359** into the analogous lactone **371** that contained the correct relative and absolute stereochemistry for C-6, C-7, and C-10 (Scheme 42). The sugar **359** was tosylated to give **360** which was transformed into epoxy alcohol **361** by cleavage of the acetal protecting group followed by treatment with base. Methylation provided **362** which was reacted with 3-tetrahydropyranyloxy-1-propynyldimethylalanine to open the epoxide. Attack occurred at the terminal position only, resulting in the secondary alcohol **363**. Removal of the tetrahydropyranyl ether from the alkyne-

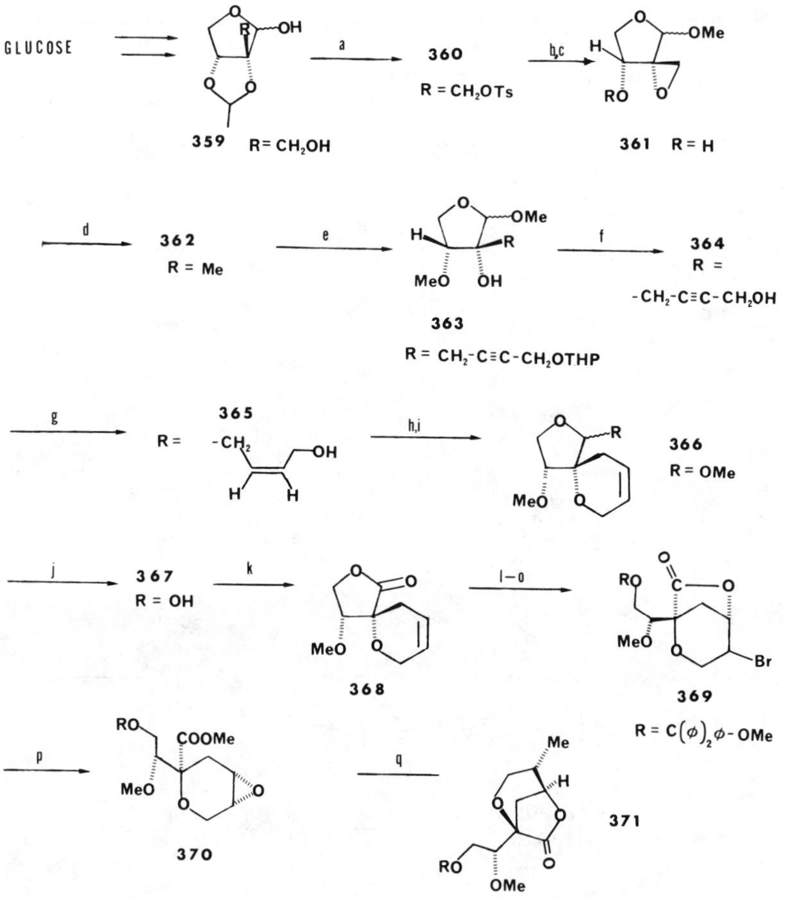

SCHEME 42. Ho's asymmetric approach to the northeast. (a) TsCl, pyridine. (b) *p*-TsOH, MeOH, refulux, 3 days. (c) KOH, MeOH. (d) NaH, MeI. (e) THPO—CH_2—C≡C-—$Al(Me)_2$, toluene, 10°C. (f) *p*-TsOH, aq MeOH, 40°C. (g) H_2, Pd–$CaCO_3$, quinoline. (h) Diethyl azodicarboxylate, triphenylphosphine, LiBr. (i) MeOH, (j) 70% aqueous TFA. (k) DMSO, Ac_2O. (l) KOH, aq MeOH. (m) *p*-Anisyldiphenylmethyl chloride, pyridine. (n) KOH, H_2O. (o) Br_2, KBr. (p) LiOH · H_2O, MeOH. (q) Me_2Li.

containing side chain gave diol **364** which was hydrogenated to the (*Z*)-olefin **365**. Cyclization of **365** to the spiroglycoside **366** was accomplished by a modification of the Mitsonubu conditions, combining diethyl azodicarboxylate–triphenylphosphine with lithium bromide. Presumably the reaction proceded through an allylic bromide derived from the initially formed oxy–phosphonium species. Conversion of the acetal to hemiacetal **367** was followed by oxidation to the γ-lactone **368**. The lactone of this spiro compound was opened with KOH to the corresponding hydroxy acid, protected as a substituted trityl ether, and then bromolactonized to **369**. Attack of methoxide at the lactone carbonyl initiated a stereospecific epoxide formation and produced **370**. This epoxide was then stereoselectively opened with dimethyl lithium cuprate to the chiral target lactone **371**.

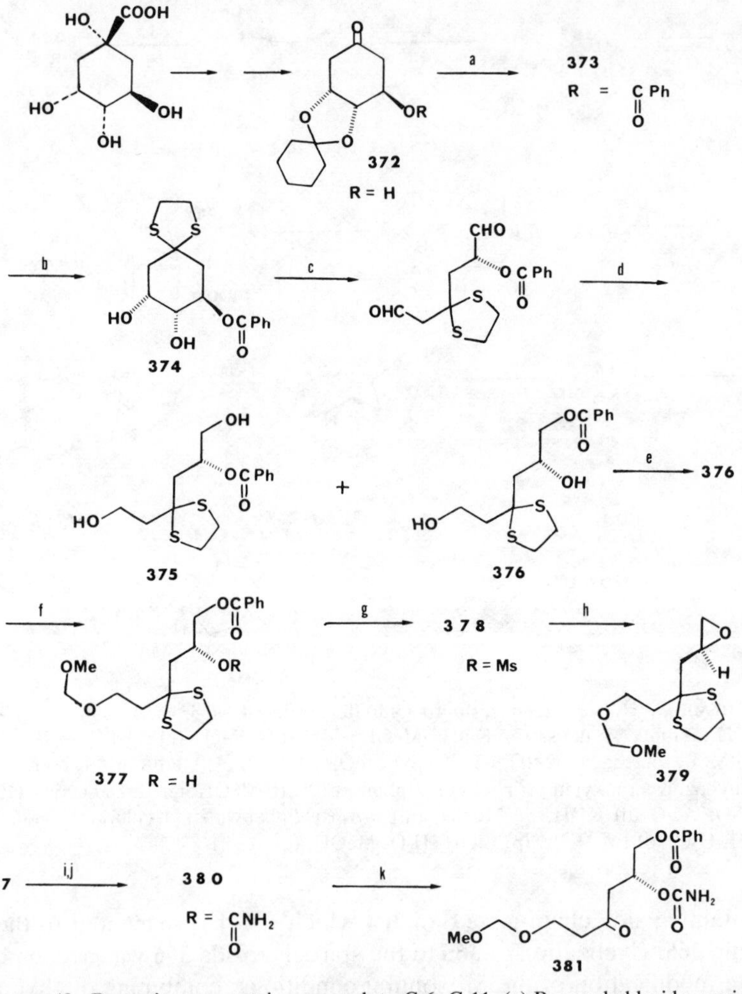

SCHEME 43. Barton's asymmetric approach to C-6–C-11. (a) Benzoyl chloride, pyridine. (b) 1,2-Ethandithiol, ZnI_2. (c) Triphenyl bismuth carbonate. (d) $NaBH_4$. (e) *N,N,N',N'*-Tetramethyl guanidine. (f) Chloromethyl methyl ether. (g) MsCl. (h) $NaOCH_3$, MeOH. (i) Nitrophenyl chloroformate, pyridine. (j) NH_3, *tert*-butanol. (k) Phenyl seleninic anydride, propylene oxide.

Barton (*32,34*) has reported an asymmetric approach to the epoxide **379** which would become the C-6–C-11 fragment of maytansine (**1**) (Scheme 43). The β-hydroxy ketone **372**, which derives from D-(−)-quinic acid, was benzoylated to give **373**. The ketone was reacted with ethanedithiol under Lewis acid catalysis and the diol was simultaneously deprotected, yielding **374**. Oxidative cleavage of the diol resulted in an unstable dialdehyde, which was reduced with sodium borohydride to the monobenzoylated triol **375**. However, the major product from this sequence was that of a benzoyl migration, **376**. Consequently, the secondary ester **375** was treated with base to complete migration to **376**. At this point the primary hydroxyl was selectively protected as its methoxy methyl ether **377**, and the secondary alcohol function was mesylated. The fully elaborated triol **378** was converted to the epoxide **379** upon methanolysis.

Alternatively, the secondary hydroxyl function of **377** was acylated with *p*-nitrophenyl chloroformate and then reacted with ammonia to give the urethane **380**. The dithioacetal was removed under mild conditions with phenylseleninic anhydride and propylene oxide to produce **381**. Unexpectedly, this ketourethane did not cyclize to the cyclic carbinolamide. Barton noted that the stereochemistry of the secondary alcohol of **376** was incorrect for the maytansinoids, so he then demonstrated the preparation of this intermediate with the opposite configuration.

Another use of a sugar as a chiral building block was reported by Isobe (*148*) (Scheme 44). D-Mannose was modified in eleven steps to the aldehyde **382** which was then used for the previously discussed (Scheme 28) heteroconjugate addition sequence. The vinyl sulfone **383** was produced as a 1 : 2 mixture of *E* and *Z* isomers, which was used without purification.

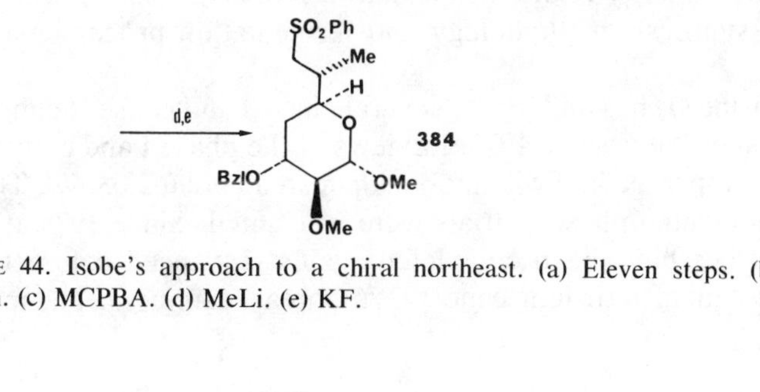

SCHEME 44. Isobe's approach to a chiral northeast. (a) Eleven steps. (b) $(TMS)_2C(Li)SO_2Ph$. (c) MCPBA. (d) MeLi. (e) KF.

Treatment with methyllithium followed by potassium fluoride in methanol provided chiral **384** in high yield. This glycoside should be compared to the racemic epoxyglycoside **218** (Scheme 39) used in Isobe's synthesis of maytansinol (**4**) (*27*) to appreciate its potential in asymmetric total synthesis.

V. Biological Activity

Several members of the maytansinoid family isolated from plants have antileukemic, cytotoxic, antitubulin, and antimitotic properties (*6,7,57, 58,59,63,149*). Maytansine (**1**) was found to have the greatest potency and has been studied extensively (*6,7,63,67,69,150–154*). The ansamitocins isolated from fermentation broths have exhibited similar biological properties (*63*).

The antitumor activity is primarily attributed to the prevention of microtubule protein polymerization due to tubulin binding, similar to the activity of the dimeric indole alkaloid vincristine (*155–159*). Work indicates that maytansine (**1**) and vincristine compete for the same or overlapping tubulin binding sites but utilize different binding mechanisms (*160*).

The ansamitocins have been shown to have antiprotozoal activity against *Tetrahymena pyriformis* and antifungal activity against *Hamigera avellanea* (*161*). Ansamitocins P-3 (**24**), P-3′ (**25**), and P-4 (**26**) were determined to be the most potent, and maytansinol (**4**) was inactive. The C-15 hydroxylated PHO-3 (**42**) and its epimer were found to have antiprotozoal activity also (*162*). The ansamitocins and maytansine (**1**) in combination with amphotericin B have produced synergistic inhibition of the growth of yeast cultures (*163*).

Structure–activity studies indicate the necessity of a C-3 ester moiety and the C-9 hydroxyl for optimal activity (*6,85,158*). Simple analogs of the carbinolamide moiety have exhibited some biological activity (*19,164–165*). Toxicological studies and clinical trials have indicated several toxicity problems even with very low doses (*69,166–168*). Hopefully, progress in structure–activity relationships and in reducing toxicity will be made now that synthesis methodology and fermentation processes are available.

Maytansine (**1**) has undergone several phase I and phase II clinical trials in the United States since 1975. Reviews of the phase I and early phase II trials have appeared and will not be duplicated here (*62,63,69*). The phase I results indicated phase II trials were warranted. Since 1978 at least 13 phase II trials have been completed against advanced colorectal carcinoma (*169*), metastatic lung cancer (*170*), breast carcinoma and melanoma

(*171*), advanced head and neck cancer (*172*), breast cancer (*173*), disseminated malignant melanoma (*174*), advanced lymphoma (*175*), small cell lung carcinoma (*176*), non-Hodgkin's lymphoma (*177*), cervical carcinoma (*178*), advanced solid tumors (*179*), soft tissue sarcoma and mesothelioma (*180*), and advanced sarcomas (*181*). These trials unfortunately indicated that **1** was not an effective antitumor agent of clinical use. Although there have been some positive clinical observations with **1**, its future in chemotherapy is doubtful (*182,183*).

Maytansine (**1**) has been reported to have antimitotic, anti-gibberellin, and auxin activity in plants (*184*). Extracts of *Trewia nudiflora* have been shown in several insect species to act as antifeedants, to reduce or suppress adult emergence or progeny, and to produce morphogenic changes (*60*). The precise modes of action and their possible use in this area remain to be defined.

References

1. Deceased, October 19, 1976.
2. S. M. Kupchan, Y. Komoda, W. A. Court, G. J. Thomas, R. M. Smith, A. Karim, C. J. Gilmore, R. C. Haltiwanger, and R. F. Bryan, *J. Am. Chem. Soc.* **95,** 1354 (1972).
3. *Cancer Chemother. Rep.* **25,** 1 (1962).
4. R. I. Geran, N. H. Greenberg, M. M. MacDonald, A. M. Schumacker, and B. J. Abbott, *Cancer Chemother. Rep., Part 3* **3,** 1 (1972).
5. K. L. Rinehart, *Acc. Chem. Res.* **5,** 57 (1972).
6. S. M. Kupchan, A. T. Sneden, A. R. Branfman, G. A. Howie, L. I. Rebhun, W. E. McIvor, R. W. Wang, and T. C. Schnaitman, *J. Med. Chem.* **21,** 31 (1978).
7. S. M. Kupchan, Y. Komoda, A. R. Branfman, A. T. Sneden, W. A. Court, G. J. Thomas, H. P. J. Hintz, R. M. Smith, A. Karim, G. A. Howie, A. K. Verma, Y. Nagao, R. G. Dailey, V. A. Zimmerly, and W. C. Sumner, *J. Org. Chem.* **42,** 2349 (1977).
8. A. I. Meyers and C. C. Shaw, *Tetrahedron Lett.* 717 (1974).
9. E. J. Corey and M. G. Bock, *Tetrahedron Lett.* 2643 (1975).
10. E. J. Corey, H. F. Wetter, A. P. Kozikowski, and A. V. Rama Rao, *Tetrahedron Lett.* 777 (1977).
11. E. J. Corey, M. G. Bock, A. P. Kozikowski, A. V. Rama Rao, D. Floyd, and B. Lipshutz, *Tetrahedron Lett.* 1051 (1978).
12. A. I. Meyers, C. C. Shaw, D. A. Horne, L. M. Trefonas, and R. J. Majeste, *Tetrahedron Lett.* 1745 (1975).
13. A. I. Meyers and R. S. Brinkmeyer, *Tetrahedron Lett.* 1749 (1975).
14. J. M. Kane and A. I. Meyers, *Tetrahedron Lett.* 771 (1977).
15. A. I. Meyers, K. Tomioka, D. M. Roland, and D. Comins, *Tetrahedron Lett.* 1375 (1978).
16. A. I. Meyers and J. P. Hudspeth, *Tetrahedron Lett.* 3925 (1981).
17. S. Stinson, *Chem. Eng. News* **58,** 54 (1980).
18. A. I. Meyers, *Chimia* **34,** 281 (1980).
19. A. H. Beaulieu, *Diss. Abstr. Int. B* **41,** 4115 (1981).

20. E. J. Corey, L. O. Weigel, D. Floyd, and M. G. Bock, *J. Am. Chem. Soc.* **100,** 2916 (1978).
21. A. I. Meyers, D. M. Roland, D. L. Comins, R. Henning, M. P. Fleming, and K. Shimizu, *J. Am. Chem. Soc.* **101,** 4732 (1979).
22. E. J. Corey, L. O. Weigel, A. R. Chamberlain, and B. J. Lipshutz, *J. Am. Chem. Soc.* **102,** 1439 (1980).
23. A. I. Meyers, D. L. Comins, D. M. Roland, R. Henning, and K. Shimizu, *J. Am. Chem. Soc.* **101,** 7104 (1979).
24. A. I. Meyers, K. Babiak, A. L. Campbell, D. L. Comins, M. P. Fleming, R. Henning, M. Heuschmann, J. P. Hudspeth, J. M. Kane, P. J. Reider, D. M. Roland, K. Shimizu, K. Tomioka, and R. D. Walkup, *J. Am. Chem. Soc.* **105,** 5015 (1983). We are indebted to Prof. Meyers for providing a preprint of this manuscript.
25. A. I. Meyers, P. J. Reider, and A. L. Campbell, *J. Am. Chem. Soc.* **102,** 6597 (1980).
26. E. J. Corey, L. O. Weigel, A. R. Chamberlain, N. Cho, and D. H. Hua, *J. Am. Chem. Soc.* **102,** 6613 (1980).
27. M. Isobe, M. Kitamura, and T. Goto, *J. Am. Chem. Soc.* **104,** 4997 (1982).
28. M. Isobe, M. Kitamura, and T. Goto, *Chem. Lett.* 1907 (1982).
29. M. Isobe, M. Kitamura, and T. Goto, *Tetrahedron Lett.* 3465 (1979).
30. M. Isobe, M. Kitamura, and T. Goto, *Chem. Lett.* 331 (1980).
31. M. Isobe, M. Kitamura, and T. Goto, *Tetrahedron Lett.* 4287 (1981).
32. D. H. R. Barton, S. D. Gero, and C. Maycock, *J. Chem. Soc., Perkin Trans. 1* 1541 (1982).
33. D. H. R. Barton, M. Benechie, F. Khuong-Huu, P. Potier, and V. Reyna-Pinedo, *Tetrahedron Lett.* **23,** 651 (1982).
34. D. H. R. Barton, S. D. Gero, and C. Maycock, *J. Chem. Soc., Chem. Commun.* **22,** 1089 (1980).
35. D. Bai, Q. Zhou, P. Pan, H. Sun, H. Zhang, Y. Shu, Y. Xu, C. Pan, Y. Du, and Y. Gao, *Chem. Nat. Prod., Proc. Sino–Am. Symp., 1980* p. 115 (1982); *CA* **98,** 198493 (1983).
36. B. Pan, H. Zhang, C. Pan, Y. Shu, Z. Wang, and Y. Gao, *Hua Hsueh Hsueh Pao* **38,** 502 (1980).
37. Q. Zhou, D. Bai, H. Sun, Y. Yang, X. Du, Y. Xu, M. Chen, and Y. Gao, *Hua Hsueh Hsueh Pao* **38,** 507 (1980).
38. W. J. Elliott and J. Fried, *J. Org. Chem.* **41,** 2469 (1976).
39. G. Gormley, Y. Chan, and J. Fried, *J. Org. Chem.* **45,** 1447 (1980).
40. J. E. Foy and B. Ganem, *Tetrahedron Lett.* 775 (1977).
41. R. Bonjouklian and B. Ganem, *Tetrahedron Lett.* 2835 (1977).
42. B. Ganem, *Proc. Asian Symp. Med. Plants Spices, 4th, 1980,* Vol. 1, p. 235 (1981).
43. O. E. Edwards and P. T. Ho, *Can. J. Chem.* **55,** 371 (1977).
44. P. T. Ho, *Can. J. Chem.* **58,** 858 (1980).
45. P. T. Ho, *Can. J. Chem.* **58,** 861 (1980).
46. H. Sirat, E. J. Thomas, and J. D. Wallis, *J. Chem. Soc., Perkin Trans. 1* 2885 (1982).
47. R. H. Wollenberg, *Tetrahedron Lett.* 717 (1978).
48. C. W. Cheng, *Diss. Abstr. Int. B* **42,** 2372 (1981).
49. E. Götschi, F. Schneider, H. Wagner, and K. Bernauer, *Helv. Chim. Acta* **60,** 1416 (1977).
50. E. Götschi, F. Schneider, H. Wagner, and K. Bernauer, *Org. Prep. Proced. Int.* **13,** 23 (1981).
51. M. Samson, P. DeClercq, H. DeWilde, and M. Vanderwalle, *Tetrahedron Lett.* 3195 (1977).
52. W. A. White, *Diss. Abstr. Int. B* **37,** 2867 (1976).

53. E. Higashide, M. Asai, K. Ootsu, S. Tanida, Y. Kozai, T. Hasegawa, T. Kishi, Y. Sugino, and M. Yoneda, *Nature (London)* **270,** 721 (1977).
54. M. Asai, E. Mizuta, M. Izawa, K. Haibara, and T. Kishi, *Tetrahedron* **35,** 1079 (1979).
55. S. Tanida, M. Izawa, and T. Hasegawa, *J. Antibiot.* **34,** 489 (1981).
56. M. Izawa, S. Tanida, and M. Asai, *J. Antibiot.* **34,** 496 (1981).
57. R. G. Powell, D. Weisleder, and C. R. Smith, *J. Org. Chem.* **46,** 4398 (1981).
58. R. G. Powell, D. Weisleder, C. R. Smith, J. Kozlowski, and W. K. Rohwedder, *J. Am. Chem. Soc.* **104,** 4929 (1982).
59. R. G. Powell, C. R. Smith, R. D. Plattner, and B. E. Jones, *J. Nat. Prod.* **46,** 660 (1983). We are indebted to Dr. Powell for providing a preprint of this manuscript.
60. B. Freedman, D. K. Reed, R. G. Powell, R. V. Madrigal, and C. R. Smith, *J. Chem. Ecol.* **8,** 409 (1982).
61. Y. Komoda, *Kagaku no Ryoiki* **28,** 887 (1974).
62. J. Douros, M. Suffness, D. Chiuten, and R. Adamson, *Adv. Med. Oncol. Res. Educ. Proc. Int. Cancer Congr. 12th, 1978,* Vol. 5, p. 59 (1979).
63. Y. Komoda and T. Kishi, *Med. Chem. (Academic)* **16,** 353 (1980).
64. K. Tomioka, *Yuki Gosei Kagaku Kyokaishi* **37,** 320 (1979).
65. M. Isobe, *Yuki Gosei Kagaku Kyokaishi* **41,** 51 (1983).
66. D. L. Bai, *Yu Chi Hua Hsueh* **1,** 1 (1981); *CA* **95,** 80766 (1981).
67. M. LeBoeuf, *Plant. Med. Phytother.* **12,** 53 (1978).
68. N. Cho, *Yuki Gosei Kagaku Kyokaishi* **40,** 296 (1982).
69. B. F. Issell and S. T. Crooke, *Cancer Treat. Rev.* **5,** 199 (1978).
70. S. A. Schepartz, *Cancer Treat. Rep.* **60,** 975 (1976).
71. R. W. Spjut and R. E. Perdue, *Cancer Treat. Rep.* **60,** 979 (1976).
72. R. E. Perdue, *Cancer Treat. Rep.* **60,** 987 (1976).
73. D. Statz and F. B. Coon, *Cancer Treat. Rep.* **60,** 999 (1976).
74. B. J. Abbott, *Cancer Treat. Rep.* **60,** 1007 (1976).
75. M. E. Wall, M. C. Wani, and H. Taylor, *Cancer Treat. Rep.* **60,** 1011 (1976).
76. J. Hartwell, *Cancer Treat. Rep.* **60,** 1031 (1976).
77. J. Douros, *Cancer Treat. Rep.* **60,** 1069 (1976).
78. A. S. Barclay and R. E. Perdue, *Cancer Treat. Rep.* **60,** 1081 (1976).
79. S. M. Kupchan, *Cancer Treat. Rep.* **60,** 1115 (1976).
80. S. M. Sieber, J. A. R. Mead, and R. H. Adamson, *Cancer Treat Rep.* **60,** 1127 (1976).
81. S. K. Carter and R. B. Livington, *Cancer Treat. Rep.* **60,** 1141 (1976).
82. C. R. Smith, R. G. Powell, and K. L. Mikolajczak, *Cancer Treat. Rep.* **60,** 1157 (1976).
83. N. R. Farnsworth, A. S. Bingel, H. S. Fong, A. Saleh, G. M. Christenson, and S. M. Saufferer, *Cancer Treat. Rep.* **60,** 1171 (1976).
84. S. M. Kupchan, Y. Komoda, G. J. Thomas, and H. P. Hintz, *J. Chem. Soc., Chem. Commun.* 1065 (1972).
85. S. M. Kupchan, Y. Komoda, A. R. Branfman, R. G. Dailey, and V. A. Zimmerly, *J. Am. Chem. Soc.* **96,** 3706 (1974).
86. A. T. Sneden and G. Beemsterboer, *J. Nat. Prod.* **43,** 637 (1980).
87. S. M. Kupchan, A. R. Branfman, A. T. Sneden, A. K. Verma, R. G. Daily, Y. Komoda, and Y. Nagao, *J. Am. Chem. Soc.* **97,** 5294 (1975).
88. A. T. Sneden, W. C. Sumner, and S. M. Kupchan, *J. Nat. Prod.* **45,** 624 (1982).
89. M. C. Wani, H. Taylor, and M. E. Wall, *J. Chem. Soc., Chem. Commun.* 390 (1973).
90. Jpn. Pat. 57/192,388.
91. Jpn. Pat. 57/193,478.
92. Jpn. Pat. 57/193,479.
93. Eur. Pat. GO 25,108

94. Eur. Pat. Appl. 65,730.
95. U.S. Pat. 4,362,663.
96. U.S. Pat. 4,371,553.
97. Jpn. Pat. 57/142,987.
98. U.S. Pat. 4,364,866.
99. Eur. Pat. Appl. 31,430; *CA* **95,** 185558.
100. Eur. Pat. Appl. 28,683; *CA* **95,** 185550.
101. Jpn. Pat. 81/20,592; *CA* **95,** 132912.
102. Eur. Pat. Appl. 25,496; *CA* **95,** 115624.
103. Eur. Pat. Appl. 25,108; *CA* **95,** 97867.
104. Eur. Pat. Appl. 25,898; *CA* **95,** 59919.
105. Eur. Pat. Appl. 21,173; *CA* **95,** 7357.
106. Eur. Pat. Appl. 21,178; *CA* **94,** 208922.
107. Eur. Pat. Appl. 21,177; *CA* **94,** 208923.
108. Eur. Pat. Appl. 11,302; *CA* **94,** 47372.
109. Eur. Pat. Appl. 11,277; *CA* **94,** 47371.
110. Eur. Pat. Appl. 11,276; *CA* **94,** 47370.
111. Eur. Pat. Appl. 10,735; *CA* **94,** 47369.
112. Jpn. Pat. 80/53,294; *CA* **93,** 186427.
113. Eur. Pat. Appl. 14,402; *CA* **95,** 43191.
114. Eur. Pat. Appl. 4466; *CA* **92,** 111076.
115. Ger. Offen. 2,911,248; *CA* **92,** 94449.
116. Eur. Pat. Appl. 65,730; *CA* **98,** 143193.
117. K. Nakahama, M. Izawa, M. Asai, M. Kida, and T. Kishi, *J. Antibiot.* **34,** 1581 (1981).
118. M. Izawa, K. Nakahama, F. Kasahara, M. Asai, and T. Kishi, *J. Antibiot.* **34,** 1587 (1981).
119. M. Izawa, Y. Wada, F. Kasahara, M. Asai, and T. Kishi, *J. Antibiot.* **34,** 1591 (1981).
120. D. E. Nettleton, D. M. Balitz, M. Brown, J. E. Moseley, and R. W. Myllymaki, *J. Nat. Prod.* **44,** 340 (1981).
121. M. S. Ahmed, H. H. S. Fong, D. D. Soejarto, R. H. Dobberstein, D. P. Waller, and R. Morenzo-Azorero, *J. Chromatogr.* **213,** 340 (1981).
122. K. H. Lee, H. Nozaki, I. Hall, R. Kasai, T. Hirayama, H. Suzuki, R. Y. Wu, and H. C. Huang, *J. Nat. Prod.* **45,** 509 (1982).
123. Z. He, Y. Zhou, G. Ma, R. Xu, and Q. He, *Zhiwu Xuebao* **24,** 360 (1982); *CA* **97,** 178714.
124. Y. Zhou, L. Huang, Q. Zhou, F. Jiang, X. He, C. Li, C. Wang, and B. Li, *Huaxue Xuebao* **39,** 427 (1981); *CA* **96,** 45973.
125. X. Wang, J. Chen, R. Wei, D. Jiang, and D. Djiang, *Yaoxue Xuebao* **16,** 628 (1981); *CA* **96,** 31684.
126. C. Li, B. Li, C. Wang, Y. Zhou, and L. Huang, *Yaoxue Xuebao* **16,** 635 (1981); *CA* **96,** 31685.
127. Y. Chou, L. Wang, C. Chou, C. Chiang, H. Ho, Y. Li, C. Wang, and P. Li, *K'o Hsueh T'ung Pao* **25,** 427 (1980); *CA* **93,** 210149.
128. H. Chi, W. Chiang, C. Tsai, and C. Pu, *Chung Ts'ao Yao* **11,** 173 (1980); *CA* **94,** 109167.
129. X. Wang, R. Wei, J. Chen, and D. Jiang, *Yao Hsueh Hsueh Pao* **16,** 59 (1981); *CA* **95,** 138482.
130. X. Wang, J. Chen, R. Wei, D. Jiang, and D. Djiang, *Yaoxue Xuebao* **16,** 628 (1981); *CA* **96,** 31684.
131. M. Izawa, K. Haibara, and M. Asai, *Chem. Pharm. Bull.* **28,** 789 (1980).

132. W. A. Wallace and A. T. Sneden, *Org. Magn. Reson.* **19,** 31 (1982).
133. R. F. Bryan, C. J. Gilmore, and R. C. Haltiwanger, *J. Chem. Soc., Perkin Trans. 2* 897 (1973).
134. A. I. Meyers and M. P. Fleming, *J. Org. Chem.* **44,** 3405 (1979).
135. L. A. Carpino and J. H. Tsao, *J. Chem. Soc., Chem. Commun.* 358 (1978).
136. R. S. Brinkmeyer, *Diss. Abstr. Int. B* **36,** 1704 (1975).
137. D. A. Evans and G. C. Andrews, *Acc. Chem. Res.* **7,** 147 (1974).
138. S. Tanaka, H. Yamamoto, H. Nozaki, K. B. Sharpless, R. C. Michaelson, and J. D. Cutting, *J. Am. Chem. Soc.* **96,** 5254 (1974).
139. B. Ganem, N. Takamura, and F. Mayerl, private communication.
140. M. Isobe, M. Kitamura, and T. Goto, *Tetrahedron Lett.* **22,** 239 (1981).
141. A. I. Meyers and D. L. Comins, *Tetrahedron Lett.* 5179 (1978).
142. K. Narasaka, A. Moriicawa, K. Saigo, and T. Mukaiyama, *Bull. Chem. Soc. Jpn.* **50,** 2773 (1977).
143. K. Maruoka, S. Hashimoto, Y. Kitagawa, H. Yamamoto, and H. Nozaki, *J. Am. Chem. Soc.* **99,** 7705 (1977).
144. T. Katsuki and K. B. Sharpless, *J. Am. Chem. Soc.* **102,** 5974 (1980).
145. C. T. Goodhue and J. R. Shaeffer, *Biotechnol. Bioeng.* **13,** 203 (1971).
146. N. Cohen, W. F. Eichel, R. J. Lopresti, C. Neukom, and G. Saucy, *J. Org. Chem.* **41,** 3505 (1976).
147. A. Horeau, *in* "Stereochemistry: Fundamental Methods" (H. B. Kagan, ed.), Vol. 3, p. 51. Thieme, Stuttgart, 1977.
148. M. Isobe, Y. Ichikawa, M. Kitamura, and T. Goto, *Chem. Lett.* 457 (1981).
149. R. Chen, Z. Hua, Z. Lu, L. Wang, and B. Xu, *Yaoxue Tongbao* **17,** 303 (1982); *CA* **97,** 192794.
150. *Drugs Fut.* **4,** 35 (1979).
151. *Drugs Fut.* **5,** 50 (1980).
152. P. N. Rao, E. J. Freireich, M. L. Smith, and T. L. Loo, *Cancer Res.* **39,** 3152 (1979).
153. D. V. Jackson and R. A. Bender, *Appl. Methods Oncol.* **1,** 277 (1978).
154. D. G. Johns, M. K. Wolpert-Defilippes, F. Mandelbaum-Shavit, B. A. Chabner, V. H. Bono, and R. H. Adamson, *Curr. Chemother., Proc. Int. Congr. Chemother., 10th, 1977* Vol. 2, p. 1180 (1978).
155. F. Mandelbaum-Shavit, M. K. Wolpert-Defilippes, and D. G. Johns, *Biochem. Biophys. Res. Commun.* **72,** 47 (1976).
156. K. Ootsu, Y. Kozai, M. Takeuchi, S. Ikeyama, K. Igarashi, K. Tsukamoto, Y. Sugino, T. Tashiro, S. Tsukagoshi, and Y. Sakurai, *Cancer Res.* **40,** 1707 (1980).
157. S. Kim, M. Tomonaga, and B. Ghetti, *Acta Neuropathol.* **52,** 161 (1980).
158. J. York, M. K. Wolpert-DeFilippes, D. G. Johns, and V. S. Sethi, *Biochem. Pharmacol.* **30,** 3239 (1981).
159. S. Ikeyama and M. Takeuchi, *Biochem. Pharmacol.* **30,** 2421 (1981).
160. R. F. Luduena and M. C. Roach, *Arch. Biochem. Biophys.* **210,** 498 (1981).
161. S. Tanida, T. Hasegawa, K. Hatano, E. Higashide, and M. Yoneda, *J. Antibiot.* **33,** 192 (1980).
162. M. Izawa, Y. Wada, F. Kasahara, M. Asai, and T. Kishi, *J. Antibiot.* **34,** 1591 (1981).
163. S. Tanida and T. Hasegawa, *Agric. Biol. Chem.* **45,** 481 (1981).
164. J. W. Lown, K. Majumdar, A. I. Meyers, and A. Hecht, *Bioorg. Chem.* **6,** 453 (1977).
165. R. J. Baker, R. J. Majeste, and L. M. Trefonas, *J. Heterocycl. Chem.* **18,** 1541 (1981).
166. D. Thake, M. Naylor, R. Denlinger, A. Guarino, and D. Cooney, *U.S. NTIS PB Rep.* **PB-244628** (1975); *CA* **84,** 54226.

167. D. Thake, M. Naylor, R. Denlinger, A. Guarino, and D. Cooney, *U.S. NTIS, PB Rep.* **PB-244566** (1975); *CA* **84,** 54227.
168. R. Meeks, E. Denine, L. Stout, J. Peckham, and C. Litterst, *Gov. Rep. Announce. Index (U.S.)* **82,** (1982); *CA* **97,** 1214886.
169. M. J. O'Connell, A. Shani, J. Rubin, and C. G. Moertel, *Cancer Treat. Rep.* **62,** 1237 (1978).
170. R. T. Eagan, E. T. Creagan, J. N. Ingle, S. Frytak, and J. Rubin, *Cancer Treat. Rep.* **62,** 1577 (1978).
171. F. Cabanillas, G. P. Bodey, M. A. Burgess, and E. J. Freireich, *Cancer Treat. Rep.* **63,** 507 (1979).
172. E. T. Creagan, T. R. Fleming, J. H. Edmonson, and J. N. Ingle, *Cancer Treat. Rep.* **63,** 2061 (1979).
173. J. A. Neidhart, L. R. Laufman, C. Vaughn, and J. D. McCracken, *Cancer Treat. Rep.* **64,** 675 (1980).
174. D. L. Ahmann, S. Frytak, L. K. Kvols, R. G. Hahn, J. H. Edmonson, H. F. Bisel, and E. T. Creagan, *Cancer Treat. Rep.* **64,** 721 (1980).
175. S. Rosenthal, D. T. Harris, J. Horton, and J. H. Glick, *Cancer Treat. Rep.* **64,** 1115 (1980).
176. R. H. Creech, K. Stanley, S. E. Vogl, D. S. Ettinger, P. D. Bonomi, and O. Salazar, *Cancer Treat. Rep.* **66,** 1417 (1982).
177. V. Ratanatharathorn, N. Gad-el-Mawla, H. E. Wilson, J. D. Bonnet, S. E. Rivkin, and R. Mass, *Cancer Treat. Rep.* **66,** 1587 (1982).
178. T. Thigpen, C. Ehrlich, and J. Blessing, *Proc. Am. Assoc. Cancer Res. Am. Soc. Clin. Oncol.* **21,** 424 (1980).
179. M. J. R. Ravry, S. Lowenbraun, and R. Birch, *Proc. Am. Assoc. Cancer Res. Am. Soc. Clin. Oncol.* **21,** 154 (1980).
180. E. Borden, A. Ash, C. Rosenbaum, and H. Lerner, *Proc. Am. Assoc. Cancer Res. Am. Soc. Clin. Oncol.* **21,** 479 (1980).
181. J. H. Edmonson, R. G. Hahn, E. T. Creagan, and M. J. O'Connell, *Cancer Treat. Rep.* **67,** 401 (1983).
182. A. P. Chahinian, C. Nogeire, and T. Ohnuma, *Cancer Treat. Rep.* **63,** 1953 (1979).
183. I. S. Jaffrey, J. M. Denefrio, and P. Chahinian, *Cancer Treat. Rep.* **64,** 193 (1980).
184. Y. Komoda and Y. Isogai, *Sci. Pap. Coll. Gen. Educ., Univ. Tokyo* **28,** 129 (1978); *CA* **89,** 158672.

—CHAPTER 3—

CEPHALOTAXUS ALKALOIDS

LIANG HUANG AND ZHI XUE

Institute of Materia Medica Chinese
Academy of Medical Sciences
Beijing, People's Republic of China

I. Introduction

A. BOTANY AND DISTRIBUTION OF *CEPHALOTAXUS*

Cephalotaxus once was regarded as a genus of the family Taxaceae. On the basis of morphology, anatomy, embryology, geographical distribu-

THE ALKALOIDS, VOL. XXIII

ISBN 0-12-469523-X

tion, and the recent findings of chemical constituents in the plant, the classification of the genus *Cephalotaxus* to a separate family, the Cephalotaxaceae of the order Coniferae, has been supported. *Cephalotaxus,* the only genus of this family, consists of eight species, namely, *C. harringtonia* (Forbes) Koch, *C. fortunei* Hook. f., *C. hainanensis* Li, *C. wilsoniana* Hay., *C. mannii* Hook. f., *C. oliveri* Mast., *C. lanceolata* K. M. Feng, *C. sinensis* Li, and possibly 2–4 varieties. *Cephalotaxus* is native to eastern Asia; *C. harringtonia* occurs in Japan and *C. mannii* in India and the southwestern part of China. The remaining six species are indigenous to China. *Cephalotaxus* plants are evergreen coniferous trees with yewlike leaves or shrubs that grow in humid valleys or in forests at an elevation of 100–2000 m (*1,2*).

B. Background and Current Status of Chemical Studies on *Cephalotaxus*

In 1954 Wall *et al.* (*3*) first reported the presence of alkaloids in *Cephalotaxus,* and Paudler *et al.* (*4*) in 1963 isolated a crystalline alkaloid named cephalotaxine from *C. drupacea,* which is now considered to be *C. harringtonia* var. *drupacea* (Sieb & Zucc.), and *C. fortunei.* Six years later the structure of cephalotaxine was finally determined by X-ray crystallographic analysis (*5*). The structure is unique to the genus *Cephalotaxus.*

Antitumor activity against leukemia P388 and L1210 in mice of the four esters of cephalotaxine, i.e., harringtonine, homoharringtonine, deoxyharringtonine, and isoharringtonine, reported by Powell *et al.* in 1972 prompted investigation of *Cephalotaxus* (*6*). Six out of the eight species of *Cephalotaxus* have since been studied. They are *C. harringtonia, C. fortunei, C. hainanensis, C. wilsoniana, C. sinensis,* and *C. oliveri,* with the first three being studied in more detail. Two classes of alkaloids have been isolated and identified. Of 28 compounds, 18 relate to the cephalotaxine series. The other 10 compounds belong to the class of homoerythrina alkaloid, which is also present in *Schelhammera pedunculata, S. multiflora,* and *Phelline comosa.* In most species of *Cephalotaxus,* the homoerythrina alkaloids are present as minor alkaloids with the exception of *C. wilsoniana.* Isolation of *Cephalotaxus* and homoerythrina alkaloids from the same plant suggested that they might be formed through similar biogenetic routes. Tables I and II summarize the distribution of *Cephalotaxus* and homoerythrina alkaloids in various species of *Cephalotaxus* reported up to 1981.

For extraction of alkaloids from *Cephalotaxus* plants the procedure of percolation of the plant material with ethanol, followed by acidification of the concentrate, removal of the nonalkaloid constituents with chloroform,

TABLE I
DISTRIBUTION OF *CEPHALOTAXUS* ALKALOIDS

Compounds	*Cephalotaxus* spp.						
	harringtonia						
	var. *harringtonia*	var. *drupacea*	*fortunei*	*hainanensis*	*wilsoniana*	*sinensis*	*oliveri*
Cephalotaxine	+ (*6,7*)	+ (*4,9*)	+ (*11–13*)	+ (*15*)	+ (*18,19*)	+ (*20,20a*)	+ (*21*)
Epicephalotaxine		—	+ (*11*)	+ (*15*)	—	—	—
Cephalotaxinone	+ (*7*)	—	+ (*11*)	+ (*16*)	—	—	—
11-Hydroxycephalotaxine	—	+ (*10*)	+ (*13,14*)	—	—	+ (*20,20a*)	—
Drupacine	—	+ (*9,10*)	+ (*13*)	+ (*15*)	+ (*18*)	+ (*20,20a*)	—
Demethylcephalotaxine	—	+ (*9*)	+ (*11*)	—	—	—	—
Demethylcephalotaxinone	+ (*8*)	—	+ (*12*)	+ (*15*)	—	+ (*20*)	—
Demethylneodrupacine	—	—	—	+ (*16*)	—	—	—
Cephalotaxinamide	—	—	—	+ (*16*)	—	—	—
Acetylcephalotaxine	—	—	+ (*11*)		+ (*18,19*)	—	—
(+)-Acetylcephalotaxine	—	—	+ (*12*)	+ (*16*)	—	—	—
Harringtonine	+ (*6,7*)	—	+ (*12,13*)	+ (*15*)	—	+ (*20*)	+ (*21*)
Homoharringtonine	+ (*6,7*)	+ (*9*)	+ (*12,13*)	+ (*15*)	—	+ (*20,20a*)	—
Deoxyharringtonine	+ (*6,7*)	—	—	+ (*15*)	—		—
Isoharringtonine	+ (*6,7*)	+ (*9*)	—	+ (*15*)	+ (*18*)	+ (*20a*)	—
Isoharringtonic acid	—	—	—	+ (*16*)	—	—	—
Deoxyharringtonic acid	—	—	—	+ (*16*)	—	—	—
Hainanensine	—	—	—	+ (*17*)	—	—	—

TABLE II

DISTRIBUTION OF HOMOERYTHRINA ALKALOIDS

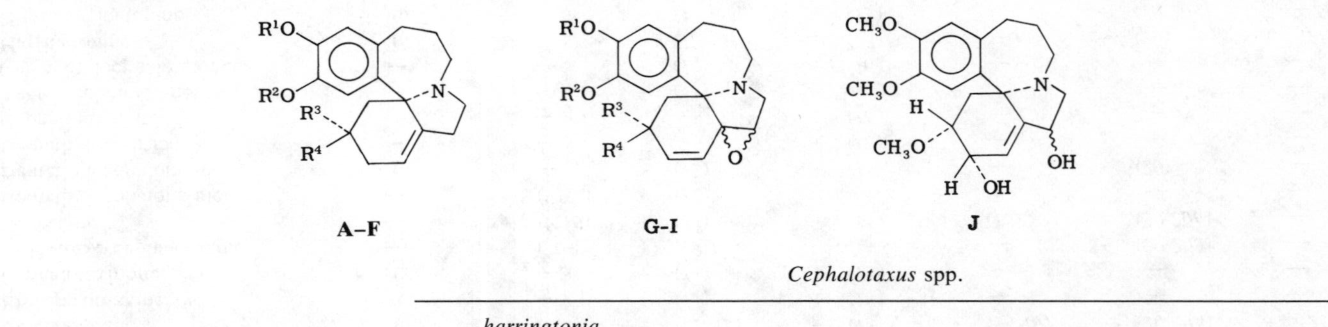

Compound	*Cephalotaxus* spp.						
	harringtonia						
	var. *harringtonia*	var. *drupacea*	*fortunei*	*hainanensis*	*wilsoniana*	*sinensis*	*oliveri*
Schelhammericine (**A**)							
R^1, R^2 = —CH_2—							
R^3 = H							
R^4 = OCH_3							
Epischelhammericine (**B**)	+(*7*)	+(*9*)		+(*15*)	+(*18*)		
R^1, R^2 = —CH_2—							
R^3 = OCH_3							
R^4 = H							

Schellhammericine B (**C**)	+(*7*)						
$R^1 = CH_3$ $R^2 = H$							
$R^3 = H$							
$R^4 = OCH_3$							
Epischellhammericine B (**D**)	+(*7*)	+(*9*)					+(*21*)
$R^1 = CH_3$ $R^2 = H$							
$R^3 = OCH_3$							
$R^4 = H$							
Methylschelhammericine B (**E**)	+(*7*)						
$R^1 = R^2 = CH_3$							
$R^3 = H$							
$R^4 = OCH_3$							
Epimethylschelhammericine B (**F**)	+(*7*)	+(*9*)			+(*18*)		
$R^1 = R^2 = CH_3$							
$R^3 = OCH_3$							
$R^4 = H$							
Wilsonine (**G**)					+(*19*)	+(*20,20a*)	
$R^1 = R^2 = CH_3$							
$R^3 = H$							
$R^4 = OCH_3$							
Epiwilsonine (**H**)			+(*14*)		+(*18,19*)	+(*20*)	
$R^1 = R^2 = CH_3$							
$R^3 = OCH_3$							
$R^4 = H$							
Cycloxyschelhammericine (**I**)				+(*15*)		+(*20a*)	
$R^1, R^2 = —CH_2—$							
$R^3 = H$							
$R^4 = OCH_3$							
Cephalofortuneine (**J**)			+(*14*)				

and extraction of the basified solution with chloroform is most commonly used. The total alkaloid is also obtained by extraction with dilute hydrochloric acid and then exchange on a column of sulfonic acid resin. The resin is then made alkaline with ammonium hydroxide and eluted with chloroform. The latter method has the advantage of avoiding the formation of a troublesome emulsion but requires more careful manipulation. The alkaloid mixture is then separated into its constituents by a combination of column chromatography with countercurrent extraction and/or partition chromatography monitored by TLC. The partition chromatography on silica gel or celite, using a buffer (pH 5) as the stationary phase and chloroform as the mobile phase, is especially useful for separation of the closely related compounds harringtonine and homoharringtonine.

Alkaloids were found in various parts of the plants, including leaves, stems, bark, roots, and seeds. In the case of *C. hainanensis,* the content of total alkaloid as well as the amounts of harringtonine and homoharringtonine were found to be much higher in bark than in leaves and branches (*2*).

Powell's group (*21a*) developed a quantitative gas chromatographic method for analysis of crude alkaloid mixtures from *Cephalotaxus* plants. The alkaloids were transformed to trimethyl silyl (TMS) derivatives with Regisil RC-1 and chromatographed on a glass column packed with 3% Dexsil 300. Most TMS alkaloids showed different relative retention time (RRT) values and were separable from each other. Acetylcephalotaxine, 11-hydroxycephalotaxine, and demethylcephalotaxine exhibited insufficient differences in RRT (between 0.79 and 0.80) to be resolved. The alkaloids in the mixture could be further identified by gas chromatography–mass spectrometry (GC–MS) in the form of TMS derivatives. Different parts of selected *Cephalotaxus* plants were analyzed. Cephalotaxine is the major alkaloid in all the samples tested except *C. wilsoniana,* which contains more homoerythrina alkaloid than *Cephalotaxus* alkaloid (45 versus 29%). The percentage of the four esters, harringtonine, homoharringtonine, isoharringtonine, and deoxyharringtonine, in the total alkaloid extract ranges from 0.7% in leaves of *C. fortunei* to 35.5% in roots of *C. harringtonia* var. *harringtonia*. The latter plant appeared to be a better source for the esters.

Delfel (*21b*) reported that young plants of *C. harringtonia* grown in a controlled environment contained much more "harringtonines" than cephalotaxine. The concentration of esters, moreover, increased with age. Differences in alkaloid composition between laboratory- and field-grown trees might be ascribed mainly to the influence of the environment on plant growth. In addition, it was found that physiological stress caused hydrolysis of the stored alkaloid esters to free cephalotaxine.

II. Structural Studies on *Cephalotaxus* Alkaloids

The 18 *Cephalotaxus* alkaloids may be divided structurally into two series: those compounds with structural variation limited to the cephalotaxine moiety and those with variation at the ester of cephalotaxine.

A. Cephalotaxine

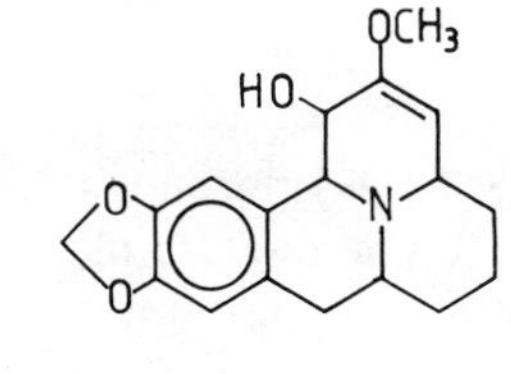

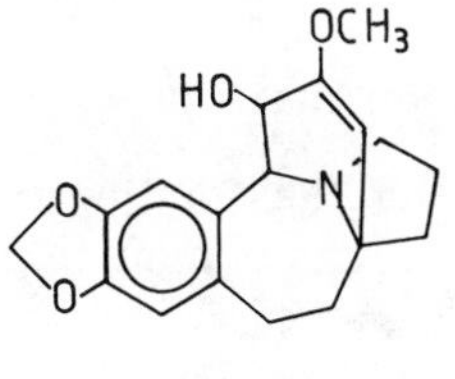

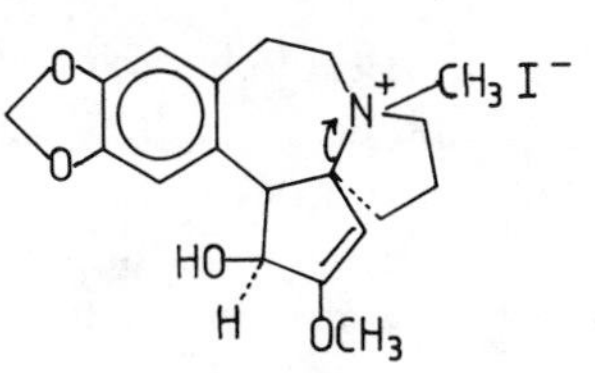

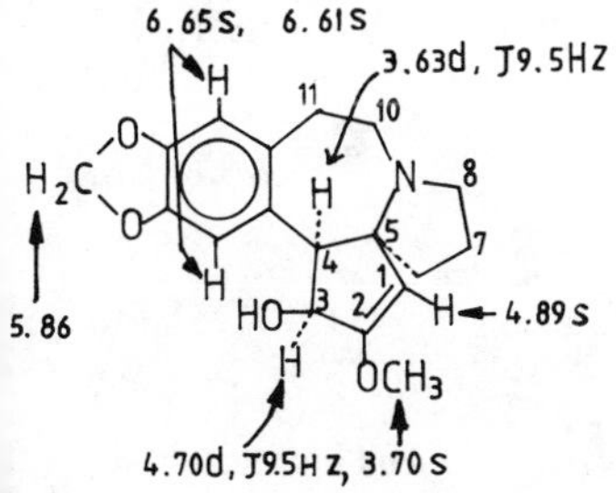

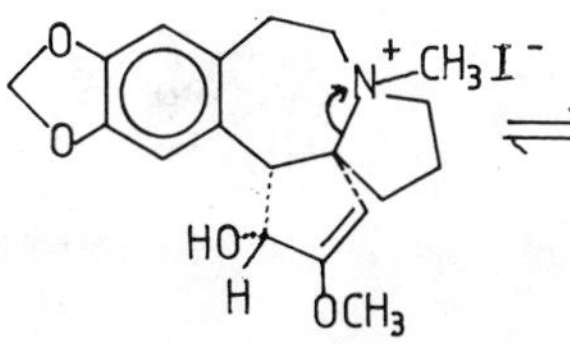

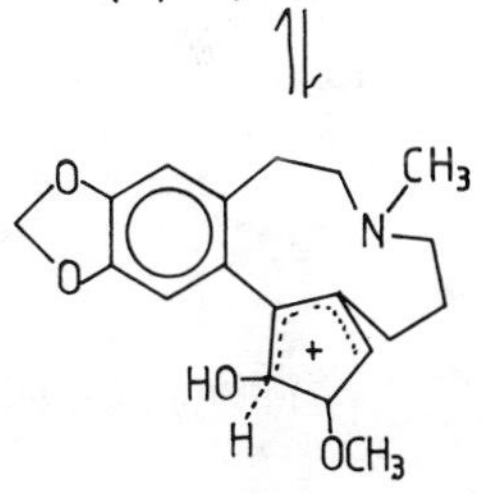

Cephalotaxine (*4,5*), $C_{18}H_{21}NO_4$, mp 135–136°C, $[\alpha]_D$ −189° (0.51, $CHCl_3$), $[\alpha]_D$ −219° (0.04, ethanol), pK_a 8.95 in 95% ethanol, showed a UV absorption spectrum [λ_{max},nm (log e)] of 290 (3.55), 260 (2.79), 238 (3.56), which is indicative of the presence of benzene ring with a substitution pattern similar to lycorine. Bands of 1040 and 934 cm^{-1} in its IR spectrum represented the methylenedioxyphenyl group, and the 1650-cm^{-1} band indicated the presence of a double bond in the molecule. Three tentative structures—**1**, **2**, and **3** (*22*)—were suggested. The structure, studied in 1969 by X-ray crystallographic analysis of cephalotaxine methiodide (±)-**4**, has been assigned as **3**. The methiodide used in the analysis, however, was a racemic mixture whereas cephalotaxine is optically active. The racemization of a molecule with four chiral centers is unusual; this might be explained by C-5–N bond breaking to **5** and subsequent reforming. The X-ray study on cephalotaxine *p*-bromobenzoate in 1974 (*23*) revealed the absolute configuration as 3*S*,4*S*,5*R*. The crystal molecular structure of cephalotaxine itself was obtained by vector search methods of X-ray study (*24*). The ^{1}H-NMR (δ) assignment of individual protons is marked directly on the structural formula (**5**). It is worthwhile to mention that ~15% racemic cephalotaxine was reportedly isolated from *C. fortunei* grown in Fujian province in China (*24a*).

Cephalotaxine is resistant to tosylation although it is readily acetylated. The double bond is inert to catalytic hydrogenation; it is sensitive to acid, which hydrolyzes the enol ether to give the demethyl compound.

B. Esters of Cephalotaxine with Antitumor Activity

Cephalotaxine itself is inactive against experimental tumors. Harringtonine, homoharringtonine, isoharringtonine, and deoxyharringtonine (**6a–6d**) are the only four esters in this series that show significant antitumor activity in animal tests.

1. Gross Structures of the Harringtonines

Although most reports stated that the "harringtonines" are amorphous solids, crystalline forms with definite melting points were obtained for all except deoxyharringtonine (*15*). The four esters are closely related. Mild transesterification yielded cephalotaxine and the respective dimethyl esters of hydroxydicarboxylic acids (**7–10**) (*25*).

Harringtonine (**6a**): $C_{28}H_{37}NO_9$, mp 73–75°C, $[\alpha]_D$ −104.6° (1.0, $CHCl_3$)

Homoharringtonine (**6b**): $C_{29}H_{39}NO_9$, mp 144–146°C, $[\alpha]_D$ −119°(0.96, $CHCl_3$)

Isoharringtonine (**6c**): $C_{28}H_{37}NO_9$, mp 69–72.5°C, $[\alpha]_D$ −99.6° (1.06, $CHCl_3$)

Deoxyharringtonine (**6d**): $C_{28}H_{37}NO_8$, amorphous, $[\alpha]_D$ − 125.4° (1.76, $CHCl_3$)

The structures of the dimethyl esters were proposed on the basis of NMR data (*25*), and the structures of **7**, **9**, and **10** were further verified by regiospecific syntheses. Isoamyl methyl ketone was used to synthesize **10** (route 1) (*26*), and an alternate route started with methyl itaconate (route 2) (*27*). Synthesis of diester **7** is shown in route 3 (*28*). The question of which carboxyl group was originally esterified in cephalotaxine remained. Mikolajczak *et al.* (*26*) compared the NMR signals (δ 3.77, δ 3.64) of the carbomethoxy groups in **10** with those of the two corresponding monoes-

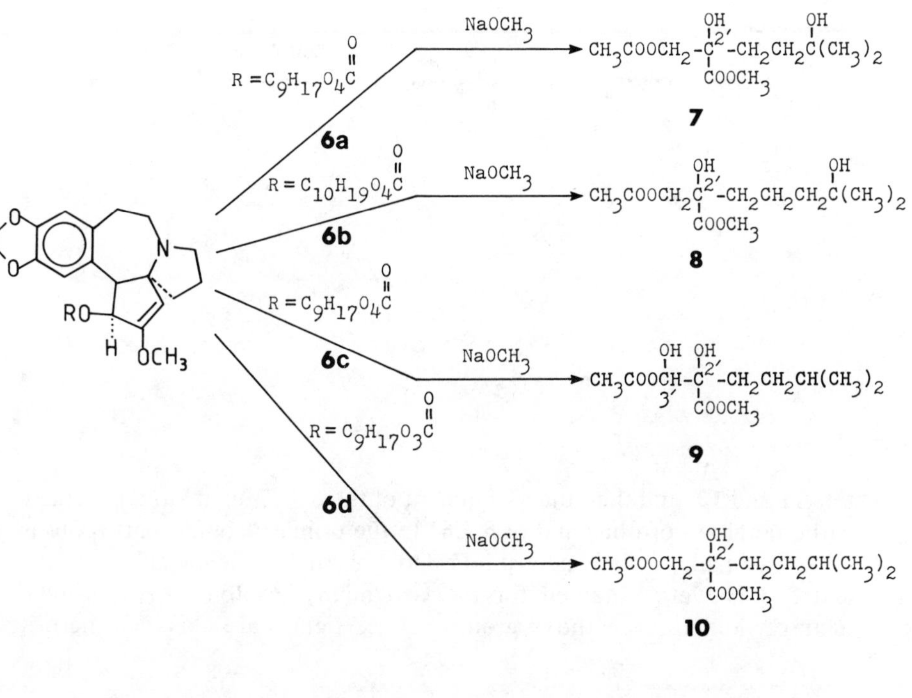

$$(CH_3)_2CHCH_2CH_2\overset{O}{\overset{\|}{C}}CH_3 \xrightarrow{(C_2H_5O)_2CO} (CH_3)_2CHCH_2CH_2\overset{O}{\overset{\|}{C}}CH_2COOC_2H_5 \xrightarrow{HCN}$$

$$(CH_3)_2CHCH_2CH_2\underset{CN}{\overset{OH}{C}}CH_2COOC_2H_5 \longrightarrow (CH_3)_2CHCH_2CH_2\underset{COOCH_3}{\overset{OH}{C}}CH_2COOCH_3$$

10

Route 1

$$H_3CO\overset{O}{\overset{\|}{C}}-\overset{CH_2}{\overset{\|}{C}}CH_2COOCH_3 \longrightarrow H_3CO\underset{O}{\underset{\|}{C}}-\overset{H_2C-O}{C}-CH_2COOCH_3 \longrightarrow H_3CO\overset{O}{\overset{\|}{C}}-\underset{CH_2COOCH_3}{\overset{OH}{C}}CH_2CH_2CH(CH_3)_2$$

10

Route 2

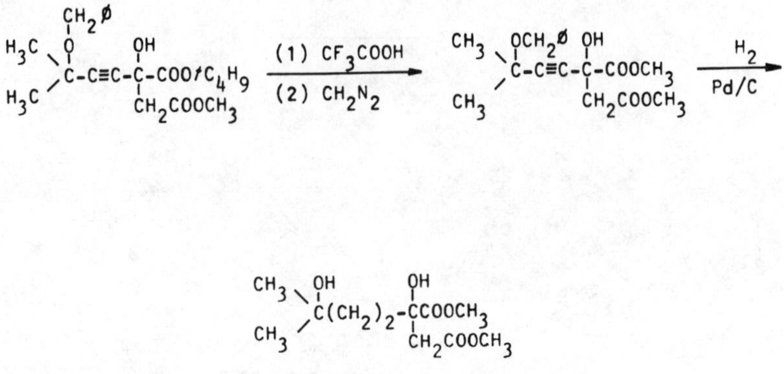

$$\begin{matrix} H_3C \\ H_3C \end{matrix} \!\!>\! \overset{\overset{CH_2\phi}{|}}{\overset{O}{|}}C\text{-}C{\equiv}C\text{-}\overset{OH}{\underset{CH_2COOCH_3}{C}}\text{-}COO\,t\,C_4H_9 \xrightarrow[(2)\ CH_2N_2]{(1)\ CF_3COOH} \begin{matrix} CH_3 \\ CH_3 \end{matrix}\!\!>\! \overset{OCH_2\phi}{C}\text{-}C{\equiv}C\text{-}\overset{OH}{\underset{CH_2COOCH_3}{C}}\text{-}COOCH_3 \xrightarrow[Pd/C]{H_2}$$

$$\begin{matrix} CH_3 \\ CH_3 \end{matrix}\!\!>\! \overset{OH}{C}(CH_2)_2\text{-}\overset{OH}{\underset{CH_2COOCH_3}{C}}COOCH_3$$

7

Route 3

ters **11** and **12**, and then the assignment of the δ 3.77 signal to the tertiary carbomethoxyl protons and of δ 3.64 to the primary counterpart protons in **10** was made. In the spectrum of deoxyharringtonine, signals at δ 3.64 and δ 3.53 were obtained for the two methyl protons in the vinylic methoxy and carbomethoxy groups. These signals are closer to the pri-

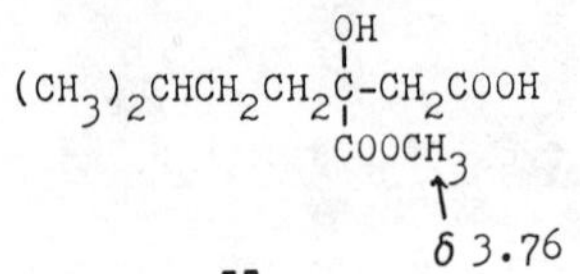

$$(CH_3)_2CHCH_2CH_2\overset{OH}{\underset{COOCH_3}{C}}\text{-}CH_2COOH \quad (\uparrow\ \delta\ 3.76)$$

11

$$(CH_3)_2CHCH_2CH_2\overset{OH}{\underset{COOH}{C}}\text{-}CH_2COOCH_3 \quad (\nearrow\ \delta\ 3.68)$$

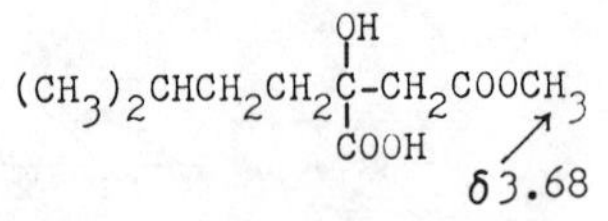

12

mary carbomethoxy found in **10**; therefore, linkage of the tertiary carboxyl group of the acid to cephalotaxine was suggested for deoxyharringtonine, as well as the parallel for the other three congeners, harringtonine, homoharringtonine, and isoharringtonine (*26*). Reaction of the acyl chloride of compound **11** with cephalotaxine yielded a mixture of diastereomers whose spectrum differed from that of the natural deoxyharringtonine. This substantiated the above assumption, i.e., the tertiary carboxyl group is involved in the acylation of cephalotaxine.

2. Stereochemistry of the Acid Moieties

The chirality of C-2′ in the acid chains of harringtonine, homo- and deoxyharringtonine has been assigned to the *R* configuration by comparing the CD spectra of molybdate complexes of the corresponding free dibasic acids of **8** and **10** with those of (*S*)-malic acid and (*S*)-citramalic acid (**13**). The Cotton effects of complexes of known (*S*)-malic acids are

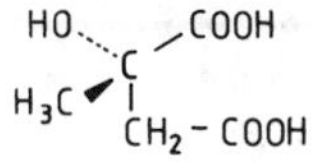

13

similar to each other while the complexes of the free acids of **8** and **10** show Cotton effects opposite (*29*) to that of the malic acids.

The acid chain of isoharringtonine has two chiral centers (C-2′ and C-3′), and therefore the stereochemistry is more complicated than the other congeners. The relative position of the hydroxyl groups was solved by stereospecific syntheses of the methyl dihydroxydicarboxylates **14** and

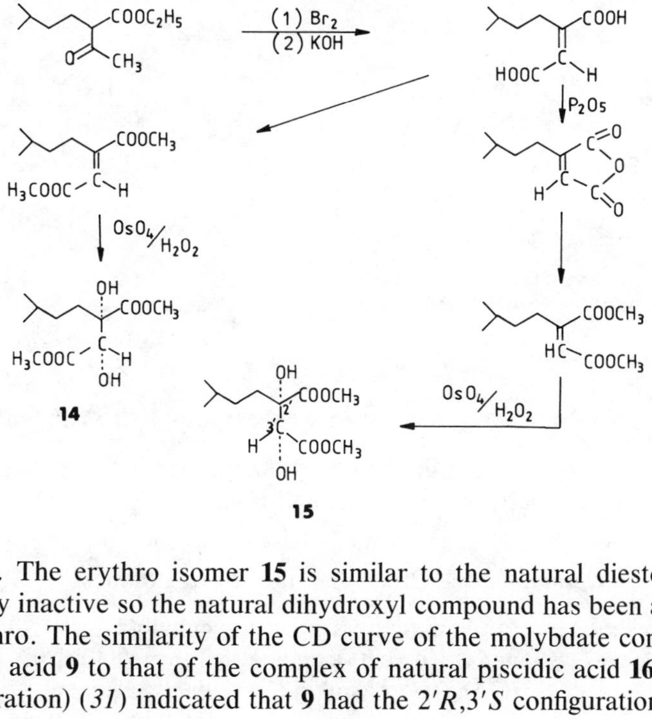

(1) Br_2 (2) KOH

P_2O_5

OsO_4/H_2O_2

14

15

15 (*30*). The erythro isomer **15** is similar to the natural diester **9** but optically inactive so the natural dihydroxyl compound has been assigned as erythro. The similarity of the CD curve of the molybdate complex of the free acid **9** to that of the complex of natural piscidic acid **16** (2*R*,3*S* configuration) (*31*) indicated that **9** had the 2′*R*,3′*S* configuration.

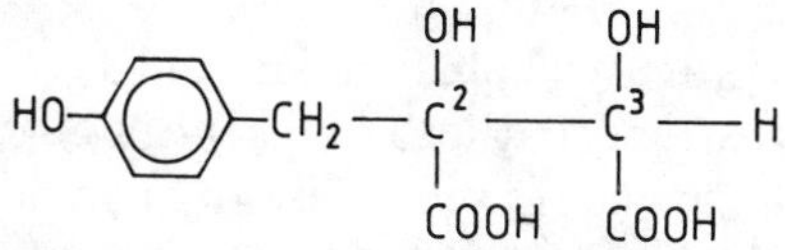

16

Thus the molecular structures, including the absolute configurations of the four antitumor esters, were completed (**17–20**).

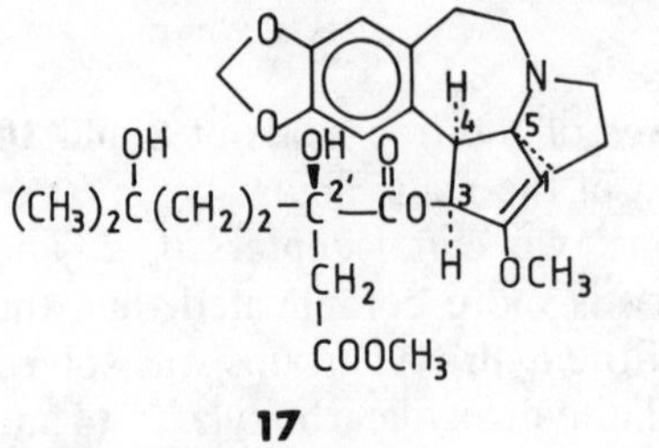

17
Harringtonine
3*S*, 4*S*, 5*R*, 2′*R*

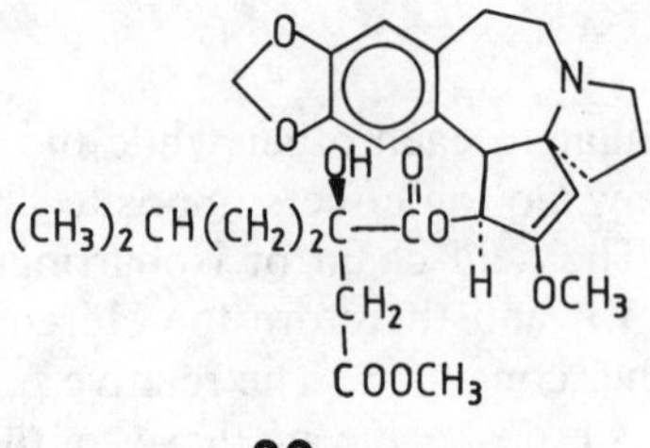

20
Deoxyharringtonine
3*S*, 4*S*, 5*R*, 2′*R*

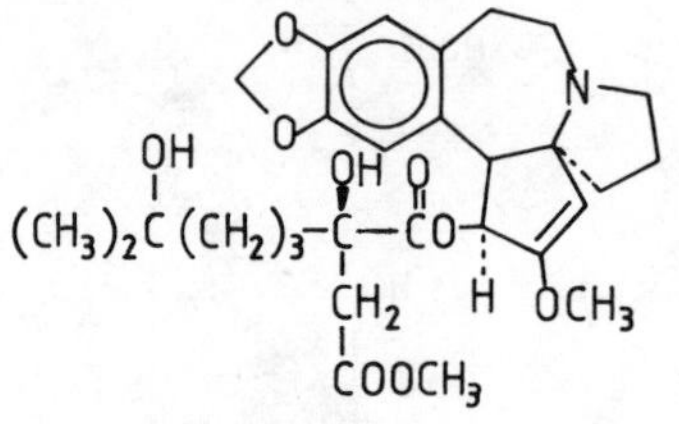

18
Homoharringtonine
3*S*, 4*S*, 5*R*, 2′*R*

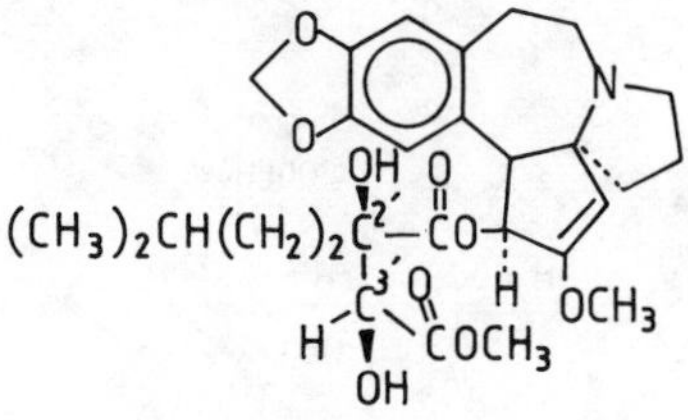

19
Isoharringtonine
3*S*, 4*S*, 5*R*, 2′*R*, 3′*S*

C. Other *Cephalotaxus* Alkaloids Isolated

Most structures were assigned based on spectral data combined with chemical transformations (Tables III and IV).

1. Esters of Cephalotaxine Other than the Harringtonines

The presence of both enantiomers of acetylcephalotaxine ($C_{20}H_{23}NO_5$, **21** and **22**) was reported. (+)-Acetylcephalotaxine, mp 140°C, $[\alpha]_D$ +102° ($CHCl_3$), $[\alpha]_D$ +130° (C_2H_5OH), found in *C. fortunei* (*12*) and *C. hainanensis* (*16*), is identical in every respect with the acetylation product of natural cephalotaxine, mp 140–142°C, $[\alpha]_D$ −97° (2.2, $CHCl_3$) (*4*) except for opposite rotation. (−)-Acetylcephalotaxine (mp 141–143°C), the acetate of natural isomer, was reported to be present in *C. fortunei* by Paudler and McKay (*11*), but specific rotation was not mentioned. The compound was isolated from *C. wilsoniana* in the form of an oil by Furukawa *et al.* (*18*), $[\alpha]_D$ −186° (0.43, C_2H_5OH), as well as by Powell (*19*), who gave no physical constants. Although no significantly substantial evidence for the

TABLE III

^{1}H-NMR VALUES FOR *CEPHALOTAXUS* ALKALOIDS[a]

Proton	Compound					
	Cephalotaxine	Epicephalotaxine	Cephalotaxinone	Demethylcephalotaxine	Demethylcephalotaxinone	Hainanensine
H-1	4.89 s	4.88 s 4.79 s	6.44 s 6.38 s	2.51	2.60 s 2.54 s	
H-3	4.70 d	4.68 d 4.56 d	— —	3.48	— —	
H-4	3.63 d	3.15 d 3.06 d	3.53 s 3.49 s	3.70	— —	
$J_{3,4}$(Hz)	9.5	5.0 5.0	— —	—	— —	
H-14 }	6.65 s	6.67 s 6.64 s	6.72 s 6.67 s	6.92	6.95 s 6.90 s	
H-17 }	6.61 s	6.62 s 6.59 s	6.66 s 6.60 s	6.65	6.68 s 6.63 s	6.53 s
OCH_3	3.70 s	3.78 s 3.64 s	3.82 s 3.78 s	—	— —	
$—OCH_2O—$	5.86 s	5.90 s 5.82 s	5.95 s 5.88 s	5.92	5.97 s 5.91 s	6.00 s
$C—CH_3$						1.60 s

[a] For more compounds, see Table IV.

TABLE IV
^{1}H NMR OF MORE CEPHALOTAXUS ALKALOIDS

	Compound			
Proton	11-Hydroxy-cephalotaxine	Drupacine	Demethylneodrupacine	Cephalotaxinamide
H-1	4.68 s	1.49 d		4.60 s
H-1′	—	2.65 d		—
$J_{1,1'}$(Hz)	—	14.0		—
H-2	—	—	4.21 t	—
H-3	4.48 d	3.99 d	—	4.67 d
H-4	3.48 d	3.45 d	3.05 s	3.52 s
$J_{3,4}$(Hz)	8.0	9.0	—	9.0
H-10	3.21 m	3.05 m	2.64 q, 2.89 br	—
H-11	4.78 t	4.87 q	4.97 bd	—
H-14	6.62 s	6.65 s	6.73 s	6.55 s
H-17	6.88 s	6.65 s	6.71 s	6.58 s
—OCH_2O—	5.91 s	5.82 s	6.00 s	5.85 s
—OCH_3	3.71 s	3.47 s	—	3.70 s

existence of (−)-acetylcephalotaxine in nature was obtained, the presence of (−)-free base cephalotaxine made the occurrence of the (−)-acetate most probable. Encountering different enantiomers in the same family of plants, however, is unusual.

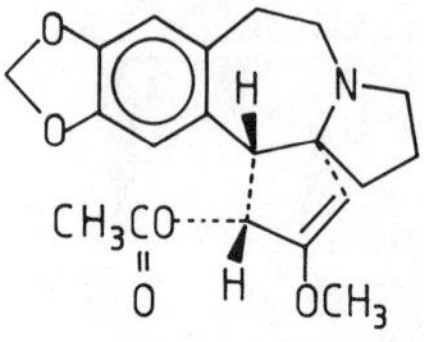

21 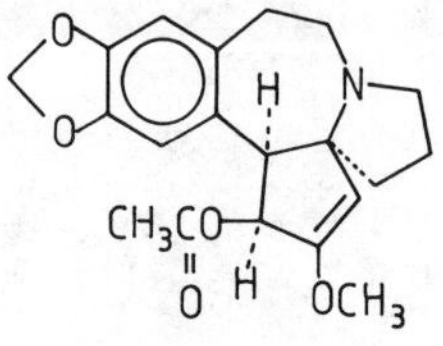 **22**

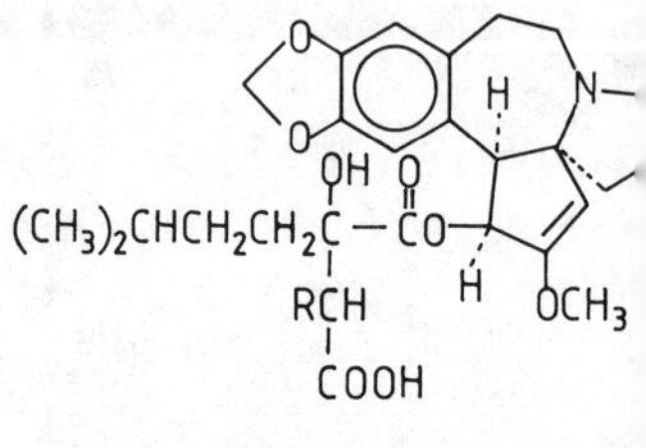

23 R = OH
24 R = H

Two acidic esters, isoharringtonic acid (**23**, $C_{27}H_{35}NO_9$, mp 224–228°C) and deoxyharringtonic acid (**24**, $C_{27}H_{35}NO_8$, mp 219–220°C), were isolated along with the corresponding diesters, iso- and deoxyharringtonine. Their spectral data and acidic property coincided nicely with the proposed structures, which were confirmed by transformation to the known esters with diazomethane (*16*).

2. Derivatives of Cephalotaxine Other than Esters

a. Cephalotaxine Derivatives with No Further Oxygenation. Epicephalotaxine (*11*) **25**, $C_{18}H_{21}NO_4$, mp 136–137°C, $[\alpha]_D$ −150° (0.8, $CHCl_3$), showed the same empirical formula but a different NMR spectrum than cephalotaxine. Melting point depression was obtained upon admixture with cephalotaxine. Compound **25** is identical to the minor

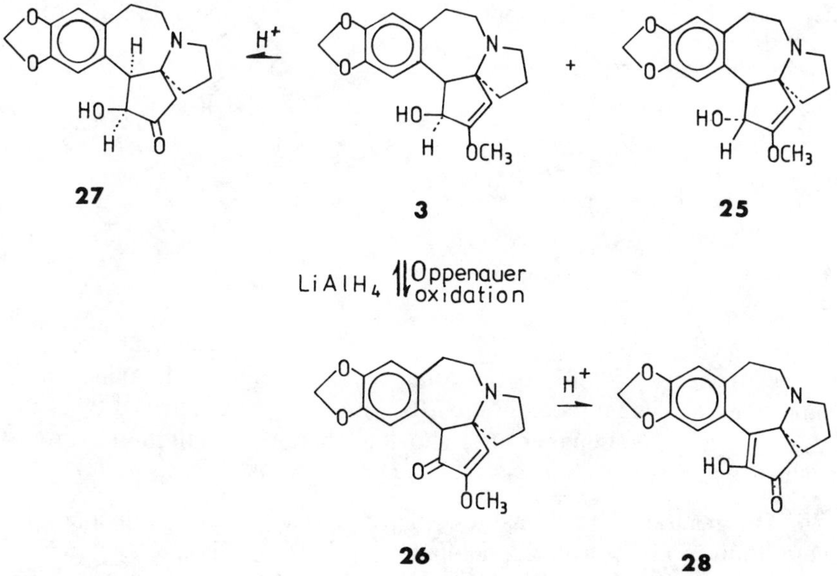

27 **3** **25**

$LiAlH_4$ Oppenauer oxidation

26 **28**

product of $LiAlH_4$ reduction of cephalotaxinone. Therefore the structure with the C-3 hydroxyl group at the α-position, epimeric to cephalotaxine, is most reasonable.

The structural assignment of cephalotaxinone, **26**, $C_{18}H_{19}NO_4$, mp 198–200°C, $[\alpha]_D$ −146° (0.63, $CHCl_3$), was based on its spectral data and confirmed by its identity to the Oppenauer oxidation product of cephalotaxine (*7,11*).

Demethylcephalotaxine (*11*), **27**, $C_{17}H_{19}NO_4$, mp 109–111°C, $[\alpha]_D$ −110° (0.28, $CHCl_3$), showed no signal of vinylic or methoxy protons in its NMR spectrum. A compound with identical properties was obtained chemically by acid hydrolysis of cephalotaxine under more strenuous conditions than usually used for a vinylic ether.

The identity of demethylcephalotaxinone (*8,15*), **28**, $C_{17}H_{17}NO_4$, mp 102–107°C, $[\alpha]_D$ +2.3° (0.52, CH_3OH), with the product obtained by acid hydrolysis of cephalotaxinone confirmed the proposed structure. The synthetic compound, however, showed an $[\alpha]_D$ +40° whereas the natural compound is nearly racemic. The enolic form was assigned based on the absence of C-3 and C-4 protons in its NMR spectrum. Conjugation of the double bond with the benzene ring and the carbonyl group seemed to be preferable to the other isomers.

Hainanensine, **29**, $C_{17}H_{17}NO_4$, mp 240–244°C, $[\alpha]_D$ 0° (CH_3OH), was isolated from *C. hainanensis*. Its UV, IR, and NMR spectra as well as its mass fragmentation pattern differed from those of compounds with a cephalotaxine skeleton. Compound **29** showed only one aromatic proton

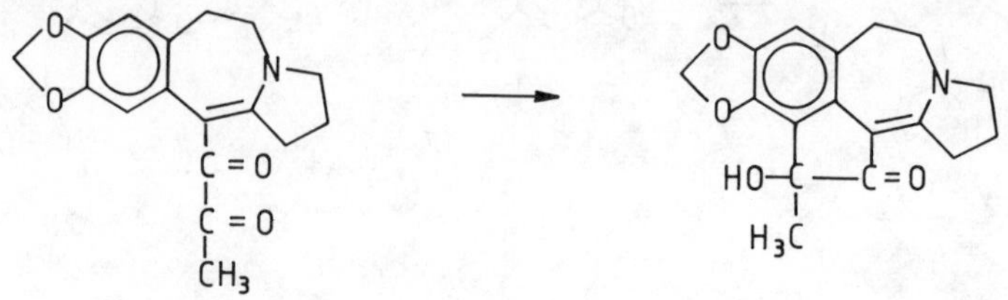

C=O C=O CH₃ HO–C–C=O H₃C

42 **29**

and a *C*-methyl group in its NMR spectrum. It is identical to a by-product of the cyclization of **42** in the total synthesis of cephalotaxine under improper experimental conditions. Its structure was confirmed by X-ray crystallography. Hainanensine showed marginal antitumor activity against experimental tumor systems (*17*).

b. Oxygenated Cephalotaxines. Three 11-oxygenated cephalotaxines were found. 11-Hydroxycephalotaxine (*10*) **30**, $C_{18}H_{21}NO_5$, mp 235–242°C, $[\alpha]_D$ −139° (0.56, $CHCl_3$); mp 260–262°C, $[\alpha]_D$ −186° (0.242, ethanol) (*13*), showed a triplet signal at δ 4.78 in its NMR, indicative of a proton on a carbon bearing both a hydroxyl and a phenyl group and next to methylene, i.e., C-11 H. On treatment with acid, **30** cyclized to give the oxygen-bridged ring compound **31**, drupacine. Other chemical transformation, e.g., **30** → **33**, **30** → **34** etc., also supported the proposed structure.

Drupacine (*10*) **31**, $C_{18}H_{21}NO_5$, mp 70–72°C, $[\alpha]_D$ −137° (0.79, $CHCl_3$), showed no vinylic proton but exhibited a pair of coupled doublets at δ 1.49 and 2.65 in its NMR spectrum. As mentioned above drupacine was formed by cyclization of compound **30**, possibly through condensation between the 11-hydroxyl group with the C-1 double bond.

Demethylneodrupacine (*16*) (**32**, $C_{17}H_{19}NO_5$, mp 197–200°C) was proposed to be a hemiketal bridged between the 11-hydroxyl group and the 3-ketone of 11-hydroxy-isodemethylcephalotaxine. The proximity of the C-11 OH and C-3 was indicated by the molecular model and was also revealed in the IR spectrum of 11-hydroxycephalotaxine—a broad band at 3500 cm^{-1} indicated the strong intramolecular bonding of OH groups.

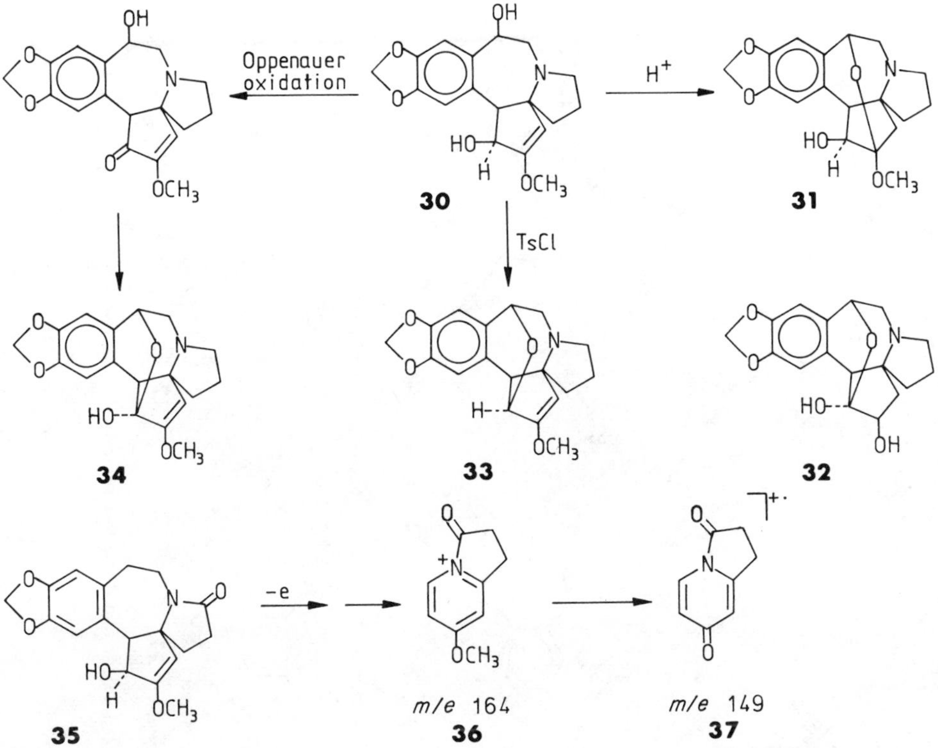

Allied hemiketal **34** was indeed obtained by oxidation of 11-hydroxycephalotaxine (*10*).

Cephalotaxinamide (*16*) (**35**, $C_{18}H_{19}NO_5$, mp 230°C) exhibited a lack of basicity compared to the rest of the compounds. The oxygen was presumed to be involved in a nonbasic amide group. The absence of isolated C-11 protons in the NMR spectrum prompted the proposal of structure **35** with oxygen at C-8. This was supported by the presence of *m/e* 164 and 149 ions in its mass spectrum that correspond to **36** and **37**.

Scheme 1 shows the interrelationship of the cephalotaxine derivatives isolated from *Cephalotaxus* but is by no means a proposal for the biogenetic pathway.

III. Syntheses of *Cephalotaxus* Alkaloids

A. Total Synthesis of Cephalotaxine

Several methods of total synthesis of cephalotaxine have been reported. They can be classified into three categories according to their strategies.

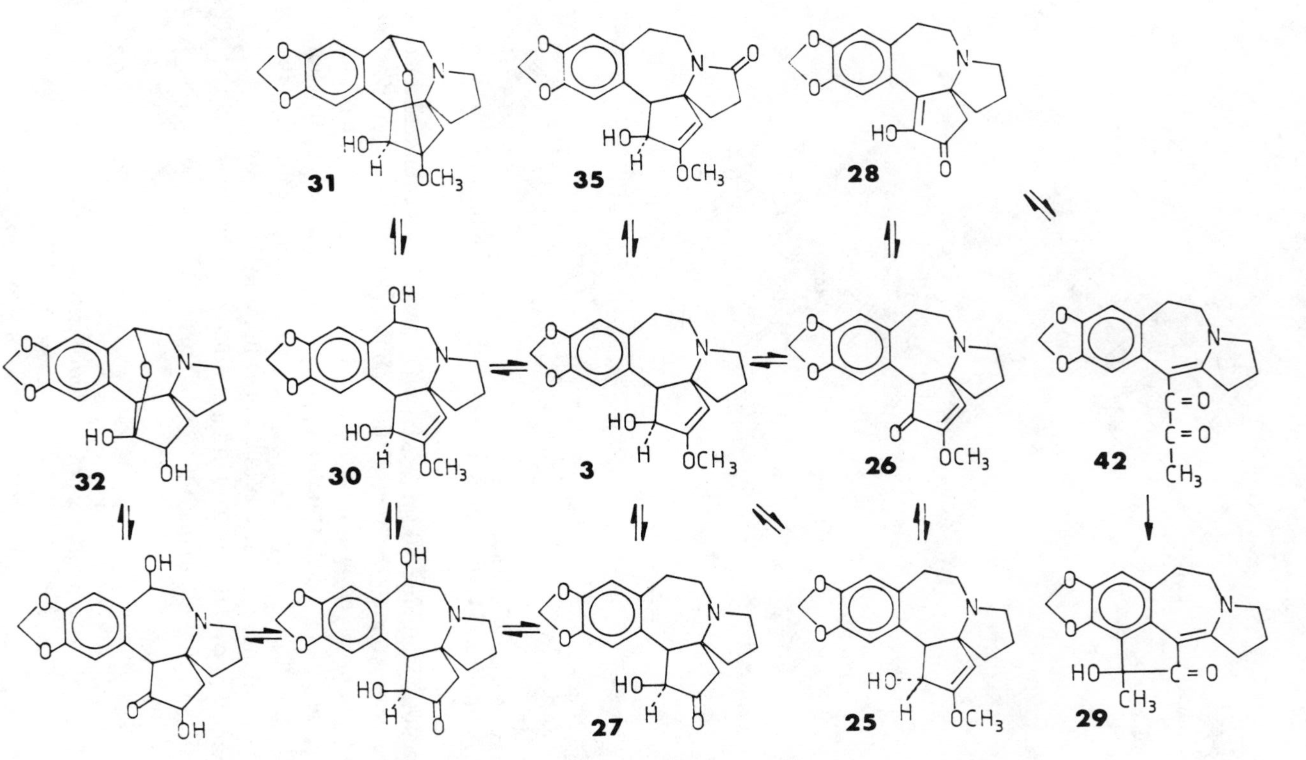

SCHEME 1. Interrelations of cephalotaxine and its natural nonesterified congeners.

1. Strategy A

The specific feature of strategy A is the connection of ring A and C to form the key intermediate **39** first. The synthesis is then completed with construction of ring D.

Auerbach and Weinreb reported the total synthesis of cephalotaxine in 1972 (*32*) and described the details in 1975 (*33*). Dolby *et al.* (*34*), Tse and Snieckus (*35*), and Weinstein and Craig (*36*) synthesized the three-ring intermediate **38** from different starting materials. Routes of various syntheses are summarized in Scheme 2. Compound **39** is quite unstable, and it decomposed at room temperature within a few days. Three of the four methods required formation of a bridge to link rings A and C followed by cyclization through bond b formation, while Dolby's approach formed bond b first. The yields of most reactions were reasonably good excepting Dolby's photocyclization step. Weinstein attempted to use pyrrolidine carboxylic acid as the starting material and go through its derivative **40**. However, an unexpected rearrangement took place during cyclization to give **41**, possibly by the formation of endocyclic amidonium ion **41′**.

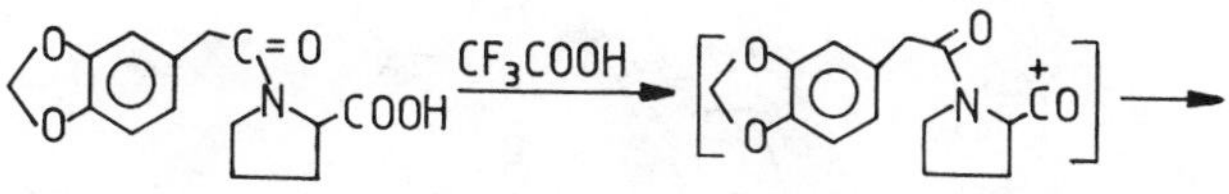

40 $\xrightarrow{CF_3COOH}$ **41′** →

41′ $\xrightarrow{CF_3COO^-}$ **41**

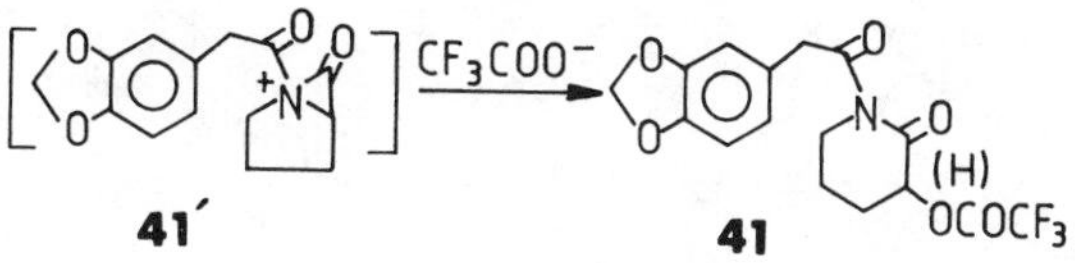

Weinreb had tried various methods to achieve the endocyclic enamine annellation to introduce ring D with the proper substitutions. Finally the methyl dicarbonyl compound **42** was cyclized smoothly in the presence of magnesium methoxide in methanol, resulting in the formation of demethylcephalotaxinone. The rigid chelated compound **43** was postulated as the intermediate. The enol form was suggested to have structure **44**, as indicated by its NMR spectrum, instead of the alternate (2-enol). Structure **44** is favored in spite of the stereo-interaction between the aromatic H and C-3 OH and despite the difficult coplanarity of cyclopentenone with the aromatic ring that precludes full overlap of π orbitals of the α,β-unsaturated carbonyl group in ring D with aromatic ring A. On methylation with diazomethane, the kinetically controlled C-3 methoxyl compound **45** was obtained. When 2,2-dimethoxypropane and *p*-toluene sulfonic acid were

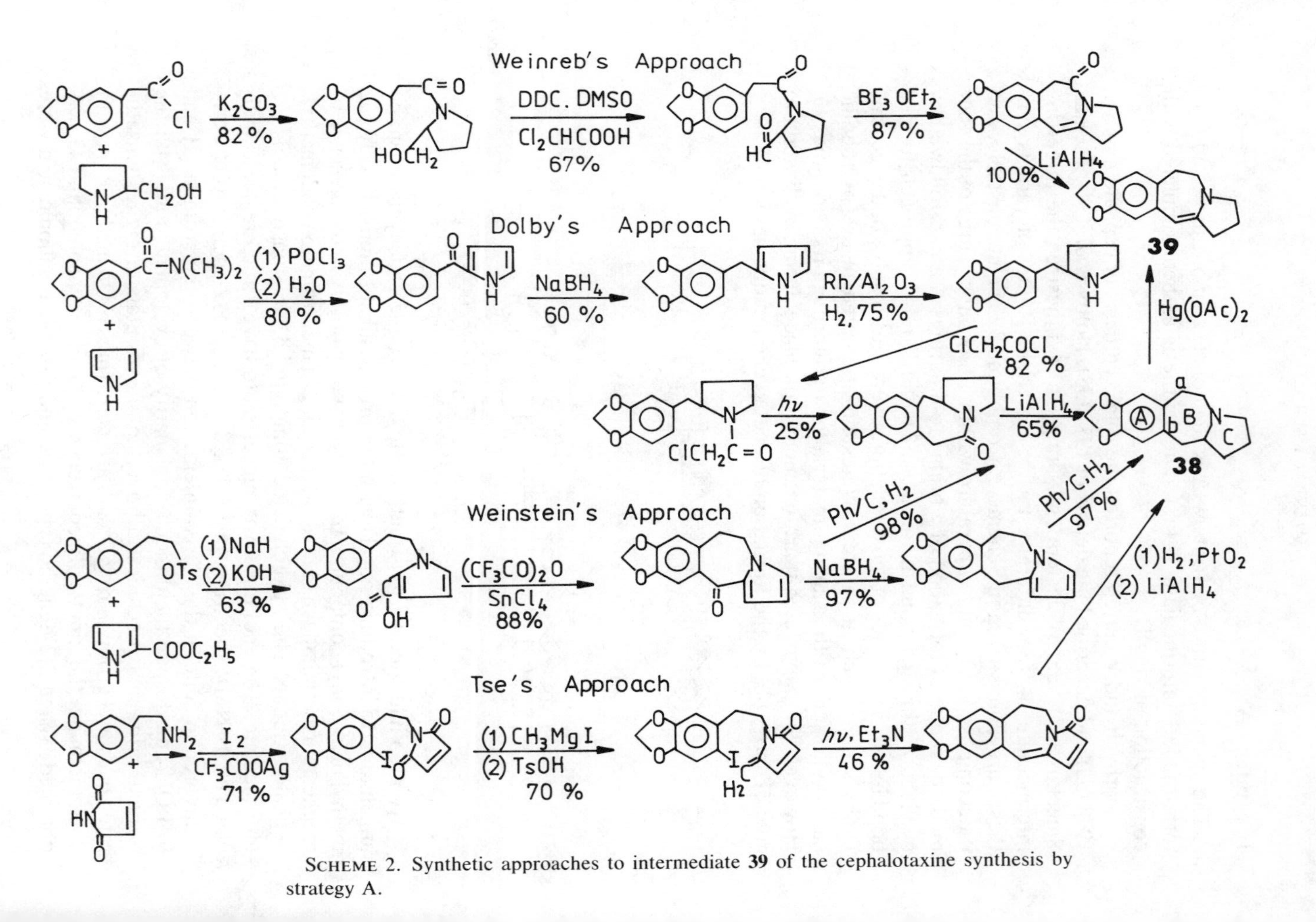

SCHEME 2. Synthetic approaches to intermediate **39** of the cephalotaxine synthesis by strategy A.

39 $\xrightarrow[\text{ClCOOC}_2\text{H}_5\ \ 73\%]{\text{CH}_3\text{COCOOH}}$ **42** ($C=O$, $C=O$, CH_3)

42 $\xrightarrow[\text{CH}_3\text{OH}\ \ 58\%]{\text{Mg(OCH}_3)_2}$ [**43**] (CH_3O-Mg) → **29** (CH_3, OH, O)

43 → **44** (HO, O)

44 $\xrightarrow[\text{TsOH}\ \ 45\%]{(\text{CH}_3\text{O})_2\text{C(CH}_3)_2}$ (O=, OCH_3) $\xrightarrow[85\%]{\text{NaBH}_4}$ **3** (HO, OCH_3)

44 $\xrightarrow[59\%]{\text{CH}_2\text{N}_2}$ **45** (CH_3O, O)

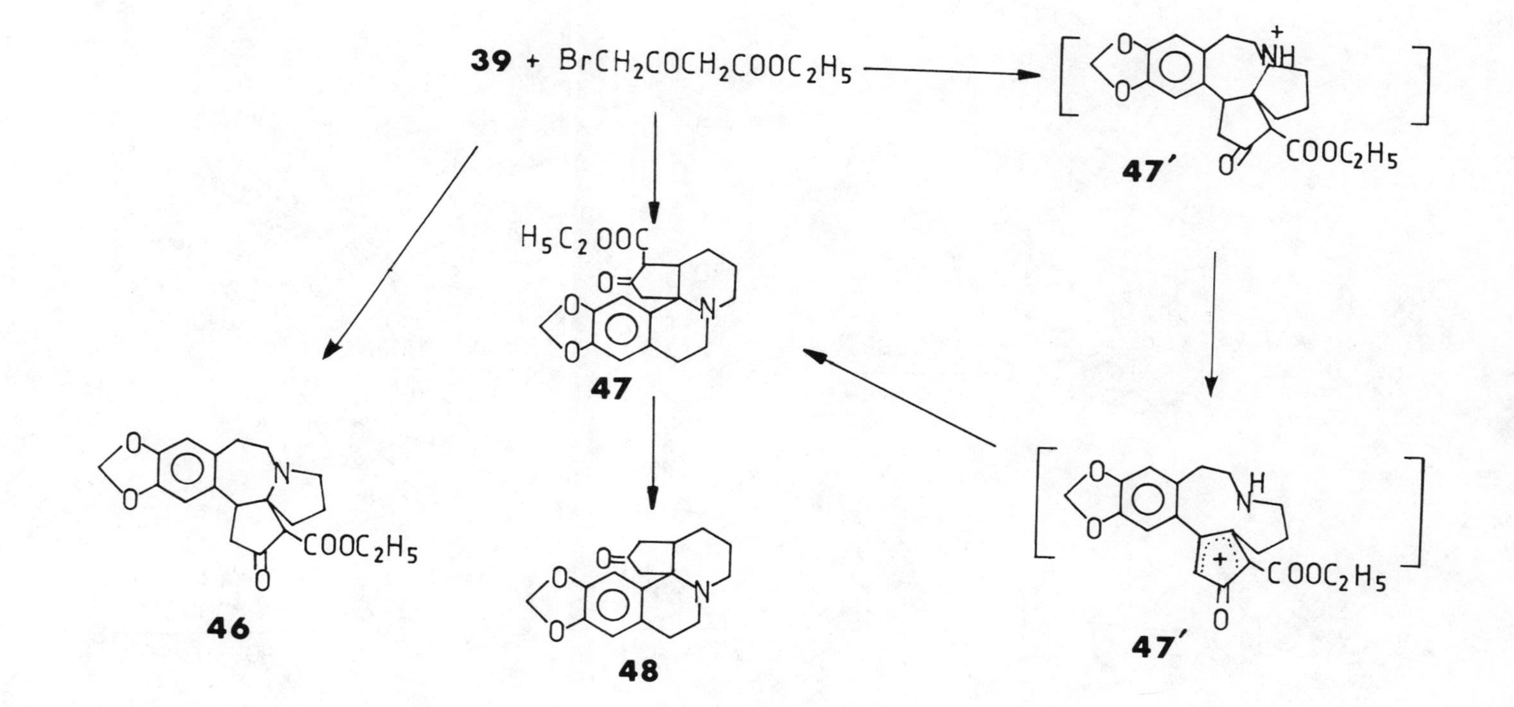
39 + BrCH2COCH2COOC2H5
47′
COOC2H5
H5C2OOC
47
46
COOC2H5
48
47′
COOC2H5

applied, the desired cephalotaxinone was obtained in 45% yield. On further reduction with sodium borohydride it yielded the final product, cephalotaxine, but in racemic form. If, in the Michael addition step from **43** to **44**, improper conditions were used, a by-product with the same structure as the naturally occurring hainanensine (*17*) was obtained.

Dolby tried to convert the enamine **39** to **46** by treating **39** with ethyl bromoacetoacetate. The cyclization, however, unexpectedly proceded to form the rearranged compound **47** and then **48**. This transformation was rationalized by the C-5—N bond breaking with subsequent ring closure between C-4 and nitrogen through the hypothetical compound **47**′.

2. Strategy B

Semmelhack *et al.* (*37–40*) designed a synthetic scheme that connected the pre-formed spiro CD ring moiety **49** to the aromatic ring through bridge a and then completed the ring closure to form cephalotaxinone by bond b formation. Spiro compound **49** was prepared from pyrrolidone through the sequence described below. Ozonolysis of the side chains of **50** required protection of the amino group, which was transformed to the *tert*-butoxycarbonyl amino in compound **51**. The protecting group was cleaved during the esterification stage in the formation of **52**. The unanticipated difficulty of acyloin cyclization was overcome by using a sodium–potassium alloy with an excess of chlorotrimethylsilane to afford the unstable product **53** which was oxidized with bromine to **54**. Compound **49** was then obtained from **54** by methylation with diazomethane.

The aromatic compound **55** reacted smoothly with the spiro CD ring to form **56**. Various approaches for ring closure of the iodo compound **56**, such as a benzyne approach, via a σ-aryl-nickel intermediate, via a copper (I) enolate, and via the SR_N1 reaction, were tried. Ring closure proceded with highest yield (94%) via a photostimulated SR_N1 reaction in liquid ammonia in the presence of solid potassium *tert*-butoxide upon external irradiation with a 450-W medium-pressure mercury arc. The cephalotaxinone thus obtained was reduced with diisobutylaluminum hydride, which produced a clean product of racemic cephalotaxine in 76% yield. The elaborate sequence of synthesis overall yielded 10% cephalotaxine from pyrrolidone.

The construction of the cephalotaxine skeleton introduced two chiral centers, C-4 and C-5. The two carbon atoms in the products obtained through ring closure as described above exhibited exclusively the same correct relative stereochemistry as the natural compound, which apparently acquires the stable form. Reduction of cephalotaxinone with sodium borohydride also afforded the C-3 OH at the proper position since the

$Et_3O\ BF_4$, 73 %

$CH_2{=}CHCH_2MgBr$, 84 %

50

$(CH_3)_3COC(O)N_3$, MgO, 95 %

$R = (CH_3)_3COCO$

51

(1) O_3, (2) Ag_2O, (3) $HC(OCH_3)_3$, H^+, 55–61 %

52

K – Na, $(CH_3)_3SiCl$

$(CH_3)_3SiN$, $(CH_3)_3SiO$, $OSi(CH_3)_3$

53

Br_2, – 78°C

54

CH_2N_2

49

52–49 (44–55 % Yield)

55

OSO_2, NO_2

70 %

56

$h\nu$, 94 %

a, b

$(ISOC_4H_9)_2AlH$, 76 %

Cephalotaxine

hydride ion comes from the less hindered side of the folding molecule as expected. As a result, the whole process yields the three chiral centers with their relative positions correlated exactly to the natural product. Therefore only racemic cephalotaxine was obtained without any other stereoisomers by these reported total syntheses.

3. Other Strategies

Two new approaches to synthesis of cephalotaxine have been attempted. One is photocyclization to construct ring C, and the other is a biogenetical type of approach. Although both of them have not been so well developed as those in strategies A and B, they are still worthy of description.

Based on the successful, exceptionally high yield (84–95%) photocyclization of the model compound **57** to the spiro system **58**, Tiner-Harding *et al.* (*41*) suggested the utility of the electron transfer-initiated iminium salt photospirocyclization method for the synthesis of cephalotaxine. This approach concerning the enclosure of ring C in the later stage of construction of molecular skeleton will be different from the other approaches discussed.

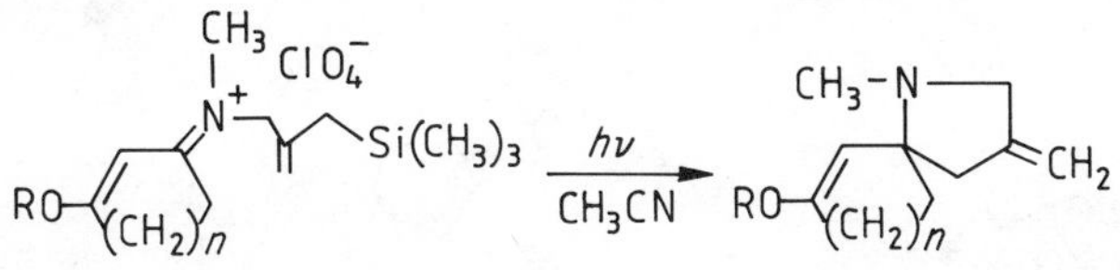

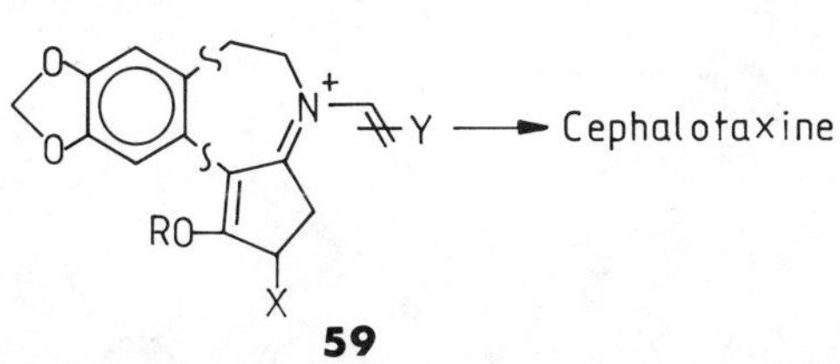

A new biogenetical type of approach to cephalotaxine has been proposed by Kupchan *et al.* (*42,43*). The trifluoroacetyl derivative of the hypothetical biogenetical intermediate, phenethyltetrahydroisoquinoline (**60**), was treated with VOF_3–CF_3COOH in CH_2Cl_2 to give the homoneospirinedienone **61** by nonphenolic oxidative coupling. Two more steps

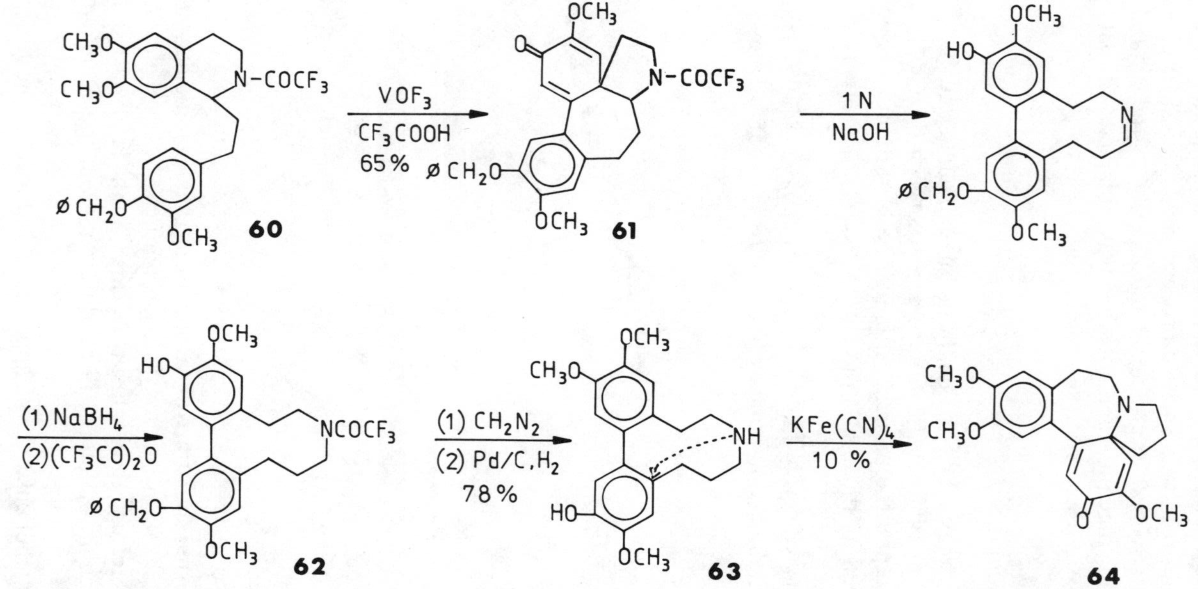

CH_3O CH_3O N–$COCF_3$ øCH_2O OCH_3 **60**

VOF_3 / CF_3COOH / 65%

OCH_3 O N–$COCF_3$ øCH_2O OCH_3 **61**

1 N / NaOH

OCH_3 HO N øCH_2O OCH_3

(1) $NaBH_4$ / (2) $(CF_3CO)_2O$

OCH_3 HO $NCOCF_3$ øCH_2O OCH_3 **62**

61–62, 70 % Yield

(1) CH_2N_2 / (2) Pd/C, H_2 / 78 %

OCH_3 CH_3O NH HO OCH_3 **63**

$KFe(CN)_4$ / 10 %

CH_3O CH_3O N O OCH_3 **64**

gave compound **62**. Methylation of the free phenolic group followed by debenzylation of the benzyl ether gave the desired monophenolic dibenzo [*d,f*]azocine **63** which was oxidized with potassium ferricyanide to yield the cephalotaxine precursor **64**.

B. Partial Syntheses of Deoxyharringtonine, Harringtonine, Homoharringtonine, and Isoharringtonine

In *Cephalotaxus* plants, the four antitumor esters, harringtonine, homoharringtonine, isoharringtonine, and deoxyharringtonine, are much less available than the major alkaloid cephalotaxine. Therefore much effort has been devoted to the partial syntheses of "harringtonines" from cephalotaxine.

1. Methods of Partial Syntheses of Mixtures of Diastereomers of Harringtonines

Attempts (*26*) to acylate cephalotaxine directly with fully pre-formed acid gave no satisfactory results due to severe steric hindrance. The partial syntheses of all four alkaloids have been completed with the last two reports on the synthesis of isoharringtonine in 1983. α-Keto acids were most often used as the starting materials. Nonstereoselective formation of the chiral center C-2′ in the acyl chain rendered a mixture of compounds with C-2′*R* and C-2′*S*, i.e., "harringtonines" and "epiharringtonines," respectively.

a. Deoxyharringtonine. The first ester synthesized was deoxyharringtonine, which has only one free hydroxyl group and which is the simplest in the series. Methods of synthesis were reported by Mikolajczak *et al.* (*44*), Huang and Pan (*24a*), and Li and Dai (*45*) independently. They all used 2-oxo-5-methyl-hexanoic acid **65** to start. The introduction of the residual moiety to ester **66** was carried out either by treating **66** with lithium methyl acetate or by Reformatsky reaction to give the desired deoxyharringtonine along with its epimer. When improper conditions were applied, a considerable amount of cephalotaxine was found in the product. Huang reported a 62% overall yield of the mixture of deoxyharringtonine and its epimer from cephalotaxine.

b. Harringtonine. Harringtonine differs from the deoxyisomer in having one extra penultimate hydroxyl group. Thus it raised the problem of proper protection of the hydroxyl group. The first successful synthesis was reported in 1975 (*46*) and described in detail in 1979 (*47*). 4,4-Di-

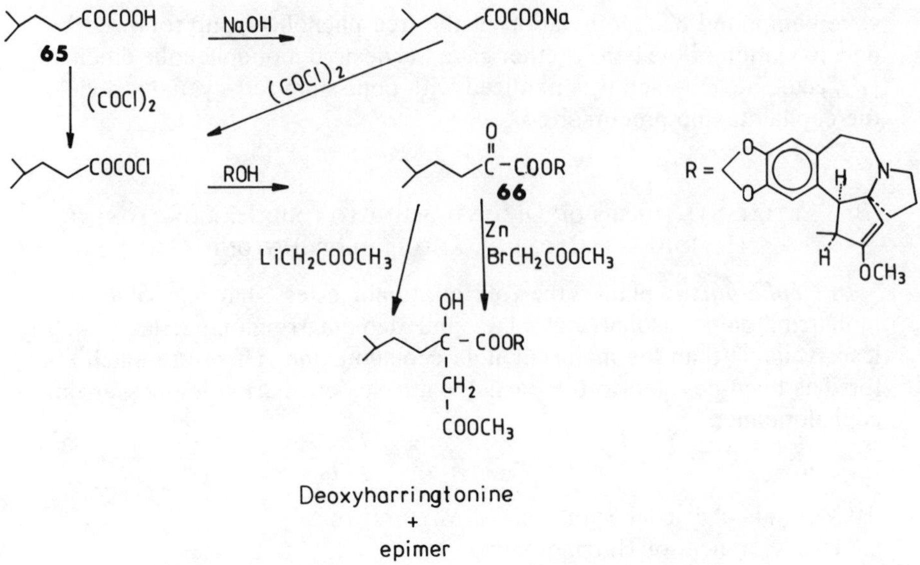

COCOOH **65** NaOH COCOONa
$(COCl)_2$ $(COCl)_2$
COCOCl ROH C–COOR **66**
$LiCH_2COOCH_3$ Zn $BrCH_2COOCH_3$
R =
OH C – COOR CH_2 $COOCH_3$

Deoxyharringtonine
+
epimer

methylbutyrolactone **67** was used as the starting material. The formation of the pivotal cyclic compound **68** protected the hydroxyl group, and at the same time the compound was accessible to reactions characteristic of carbonyl group. Acylation of cephalotaxine with **68** or its chloride gave a more or less complex mixture due to the labile properties of **68**. Dehydration and dimerization might take place under the conditions used. In order to make the reaction more controllable, **68** was dehydrated first to **69** or transferred to ketal **73**. Acylation of cephalotaxine was carried out with acyl chloride **70** or directly with the ketal in the presence of DDC followed by acid treatment to give the desired ester **72**. The introduction of the branch chain of the acid moiety by Reformatsky reaction in this case was made possible by using Ricke's active metallic zinc liberated from zinc chloride with metallic potassium (*48*). The yield increased in the presence of trimethyl borate. An overall yield of 30–40% from cephalotaxine to the mixture of harringtonine and its epimer was obtained.

Mikolajczak and Smith (*49*) reported an almost identical synthetic route in 1978. They prepared the hemiketal **68** from ethyl-4-methyl-Δ^3-valerate. A lower yield from Reformatsky reaction was reported by these authors. The difference in yields of the last step between the two reports probably was due to the difference in activities of the metallic zinc used.

Kelly *et al.* (*50*) succeeded in the synthesis of harringtonine with a different approach by using **75** as the starting material. In the key intermediate **76** the primary carboxyl group and the penultimate hydroxyl group

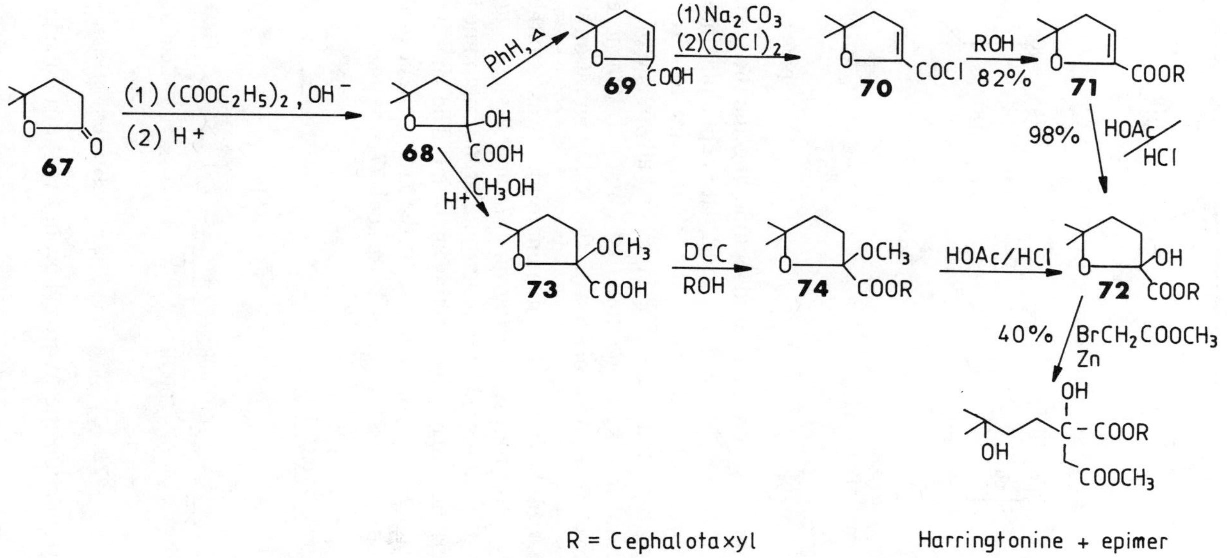

67

(1) $(COOC_2H_5)_2$, OH^-

(2) H^+

68 OH COOH

PhH, Δ

69 COOH

(1) Na_2CO_3

(2) $(COCl)_2$

70 COCl

ROH

82%

71 COOR

98% HOAc/HCl

H^+ CH_3OH

73 OCH_3 COOH

DCC

ROH

74 OCH_3 COOR

HOAc/HCl

72 OH COOR

40% $BrCH_2COOCH_3$ Zn

OH OH C–COOR $COOCH_3$

R = Cephalotaxyl

Harringtonine + epimer

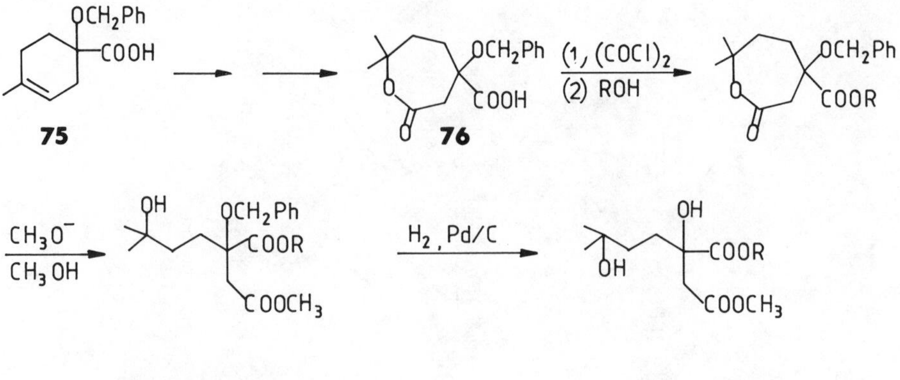

75 → → **76** (1) $(COCl)_2$ (2) ROH →

CH_3O^- / CH_3OH → H_2, Pd/C →

R = Cephalotaxyl

Harringtonine + epimer

were tied up in the seven-membered lactone ring and the tertiary carboxyl group was left free to be available to acylate cephalotaxine. The overall yield of the three-step conversion of cephalotaxine to harringtonine and its epimer was reported to be 35%. A special advantage of this synthesis was that natural harringtonine could be obtained with no epimer if the optically active enantiomer of **76** with the *R* configuration was used.

c. Homoharringtonine. Homoharringtonine was synthesized independently by two groups, Zhao *et al.* (*51*) and Wang *et al.* (*52*). Zhao attempted first to mimic the synthetic route of harringtonine by using 5,5-dimethylvalerolactone as the starting material. The ring homolog of **71** showed resistance to hydration due to the steric effect in the six-membered ring, so the unsaturated α-keto acid **77** was used instead. After

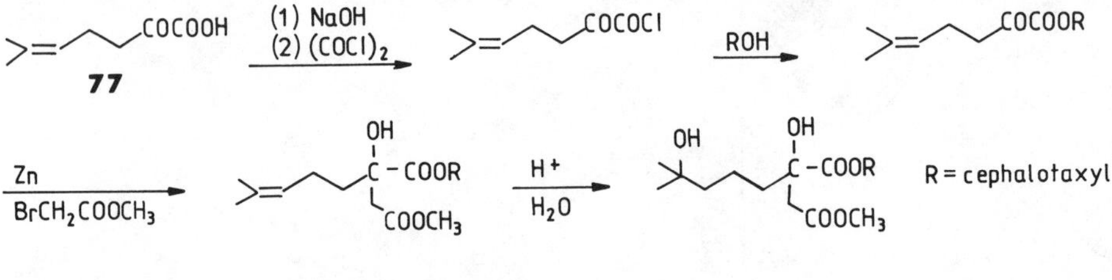

77 (1) NaOH (2) $(COCl)_2$ → ROH →

Zn / $BrCH_2COOCH_3$ → **78** H^+ / H_2O → R = cephalotaxyl

Homoharringtonine + epimer

Reformatsky reaction the hydroxyl group was introduced at the double bond in **78**.

Wang started with the oxoheptanoic acid **79**. The geminate methyl group was formed at the last step by treating **80** with Grignard reagent. Similar approaches were attempted by Hiranuma (*52a*).

79 $\xrightarrow[(3)\ ROH]{(1)\ NaOH,\ (2)\ (COCl)_2}$ C-COOR $\xrightarrow[BrCH_2COOCH_3]{Zn}$ 80

$\xrightarrow[(2)\ CH_3MgI]{(1)\ H^+}$ Homoharringtonine + epimer   R = Cephalotaxyl

d. Isoharringtonine. Isoharringtonine has two vicinal dihydroxyl groups and two chiral centers in the acid moiety. This made the synthesis more complicated than that for its congeners. Pan *et al.* (*53*) and Li *et al.* (*54*) reported the synthesis of isoharringtonine independently in 1982. They used the same intermediate **66** as in the deoxyharringtonine synthesis and introduced the side chain by two different methods. Li used methyl-α-bromo-α-benzyloxyacetate in the Reformatsky reaction in place of the α-bromoacetate used for synthesis of the other three congeners, which was followed by hydrogenolysis of **81** to yield isoharringtonine with its three isomers. Condensation of the lithium compound of the protected methyl α-hydroxyacetate **82** with **66** and subsequent hydrolysis were carried out by Pan to give a mixture of four isomeric compounds.

2. Separation of Isomers

The partial syntheses of "harringtonines" from (−)-cephalotaxine provided mixtures of isomers differing at the chiral center C-2′ in the acid moiety in the three congeners, except isoharringtonine, which yielded four isomers with different absolute configurations at C-2′ and C-3′. Separation of the isomers is a rather tedious task and varied with different compounds. Deoxyharringtonine and epideoxyharringtonine were separated by fractional recrystallization of their picrates (*24a*). Both isomers were obtained as amorphous solids and showed different NMR spectra (see Section IV) and specific rotations. They were also separated by preparative TLC on silica gel through double development with 20% methanol in benzene as reported by Mikolajczak (*44*). Harringtonine and its epimer can be separated by (1) partition column chromatography on silica gel using a pH 5 buffer as the stationary phase and $CHCl_3$ as the mobile phase (*55*); (2) preparative TLC of the hydrochlorides on acidic silica gel using $CH_2Cl_2 : CH_3OH$ (4 : 1) as the developer (*56*); (3) preparative HPLC using a reverse-phase μBondapak C18 column and $H_2O : CH_3OH : (C_2H_5)_2NH$ (55 : 45 : 0.05) as eluant (*50*); and (4) preparative HPLC on silica gel H, eluted with $CHCl_3 : CH_3OH$ (4 : 1) at pH 3 adjusted

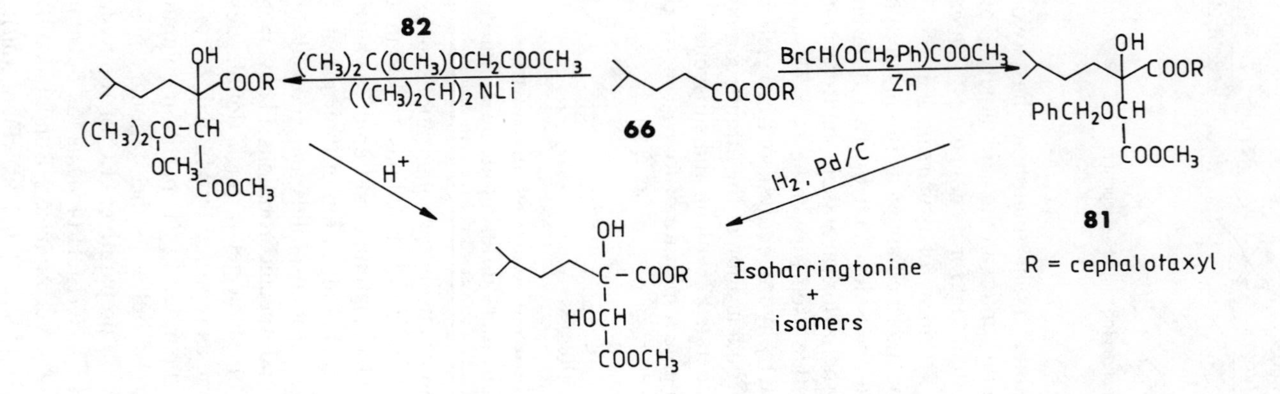
82
(CH3)2C(OCH3)OCH2COOCH3
((CH3)2CH)2NLi
OH
COOR
(CH3)2CO-CH
OCH3
COOCH3
COCOOR
66
BrCH(OCH2Ph)COOCH3
Zn
PhCH2OCH
81
R = cephalotaxyl
H+
H2, Pd/C
C - COOR
HOCH
Isoharringtonine
+
isomers

by HCl (*57*). From method 4 both harringtonine and epiharringtonine were obtained in crystalline forms.

The separation of the four stereoisomers of isoharringtonine with configurations 2′*R*3′*R*, 2′*S*3′*S*, 2′*R*3′*S*, and 2′*S*3′*R* were achieved by Li (*58*). The benzylether **81** was first separated to give two pairs, erythro and threo isomers, by chromatotron on silica gel. These were hydrogenated separately, and the debenzylated products were subjected to further chromatotron to provide the four individual isomers with only the natural product as a crystalline solid. The following is a tabulation showing some physical constants for synthetic harringtonines:

	mp (°C)	$[\alpha]_D$	Ref. and notes
Harringtonine	82–84	−108.6° (1.03, $CHCl_3$)	(*57*)
Epiharringtonine	124–126	−104.8° (1.03, $CHCl_3$)	(*57*)
Deoxyharringtonine	—	−113° (1.06, $CHCl_3$)	(*24a*)
Epideoxyharringtonine	—	−109° (1.18, $CHCl_3$)	(*24a*)
Isoharringtonine			
2′*R*,3′*S*	68–72	−99.0° (0.52, $CHCl_3$)	(*58*)
2′*S*,3′*R*	—	−113.0° (0.46, $CHCl_3$)	Configuration
2′*R*,3′*R*	—	−138° (0.55, $CHCl_3$)	tentatively
2′*S*,3′*S*	—	−103.1° (0.64, $CHCl_3$)	assigned based
			on NMR studies

C. Asymmetric Syntheses of Deoxyharringtonine and Homoharringtonine

Asymmetric partial syntheses of deoxyharringtonine and homoharringtonine were achieved by Cheng *et al.* (*59*) by the method of stereoselective condensation of (+)-(*R*)-α-sulfinylester **83** with a carbonyl compound (*60*). The cephalotaxyl α-ketocarboxylate **66** or **84**, the intermediates in the synthesis of deoxy- and homoharringtonine, respectively, was treated with the chiral sulfinylester to afford **85** or **86**. Subsequent desulfurization, hydrolysis, and methylation with diazomethane led to the desired deoxy compound. For homoharringtonine, desulfurization with aluminum amalgam was preceded by hydration of **86** to **87** in order to avoid saturation of the double bond during desulfurization. The products obtained by these two preparations were exclusively the natural isomers as indicated by the HPLC analysis.

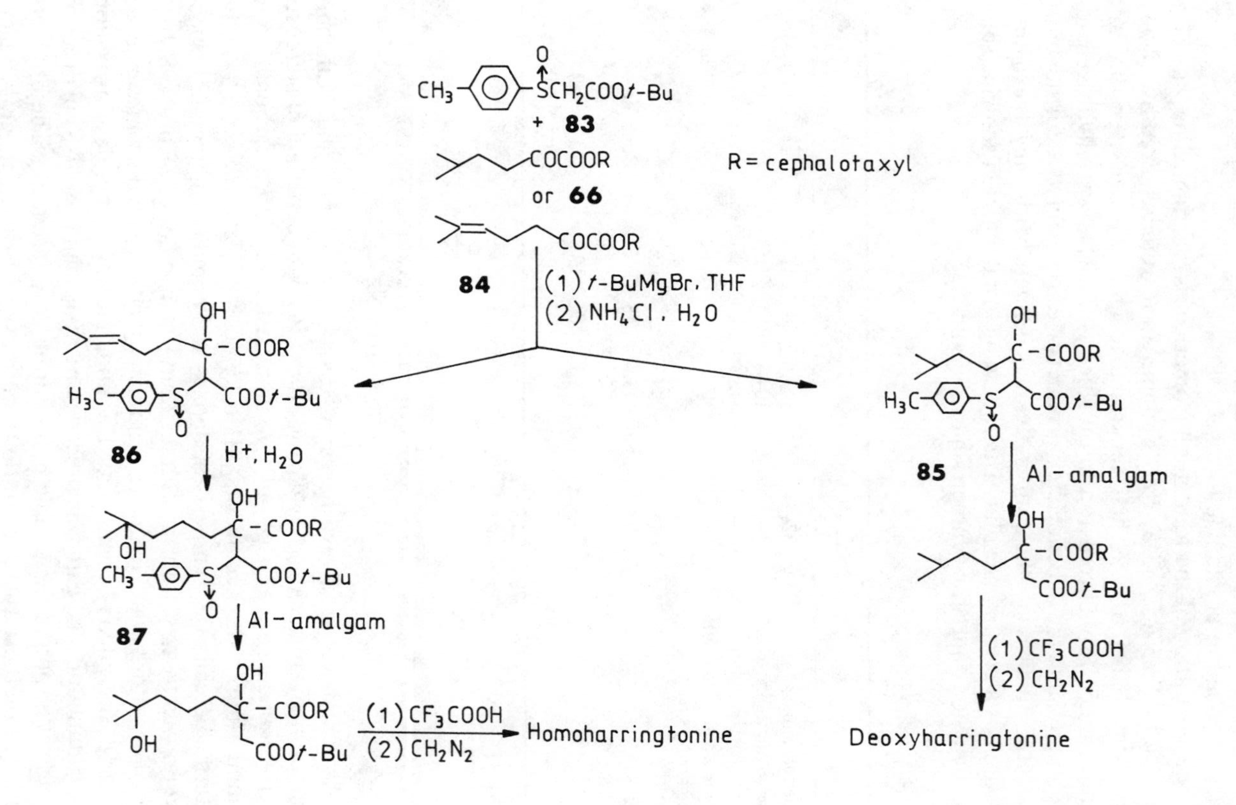

CH3 SCH2COOt-Bu
+ 83
COCOOR
or 66
COCOOR
84
R = cephalotaxyl
(1) t-BuMgBr, THF
(2) NH4Cl, H2O
OH
C-COOR
H3C S COOt-Bu
O
86
H+, H2O
OH
C-COOR
OH
CH3 S COOt-Bu
O
87
Al-amalgam
OH
C-COOR
OH
COOt-Bu
(1) CF3COOH
(2) CH2N2
Homoharringtonine
OH
C-COOR
H3C S COOt-Bu
O
85
Al-amalgam
OH
C-COOR
COOt-Bu
(1) CF3COOH
(2) CH2N2
Deoxyharringtonine

IV. Nuclear Magnetic Resonance Spectroscopy and Mass Spectrometry of *Cephalotaxus* Alkaloids

A. Nuclear Magnetic Resonance

1. Proton Nuclear Magnetic Resonance

The ^{1}H-NMR data of most of the *Cephalotaxus* alkaloids were listed in Section II. In this section, discussion will concentrate on the antitumor esters. The ^{1}H chemical shift assignments of cephalotaxine, cephalotaxyl α-hydroxyl-α-methylbutyrate (II), natural "harringtonines," and the epimers of harringtonine and deoxyharringtonine are listed in Table V for comparison. Acylation of the C-3 OH of cephalotaxine with α-hydroxy-α-methylbutyric acid causes the downfield shift of δ 0.17 for the C-1 H, 1.20 for the C-3 H and 0.15 for the C-4 H. In a comparison of the δ values of "harringtonines" and II with that of cephalotaxine, the signal at δ 6.61–6.64 in the esters may be assigned the C-17 H which eventually would be less influenced by acylation than the C-14 H (δ 6.54). The signal of vinylic OCH_3 around δ 3.65–3.70 may be distinguished from δ 3.58 of the CH_3 protons of the carbomethoxy group in the natural ester series in correlating with the signal of OCH_3 in cephalotaxine and II (δ 3.70 and δ 3.68, respectively). The differences in chemical shifts of the C-3 H between the natural "harringtonine" and II ($\Delta\delta$ 0.07–0.10) possible denotes the effect of the branched methylene carbomethoxy group.

In the natural series of harringtonines, the spectra of the four esters show good correlation among them but they differ from those of their epimers (epiharringtonine and epideoxyharringtonine) in the δ values of C-3 H, C-14 H, CH_3OCO, geminate methyl groups, and C-3′ H_2 as listed in Table V. The differences appear to be consistent. Changing to C-2′ S effects characteristic changes in those values which may serve as a valuable tool in distinguishing the absolute configuration of C-2′. The crowded cavity formed by C-14, C-4, C-3, and C-2 OCH_3 as well as the bulky acid chain provides less freedom of rotation of the side chain and nearby groups. Therefore different configurations of C-2′ may exert and reflect the more obvious differences in their long range shielding effects.

Table VI lists the ^{1}H-NMR values of isoharringtonine and its three isomers. The configurations of the latter were tentatively assigned based on NMR data in conjunction with conformational analysis and molecular models. The effects of the C-2′ configuration on the chemical shifts of the C-3 H, the C-3′H, and the dimethyl group are obvious and are in alignment with the other esters. But the chemical shifts of the C-14 H and $OCOCH_3$ do not correlate so well to its congeners probably due to the

TABLE V

^1H NMR of Cephalotaxine and Some of Its Esters[a]

^1H (δ)	Compound							
	I	II (*61*) 100MC	III 250MC	III′ 90MC	IV 100MC	IV′ 100MC	V 100MC	VI 100MC
C-1 H	4.89(s,1*H*)	5.06(s,1*H*)	5.08(s,1*H*)	5.06(s,1*H*)	5.05(s,1*H*)	5.05(s,1*H*)	5.05(s,1*H*)	5.06(s,1*H*)
C-3 H	4.70(d,1*H*)	5.90, 5.93 (2d,1*H*)	6.00(d,1*H*)	5.93(d,1*H*)	6.00(d,1*H*)	5.90(d,1*H*)	5.99(d,1*H*)	6.02(d,1*H*)
			Δδ(2′*R* − 2′*S*) = 0.07		Δδ(2′*R* − 2′*S*) = 0.10			
C-4 H	3.63(d,1*H*)	3.78(d,1*H*)	3.78(d,1*H*)	3.78(d,1*H*)	3.78(d,1*H*)	3.80(d,1*H*)	3.77(d,1*H*)	3.77(d,1*H*)
$J_{3,4}$ c/s	9.2	9	10	10	10	10	10	10
C-14 H, C-17 H	6.65(s,1*H*) 6.61(s,1*H*)	6.59, 6.58, 6.62 (3s,2*H*)	6.56(s,1*H*) 6.64(s,1*H*)	6.64(s,2*H*)	6.54(s,1*H*) 6.62(s,1*H*)	6.62(s,2*H*)	6.54(s,1*H*) 6.61(s,1*H*)	6.52(s,1*H*) 6.64(s,1*H*)
			Δδ(2′*R* − 2′*S*) = −0.08		Δδ(2′*R* − 2′*S*) = −0.08			
—OCH₂O—	5.86	5.85	5.87(s,2*H*)	5.91, 5.95 J_{AB} = 3c/s	5.88(s,2*H*)	5.89(s,2*H*)	5.85(s,2*H*)	5.82(*m*,2*H*)
CH_3O—, CH_3OCO—	3.70(*S*,6*H*)	3.68(*S*,3*H*)	3.70(s,3*H*) 3.58(s,3*H*)	3.65(s,6*H*)	3.70(*S*,3*H*) 3.59(*S*,3*H*)	3.70(*S*,3*H*) 3.68(*S*,3*H*)	3.67(s,3*H*) 3.58(s,3*H*)	3.67(s,3*H*) 3.60(s,3*H*)
			Δδ(2′*R* − 2′*S*) = −0.07		Δδ(2′*R* − 2′*S*) = −0.09			
$(CH_3)_2$>C	—	—	1.15(s,3*H*) 1.17(s,3*H*)	1.04(s,6*H*)	0.85(d,6*H*) *J* = 6c/s	0.80(d,6*H*) *J* = 6c/s	1.18(s,6*H*)	0.86(d,6*H*) *J* = 4c/s
			Δδ(2′*R* − 2′*S*) = 0.11–0.13		Δδ(2′*R* − 2′*S*) = 0.05			
C-3′ H	—	—	2.10(2*H*, J_{AB} = 16c/s)	2.58(2*H*) J_{AB} = 16c/s	1.90, 2.27 (q,2*H*) J_{AB} = 16c/s	2.64, 2.54 (q,2*H*) J_{AB} = 16c/s	2.10 (2*H*,J_{AB} = 16c/s)	—
			Δδ(2′*R* − 2′*S*) = 0.48		Δδ(2*R*′ − 2′*S*) = ~0.50			

[a] Key to compounds: I, cephalotaxine; II, α-OH-α-methylbutyrylcephalotaxine; III, harringtonine; III′, epiharringtonine; IV, deoxyharring-

TABLE VI
^{1}H NMR of Isoharringtonines (200MC)

H	Compound			
	C-2′*R*, C-3′*R*	C-2′*S*, C-3′*S*	C-2′*R*, C-3′*S* natural isomer	C-2′*S*, C-3′*R*
C-1 H	5.10(s,1*H*)	5.11(s,1*H*)	5.08(s,1*H*)	5.05(s,1*H*)
C-3 H	6.00(d,1*H*)	5.91(d,1*H*)	6.04(d,1*H*)	5.91(d,1*H*)
	$\Delta\delta$(C-2′*R* − C-2′*S*)0.09		$\Delta\delta$(C-2′*R* − C-2′*S*)0.13	
C-4 H	3.82(d,1*H* J = 9c/s)	3.83(d,1*H* J = 8c/s)	3.78(d,1*H* J = 10c/s)	3.80(d,1*H* J = 10c/s)
C-14 H / C-17 H	6.63(s,1*H*) 6.66(s,1*H*)	6.64(s,2*H*)	6.52(s,1*H*) 6.66(s,1*H*)	6.59(s,1*H*) 6.61(s,1*H*)
$—OCH_2O—$	5.86(s,1*H*) 5.88(s,1*H*)	5.92(s,2*H*)	5.82 5.87 2*H* J_{ab} = 2.4	5.89(s,2*H*)
$CH_3O—$	3.70(s,3*H*)	3.75(s,3*H*)	3.70(s,3*H*)	3.70(s,3*H*)
$CH_3OCO—$	3.75(s,3*H*)	3.77(s,3*H*)	3.61(s,3*H*)	3.72(s,3*H*)
$(CH_3)_2>C$	0.85(d,6*H* J = 5c/s)	0.76(d,6*H* J = 6c/s)	0.86(d,6*H* J = 4c/s)	0.75(d,6*H* J = 6c/s)
	$\Delta\delta$(C-2′*R* − C-2′*S*)0.09		$\Delta\delta$(C-2′*R* − C-2′*S*)0.11	
C-3′ H	3.43(d,1*H* J = 9c/s)	4.13(d,1*H* J = 10c/s)	3.36(d,1*H* J = 8c/s)	4.18(d,1*H* J = 8c/s)
	$\Delta\delta$(C-2′*R* − C-2′*S*)−0.70		$\Delta\delta$(C-2′*R* − C-2′*S*)−0.82	
C-3′ OH	3.43	3.30(d,1*H* J = 10c/s)	2.86(d,1*H* J = 8c/s)	2.92(d,1*H* J = 8c/s)
		$\Delta\delta$ threo − erythro 0.57–0.38		

influence of the C-3′ OH. From the values of the C-3′ OH protons of the four isomers, the threo form might be distinguished from the erythro form by $\Delta\delta$ 0.38–0.57 (*58*).

2. [13]C Nuclear Magnetic Resonance

The only ^{13}C-NMR data for the *Cephalotaxus* alkaloids were reported by Weisleder *et al.* (*62*). The chemical shift assignments of cephalotaxine, cephalotaxinone, drupacine, acetylcephalotaxine, harringtonine, and isoharringtonine are listed in Table VII, and the assignment for the acid protons of the esters in Table VIII. The spectra seem to be straightforward.

B. Mass Spectrometry

The characteristics of mass spectra of *Cephalotaxus* alkaloids are valuable in the elucidation of their structures. Here more emphasis will be

TABLE VII
CHEMICAL SHIFT ASSIGNMENTS[a] (^{13}C)

	Ceph.[c]	Ac-Ceph.[c]	H.[c]	Iso-H.[c]	Drup.[c]	Ceph-one[c]
C-1	97.6d	100.5	100.9	100.7d	35.7t	123.9d
C-2	160.5s	157.6	157.6	157.6s	108.6s	158.3s
C-3	73.2d	74.8	74.8	74.8d	73.6d	200.8s
C-4	57.1d	57.3	57.3	57.1d	59.7d	60.8d
C-5	70.5s	70.6	70.7	70.6s	65.2s	65.4s
C-6	43.6t	43.5	43.5	43.5t	43.4t	39.0t
C-7	20.3t	20.4	20.5	20.4t	22.4t	20.1t
C-8	53.8t	53.9	53.9	53.9t	52.1t	52.9t
C-10	48.5t	48.5	48.6	48.5t	56.6t	47.6t
C-11	31.7t	31.7	31.5	31.5t	78.3d	31.4t
C-12	128.2s	128.7	128.5	128.5s	130.2s	128.6s
C-13	134.3s	133.9	133.4	133.6s	131.2s	130.7s
C-14	112.6d	109.5	112.8	112.7d	111.9d	112.5d
C-15	146.8s[b]	146.7[b]	146.8[b]	146.7s[b]	147.8s[b]	147.3s[b]
C-16	146.0s[b]	146.0[b]	146.0[b]	145.8s[b]	146.5s[b]	146.3s[b]
C-17	110.2d	109.5	109.8	110.0d	107.7d	110.3d
C-18	100.8t	100.8	100.9	100.9t	101.3t	101.1t
C-19	58.1q	56.9	56.2	56.1q	53.9q	57.3q

[a] The δ values are in ppm downfield from TMS. From Weisleder *et al.* (*62*).
[b] Signals thus identified within any vertical column may be reversed.
[c] Ceph., cephalotaxine; Ac-Ceph., acetylcephalotaxine; H., harringtonine; Drup., Drupacine; Ceph-one., cephalotaxinone.

TABLE VIII
CHEMICAL SHIFT ASSIGNMENTS[a] FOR THE ACID PORTIONS OF THE ALKALOIDS (^{13}C)

	H.	Iso-H.	Compd. **10**
C-1	173.9[b]	173.1s[b]	175.7s[b]
C-2	74.8	79.2s	75.4s
C-3	33.1	33.0t	37.2t
C-4	37.0	31.5t	32.0t
C-5	70.1	28.1d	28.1d
C-6	29.6[c]	22.7q[c]	22.4q
C-7	28.8[c]	22.2q[c]	22.4q
C-8	42.8	75.1d	43.5t
C-9	170.3[b]	171.7s[b]	171.4s[b]
C-10	51.5	52.3q	52.8q[c]
C-11	—	—	51.8q[c]

[a] The δ values are in ppm downfield from TMS. From Weisleder *et al.* (*62*).
[b,c] Signal identified by the same letter within any vertical column may be reversed.

given to discussion of the major alkaloids, cephalotaxine and its four esters with antitumor activity. In the mass spectrum of cephalotaxine, molecular ion M^+ 315 is the base peak. The ions in the mass spectrum are *m/z* 315(100), 314(32), 300(63), 298(88), 284(81), 282(16), 272(19), 268(7), 254(17), 242(6), 228(15), 214(25), 166(51), 150(33), 137(39), and 115(17). In the higher mass region, M − 15 (*m/z* 300), M − 17 (*m/z* 298), and M − 31 (*m/z* 284) are comparatively abundant. The relatively large peak at *m/z* 298 and no visible peak corresponding to M − 18 indicate the easy cleavage of OH from C-3 and no elimination of H_2O. This is also true in other C-3-substituted cephalotaxines such as the C-3 Cl, the C-3 $OCOCH_3$, etc., which give peaks of M − Cl or M − $OCOCH_3$. In the medium mass region, *m/z* 166, 150, and 137 are characteristic of cephalotaxine skeleton. Possible mechanistic pathways to rationalize the characteristic fragmentation have been postulated by Cong (*63*) and by Yu *et al.* (*64*). Breakage of the labile C-4—C-5 bond that is activated by the C-5—N bond, the phenyl group, the C-1 double bond, and the quaternary C-5 to form the intermediate ion M′ was assumed by both groups. The rest of the pathways postulated by the two reports, however, differed from each other. Cong's mechanistic fragmentation is shown in Figs. 1 and 2.

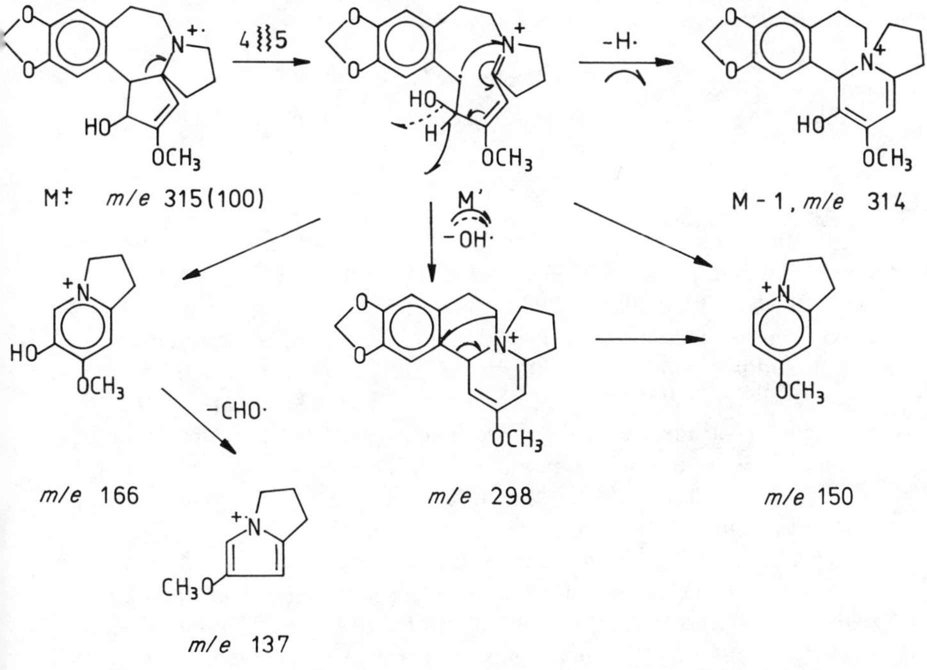

FIG. 1. Hypothetical mechanistic pathway of fragmentation of cephalotaxine, part 1.

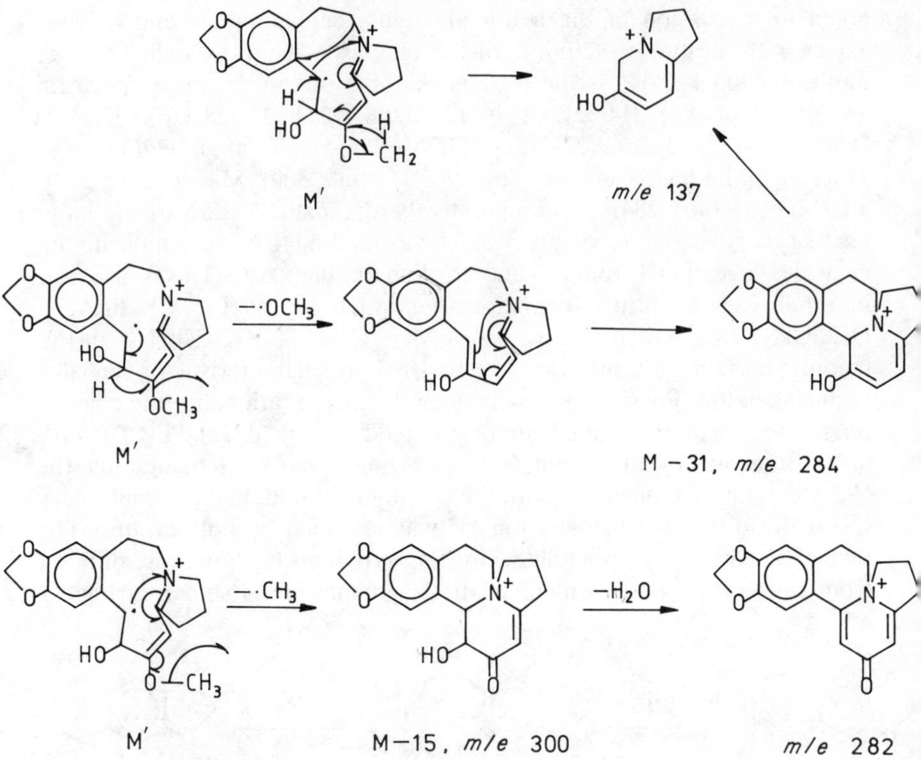

FIG. 2. Hypothetical mechanistic pathway of fragmentation of cephalotaxine, part 2.

The molecular ion of cephalotaxine esters usually are of medium abundance. The ion peaks of M − CH_3 and M − OCH_3 are weak (M − CH_3 < M − OCH_3). The peak *m/z* 298 is present as the base peak in the medium mass region. Peaks at *m/z* 315 < 314 and 284 < 282 in the medium mass region and peaks *m/z* 166, 150, and 137 in the lower mass region are also characteristic for this type of compound. The main peaks of the "harringtonines" are listed in Table IX. The presence of the substantial peak *m/z* 314 is especially useful in identifying the mass of the acyl chain (M − 314). Those compounds containing a $(CH_3)_2COH$ side chain, e.g., harringtonine and its homolog, yield peaks of M − 59 and *m/z* 59. The presence of a vicinal dihydroxyl as in isoharringtonine caused the appearance of an M − $CH(OH)COOCH_3$ peak. The hypothetical fragmentation pathway of homoharringtonine is illustrated in Fig. 3. The mass spectrum of harringtonine by chemical ionization was also reported (*64a*).

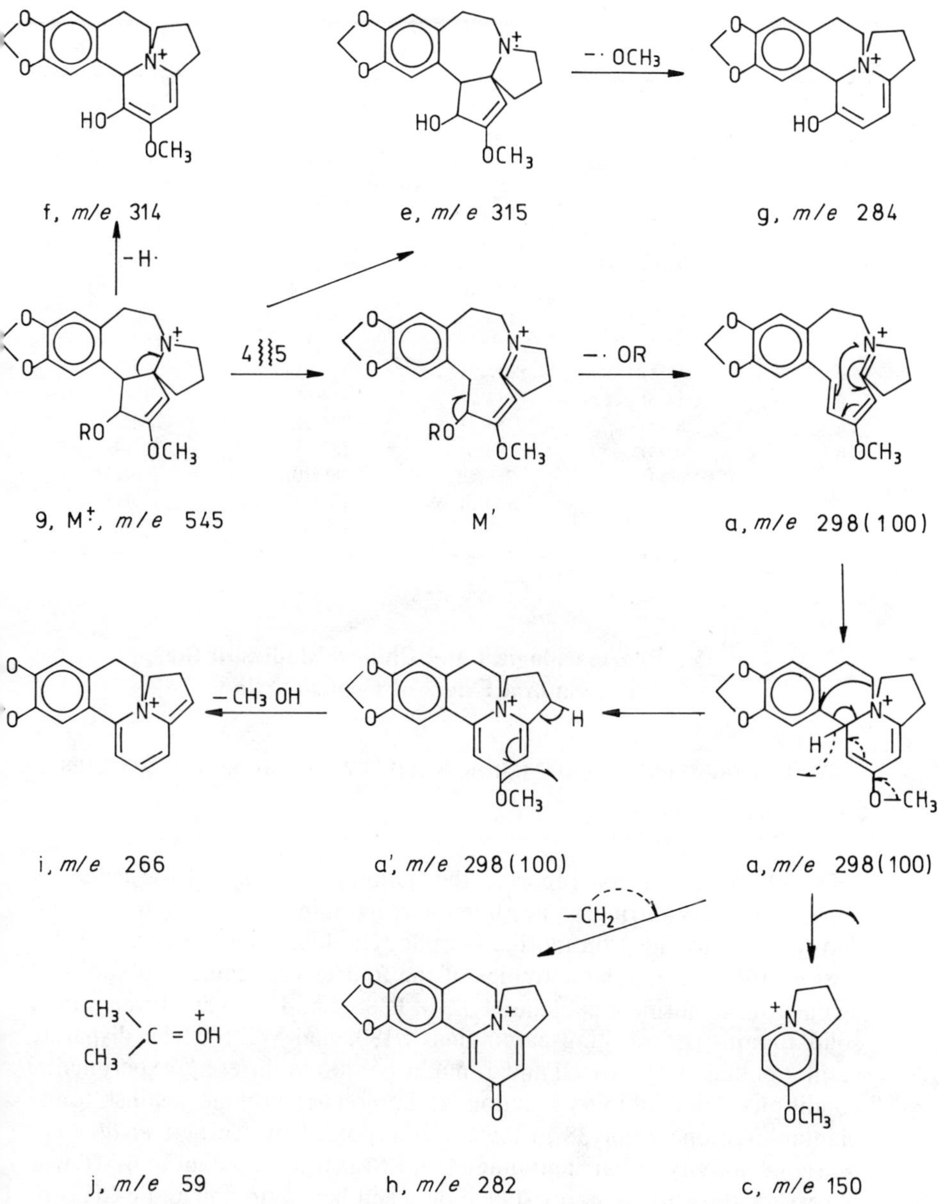

f, *m/e* 314

e, *m/e* 315

g, *m/e* 284

9, $M^{+\cdot}$, *m/e* 545

M′

a, *m/e* 298 (100)

i, *m/e* 266

a′, *m/e* 298 (100)

a, *m/e* 298 (100)

j, *m/e* 59

h, *m/e* 282

c, *m/e* 150

FIG. 3. Hypothetical mechanistic pathway of fragmentation of homoharringtonine.

TABLE IX
MAJOR PEAKS IN MASS SPECTRA OF HARRINGTONINES

	Compound			
Ion	Deoxyharringtonine	Harringtonine	Isoharringtonine	Homoharringtonine
M	515(33)	531(36)	531(13)	545(20)
M − 15	500(3)	516(3)	516(1)	530(1)
M − 31	484(8)	500(7)	500(3)	514(4)
a	298(100)	298(100)	298(100)	298(100)
c	150(8)	150(8)	150(6)	150(2)
d	137(2)	137(3)	137(1)	137(1)
e	315(4)	315(5)	315(4)	315(3)
f	314(6)	314(7)	314(6)	314(5)
g	284(2)	284(4)	284(2)	284(2)
h	282(4)	282(6)	282(4)	282(4)
i	266(11)	266(10)	266(10)	266(11)
i^{2+}	133(4)	134(6)	133(3)	133(1)
j	—	59(6)	—	59(6)

V. Pharmacological and Clinical Studies of the Four Natural Esters of Cephalotaxine

A. Experimental Antitumor Activity and Toxicity Studies

1. Antitumor Activity

Powell *et al.* (*6*) first reported the inhibitory actions of the esters of cephalotaxine—harringtonine, homoharringtonine, isoharringtonine and deoxyharringtonine—on murine lymphocytic leukemia P388 and L1210. Ji *et al.* (*65*) studied the activities of the four esters (equal toxic dosage) against three murine leukemias (L1210, P388, and L615) and three rodent solid tumors (Lewis lung carcinoma, S180, and W256). The disparate results obtained by Powell and Ji might be due to different experimental conditions. The inhibitory action of homoharringtonine against transplantable colon tumor 38 in mice was reported by Corbett *et al.* (*66*). Marginal activity of harringtonine in an L615 strain resistant to 6-MP was observed, thus no cross-resistance between harringtonine and 6-MP was suggested (*67*). These results are summarized in Tables X and XI.

Harringtonine and homoharringtonine showed higher antitumor activity against leukemia L-1210 and P-388 and B16 melanoma in mice than vinca alkaloids, vincristine (VCR), and vinblastine (VLB) and both dem-

TABLE X

INHIBITORY EFFECT OF HARRINGTONINES ON MURINE LEUKEMIA

Compound	P388				L1210				L615	
	Dose (mg/kg)	ILS[a] (%)	Dose (mg/kg)	ILS[b] (%)	Dose (mg/kg)	ILS[a] (%)	Dose (mg/kg)	ILS[b] (%)	Dose (mg/kg)	ILS[a] (%)
Harringtonine	0.70[c]	36.6	0.50	194	0.70[c]	82	0.50	31	1.40[d]	20.6
			1.00	305			1.00	35	1.05[e]	14.7
			2.00	105			2.00	37	0.70[c]	8.8
Homoharringtonine	0.55[c]	36.1	0.25	144	0.55[c]	82	0.25	23	1.10[d]	23.5
			0.50	172			0.50	20	0.83[e]	17.6
			1.00	238			1.00	42	0.55[c]	14.7
Isoharringtonine	2.40[c]	38.2	1.87	50	2.40[c]	>65.3	1.87	9	4.80[d]	46.0
			3.75	72			3.75	24	3.60[e]	35.0
			7.50	172			7.50	26	2.40[c]	18.2
Deoxyharringtonine	2.70[c]	44.4	0.50	45	2.70[c]	85.7		—	5.40[d]	49.0
			1.00	55				—	4.05[e]	38.0
			2.00	80				—	2.70[c]	32.0

[a] Ref. *65*, route of administration ip.

[b] Ref. 6.

[c] 1/6 LD_{50}.

[d] 1/3 LD_{50}.

[e] 1/4 LD_{50}.

TABLE XI
EFFECT OF "HARRINGTONINES" ON SOLID TUMORS

	W256		Lewis lung carcinoma		S180		Colon-38	
Compound	Dose (mg/kg)	C-T/C[a] (%)	Dose (mg/kg)	C-T/C[a] (%)	Dose (mg/kg)	C-T/C[a] (%)	MTD (mg/kg)	ILS (%)
Harringtonine	1.05[c]	10.6	0.70[d]	37.1	1.40[e]	41.0 (4/10 died)		
					1.05[c]	54.0		
					0.70[d]	27.0		
Homoharringtonine	0.83[c]	23.4	0.55[d]	38.6	1.10[e]	17.0 (2/10 died)	2.9 (iv 3,10,17)	86.0
					0.83[c]	41.0	7.2 (iv 14,21,18)	50.0
					0.55[d]	39.0		
Isoharringtonine	3.60[c]	61.7	2.40[d]	47.8	4.80[e]	44.0		
					3.60[c]	27.0		
					2.40[d]	9.0		
Deoxyharringtonine	4.05[c]	53.2	2.70[d]	59.4	5.40[e]	50.0		
					4.05[c]	38.0		
					2.70[d]	20.5		

[a] Ref. *65*, route of administration ip.
[b] Ref *66*; MTD, maximum tolerated dosage.
[c] 1/4 LD_{50}.
[d] 1/6 LD_{50}.
[e] 1/3 LD_{50}.

onstrated moderate activity toward P-388 resistant to VCR. (*67a*). Study on the antitumor activity in *in vitro* culture of fresh tumor cells of patients indicated significant activity of harringtonine in sarcoma, breast cancer, and ovarian and endometrial carcinoma. Homoharringtonine is more active in the continuous exposure studies (*67b,c*). The activity of homoharringtonine against fresh surgical explants of 80 human tumors representing different histologic types was tested in the 6-day *in vivo* subrenal capsule assay. The agent was active against 13 tumors. Furthermore, 32 tumors were unresponsive to all drugs tested and 2 of these unresponsive tumors were inhibited by homoharringtonine (*67d*).

In vitro, harringtonine is active toward cultures of KB, HeLa, and L cells, and KB cells are most sensitive (*68*). The relative degrees of activity of the four closely related esters varied with tumor lines. The esters showed comparable activity against L1210 and P388. In leukemia L615, solid tumor W256, and Lewis lung carcinoma, isoharringtonine and deoxyharringtonine showed higher inhibitory effects than the other two,

TABLE XII
ACUTE TOXICITY IN MICE

Compound	LD_{50} (mg/kg)	
	ip	iv
Harringtonine	4.3 ± 0.50	4.5 ± 0.21
Homoharringtonine	3.3 ± 0.44	2.4 ± 0.25
Isoharringtonine	14.6 ± 0.66	13.3 ± 0.05
Deoxyharringtonine	16.0 ± 2.40	8.8 ± 0.52

although with larger doses. In sarcoma S180, the difference between the esters varied with dose. It appeared that the penultimate hydroxyl group in the acyl chain might be the cause of dissimilarity.

2. Toxicity Studies

The acute toxicities of the "harringtonines" in mice are compared in Table XII. Harringtonine and its homolog are 2–4 times more toxic than the other two allied esters. A study on the subacute toxicities of the four esters in dogs indicated that bone marrow depression was the main toxic effect; gastrointestinal disturbance was also observed. No significant effect on liver and renal function was noted (*65*).

Semisynthetic harringtonine is a mixture of harringtonine and its epimer in a ratio of about 1 : 1.1. The epiharringtonine is almost biologically inactive compared to harringtonine. The LD_{50} of the mixture is about twice of that of the natural isomer. The semisynthetic mixture demonstrated antitumor activity comparable to natural harringtonine at a dosage twice that of the latter. This was verified by the inhibitory activity shown by combined administration of pure harringtonine and its pure epimer in the ratio and doses corresponding to that of the synthetic mixture (*69*).

Zhang *et al.* (*70*) reported that the major target organs involved in toxicity in rats, rabbits, and dogs by treatment with harringtonine and homoharringtonine were limited to the gastrointestinal tract, heart, and hematopoietic organs. Most deaths were attributed to cardiac dysfunction. Hepatic and renal toxicities were seen only in individual cases in mild and moderate degrees at the lethal dose. The cardiac and hematopoietic toxicity appeared to be moderately cumulative. Electronmicroscopic study indicated that homoharringtonine doses of 1.5 mg/kg induced a series of ultrastructural alterations in cardiac muscle cells of mice (*71*).

B. Pharmacological Studies

1. Influence on the Immune Response (*67*)

In plaque forming cell (PFC) assays, harringtonine caused a decrease in the number of PFCs in inbred 615 mice compared to the control. The compound also exerted a marked inhibitory effect on graft-versus-host (GVH) reaction as demonstrated by the reduction of spleen indexes. This indicated that harringtonine, like most cancer chemotherapeutic agents, has suppressive effects on humoral and cellular immunity as well.

2. Physiological Disposition

Absorption, distribution, and excretion patterns of [^{3}H]harringtonine (*72*) and [^{3}H]homoharringtonine (*73*) in normal rats and mice and in tumor-bearing mice were studied. After intravenous injection of the labeled drugs, the blood radioactivity level decreased rapidly with two biological phases of the half-life. Fifteen min after intravenous injection of [^{3}H]harringtonine to normal rats, the highest level of radioactivity was present in kidney, and the levels of the radioactivity in the other organs were in the following order: liver, bone marrow, lung, heart, gastrointestinal tract, spleen, and muscle. The levels in the testis, whole blood, and brain were low. After 2 hr, radioactivity levels in these organs had decreased significantly except that in bone marrow, which retained more than half of the original level.

In the case of homoharringtonine, bone marrow showed the highest level of radioactivity 15 min after administration. The distribution patterns in other organs were similar to that of harringtonine. Within 2 hr of injection, the radioactivity in various tissues decreased rapidly, but the decrease was slower in the gastrointestinal tract. Twenty-four hr later, only very low levels were found in all tissues for both alkaloids. In tumor-bearing mice, a similar distribution pattern to that of the normal rat was observed for both agents. A considerable amount of radioactivity in the form of metabolites and intact drugs was excreted in urine and less in feces in the 24 hr after administration of harringtonine or homoharringtonine. Biliary excretion was also an important route of clearance. The above results indicated that both drugs were distributed widely in the body and eliminated at a fairly rapid rate. Serious accumulation is unlikely. Determination of harringtonine and homoharringtonine in serum by GC–MS, with the limit of detection 10 ng/ml, has been achieved. This will facilitate the study of metabolism of these agents (*73a,b*).

3. Cytokinetic Studies

Different methods and techniques were used to study the cytokinetic effects of harringtonine and homoharringtonine. Baaske and Heinstein (*68*) reported that in synchronized KB cells, G_1 cells are the most sensitive to the inhibitory action of homoharringtonine on the incorporation of [^{3}H]lysine into the acid soluble material of protein. Otherwise only limited cell-cycle specificity was found by measuring protein synthesis.

From radiographic studies on the effect of harringtonine on ascitic cells of leukemia L1210-bearing mice by a [^{3}H]TdR pulse labeling technique, Pan *et al.* (*74*) found sensitivity in cells of all cell-cycle phases to the agent, with S-phase cells being manifested more. Retardation of the progression of G_2 cells into M phase was observed by the colcemide blocking method. Continuous labeling with [^{3}H]TdR demonstrated that 37% of G_0 cells suffered serious morphological damaged from harringtonine. Wang *et al.* (*75*) observed that harringtonine caused a decrease in the number of S cells and blocked the transition of cells from G_1 phase to S phase, as revealed by microscopic photometric study. The above studies came to the conclusion that harringtonine is a cell-cycle nonspecific agent, affecting cells in G_0 and S phases as well as the passage of cells from G_1 to S phase and from G_2 to M phase. The action on cells in S phase is more obvious.

4. Mode of Action

Huang (*76*) reported that harringtonine inhibited the biosyntheses of protein and DNA in HeLa cells while synthesis of RNA was unaffected. The inhibitory actions are partially reversible. The drug induced breakdown of polyribosomes to monosomes with release of completed globin chains. Therefore Huang postulated that the principal effect of the agent on protein synthesis appeared to be directed against the initiation process but not elongation. Tscherne and Pestka (*77*) also suggested the inhibitory effects of harringtonine, homoharringtonine, and isoharringtonine on the initiation of protein synthesis in intact HeLa cells. The disappearance or run-off of the polyribosomes and appearance of monosomes and subunits were observed.

Later Fresno *et al.* (*78*) elucidated the specific steps of protein biosynthesis that are inhibited by harringtonine, homoharringtonine, and isoharringtonine by using a more resolved cell-free eukaryotic system. They reported that the three alkaloids did not inhibit the initiation process of protein biosynthesis but affected its initial cycle of elongation in a double action that blocked both the peptide bond formation and aminoacyl-tRNA binding.

The action of harringtonine on DNA and protein syntheses were further supported by the related ultrastructural changes of L1210 cells under the influence of the agent (*79*). From electron-microscopic study of the effect of homoharringtonine on hepatocyte nucleoli, an inhibitory effect at the ribosomal level of protein synthesis was suggested (*71*).

The study of Wu *et al.* (*80,81a–d*) indicated that the inhibitory effect of harringtonine on DNA synthesis in L1210 cells did not appear to be caused by direct damage to the DNA template. An action on DNA polymerase was suggested.

5. Comparative Studies of the Effect of the Four Esters on Protein and DNA Syntheses*

The concentrations of the four esters required for 50% inhibition of protein synthesis in HeLa cells were different from each other but were of the same order of magnitude (0.04–0.09 μM). This was also true in reticulocytes (0.10–0.19 μM) and lysate (2–7.5 μM) (*76*), which were less sensitive.

Xu *et al.* (*82*) compared the effects of harringtonine and its three allied esters on the biosynthesis of protein in leukemia L615 and P388 cells. The inhibitory effects (T/C, %) on the incorporation of DL-[^{14}C]phenylalanine into protein of ascitic cells in mice bearing P388 were 3.9, 3.4, 11.5, and 4.7 for H, HH, IH, and DH respectively. The actions of the four esters on the incorporation of DL-[^{3}H]leucine are illustrated in Table XIII. At 1/5 of the LD_{50} for the esters, the percentages of inhibition were close to each

TABLE XIII
EFFECT OF HARRINGTONINES ON THE INCORPORATION OF DL-[^{3}H]LEUCINE INTO PROTEIN (P388)

	Control	1/5 LD_{50}		1/20 LD_{50}		1/50 LD_{50}	
Drugs	cpm/mg pr.	cpm/mg pr.	T/C (%)	cpm/mg pr.	T/C (%)	cpm/mg pr.	T/C (%)
Harringtonine	568.7	25.5	4.5	70.9	12.5	115.1	20.2
Homoharringtonine	997.2	30.7	3.2	63.2	6.5	457.8	46.9
Isoharringtonine	3614.1	91.2	2.5	157.5	4.4	341.1	9.4
Deoxyharringtonine	3311.0	116.7	3.5	139.9	4.2	205.8	6.2

* In this section H, HH, IH, and DH have been used in place of harringtonine, homoharringtonine, isoharringtonine and deoxyharringtonine, respectively.

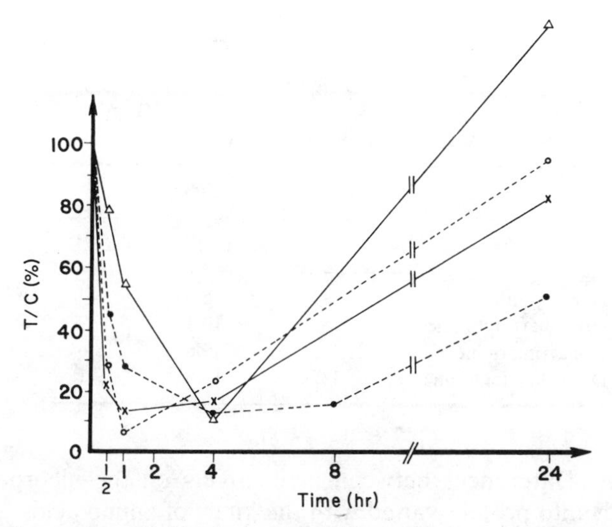

FIG. 4. Effects of harringtonines on incorporation of L-[^{3}H]asparagine into protein of leukemia P388 cells. ●----●, Harringtonine; △——△, homoharringtonine; ○----○, isoharringtonine; ×——×, deoxyharringtonine.

other. When the dosages were reduced to 1/50 of the LD_{50}, IH and DH still retained significant inhibitory activity, but the activity of H and HH decreased markedly. The effects on the incorporation of L-[^{3}H]asparagine into protein of ascitic cells of leukemia P388 (Fig. 4) and into spleen cells of mice bearing L615 (Fig. 5) showed somewhat different patterns for the

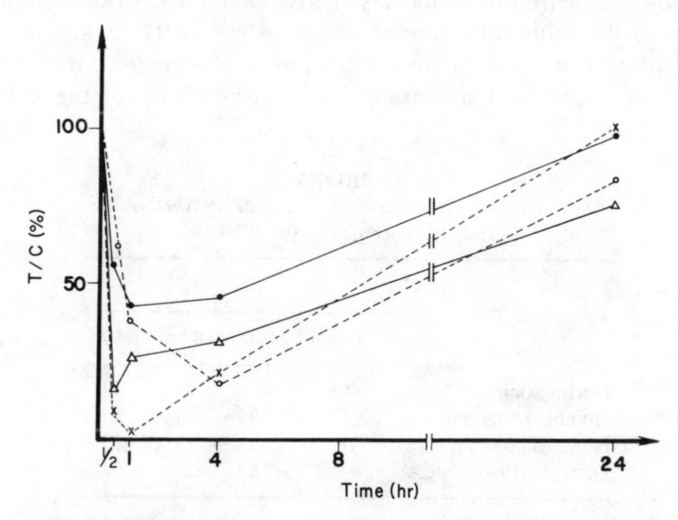

FIG. 5. Effects of harringtonines on incorporation of L-[^{3}H]asparagine into protein of spleen tissue of mice bearing leukemia L615. ●——●, Harringtonine; △——△, homoharringtonine; ○----○, isoharringtonine; ×----×, deoxyharringtonine.

TABLE XIV
EFFECT ON INCORPORATION OF [^{3}H]TdR INTO DNA OF LEUKEMIA P388 CELLS (ip)

	T/C (%)		
	0.02 LD_{50}	0.05 LD_{50}	0.2 LD_{50}
Harringtonine	21.7	6.0	6.7
Homoharringtonine	18.4	10.3	5.9
Isoharringtonine	18.9	9.6	8.0
Deoxyharringtonine	13.4	11.8	9.1

four agents. Differences between their effects on the incorporation of amino acids into protein varied with the kinds of amino acids and tumor cells used.

The degrees of inhibition of the incorporation of [^{3}H]TdR into DNA of leukemia P388 cells *in vivo* by the esters with various equal toxic dosages are compared in Table XIV. No significant difference between the esters was observed 1 hr after ip administration of the agents. When the esters were given by iv at 0.2 LD_{50}, IH and DH differed from the other two in their stronger inhibitory actions (Table XV). All four esters inhibited the [^{3}H]TdR incorporation into DNA in cells of spleen, intestinal mucosa, and bone marrow of both normal and tumor bearing mice, but differences in rates and extents of inhibition were observed. The bone marrow cells of normal mice appeared to be more sensitive than those of the tumor bearing mice to the inhibitory actions of the esters on DNA synthesis. No obvious difference between normal and tumor-bearing mice was observed in spleen and intestinal mucosa cells. Nonspecificity of the inhibitory

TABLE XV
EFFECT ON INCORPORATION OF [^{3}H]TdR INTO DNA OF LEUKEMIA P388 CELLS (iv, 0.2 LD_{50})

	T/C (%)			
	30 min	1 hr	4 hr	24 hr
Harringtonine	63	28	39	74
Homoharringtonine	53	39	63	87
Isoharringtonine	12	9	5	82
Deoxyharringtonine	24	8	16	71

action of the esters on DNA synthesis toward leukemic mice was suggested (*83*).

C. Clinical Trails of Harringtonine and Homoharringtonine

The first report of a clinical trail of harringtonine appeared in 1977 (*84*). The result of 31 cases of different types of leukemia treated with a daily dose of 0.2–0.3 mg/kg of harringtonine for 5–7 days with a 1–2 week interval by slow intravenous drip is listed in Table XVI. Another 165 cases of nonlymphocytic leukemia were evaluated later (*85*), and results correlated well with the first study, to give 20% complete remission and 72.7% total remission. Acute myelocytic leukemia was also shown to be the most sensitive leukemia. Of 102 cases of acute myelocytic leukemia, 21 showed complete remission and 57 cases showed various degrees of partial remission. The results obtained for acute monocytic leukemia and erythroleukemia were also promising; they showed 5 of 33 and 4 of 10 complete remission and 15 of 33 and 3 of 10 partial remission for monocytic leukemia and erythroleukemia, respectively.

Disturbance of the gastrointestinal system is the main side effect. About 44% showed tachycardia during drug administration. Of 102 patients, 6 demonstrated S–T changes and 5 showed arrhythmia. Varying degrees of alopecia were observed in a small number of patients.

In a clinical trail of homoharringtonine on 94 patients with nonlymphocytic leukemia similar results were obtained (22.3% complete remission and 63.8% total remission) (*85*). Acute myelocytic leukemia and erythroleukemia were also most sensitive to the agent. Its toxic effects are similar to that of harringtonine. Harringtonine and homoharringtonine are being

TABLE XVI
Results of Clinical Trials of Harringtonine

Diagnosis	Number of cases	Complete remission	Partial remission	No response
Acute granulocytic leukemia	26	7	14	5
Acute lymphocytic leukemia	3		2	1
Acute monocytic leukemia	1		1	
Erythroleukemia	1		1	
	31	7	18	6

used in hematology clinics in China for treatment of nonlymphocytic leukemias. They also are used as a constituent in the combination therapy HAOP (harringtonine or homoharringtonine, arabinosyl cytosine, vincristine, and prednisone) and HOP (without arabinosyl cytosine) for leukemia with satisfactory results (*86*).

A phase I study of homoharringtonine has been summarized by the National Cancer Institute (NCI) (*87*). Hypotension has been found to be the limiting toxicity, and other side effects, such as myelosuppression, disturbance of gastrointestinal system, fever, chills, and fatigue, were also observed. A dosage regimen of 3.5–4.0 mg/m^2/day has been suggested for phase II trails.

A total of 161 cases in a clinical trail of semisynthetic harringtonine (*88*) in the treatment of nonlymphocytic leukemia were evaluated. Daily doses of 2–20 mg for 5–10 days were applied. Comparable results to that of the natural harringtonine were obtained.

Neither clinical data of harringtonine and homoharringtonine in the treatment of solid tumors nor clinical data of isoharringtonine and deoxyharringtonine are available. The effect of harringtonine on cells in G_0 phase warrants its trail on slow growing tumors.

VI. Unnatural Cephalotaxine Esters and Their Antitumor Activity

The revelation of antitumor activity of the four esters of cephalotaxine by Powell (*6*) prompted study on the structural feature essential to their biological activity and on the preparation of derivatives in an attempt to improve their antitumor activity. As stated above, the free cephalotaxine alcohol and the dimethyl ester of the acid obtained from hydrolysis of deoxyharringtonine are both void of activity. This led to the syntheses of ester derivatives primarily with modifications of the acyl moiety. The derivatives reported (*45,61,89–91*) are listed in Table XVII and are classified roughly according to the acyl structures. Most of the derivatives were prepared simply by treating cephalotaxine with the corresponding acyl chloride. The epiharringtonine and epideoxyharringtonine were obtained from the mixtures of partial synthetic products. The rearranged ester derivative (deri.* **48**) was formed from reaction of the silver salt of methyl 3-carboxy-3-hydroxy-6-methylheptanoate with chloride **88** accompanied by an allylic rearrangement.

* "Deri." preceding the number denotes the compound listed in Table XVII.

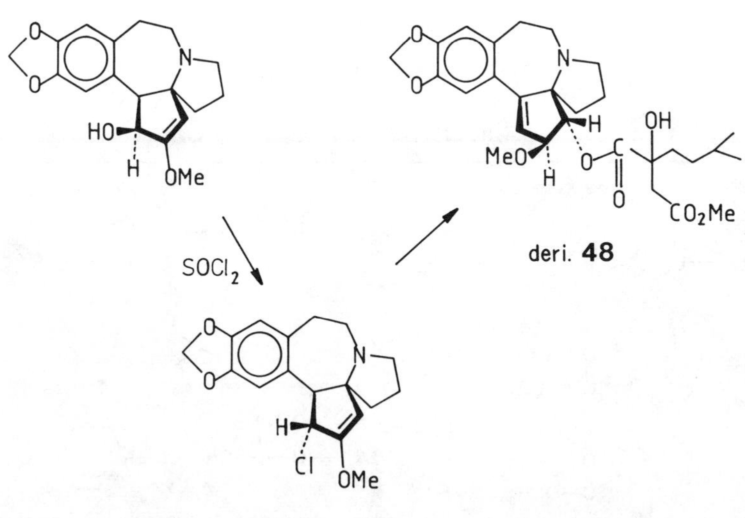

SOCl$_2$

deri. **48**

88

TABLE XVII
DERIVATIVES SYNTHESIZED

1 Epideoxyharringtonine C-2′S **2** Epiharringtonine C-2′S	**6** —CCH(OH)—Ph
R with main feature of natural acyl group	**7** —C(O)—CH(Ph)—OCO(O)CH$_2$CCl$_3$
3 —C(O)—C(OH)(CH$_2$COOCH$_3$)—(CH$_2$)$_4$CH$_3$	**8** —C(O)—C(OH)(CH$_3$)—CH$_2$CH$_3$
4 —C(O)—C(OH)(Ph)—CH$_2$COOCH$_3$	**9** —C(O)—C(OCO(O)CH$_2$CCl$_3$)(CH$_3$)—CH$_2$CH$_3$
5 —C(O)—CH$_2$—C(OH)(CH$_2$COOCH$_3$)—CH$_2$CH$_2$CH(CH$_3$)$_2$	**10** —C(O)—CH$_2$O—(2-ClC$_6$H$_4$)
α-Hydroxyl acyl type and its derivatives	**11** —C(O)—CH$_2$O—(2,4-Cl$_2$C$_6$H$_3$)
	Methyl succinyl type

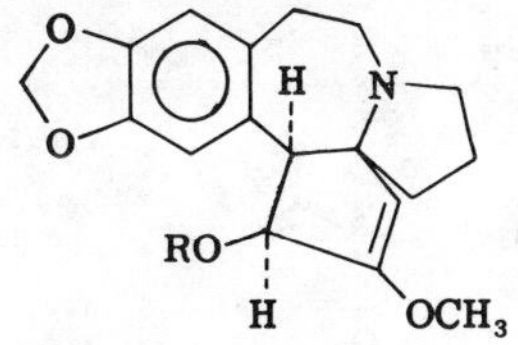

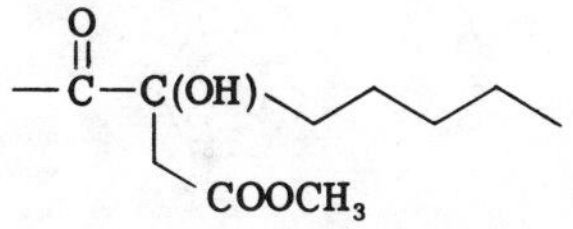

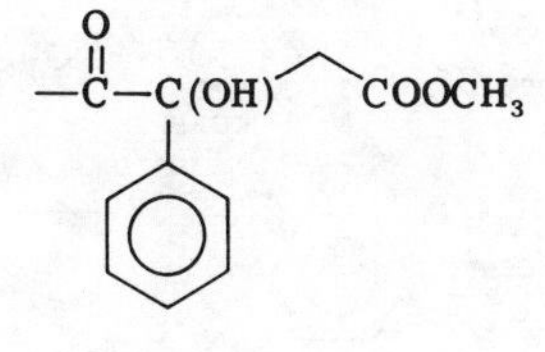

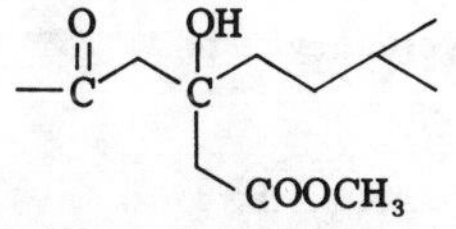

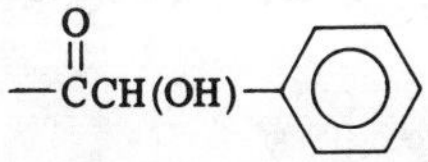

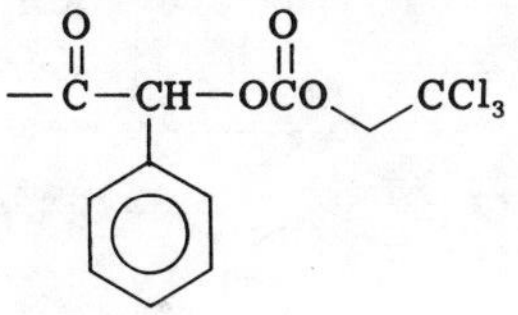

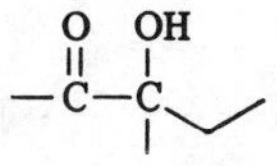

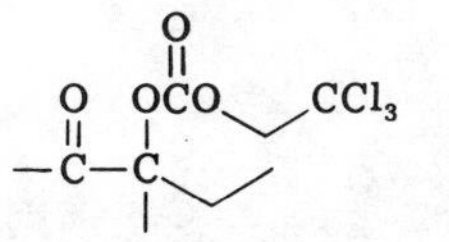

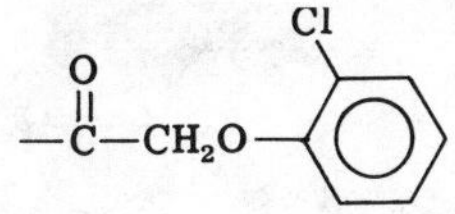

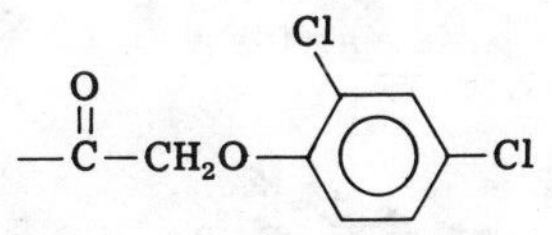

(*Continued*)

TABLE XVII (*Continued*)

12 $-\overset{O}{\overset{\|}{C}}-CH_2CH_2-\overset{O}{\overset{\|}{C}}-OCH_3$

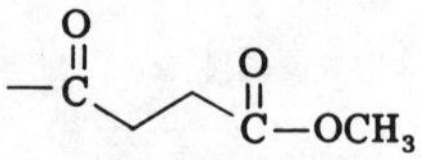

13 $-\overset{O}{\overset{\|}{C}}-$ (branched) $-\overset{O}{\overset{\|}{C}}-OCH_3$

14 $-\overset{O}{\overset{\|}{C}}-$ (branched) $-\overset{O}{\overset{\|}{C}}-OCH_3$

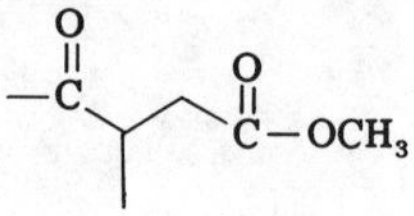

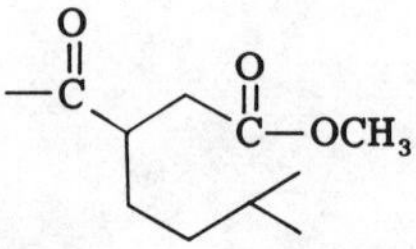

Methyl fumaryl and its analogs

15 $-\overset{O}{\overset{\|}{C}}-CH=CH-\overset{O}{\overset{\|}{C}}-OCH_3$

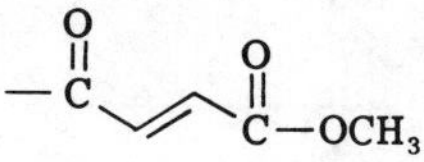

16 $-\overset{O}{\overset{\|}{C}}-$ (branched) $-\overset{O}{\overset{\|}{C}}-OCH_3$

17 $-\overset{O}{\overset{\|}{C}}-$ (branched) $-\overset{O}{\overset{\|}{C}}-OCH_3$

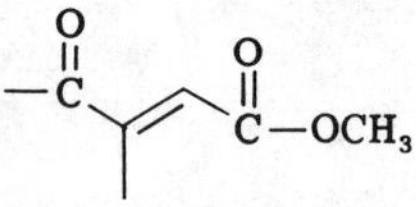

18 $-\overset{O}{\overset{\|}{C}}-C$ $CH_3OOC\overset{\|}{C}H$

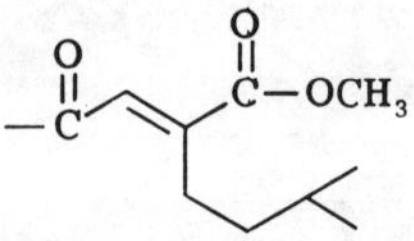

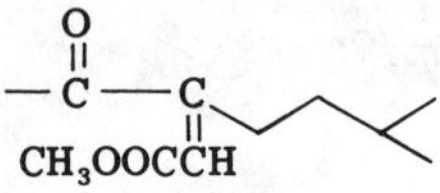

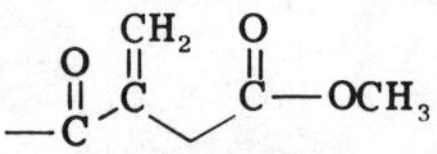

19 $-\overset{O}{\overset{\|}{C}}-\overset{CH_2}{\overset{\|}{C}}-CH_2-\overset{O}{\overset{\|}{C}}-OCH_3$

20 $-\overset{O}{\overset{\|}{C}}-CH=CH-CH=CH-\overset{O}{\overset{\|}{C}}-OCH_3$

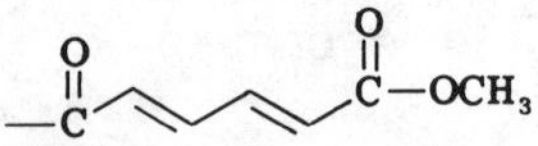

α-Keto acyl type

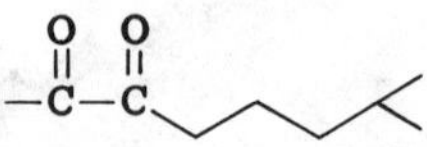

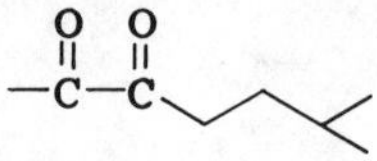

21 $-\overset{O}{\overset{\|}{C}}-\overset{O}{\overset{\|}{C}}-$

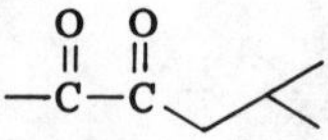

22 $-\overset{O}{\overset{\|}{C}}-\overset{O}{\overset{\|}{C}}-$

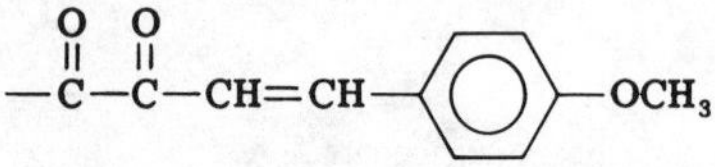

23 $-\overset{O}{\overset{\|}{C}}-\overset{O}{\overset{\|}{C}}-$

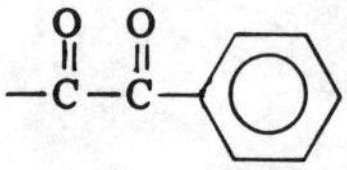

24 $-\overset{O}{\overset{\|}{C}}-\overset{O}{\overset{\|}{C}}-CH=CH-C_6H_4-OCH_3$

25 $-\overset{O}{\overset{\|}{C}}-\overset{O}{\overset{\|}{C}}-C_6H_5$

26 $-\overset{O}{\overset{\|}{C}}-\overset{O}{\overset{\|}{C}}-OC_2H_5$

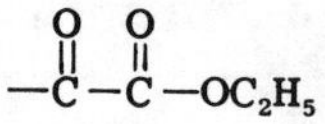

Unsaturated acyl type

27 $-\overset{O}{\overset{\|}{C}}-$

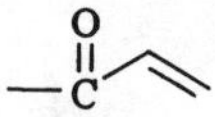

28 $-\overset{O}{\overset{\|}{C}}-$

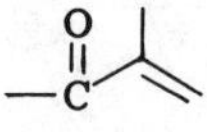

29 $-\overset{O}{\overset{\|}{C}}-$

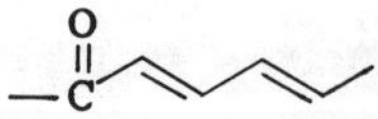

30 $-\overset{O}{\overset{\|}{C}}-CH=CH-C_6H_5$

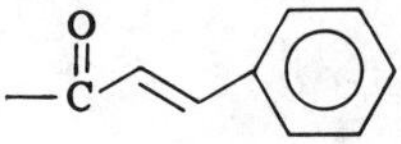

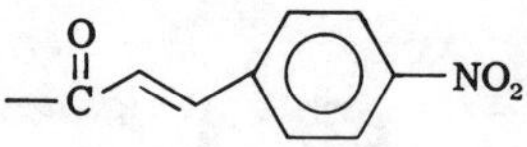

31 $-\overset{O}{\overset{\|}{C}}-CH=CH-C_6H_4-NO_2$

32 $-\overset{O}{\overset{\|}{C}}-C_6H_5$

$_2Cl$

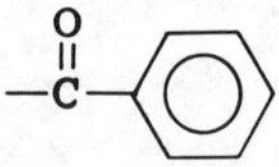

Heterocyclic acyl type

33 $-\overset{O}{\overset{\|}{C}}-$ (pyridyl, N)

$_2H_5$

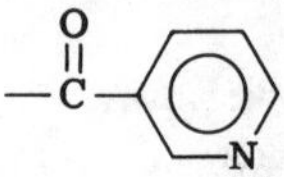

TABLE XVII (*Continued*)

34 —C(=O)— (furan ring)

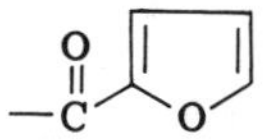

35 —C(=O)— (furan ring), CH_3OOC

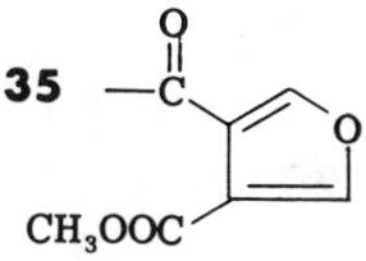

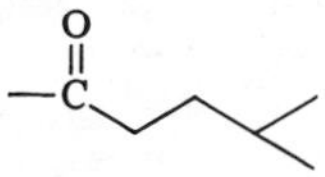

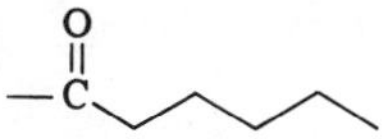

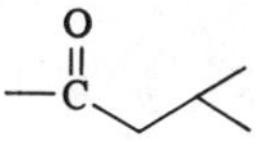

Alkyl acyl type

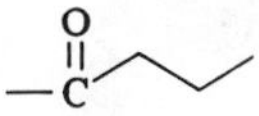

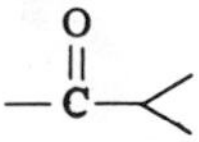

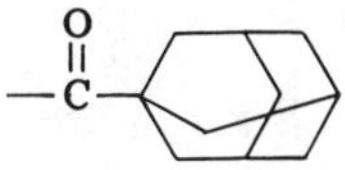

36 —C(=O)—

37 —C(=O)—

38 —C(=O)—

39 —C(=O)—

40 —C(=O)—

41 —C(=O)—

42 —C(=O)—CH_2Cl

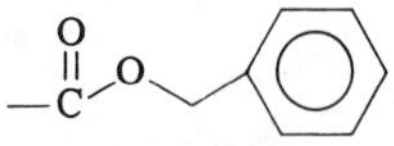

Carbonate

43 —C(=O)—OC_2H_5

44 —C(=O)—O—

45 —C(=O)—O— CCl_3

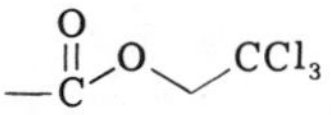

Sulfonate

46 —SO_2CH_3

Different nucleus

47 (+)-Cephalotaxyl-C(=O)—C(=)—C(=O)—OCH_3

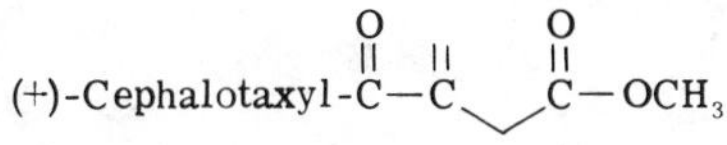

48 O, O, N, H, CH_3O, H, O—C(=O), OH, $COOCH_3$

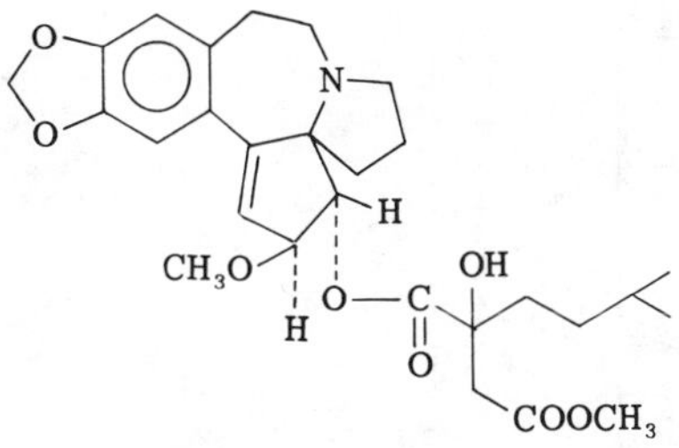

Among nearly 50 compounds tested, only 7 demonstrated some activity against P388 lymphocytic leukemia in mice (Table XVIII). Two (deri. **15** and **17**) of the four methyl cephalotaxyl fumarates were found to be active. The itaconate derivative **19** showed higher inhibitory action than the fumarate series. In the three trichloroethyl carbonates, two compounds (deri. **7** and **45**) exhibited activity while the unesterified precursor of deri. **7**, cephalotaxyl mandelate (**6**) appeared to be inactive. In derivative **9** the

TABLE XVIII

ANTITUMOR ACTIVITY (AGAINST P388 LYMPHOCYTIC LEUKEMIA IN MICE) OF THE ACTIVE DERIVATIVES

Compound		Dosage	T/C	Compound		Dosage	T/C
No.	Acyl chain	(mg/kg × 9)	(%)	No.	Acyl chain	(mg/kg × 9)	(%)
Deoxyharringtonine		4	174	**29**	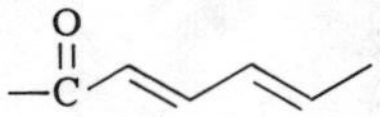 $-C(=O)-CH=CH-CH=CH-CH_3$	20	130
						40	125
7	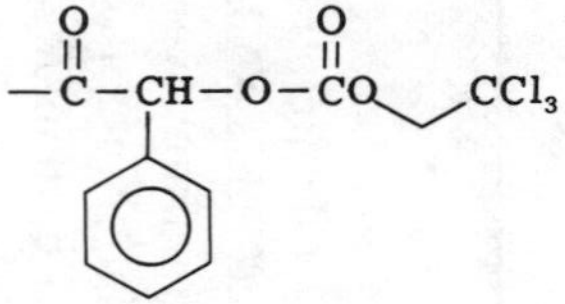 $-C(=O)-CH(C_6H_5)-O-C(=O)-O-CH_2-CCl_3$	80	138	**45**	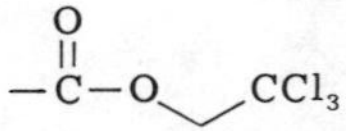 $-C(=O)-O-CH_2-CCl_3$	20	128–195
						40	140–183
						160	162
15	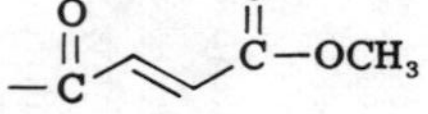 $-C(=O)-CH=CH-C(=O)-OCH_3$	10	136	Harringtonine		0.5[a]	177
		20	125	Epiharringtonine		50[a]	124
						30[a]	111
						10[a]	—
17	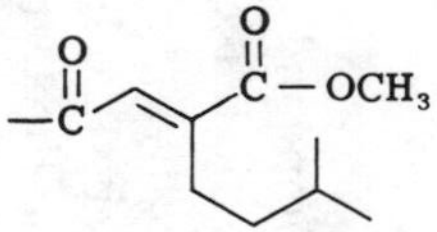 $-C(=O)-CH=C(C(=O)-OCH_3)-CH_2CH_2CH(CH_3)_2$	20	131				
19	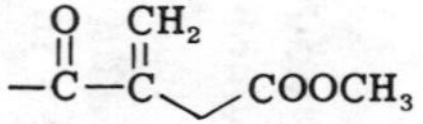 $-C(=O)-C(=CH_2)-CH_2-COOCH_3$	40	135				
		80	173				
		240	169				
26	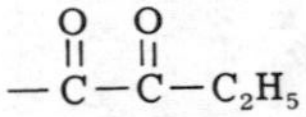 $-C(=O)-C(=O)-C_2H_5$	20	135				
		20	211				

[a] mg/kg × 7

trichloroethyl carbonate group induced no activity to the parent compound, deri. **8**. The other two esters, deri. **26** and **29**, also demonstrated some activity but with inconsistent results. Dosages of these synthetic derivatives much higher than that of the natural esters were required for revealing antitumor activity. Some of them lost their activity when aqueous 0.5% citric acid solution was used in place of the original vehicle. It is too early to draw any structure–activity correlations because of the small number of compounds so far synthesized and tested.

Epiharringtonine, which differs from harringtonine only in the configuration of the chiral center C-2′ exhibited no antitumor activity with various dosages up to 100 times (50 mg/kg × 7) of that of the natural isomer. The C-2′ epimer also showed much lower toxicity (about 1/100 that of the natural one). This preliminary work indicated the specific stereo-requirement of the structure of the acyl group in "harringtonines" for both their antitumor activity and toxicity.

Esterification of the hydroxyl group of cephalotaxine in harringtonine might not be only to perform the function of a carrier or of a protective group for the hydroxyl group in the free alkaloid. A more complicated function of the acyl moiety such as reaction with enzyme or combination with a receptor might be involved.

Other derivatives of cephalotaxine such as cephalotaxyl succinyltyrosine, which might be useful in radioimmunoassay, carbamic esters of cephalotaxine formed by treating it with alkyl isocyanate and phosphorous derivatives, cephalotaxine 2-hydroxy-2-dibutoxyphosphoryl-6-methylhaptanoate and cephalotaxine 2-dibutoxyphosphoryloxyl-6-methylheptanoate, have been prepared but no biological activity was reported (*91a,b*).

Structural modification of "harringtonines" has been limited to the acyl part; the cephalotaxine nucleus of the molecule has barely been touched. The only two compounds with changes in the cephalotaxine portion so far reported are the enantiomer of the active deri. **19**, methyl (+)-cephalotaxyliticonate, and the rearranged ester. Both showed no activity in the test. The road to modification of the nucleus is wide open. In comparison with study done on the structural modifications of other active natural products, the development of derivatives of this particular type of compound has a long way to go.

VII. Biosynthesis of *Cephalotaxus* Alkaloids

A. Cephalotaxine

Barton's phenol coupling biosynthetic pathway for *Erythrina* alkaloids via 1-benzyl-tetrahydroisoquinoline derived from two molecules of tyrosine or phenylalanine was extended by Fitzgerald to homoerythrina alkaloids by employing 1-phenethyltetrahydroisoquinoline as the precursor. On the basis of the presence of homoerythrina alkaloids with *Cephalotaxus* alkaloids in *Cephalotaxus,* Powell adopted Fitzgerald's scheme to the cephalotaxine series by suggesting an alternate path of ring closure (*7*).

The biosynthesis of *Cephalotaxus* alkaloids has been elaborately studied by Parry's group (*92–95*). Incorporation experiments using various labeled precursors followed by proper degradation of the radioactive biosynthetic products suggested that ring A of the cephalotaxine molecule is derived from tyrosine and ring D from phenylalanine. The major processes for degradation of cephalotaxine to locate the positions of the incorporated radioactive carbon atoms used by Parry are summarized in Fig. 6 (see pp. 216, 217). Labeled C-4 and C-11, labeled C-2, and labeled C-3 can be located independently by following route a, d, and c respectively.

Route b affords information about the radioactive carbons at positions 10, 11, 8, 7, and 6. DL-[3-^{14}C]Tyrosine was administered to young *C. harringtonia* plants by the cotton-wick method and worked-up after 8 weeks. Degradation of biosynthetic cephalotaxine by permanganate oxidation gave 4,5-methylenedioxyphthalic acid, which contained all the radioactivity incorporated into the alkaloid. The corresponding anthranilic acid carried 68% of the activity. This indicated that C-3 of tyrosine was incorporated into C-11 of cephalotaxine. When DL-[2-^{14}C]tyrosine was given instead, 37% of the total radioactivity incorporated in cephalotaxine was found in C-10 and no activity was found in C-8 or C-7 of the alkaloid. The low specificity of incorporation was postulated to be due to the catabolism of the amino acid. This was supported by the 68% labeling at C-10 after a shorter incorporation period. The precursor DL-[1-^{14}C]tyrosine provided only 7.6% of the cephalotaxine-incorporated radioactivity at C-8. This probably is not due to direct incorporation but is due to reentry of the radioactive carbon dioxide that was produced by decarboxylation of tyrosine. When L-[ring-^{14}C]tyrosine or L-[*p*-^{14}C]tyrosine was administered, the radioactivity incorporated into the alkaloids was found almost exclusively in ring A. These experiments disclosed that ring A, C-11 and C-10 of cephalotaxine are biogenetically derived from the aromatic ring, C-3 and C-2 of tyrosine, respectively.

The above results raised the question of the origin of rings C and D of cephalotaxine. Participation of phenylalanine in the biosynthesis of cephalotaxine was suggested on the basis of colchicine biosynthesis and also of biosynthesis of schelhammeridine in *Schelhammera*. Administration of DL-[1-^{14}C]phenylalanine to *C. harringtonia* showed that 84% of the total radioactivity incorporated in cephalotaxine was found in C-8 of the alkaloid. When double-labeled phenylalanine, DL-3-(*R*,*S*)-[3-^{3}H,2-^{14}C]-phenylalanine (^{3}H/^{14}C = 4.55) was used as the precursor, the ^{3}H/^{14}C in the incorporated cephalotaxine was reduced to 2.12, which corresponded to a 50% loss of tritium. The 50% loss of ^{3}H could be explained either by the loss of ammonia from the amino acid to form the intermediate, cinnamic acid, or by an exchange process catalyzed by the tautomerase enzyme that is present in the pathway. When DL-3-(*R*,*S*)-[3-^{3}H,*p*-^{14}C] phenylalanine was administered as the precursor, then 100% of ^{14}C in cephalotaxine was distributed at C-3 and 50% of tritium was lost. When DL-3-(*R*,*S*)-[3-^{3}H,*m*-^{14}C]phenylalanine was given to the plant, 98.5% of the radioactivity in cephalotaxine was located at C-2 and the ratio of ^{3}H to ^{14}C remained unchanged. The loss of one meta carbon atom in conjunction with a 50% loss of tritium might account for the unchanged ratio of the two radioactive atoms.

The above results not only indicated that the ring D and ring C of cephalotaxine are biogenetically derived from phenylalanine but also revealed a more intriguing feature of cephalotaxine biosynthesis about the formation of ring D. The possible paths of ring contraction with loss of one carbon from the oxidative phenolic coupling product **89** of phenethyltetrahydroisoquinoline to ring D through benzilic rearrangement are shown in Fig. 7 (see p. 218). The above experimental results that found ^{14}C at C-3 of cephalotaxine from para-labeled phenylalanine and at C-2 accompanied by loss of one carbon atom from the double-labeled [3-^{3}H,*m*-^{14}C]phenylalanine precursor coincides well with path a. The incorporation experiments also eliminated the possibility of the presence of triketone intermediate **90** and corresponding ring contraction through path c since no C-2-labeled cephalotaxine was found with the [*p*-^{14}C]phenylalanine precursor. This meant that the methoxyl enol group is present before the ring contraction and is not formed by methylation afterward. This was further verified by the following incorporation studies.

B. Interrelations between Cephalotaxine and Its Demethyl Counterparts

Incorporation experiments with DL-[3-^{14}C]cephalotaxine showed that cephalotaxine can be transferred to cephalotaxinone, demethylcephalotaxine, and demethylcephalotaxinone. Parallel results were obtained with DL-[3-^{14}C]cephalotaxinone which gave radioactive cephalotaxine and demethylcephalotaxinone. However no labeled cephalotaxine or cephalotaxinone was isolated in cases where [3-^{14}C]-demethylcephalotaxine or the corresponding labeled ketone was used as the precursor. These results demonstrated that biogenetically (1) cephalotaxine and cephalo-

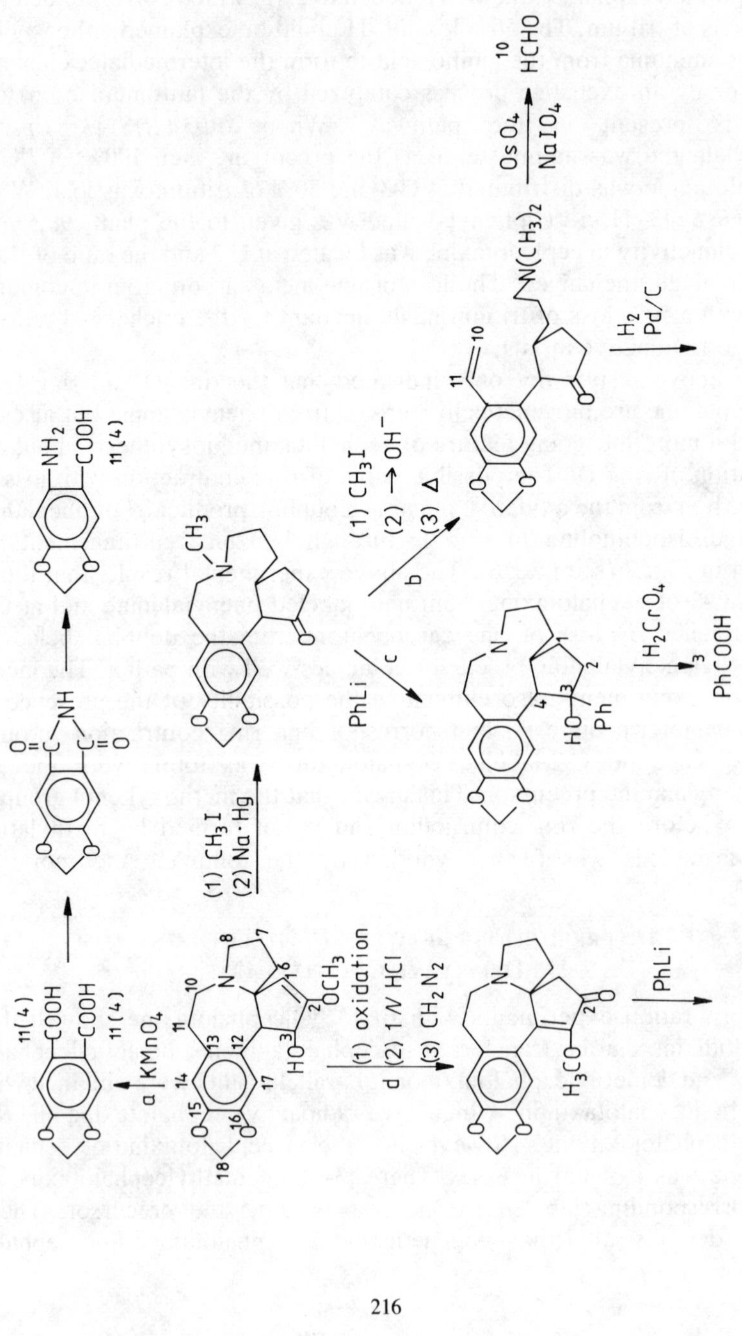

11(4)
COOH
COOH
11(4)

a | $KMnO_4$

14 11 10
15 13 N 8
18 16 12
17 7
HO 3 1 6
2
OCH_3

(1) CH_3I
(2) Na·Hg

CH_3
N

O
C
NH
C
O

NH_2
COOH
11(4)

d
(1) oxidation
(2) 1 *N* HCl
(3) CH_2N_2

N
H_3CO
2
O

PhLi

PhLi
c

(1) CH_3I
(2) → OH^-
(3) Δ
b

N
4
HO 3
Ph 2
1

H_2CrO_4
3
PhCOOH

11
10
$N(CH_3)_2$
O

OsO_4
$NaIO_4$
10
HCHO

H_2
Pd/C

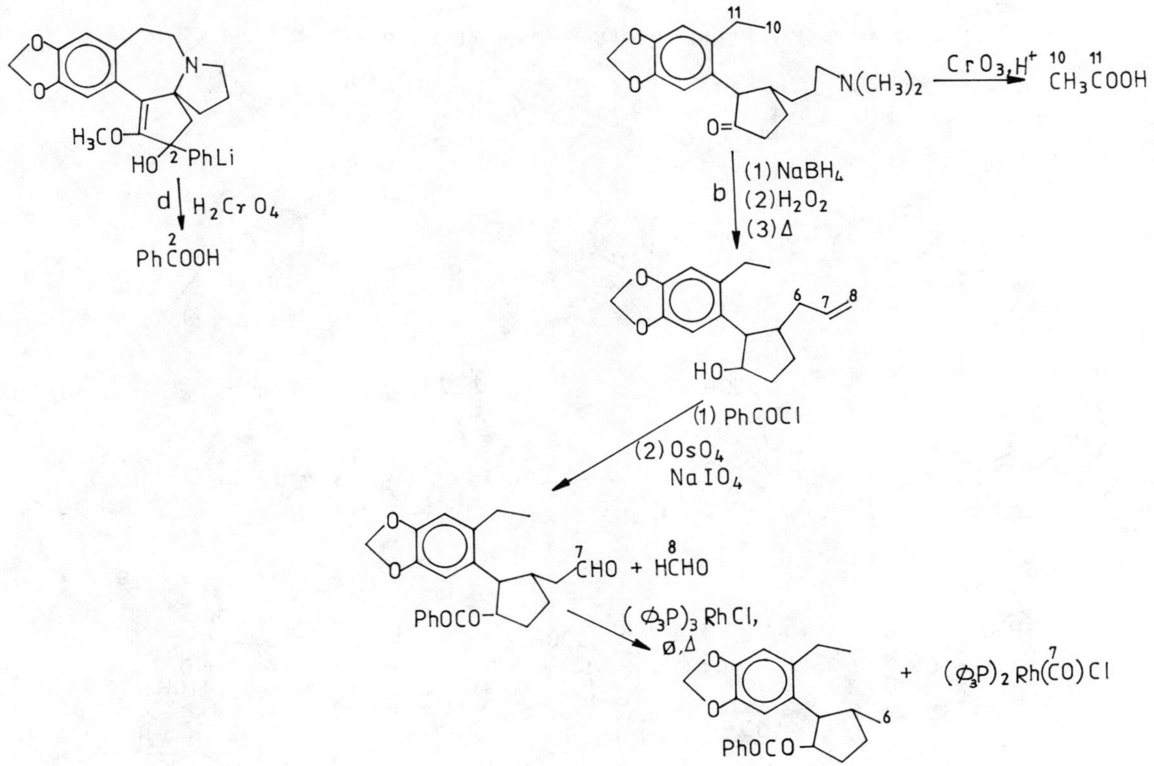

FIG. 6. Degradation processes of cephalotaxine.

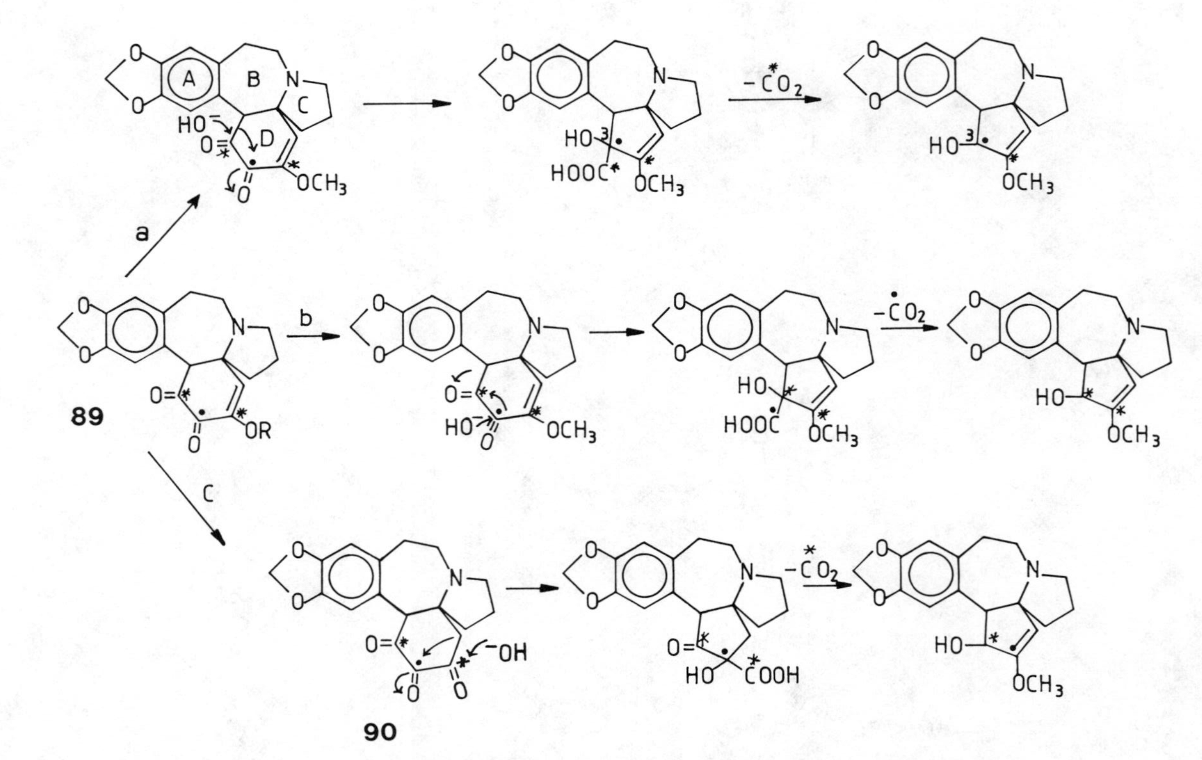

FIG. 7. Possible routes of biogenetical formation of ring D of cephalotaxine.

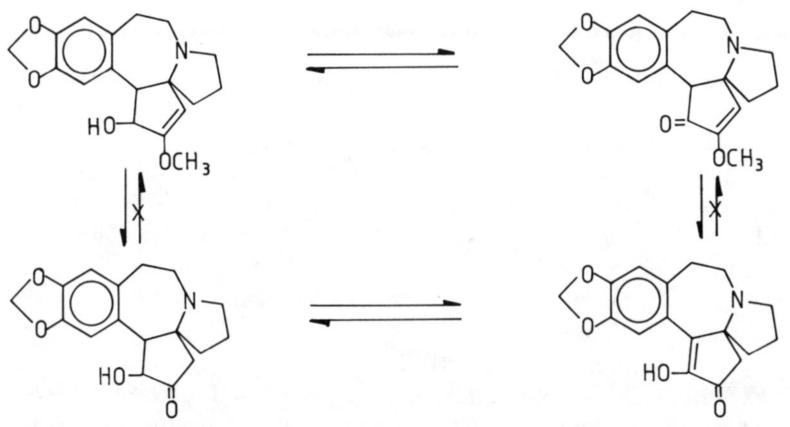

FIG. 8. Interrelations between cephalotaxine and its demethyl counterparts.

taxinone are interconvertible; (2) demethylation of cephalotaxine or cephalotaxinone is a catabolic process and is irreversible; and (3) the methyl group of the C-2 methoxy in cephalotaxine is introduced before benzilic rearrangement (Fig. 8).

The hypothetical overall biosynthetic pathway of cephalotaxine from one molecule of tyrosine and one molecule of phenylalanine is delineated in Fig. 9.

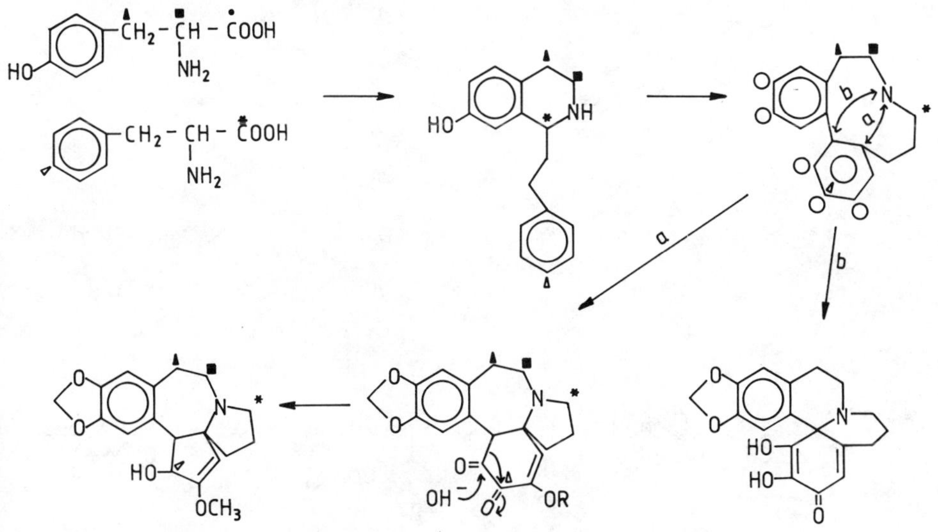

FIG. 9. Hypothetical biosynthetic pathway of cephalotaxine.

C. Ester Side Chains of the Harringtonines

The biosyntheses of the acid chains of the esters were also studied by Parry's group (*95–97*). The hypothetical biosynthetic path is shown in Fig. 10. The position of the radioactive carbons in the labeled biosynthetic products were determined by the degradation processes indicated in Figs. 11–13. The biosynthesis of α-isoamyl malic acid **91** (the dibasic acid related to deoxyharringtonine) was predicted to resemble the conversion of valine into leucine in microorganisms. This biosynthetic route was established by the experimental isolation of labeled **92** from the precursor L-[1-^{14}C]leucine and labeled acid **91** from the precursor [1-^{14}C]-**93** and also from DL-[1-^{14}C]homoleucine (**94**). The range of radioactivity residing in the expected position of the dibasic acids was found to be 84–93% of the incorporated radioactivity in the corresponding biosynthetic products. The considerably high incorporation level of DL-[1-^{14}C]homoleucine obtained was indicative of homoleucine's being an efficient and specific precursor.

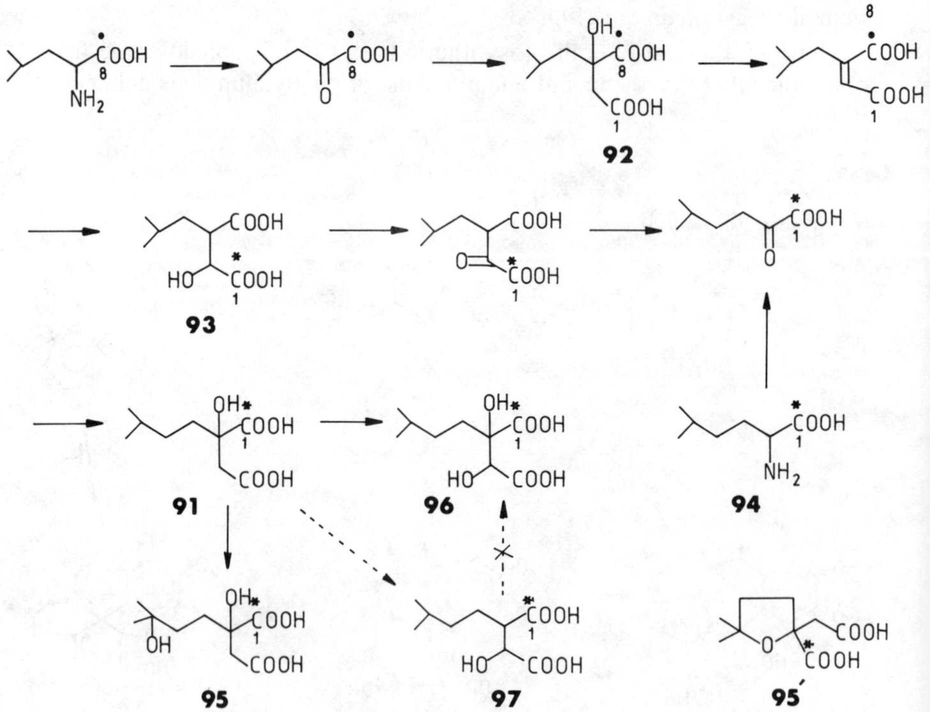

COOH 8 NH2 → COOH 8 O → OH COOH 8 COOH 1 92 → 8 COOH COOH 1

→ COOH HO COOH 1 93 → COOH O COOH 1 → COOH 1 O

→ OH COOH 1 COOH 91 → OH COOH 1 HO COOH 96 COOH 1 NH2 94

OH COOH 1 OH COOH 95 COOH 1 HO COOH 97 O COOH COOH 95′

Fig. 10. Hypothetical biosynthetic pathway of the acyl portion of the harringtonines.

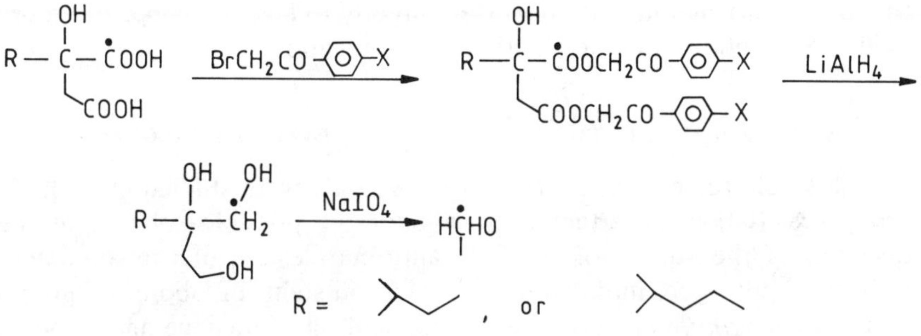

FIG. 11. Degradation route of alkyl-substituted malic acid.

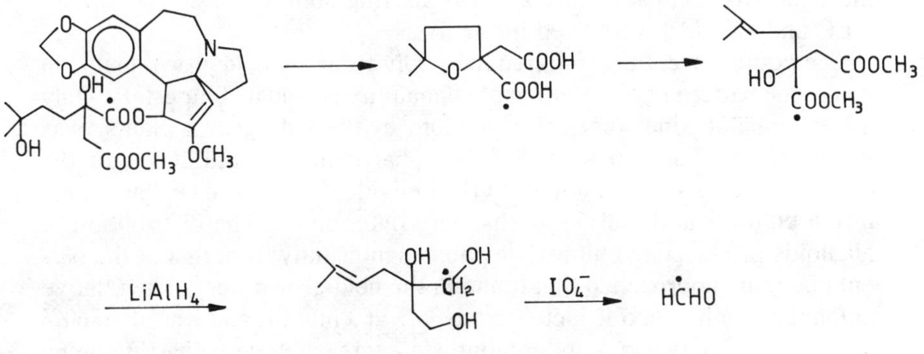

FIG. 12. Degradation route of harringtonine.

The derivation of acid **95** and acid **96** (acid portion of harringtonine and isoharringtonine, respectively) from acid **91** was speculated. Administration of [1-^{14}C]-**91** to the plant afforded labeled acid **95** with 98% of its radioactivity in C-1. The same precursor provided labeled acid **96**. However, [1-^{14}C]-**97** failed to be incorporated. This indicated direct hydroxylation at C-3 of **91** to **96**. The acid portion of homoharringtonine was postulated to be formed by homologation of acid **91** with subsequent hydroxylation. An attempt to examine whether deoxyharringtonine is converted directly to harringtonine or is first deacylated followed by hydroxylation and then reacylation was investigated by using double-labeled deoxyharringtonine. The results of the incorporation experiments evinced

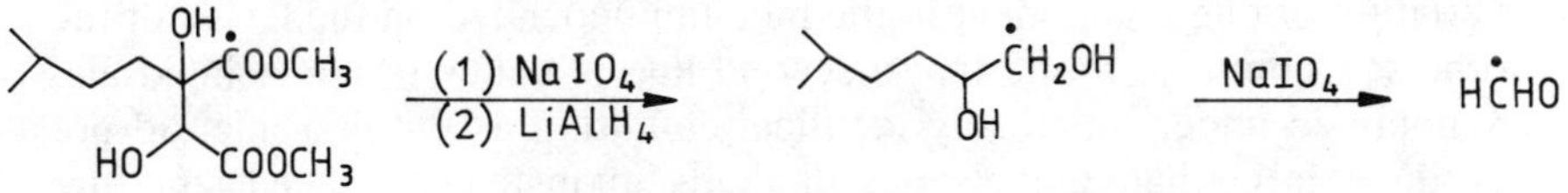

FIG. 13. Degradation route of the acyl portion of isoharringtonine.

that deoxyharringtonine is converted directly to harringtonine. But more evidence is required to substantiate this indication.

D. Tissue Culturing of *Cephalotaxus* Alkaloids

Callus cultures of *Cephalotaxus harringtonia* were studied by Delfel's group (98–100) in an attempt to look for the possible solution to the question of the source of the four antitumor esters of cephalotaxine. Callus cultures were initiated from leaves and stems of laboratory grown young *C. harringtonia* trees and maintained on Murashige and Skoog's salt mixture medium supplemented with other nutrients. The cephalotaxine alkaloids (cephalotaxine and the harringtonines) were quantitated. GLC and GC–MS were used for analysis.

The young parent tree differed markedly from the field grown plants in its unique pattern of distribution of cephalotaxine and its four esters. Only 4.6% cephalotaxine, the major alkaloid in the field grown plants, was present in conjunction with 79% homoharringtonine and 16.4% of the other three esters. The amounts of the individual esters and cephalotaxine in the cultures at 3 and 6 months were determined. The distribution of alkaloids produced in cultures deviated significantly from that of the parent plant and approached that found in the field-grown trees. The relative amount of cephalotaxine increased to 38% at 3 months and was up to 60% at 6 months. Between 3 and 6 months a decrease in deoxyharringtonine (38.4 → 8.8%) and an increase in isoharringtonine (9.9 → 19.2%) were observed, while the percentages of harringtonine and homoharringtonine remained nearly unchanged. The alkaloid/callus ratio was almost constant (10.1 mg/kg at 3 months and 11.1 mg/kg at 6 months). But the actual dried weight of callus at 6 months was twice that at 3 months. Eventually in the same period a roughly parallel increase in total production of these cephalotaxine alkaloids was obtained. At 6 months the total alkaloid production was 1–3% of the level reported for mature field-grown trees. In addition, a new alkaloid, homodeoxyharringtonine, was detected and characterized by GC–MS in the 6-month culture medium.

It was also reported that the callus cultures grew best at 25°C and that the initial pH was unimportant (in the range of 4.5–8.0). The final pH was around 4.2, possibly too acidic for optimum growth. At the end of six transfers the vitamin-free treatment still supported growth. The effect of variation of one component in the medium depended on the levels of the others. Delfel also found that a second line of callus tissue isolated and maintained under practically identical condition as the first line did not produce the cephalotaxine series alkaloids but instead produced a mixture of nine new compounds with characteristic differences between one another in GLC behavior on two liquid phases. The mass spectra of the trimethylsilyl derivatives of the new compounds showed none of the characeristic ions of the homoerythrina or *Cephalotaxus* alkaloids. Variation of the nutrient composition of the culture media did affect the relative amounts of the alkaloids, but it could not restore the biosynthesis of alkaloids to normal series. The new compounds were suggested to be intermediates in the biosynthetic pathway, and their production could be explained by differences in the choice of the starting materials or the growth conditions employed.

Addendum

Ma *et al.* (*101*) have reported the isolation of 4-hydroxy-cephalotaxine, **98**, $C_{18}H_{21}NO_5$, mp. 135–137°C, $[\alpha]_D$ +120 (C 0.025, CH_3OH), from *C.*

98

fortunei. The assigned structure was based on PMR and supported by the mass spectrum. Assignment of the chemical shifts of C-14 H and C-17 H based on NOE differed from that mentioned in Section IV,A,1 of this chapter.

Acknowledgments

We are grateful to Ms. K. T. Sun for preparing all the drawing and typing works.

References

1. W. Cheng, L. Fu, and C. Chao, *in* "Flora Reipublicae Popularis Sinicae" (Delectis Florae Reipublicae Popularis Sinicae Agendae Academiae Sinicae Edita), Vol. 7, p. 423. (1978).
2. Y. H. Chen and G. Huang, *Zhongcaoyao Tongxun* 254 (1977).
3. M. E. Wall, C. R. Eddy, J. J. Willaman, D. S. Correll, B. G. Schubert, and H. S. Gentry, *J. Am. Pharm. Assoc.* **43,** 503 (1954).
4. W. W. Paudler, G. I. Kerley, and J. Mckay, *J. Org. Chem.* **28,** 2194 (1963).
5. D. J. Abraham, R. D. Rosenstein, and E. L. McGandy, *Tetrahedron Lett.* 4085 (1969).

6. R. G. Powell, D. Weisleder, and C. R. Smith, Jr., *J. Pharm. Sci.* **61,** 1227 (1972).
7. R. G. Powell, *Phytochemistry* **11,** 1467 (1972).
8. R. G. Powell and K. L. Mikolajczak, *Phytochemistry* **12,** 2987 (1973).
9. S. Asada, *Yakugaku Zasshi* **93,** 916 (1973).
10. R. G. Powell, R. V. Madrigal, C. R. Smith, Jr., and K. L. Mikolajczak, *J. Org. Chem.* **39,** 676 (1974).
11. W. W. Paudler and J. McKay, *J. Org. Chem.* **38,** 2110 (1973).
12. G. E. Ma, L. T. Lin, T. Y. Chao, and H. C. Fan, *Acta Chim. Sin.* **35,** 201 (1977).
13. G. E. Ma, C. E. Lu, and G. J. Fan, *Chin. Pharm. Bull.* **17**, 205 (1982).
14. G. E. Ma, L. T. Lin, T. Y. Chao, and H. C. Fan, *Acta Chim. Sin.* **36,** 129 (1978).
15. Anonymous, *Acta Chim. Sin.* **34,** 283 (1976).
16. Z. Xue, L. Z. Xu, D. H. Chen, and L. Huang, *Acta Pharm. Sin.* **16,** 752 (1981).
17. N. Sun and X. Liang, *Acta Pharm. Sin.* **16,** 24 (1981).
18. H. Furukawa, M. Itogawa, M. Haruna, Y. Jinno, K. Ito, and S. T. Lu, *Yakugaku Zasshi* **96,** 1373 (1976).
19. R. G. Powell, K. L. Mikolajczak, D. Weisleder, and C. R. Smith, Jr., *Phytochemistry* **11,** 3317 (1972).
20. Anonymous, *Acta Bot. Sin.* **22,** 156 (1980).
20a. L. J. Ren and Z. Xue, *Chin. Tradit. Herb. Drugs* **12,** 241 (1981).
21. F. X. Zhang, Z. H. Wang, W. D. Pan, Y. J. Li, L. T. Mai, J. Q. Sun, and G. E. Ma, *Acta Bot. Sin.* **20,** 129 (1978).
21a. G. F. Spencer, R. D. Plattner, and R. G. Powell, *J. Chromatogr.* **120,** 335 (1976).
21b. N. E. Delfel, *Phytochemistry* **19,** 403(1980).
22. R. G. Powell, D. Weisleder, C. R. Smith, Jr., and I. A. Wolff, *Tetrahedron Lett.* 4081 (1969).
23. S. K. Arora, R. B. Bates, R. A. Grady, and R. G. Powell, *J. Org. Chem.* **39,** 1269 (1974).
24. S. K. Arora, R. B. Bates, R. A. Grady, G. Germain, J. P. DeClercq, and R. G. Powell, *J. Org. Chem.* **41,** 551 (1976).
24a. W. Huang, Y. Li, and X. Pan, *Lanzhou Daxue Xueboa, Ziran Kexueban* 148 (1974); *Sci. Sin. (Engl. Ed.)* **23,** 835 (1980).
25. R. G. Powell, D. Weisleder, C. R. Smith, Jr., and W. K. Rohwedder, *Tetrahedron Lett.* 815 (1970).
26. K. L. Mikolajczak, R. G. Powell, and C. R. Smith, Jr., *Tetrahedron* **28,** 1995 (1972).
27. J. Auerbach, T. Ipaktchi, and S. M. Weinreb, *Tetrahedron Lett.* 4561 (1973).
28. T. R. Kelly, J. C. McKenna, and P. A. Christenson, *Tetrahedron Lett.* 3501 (1973).
29. S. Brandänge, S. Josephson, and S. Vallen, *Acta Chem. Scand., Ser. B* **B28,** 153 (1974).
30. T. Ipaktchi and S. M. Weinreb, *Tetrahedron Lett.* 3895 (1973).
31. S. Brandänge, S. Josephson, S. Vallen, and R. G. Powell, *Acta Chem. Scand., Ser. B* **B28,** 1237 (1974).
32. J. Auerbach and S. M. Weinreb, *J. Am. Chem. Soc.* **94,** 7172 (1972).
33. S. M. Weinreb and J. Auerbach, *J. Am. Chem. Soc.* **97,** 2503 (1975).
34. L. J. Dolby, S. J. Nelson, and D. Senkovich, *J. Org. Chem.* **37,** 3691 (1972).
35. I. Tse and V. Snieckus, *J. Chem. Soc., Chem. Commun.* 505 (1976).
36. B. Weinstein and A. R. Craig, *J. Org. Chem.* **41,** 875 (1976).
37. M. F. Semmelhack, B. P. Chong, and L. D. Jones, *J. Am. Chem. Soc.* **94,** 8629 (1972).
38. M. F. Semmelhack, R. D. Stauffer, and T. D. Rogerson, *Tetrahedron Lett.* 4519 (1973).
39. M. F. Semmelhack, B. P. Chong, R. D. Stauffer, T. D. Rogerson, A. Chong, and L. D. Jones, *J. Am. Chem. Soc.* **97,** 2507 (1975).

40. M. F. Semmelhack and T. M. Bargar, *J. Org. Chem.* **42,** 1481 (1977).
41. T. Tiner-Harding, J. W. Ullrich, F. T. Chiu, S. F. Chen, and P. S. Mariano, *J. Org. Chem.* **47,** 3360 (1982).
42. S. M. Kupchan, O. P. Dhingra, and D. K. Kim, *J. Chem. Soc., Chem. Commun.* 847 (1977).
43. S. M. Kupchan, O. P. Dhingra, and C. K. Kim, *J. Org. Chem.* **43,** 4464 (1978).
44. K. L. Mikolajczak, C. R. Smith, Jr., D. Weisleder, T. R. Kelly, J. C. McKenna, and P. A. Christenson, *Tetrahedron Lett.* 283 (1974).
45. S. Li and J. Dai, *Acta Chim. Sin.* **33,** 75 (1975).
46. Anonymous, *Kexue Tongbao* **20,** 437 (1975); **21,** 512 (1976).
47. L. Huang, Y. Xi, J. Guo, D. Liu, S. Xu, K. Wu, J. Cheng, Y. Jiang, Y. Gao, Z. Guo, L. Li, M. Zhang, and F. Chu, *Sci. Sin.* (*Engl. Ed.*) **22,** 1333 (1979).
48. R. D. Ricke, *Synthesis* 452 (1975).
49. K. L. Mikolajczak and C. R. Smith, Jr., *J. Org. Chem.* **43,** 4762 (1978).
50. T. R. Kelly, R. W. McNutt, Jr., M. Montury, N. P. Tosches, K. L. Mikolajczak, C. R. Smith, Jr., and D. Weisleder, *J. Org. Chem.* **44,** 63 (1979).
51. Z. Zhao, Y. Xi, H. Zhao, J. Hou, J. Zhang, and Z. Wang, *Acta Pharm. Sin.* **15,** 46 (1980).
52. Y. Wang, Y. Li, X. Pan, Z. Li, and W. Huang, *Kexue Tongbao* **25,** 576 (1980).
52a. S. Hiranuma and T. Hudlicky, *Tetrahedron Lett.* **23,** 3431 (1982).
53. X. Pan, Y. Li, Z. Li, J. Tian, T. Zheng, Y. Cui, P. Chang, Y. Wang, and W. Huang, *Kexue Tongbao* **27,** 1048 (1982).
54. Y. Li, K. Wu, and L. Huang, *Acta Pharm. Sin.* **17,** 866 (1982).
55. Anonymous, *Kexue Tongbao* **23,** 696 (1978).
56. G. Zhang, Z. Zhou, Y. Guo, J. Guo, D. Liu, and F. Chu, *Fenxi Huaxue* 291 (1981).
57. L. Huang, Q. Fang, J. Cheng, and Z. Jiang, *Fenxi Huaxue* 158 (1983).
58. Y. Li, K. Wu, and L. Huang, to be published.
59. J. Cheng, J. Zhang, C. Zhang, J. Yang, and L. Huang, *Acta Pharm. Sin.* **19**, 178 (1984).
60. C. Mioskowski and G. Solladie, *Tetrahedron* **36**, 227 (1980).
61. K. L. Mikolajczak, C. R. Smith, Jr., and R. G. Powell, *J. Pharm. Sci.* **63**, 1280 (1974).
62. D. Weisleder, R. G. Powell, and C. R. Smith, Jr., *Org. Magn. Reson.* **13,** 114 (1980).
63. P. Z. Cong, *Acta Pharm. Sin.* **18,** 215 (1983).
64. Q. Yu, G. Ma, and Z. Huang, *Acta Chim. Sin.* **40,** 539 (1982).
64a. J. Greaves, J. Roboz, H. Jui, R. Suzuki, and J. F. Holland, *Anal. Chem. Symp. Ser.* **12,** 135 (1983).
65. X. Ji *et al.,* private communication.
66. T. H. Corbett, D. P. Griswold, Jr., B. J. Roberts, J. C. Peckham, and F. M. Schabel, *Cancer* **40,** 2660 (1977).
67. Anonymous, *Chin. Med. J.* **3,** 131 (1977).
67a. S. Takeda, N. Yajima, K. Kitazato, and N. Unemi, *J. Pharmacobio. Dyn.* **5,** 841 (1982).
67b. T. L. Jiang, R. H. Liu, and S. E. Salmon, *Invest. New Drug* **1,** 21 (1983).
67c. T. L. Jiang, S. E. Salmon, and R. H. Liu, *Eur. J. Cancer Clin. Oncol.* **19,** 263 (1983).
67d. W. R. Cobb, A. E. Bogden, S. D. Reich, T. W. Griffin, D. E. Kelton, and D. J. LePage, *Cancer Treat. Rep.* **67,** 173 (1983).
68. D. M. Baaske and P. Heinstein, *Antimicrob. Agents Chemother.* **12,** 298 (1977).
69. X. Ji, F. Zhang, and X. Dong, *Acta Pharm. Sin.* **18,** 299 (1983).
70. Y. Zhang, H. Yu, X. Luo, Y. Zheng, W. Li, X. Liu, and Y. Yuan, *Acta Pharm. Sin.* **14,** 135 (1979).
71. Z. Wang and B. Hsu, *Acta Biochim. Biophys. Sin.* **12,** 231 (1980).

72. X. Ji, J. Liu, and Z. Liu, *Acta Pharm. Sin.* **14,** 234 (1979).
73. X. Ji, Y. Liu, H. Lin, and Z. Liu, *Acta Pharm. Sin.* **17**, 881 (1982).
73a. J. Roboz, J. Greaves, H. Jui, and J. F. Holland, *Biomed. Mass Spectrum* **9,** 510 (1982).
73b. H. Jui and J. Roboz, *J. Chromatogr.* **233,** 203 (1982).
74. Z. Pan, R. Han, and Y. Wang, *Acta Biochim. Biophys. Sin.* **12,** 13 (1980).
75. Y. Wang, Z. Pan, and R. Han, *Chin. J. Oncol.* **2,** 247 (1980).
76. M. T. Huang, *Mol. Pharmacol.* **11,** 511 (1975).
77. J. S. Tscherne and S. Pestka, *Antimicrob. Agents Chemother.* **8,** 479 (1975).
78. M. Fresno, A. Jimenez, and D. Vazquez, *Eur. J. Biochem.* **72,** 323 (1977).
79. J. Si, K. Li, K. Son, W. Hu, S. Men, R. Han, Z. Pan, and Z. Li, *Acta Biochim. Biophys. Sin.* **11,** 333 (1979).
80. G. Wu, F. Fang, R. Han, and Z. Sun, *Acta Biochim. Biophys. Sin.* **13,** 509 (1981).
81a. G. Wu and C. Qiou, *Acta Biochim. Biophys. Sin.* **14,** 553 (1982).
81b. G. Wu and C. Qiou, *Shengwu Huaxue Yu Shengwu Wuli Xuebao* **14,** 553 (1982).
81c. S. Yang, F. Fang, and G. Wu, *Yaoxue Xuebao* **17,** 721 (1982).
81d. G. Wu, F. Fang, J. Zuo, R. Han, and Z. Sun, *Zhongguo Yixue Kexueyuan Xuebao* **4,** 78 (1982).
82. Y. Xu, C. Du, F. Zhang, and X. Ji, *Acta Pharm. Sin.* **16,** 661 (1981).
83. Y. Xu and C. Du, *Acta Pharm. Sin.* **2,** 252 (1981).
84. Anonymous, *Chin. Med. J.* **3,** 319 (1977).
85. Private communication.
86. F. Ma, B. Zheng, R. Yu, and H. Zhang, *Chin. J. Hematol.* **3,** 53 (1982).
87. Minutes of the Phase I Working Group Meeting, Nov. 9–10, 1982. Investigational Drug Branch Cancer Therapy Evaluation Program, National Cancer Institute, Bethesda, Maryland, 1982.
88. Private communication.
89. K. L. Mikolajczak, R. G. Powell, and C. R. Smith, Jr., *J. Med. Chem.* **18,** 63 (1975).
90. K. L. Mikolajczak, C. R. Smith, Jr., and D. Weisleder, *J. Med. Chem.* **20,** 328 (1977).
91. S. Li, H. Sun, X. Lu, S. Zhang, F. Lu, J. Dai, and Y. Xu, *Acta Pharm. Sin.* **16**, 821 (1981).
91a. C. Chan, X. Huang, D. Xiao, J. Zhu, and L. Lu, *J. of Hangzhou University, Sc.ed.* **10,** 87 (1983).
91b. Q. Mu, Y. Li, and W. Huang, *Kexue Tongbao* **27,** 759 (1982).
92. R. J. Parry and J. M. Schwab, *J. Am. Chem. Soc.* **97,** 2555 (1975).
93. J. M. Schwab, M. N. T. Chang, and R. J. Parry, *J. Am. Chem. Soc.* **99,** 2368 (1977).
94. R. J. Parry, M. N. T. Chang, J. M. Schwab, and B. M. Foxman, *J. Am. Chem. Soc.* **102,** 1099 (1980).
95. R. J. Parry, *Recent Adv. Phytochem.* **13,** 55 (1979).
96. R. J. Parry, D. D. Sternbach, and M. D. Cabelli, *J. Am. Chem. Soc.* **98,** 6380 (1976).
97. A. Gitterman, R. J. Parry, R. F. Dufresne, D. D. Sternbach, and M. D. Cabelli, *J. Am. Chem. Soc.* **102,** 2074 (1980).
98. N. E. Delfel and J. A. Rothfus, *Phytochemistry* **16,** 1595 (1977).
99. N. E. Delfel, *Planta Med.* **39,** 168 (1980).
100. N. E. Delfel and L. J. Smith, *Planta Med.* **40,** 237 (1980).
101. G. E. Ma, G. Q. Sun, M. A. ElSohly, and C. E. Turner, *Lloydia* **45,** 585 (1982).

—CHAPTER 4—

CONSTITUENTS OF RED PEPPER SPECIES: CHEMISTRY, BIOCHEMISTRY, PHARMACOLOGY, AND FOOD SCIENCE OF THE PUNGENT PRINCIPLE OF *CAPSICUM* SPECIES

T. SUZUKI

The Research Institute for Food Science
Kyoto University
Kyoto, Japan

AND

K. IWAI

Department of Food Science and Technology
Faculty of Agriculture
Kyoto University
Kyoto, Japan

THE ALKALOIDS, VOL. XXIII

ISBN 0-12-469523-X

I. Introduction

This chapter reviews the pungent principle of red pepper species (*Capsicum* species), known as a major component of the hot red pepper spice. The pungent principle represents a group of compounds called capsaicin and its analogs, or capsaicinoids. Their general structure has been assigned as that shown for compound **1** [*N*-(4-hydroxy-3-methoxybenzyl)alkylamides].

CH_3O / HO — benzene ring — CH_2NHCOR

1

	R	Names for acyl moieties
2	$(CH_3)_2CHCH{=}CH(CH_2)_4-$	8-methylnon-trans-6-enoic
3	$(CH_3)_2CH(CH_2)_6-$	8-methylnonanoic
4	$(CH_3)_2CH(CH_2)_5-$	7-methyloctanoic
5	$(CH_3)_2CH(CH_2)_7-$	9-methyldecanoic
6	$(CH_3)_2CHCH{=}CH(CH_2)_5-$	9-methyldec-trans-7-enoic
7	$(CH_3)_2CHCH_2CH{=}CH(CH_2)_4-$	9-methyldec-trans-6-enoic
8	$CH_3CH_2(CH_3)CH(CH_2)_6-$	8-methyldecanoic
9	$CH_3CH_2(CH_3)CHCH{=}CH(CH_2)_4-$	8-methyldec-trans-6-enoic
10	$CH_3(CH_2)_6-$	octanoic
11	$CH_3(CH_2)_7-$	nonoic
12	$CH_3(CH_2)_8-$	decanoic
13	$CH_3(CH_2)_9-$	undecanoic
14	$CH_3(CH_2)_{10}-$	dodecanoic
15	$CH_3CH_2(CH_3)CH(CH_2)_4-$	6-methyloctanoic
16	$CH_3(CH_2)_{11}-$	tridecanoic

It has long been known that the fruit of *Capsicum* species forms, accumulates, and secretes severely pungent materials in its fruits. The *Capsicum* fruits with hot taste, widely known by the name "hot chile," have been used as an important spice to enhance the palatability of food. The fruits have also been used as a medicinal drug. Studies on the pungent principle of *Capsicum* species began as early as the 1810s, however, until the 1950s, no attention was paid to pharmacological or biochemical aspects. Although the pungent principle of *Capsicum* species had been called "protoalkaloid" (*1,2*) or "alkaloid" (*3,4*), it never was regarded as a member of the alkaloids in general and was heretofore not discussed in

The Alkaloids. The recently revealed potent pharmacological activities of capsaicinoids and the fact that capsaicin and its analogs occur in *Capsicum* species and are composed of nitrogen, oxygen, carbon, and hydrogen make a review of the constituents of *Capsicum* species in this treatise appropriate.

In the present chapter, the authors intend to broaden the reader's knowledge of the various aspects of capsaicinoids. In Section II, structure, nomenclature, chemistry, and occurrence of capsaicin and its analogs in nature are described. Recent advances in the analysis of capsaicinoids will be reviewed in Section III. The synthesis of capsaicin and its analogs including related compounds will be briefly reviewed in Section IV. The accumulation and secretion in the *Capsicum* plant will be discussed in Section V. The nutritional and pharmacological effects of capsaicin and its analogs as well as their metabolism in mammalian tissues are reviewed in Section VI. Finally, Section VII, the food chemical aspects are briefly mentioned, but since this is an introductory review, detailed data will not be given. For more detailed information the reader should consult the original papers cited in the individual sections.

Beside capsaicin and its analogs, *Capsicum* species contain glycoalkaloids such as solanine and solanidine (*5–7*); however, they will not be covered in this chapter.

II. Structure, Nomenclature, Chemistry, and Occurrence

Some reviews on the history of identification of the pungent principles of red pepper species have already been published (*4,8–10*), however, a short history of capsaicin and its analogs seems necessary here. Because of its strong pungency and irritating effect, the pungent principle of *Capsicum* fruit was of interest for a long time. As early as 1810, the pungent substance of *Capsicum* fruit was being studied under the names capsicol, capsicin, capsacutin, etc. (*8,9*). In 1876 Trench first obtained the crude pungent substance as an oleoresin from paprika (*11*) and isolated the pungent substance in crystalline form (*12,13*). Thresh named it capsaicin after a compound from the genus *Capsicum*. Flückiger and Hanbury also succeeded in isolating the pungent substance in crystalline form of mp 59°C (*14*). The formula $C_9H_{14}O_2$ for capsaicin assigned by Thresh (*13*) was quite different from the exact formula, and the nitrogen atom was overlooked.

Later, Micko proposed the formula $C_{18}H_{28}NO_3$ for capsaicin, which was obtained as crystal of mp 63.5°C (*15,16*). Micko also established the presence of a phenolic hydroxy group and a methoxy group in the capsacin molecule, which consequently led to the establishment of vanillyl group. On the basis of his empirical formula, Micko proposed the chemical formula $C_{17}H_{24}ON(OH)(OCH_3)$ for capsaicin. In 1910, Nelson (*17*) isolated a crystal of "capsaicin" with mp 64.5°C. In his subsequent published works (*18,19*), the chemical structure of capsaicin was vigorously studied and in 1923 established as the vanillylamide of 8-methylnon-6-enoic acid, or *N*-(4-hydroxy-3-methoxybenzyl)-8-methylnon-6-enamide (*20*). In their work, vanillylamine and 8-methylnon-6-enoic acid were obtained by acid and alkaline hydrolyses of capsaicin isolated from *Capsicum* fruits, and capsaicin was obtained from vanillylamine and 8-methylnon-6-enoic acid chloride.

Lapworth and Royle, competing with Nelson, proposed that capsaicin had the structure of a dihydrooxyazole derivative (*21*). In 1955 Crombie *et al.* confirmed the chemical structure of capsaicin as *N*-(4-hydroxy-3-methoxybenzyl)-8-methylnon-6-*trans*-enamide (**2**) (*22*). The assignment of the chemical structure of capsaicin seemed to finish the work on identification of the "capsaicin" formula.

In 1958, however, Kosuge and co-workers reported that the chemically pure "capsaicin" was a mixture of two closely related compounds (*23,24*). The "pure capsaicin" gave two components in a ratio of 2.1 : 1 by paper chromatography and subsequent colorimetric determination (*23*). The "pungent component I" was found to be capsaicin (**2**) (*25*), and the other minor component, "pungent component II," was identified as *N*-(4-hydroxy-3-methoxybenzyl)-8-methylnonamide (**3**) (*26*). Since the double bond of the acyl residue of the latter compound could be hydrogenated, Kosuge named the new compound dihydrocapsaicin. The heterogeneity of "pure capsaicin" was also recognized in material originating from other *Capsicum* species such as *C. annuum* L. var. *parvoaccuminatum* Makino (Santaka), *C. annuum* var. *fasciculatum* Irish (*23*). Later, Friedrich and Rangoonwala (*27*) and Juhasz and Tyihak (*28*) independently reported similar results.

Bennett and Kirby (*29*), in their study using nuclear magnetic resonance (NMR), mass spectrometry (MS), and radioisotopic techniques, also speculated that natural capsaicin is composed of at least five closely related amides, three of which are now recognized as nordihydrocapsaicin (molecular weight 293; formula **4**), homodihydrocapsaicin (molecular weight 321; formula **8**), and homocapsaicin (molecular weight 319 formula **6**). Capsaicin (**2**), was recognized to be the major component, constituting 70% of total pungent acid amides, while dihydrocapsaicin (**3**) amounted to 30% or less. Remaining components were found only in trace amounts.

Bennett and Kirby's speculation was independently confirmed by Jentsch *et al.* (*30*) and by Kosuge and Furuta (*31*). The chemical structures of nordihydrocapsaicin and homodihydrocapsaicin were shown to be that of *N*-(4-hydroxy-3-methoxybenzyl)-7-methyloctanamide (**4**) (*31*) and *N*-(4-hydroxy-3-methoxybenzyl)-9-methyldecanamide (**8**) (*30*), respectively. No capsaicin with a cis configuration has been reported heretofore. The formula of homocapsaicin was assigned as *N*-(4-hydroxy-3-methoxybenzyl)-9-methyldecenamide by Bennett and Kirby (*29*); the exact position, however, of the double bond in the branched amide substituent was left open. Masada *et al.* (*32*) in their paper presented the 9-methyldec-7-enoyl residue as the acyl part of homocapsaicin (**6**). More recently, Jurenitsch *et al.* examined the double bond position of homocapsaicin by oxidation with osmium tetroxide and subsequent analysis by gas chromatography–mass spectrometry (*33*). They found that the double bond in the acyl residue was at carbon 6 and that homocapsaicin is the vanillylamide of 9-methyldec-6-*trans*-enoic acid (**7**). Beside the five analogs Bennett and Kirby speculated to occur (*29*), Jurenitsch *et al.* found and identified several more analogs as minor components of capsaicinoids in *Capsicum* fruit, represented by formulas **9** to **16** (*33–36*). "Homodihydrocapsaicin II" (**8**), "homocapsaicin II" (**9**), and "nordihydrocapsaicin II" (**15**), have anteiso acid type acyl residues in their fatty acid chains.

The term "capsaicinoid" first proposed by Kosuge *et al.* to describe capsaicin and dihydrocapsaicin is widely used for compounds having the general structure **1**. In this chapter, the term "capsaicinoids" will hereafter be used for "capsaicin and its analogs" except for special cases. Some chemical properties established for capsaicin and some of its analogs are summarized below (*8,37–39*).

Capsaicin (**2**) [*N*-(4-hydroxy-3-methoxybenzyl)-8-methylnon-6-*trans*-enamide]: $C_{18}H_{27}O_3N$, MW 305.199, odorless white needle crystal with severe burning pungency; mp 64.5°C, bp at 0.01 mm Hg = 210–220°C, sublimate at 115°C; UV A_{max} 227, 281 nm (ε = 7000, 2500); easily soluble in ethyl ether, ethyl alcohol, acetone, methyl alcohol, carbon tetrachloride, benzene, and hot alkali; slightly soluble in carbon disulfide, hot water, and conc HCl; practically insoluble in cold water; fairly resistant to acids and alkalies at ordinary temperatures. By alkaline hydrolysis with 25% sodium hydroxide at 180°C in an autoclave for 30 min, capsaicin gives 8-methylnon-6-*trans*-enoic acid and pyrocatechin or other decomposed phenolic products from vanillylamine with release of ammonia (*18*). It afforded 3-methoxy-4-hydroxybenzylamine chloride by acid hydrolysis with hydrochloric acid (*18*) and isobutyric acid plus vanillylcarbamoyl ipentanoic acid by ozonolysis (*25*). Commercial grade "capsaicin" often is a mixture of capsaicin, dihydrocapsaicin, and nordihydrocapsaicin.

Dihydrocapsaicin (**3**) [*N*-(4-hydroxy-3-methoxybenzyl)-8-methylnonanamide]: $C_{18}H_{29}O_3N$, MW 307.215; odorless white opaque crystal with strong pungency; mp 65.6–65.8°C; UV A_{max} below 230 nm, 281 nm in abs. ethyl alcohol; solubility is almost the same as that of capsaicin.

Nordihydrocapsaicin (**4**) [*N*-(4-hydroxy-3-methoxybenzyl)-7-methyloctanamide]: $C_{17}H_{27}O_3N$, MW 293.199; white crystal with strong pungency; mp 65.6°C; UV A_{max} 280.5 nm in abs. ethyl alcohol.

Detailed chemical properties of other analogs have not yet been established.

Prior to review of studies on the occurrence of capsaicinoids in the various *Capsicum* species, a taxonomical introduction to *Capsicum* species should be made. The so-called red peppers have been familiar to nearly all Spanish South Americans by the Arawakan name *aji* and by the Nahuatlan name *chilli* in Mexico and Central America (*39a*). The genus *Capsicum,* which is commonly known as "red chile," "chilli peppers," "hot red pepper," "tabasco," "paprika," "cayenne," etc., is a member of the Solanaceae (nightshade family) that originated in Central and South America. In some books the popular name "chile" or "chili" is explained by reasoning that the hot pepper species originated in the South American country of Chile; however, the name "chile" seems to have nothing to do with the country name, and, on the contrary, it apparently originated in a district of Central America. *Capsicum* species have been thought to be of Central American origin, but one species was introduced to Europe in the fifteenth century. By the middle of the seventeenth century, the *Capsicum* pepper (more correctly *C. annuum* var. *annuum*) had become cultivated throughout southern and middle Europe to the Asian and African tropical and subtropical regions as a spice and/or medicinal drug.

It is generally believed that about 20 *Capsicum* species are distributed worldwide. However, the question of exactly how many species are involved is still controversial. Heither (*40*) recognized five species as the fundamental cultivated species of *Capsicum: C. annuum, C. frutescens, C. chinense, C. pendulum,* and *C. pubescens*. On the other hand, Pickersgill (*41*) and Tanaka (*42*) independently recognized four species as the fundamental cultivated species: *C. annuum* var. *annuum, C. chinense, C. baccatum* var. *pendulum,* and *C. pubescens*. According to them, *C. frutescens* is not regarded as a cultivated variety but as a semicultivated one. The following species are regarded as the ancestral species for the individual cultivated species: *C. annuum* var. *minimum* for *C. annuum* var. *annuum, C. frutescens* for *C. chinense, C. baccatum* var. *baccatum* for

C. baccatum var. *pendulum,* and *C. eximium* for *C. pubescens. Capsicum annuum* var. *minimum* is believed to have originated in Mexico, *C. frutescens* in swamps around the higher reaches of the Amazon river in Colombia and Peru, *C. baccatum* var. *baccatum* in southern Peru and Bolivia, and *C. eximium* in mountainous areas of southern Peru and northern Bolivia. The majority of cultivated varieties of *Capsicum* belong to *C. annuum* var. *annuum*. Species other than *C. annuum* var. *annuum* have not yet been cultivated outside the New World. Any *Capsicum* species that accumulates only small amounts of or no capsaicinoid belongs to *C. annuum* var. *annuum*.

For further topographical, taxonomical, horticultural, and plant breeding information on *Capsicum* species, refer to the detailed papers (*43–55a*).

With regard to the occurrence of capsaicinoids in various *Capsicum* species, more than 20 papers have been presented. Govindarajan reviewed capsaicinoid content and composition in varieties of *Capsicum* species (*56*). When Micko isolated capsaicin from paprika (*C. annuum*) and from cayenne pepper (*C. fastigiatum*) (*15,16*), the content of crude capsaicin was 0.03 and 0.55%, respectively. Nelson extracted 0.14% capsaicinoid by weight from African *Capsicum* (*17*). However,.these values do not necessarily reflect the capsaicinoid content in the individual species because the studies were not intended for quantitation of capsaicinoid. Until 1960 few studies on the capsaicinoid content of different *Capsicum* varieties had appeared. In 1960 Ohta (*57*) first investigated the capsaicin content of several varieties of *C. annuum* and related species from the standpoints of plant physiology and genetics. He determined the capsaicin content, which should be now taken as the capsaicinoid content, of several common varieties of *C. annuum* L. in Japan: *C. frutescens* L., *C. chacoense* Hunz., *C. microcanpon* Cav., *C. pendulum* Willd., and *C. pubescens* Ruis et Pavon. In Ohta's study, *C. frutescens* was evaluated to be the most promising species for breeding more pungent hot peppers. Anantha Samy *et al.* (*58*) also reported the capsaicin content of several green peppers.

Csedo *et al.* subsequently determined the capsaicinoid content of five *Capsicum* species (*59*): 640–768 mg% in *C. annuum* var. *typicum,* 35–386 mg% in a hybrid from Chile, 188–342 mg% in *C. annuum* var. *accuminatum,* 280–258 mg% in *C. annuum* var. *cordiforme,* and 195–322 mg% in *C. annuum* var. *cerasiforme*. Deb *et al.* (*60*) analyzed the "capsaicin" content of 12 varieties of *Capsicum* species by using spectrophotometry and showed that the capsaicinoid content ranged from 0.605 to 1.498%. Similar values were reported in Thirumalachar's work (*61*). Csedo and Kopp (*62*) determined the capsaicinoid content of six different varieties of paprika (*C. annuum*) by different methods, and showed that *C. annuum* var. *typicum* cv. fasciculatum has the highest capsaicinoid content, 640–768 mg%. Meanwhile, the greatest variation was observed in *C. annuum* var. *longum,* where content ranged from 35 to 385 mg%.

Schratz and Rangoonwala also investigated the variability of capsaicinoid content in different *Capsicum* species (*63*). They showed that the capsaicinoid content of various *Capsicum* species varied widely from 0.1 to 17.0 mg%. In their study *C. annuum* contained a greater absolute amount of capsaicinoid than *C. pendulum* or *C. frutescens;* however, on the basis of relative amount, the two latter species contained higher amounts of capsaicinoid than the large fruit. Karawya *et al.* (*64*) demonstrated statistically reliable values of capsaicinoid content for *C. minimum* (0.748%) and *C. frutescens* (0.420%). Govindarajan *et al.* also reported on capsaicin content in various *Capsicum* species (*65–68*). In Pankar and Magar's paper (*68*), the capsaicin content in 10 popular varieties of red chili was 0.053 to 0.912%. The data presented by Yuste *et al.* (*69*) on capsaicinoid content of 14 varieties of *C. annuum* and *C. frutescens* ranged from 0.05 to 0.33% in *C. annuum* and from 0.35 to 0.85% in *C. frutescens.* They showed that capsaicinoid content in Mexican varieties were generally less than those of Indian, Japanese, and African varieties.

As will be discussed in Section V, the capsaicinoid content varies with environmental conditions. In general, however, variation in capsaicinoid content is greater in *C. annuum* than in other species. Although in the papers reviewed heretofore, values were expressed as "capsaicin," they actually represent the content of capsaicin and its analogs. It was, of course, Kosuge and Inagaki's paper that discussed capsaicin and its analogs individually (*70*). Later, Juhasz and Tyihak (*28*) reported that capsaicin content, designated "capsaicin a," was superior to dihydrocapsaicin, designated by "capsaicin b," in the fruits of *C. annuum, C. frutescens, C. argulosum,* and *C. pubescens.* They showed that the highest concentration of capsaicinoid was found in *C. argulosum.*

Jurenitsch *et al.* (*55*) vigorously investigated the total capsaicinoid content and its composition in various *Capsicum* varieties—using as many as 71 fresh and taxonomically highly reliable samples belonging to *C. pubescens* (2 cultivars), *C. baccatum* var. *pendulum* (3 cultivars), *C. frutescens* or *C. chinense* (5 cultivars), and *C. annuum* var. *annuum* (13 cultivars)—from a taxonomical and anatomical viewpoint to see if there is any species specificity in capsaicinoid composition. They not only quantitated the total capsaicinoid content but also determined the composition of the individual capsaicin analogs. Among the species they examined, *C. annuum* var. *accuminatum* Fingerhurth showed the highest capsaicinoid content (as high as 1.477%), the second highest was "paprika" purchased from a market in Wien, and *C. microcaprus* DC was in third place. As many as 9 out of 38 samples belonging to *C. annuum* var. *annuum* exceeded 1% in capsaicinoid content. In *C. pubescens* the composition of capsaicinoid ranged from 25.5 to 36.3% capsaicin, 48.8 to 54.2% dihydrocapsaicin, 5.1 to 15.4% nordihydrocapsaicin, 0.2 to 10.1% homodihydrocapsaicin, and 2.3 to 6.9% caprylic acid vanillylamide. *Capsicum baccatum* var. *pendulum* yielded 38.1 to 66.6% capsaicin, 26.3 to 61.3% dihydrocapsaicin, 2.5 to 9.4% nordihydrocapsaicin, and 0.2 to 0.5% caprylic acid vanillylamide. From 63.2 to 77.2% capsaicin, 21.0 to 32.0% dihydrocapsaicin, 0.8 to 8.6% nordihydrocapsaicin, 3.2 to 3.6% homodihydrocapsaicin, and 0.1 to 0.6% caprylic acid vanillylamide were obtained from *C. frutescens* or *C. chinense*. In *C. annuum* var. *annuum* the capsaicinoid was composed of 36.9 to 59.1% capsaicin, 30.4 to 51.2% dihydrocapsaicin, 7.0 to 22.2% nordihydrocapsaicin, 0.7 to 4.5% homodihydrocapsaicin, and 0.1 to 2.6% caprylic acid vanillylamide.

Pronounced differences in capsaicinoid composition were recognized in some species, so Jurenitsch *et al.* suggest that determination of capsaicinoid and its composition can be a useful method for taxonomical identification of individual species. The present authors also analyzed the capsaicinoid content and composition of more than 40 different cultivars which belong to *C. annuum* var. *annuum, C. chinense, C. baccatum,* and *C. pubescens,* and obtained results that suggest a species-specific relationship in capsaicinoid content and composition exists (K. Iwai and T. Suzuki, unpublished data). Similar results showing significant differences in capsaicinoid content between pungent and sweet types were also reported by Quagliotti and Ottaviano (*71*), who investigated genetic characteristics of capsaicinoid variability. Studies on the variability of capsaicinoid content from the hereditary aspect were reported by Kvachadze (*72*), Arya and Saini (*73*), Sooch (*74*), and by Nowaczyk (*75*) independently.

III. Analysis of Capsaicin and Analogs

Choice of suitable analytical methods is important to make the study of capsaicin and analogs successful. So far more than 150 analyses of capsaicinoids have been reported. However, with the development of more convenient techniques for capsaicinoid analysis, less convenient or unreliable methods have faded away. Methods for the analysis of capsaicin and its analogs have been reviewed in several papers (*4,8,39,56,76–81*). Jordan *et al.* (*76*) reviewed the assay methods of capsaicin, which varied from time to time and proposed a standard assay method for capsaicinoids that is based on the vanadium trioxychloride colorimetric reaction. Newman, on the other hand, reviewed the chemistry of red hot peppers by tracing works from 1816 to the early 1950s (*39*). Suzuki *et al.* (*77*) briefly reviewed the history of research of the pungent principle of *Capsicum* species. To establish a reliable assay method for capsaicinoids, a joint committee of the Pharmaceutical Society and the Society for Analytical Chemistry on the Methods of Assay of Crude Drugs in Great Britain (The Joint Committee) in 1959 carefully examined various methods for extraction, purification, and spectrophotometric and colorimetric determination of capsaicinoids, and selected some recommendable procedures for analyses (*78*). The Joint Committee in 1964 revised some procedures and also mentioned an organoleptic method, paper chromatography, gas chromatography, and infrared spectrometry for capsaicinoid determination (*80*). Csedo *et al.* reexamined conventional analytical methods carefully and proposed a photocolorimetric method by modifying Spanyar's original method (*82*). Molnar (*8*) and Mathew *et al.* (*4*) separately made brief comments on the analytical methods for capsaicinoids in their reviews on pharmacological effects of capsaicinoids. Govindarajan reviewed analytical methods of capsaicinoids by focusing on organoleptic assay methods and chromatographic analyses (*56*). Pruthi (*81*) summarized analytical methods for capsaicinoids, briefly but ingeniously, in two tables.

In spite of the effort on reviews, most methods described therein are getting out-of-date because of recent dramatic progress in analytical chemistry. Even in the latest reviews by Govindarajan (1979) and by Pruthi (1980), modern techniques using gas chromatography–mass spectrometry, high performance liquid chromatography, or high performance thin-layer chromatography are not cited. In this chapter, modern techniques will also be mentioned.

A. Extraction and Purification

1. Extraction

Among analytical procedures the extraction method may be the only one in which traditional procedures are still being used. Extraction is the first step in capsaicinoid analysis: without performing sufficient extraction of capsaicinoids from the starting materials satisfactory results cannot be obtained. The extraction procedure was the first step in the isolation of the pungent substance in pure crystalline form. In the early days of capsaicinoid research, more polar solvents such as methyl alcohol, ethyl alcohol, isopropyl alcohol, acetone, and ethyl ether were preferred for extraction of the pungent substance from *Capsicum* fruits with Soxhlet extractors. The crude *Capsicum* extracts usually contained impurities such as fats, lipids, sterols, and pigments along with capsaicinoids, and such impurities often interfered with the crystallization of the capsaicinoids. Therefore, crude capsaicinoid extracts generally designated "oleoresin" were further purified by multistage solvent extraction and/or running through appropriate column adsorbents.

Besides the polar solvents mentioned above, chloroform, methylene chloride, and aqueous sodium hydroxide or ethyl acetate have also been used. Through introduction of modern techniques with higher separability, a universal solvent system for lipid extraction, chloroform–methyl alcohol (2 : 1, v/v), is now often used for capsaicinoid extraction because of its high extractability, especially of wet materials. Throughout capsaicinoid studies, alcohols, acetone, and ethyl ether have been chosen as the major solvents for extraction. Extraction should be done until no color is left in the material. Nonpolar solvents such as petroleum ether are not generally used for capsaicinoid extraction but rather for removing lipids and pigments by solvent partition.

The Joint Committee established in 1959 two recommended procedures for the extraction of capsaicinoids from *Capsicum* fruits. The procedures are preparation of oleoresin by extracting *Capsicum* powder with 96% ethyl alcohol for 48 hr followed by purification on aluminum oxide with activated carbon and Kieselguhr, and extraction and purification by ether–alkali partition extraction (*78*).

Extraction of capsaicinoids from animal tissues is performed by acetone with a recovery of approximately 90% (*83*). Johnson *et al.* (*84*) reported trace amounts of natural capsaicinoids in animal feed, human urine, and waste water in a study in which 80% methyl alcohol–20% 0.1 *N* hydrochloric acid was used for the extraction of capsaicinoids from animal feed and dichloroethane was used for extraction of capsaicinoids from human urine and waste water. More recently ethyl acetate (*85–87*), chlorinated carbon solvents such as di- or trichloromethylene (*1*) or chloroform (*33,88,89*), and chloroform–methyl alcohol (*35*) have been preferably used for extraction of capsaicinoids from *Capsicum* fruits. For preparing oleoresin from *Capsicum* fruits for food additives or medicinal drugs, the use of ethyl acetate is proposed as an efficient and less harmful solvent (*87*). The extraction efficiency of ethyl acetate was shown to be comparable to methylene chloride and acetone. Lyophilized fresh green chili peppers extracted with acetone, potassium hydroxide, and ethyl ether gave a much higher yield than known conventional procedures (*90*).

An interesting paper examining the effect of ultrasound and heating on the extraction efficiency of capsaicinoids was presented by Adamski and

Socha (*91*). Ultrasonic treatment for 30 min showed about the same effect as heating at 30°C which yielded the highest extraction efficiency. On the other hand, Meerov *et al.* (*92*) reported that the extraction of *C. annuum* fruit by liquid carbon dioxide yielded the highest amount of capsaicinoid compared to other extraction methods.

2. Purification

Crude extracts (oleoresin) prepared from *Capsicum* fruits, ointments, tinctures, animal tissues, etc. usually contain a considerable amount of impurities such as waxes, fats, phospholipids, sterols, and coloring matters, so crude extracts must be further purified by various procedures to obtain pure crystals or to provide material for further analyses. As briefly mentioned in Section III,A,1, two major procedures were proposed for removing impurities from crude extracts, i.e., purification of crude capsaicinoid with solvents by taking advantage of the differing solubilities of capsaicinoid and impurities (*15,17,93–96*) and column chromatography through alumina, charcoal, and Kieselghur (*78*).

Lapworth and Royle removed impurities from capsaicinoid extracts by precipitating them as barium salts (*21*). The Joint Committee established column chromatography and the ether–alkali extraction method as recommendable methods for purification. In the recommended chromatographic procedure, impurities are removed by adsorption on the column adsorbents and elution with alcohol; however, materials such as fats and waxes cannot be removed by column chromatography. In such cases, a method of partition extraction between aqueous alkaline and organic solvent is applied (*78,80*). Rangoonwala reported a novel procedure modified from Schulte and Krüger's method in which the chili powder is first extracted on the column with diethyl ether to obtain a single capsaicinoid crystal (*97*). Müller-Stock *et al.* (*98*) reported a column chromatographic purification of capsaicinoid and subsequent subfractionation of capsaicin analogs. In case a sufficient gram quantity of crude capsaicinoid is available, pure capsaicinoid can be obtained by repeated crystallization from light petroleum or hexane until the melting point of the crystal becomes constant at ~64 to 65°C (*23,66,77,78,97,99–103*). Individual capsaicin analogs, however, cannot be separated by recrystallization. When sufficient capsaicinoid extract is not available, the above mentioned purification methods cannot be applied, but modern techniques such as paper chromatography (PC), thin-layer chromatography (TLC), and high performance liquid chromatography (HPLC) are recommended instead (*64,67, 86,104–106*).

B. Sensory Evaluation

Organoleptic evaluation was the first method applied to the identification and quantitation of capsaicinoids. Despite having been criticized for low accuracy and poor reproducibility, organoleptic methods have been proposed since the early days of capsaicinoid research (*17,18,39,80, 107–112*).

As early as 1912, the basic principle of pungency evaluation using an organoleptic method was established by Scoville (*107*). Scoville quantified the pungency of *Capsicum* fruits by examining diluted ethanolic-sweetened solutions of *Capsicum* extract to determine the greatest dilution at which definite pungency could be recognized (the recognition threshold). Alcohol solution was added to the sweetened water in definite proportions until a distinct but weak pungency was perceptible on the tongue. Scoville expressed the greatest dilution as the reciprocal of the dilution in Scoville heat units. Scoville heat units for pure capsaicin are reported as 15 to 17 $\times$ 10^6 (*77,113*). Scoville's method is simple and convenient; however, the accuracy of the method is low and often the reproducibility among laboratories is poor.

Beside Scoville's method, other sensory methods developed and modified by other researchers were proposed (*18,39,108,109,112,114*); nevertheless, the accuracy and reproducibility of sensory methods are definitely inferior to chemical methods using modern instruments. Thus, with the development of modern chemical methods, the organoleptic method, which had once been officially approved in the United States Pharmacopoeia and later in the National Formulary, is no longer in favor among pharmacologists and biochemists. However, because of its convenience the American Spice Trade Association (ASTA) has employed the method as an official and approved method for pungency evaluation for *Capsicum* since 1968 (*56*).

Suzuki *et al.* compared the pungency of *Capsicum* determined by the organoleptic method to that determined by chemical methods during the course of establishing and developing new determination methods (*77*). Out of 21 samples, 8 showed good consistency with chemical methods (chromatography and colorimetry). Suzuki *et al.* pointed out that the rapid fatigue encountered in the organoleptic method is the main factor that makes the organoleptic method unreliable. Hartman (*102*) compared Scoville heat units for several *Capsicum* species and correlated Scoville heat units to the proportion of capsaicinoid content as determined by GLC. Todd *et al.* also examined the pungency of *Capsicum* fruit by comparing the more convenient method to GLC (*115*). They showed that comparative data for pungency determined by the organoleptic method and by GLC on samples ranging from 50,000 to 2,000,000 Scoville heat units exhibited a correlation coefficient of 0.95 for the raw data.

Govindarajan proposed a standardized procedure for evaluation of pungency by Scoville heat units (*112*) by which a linear regression was obtained between Scoville heat units and capsaicin content of the samples. A detailed description of the organoleptic determination method is given in the review by Govindarajan (*56*). Unlike colorimetric methods, other spectrophotometric methods, or other modern methods, organoleptic methods require neither preliminary purification to remove the interfering materials nor special apparatus. Nevertheless, the organoleptic method does require skillful and experienced tasters, so it is eventually a very expensive and time-consuming method compared to chemical methods (*115*). Furthermore, the sensitivity and reliability of modern methods such as GC–MS, HPTLC, and HPLC are absolutely superior to organoleptic methods. Therefore, instrumental assay methods are becoming the major techniques for capsaicinoid determination except in special cases.

C. Determination by Colorimetry and Spectrometry

1. Colorimetric Determination

Colorimetric determination is the oldest techinique for capsaicinoid determination among chemical assay methods. Colorimetric determination methods are based on the principle of color formation by reduction of a chromophore in reaction with capsaicinoids. They are classified into three methods based on the color reagents. The first is the method using color reactions of capsaicinoids with vanadium trioxychloride or ammonium vanadate to give a greenish-blue color (*94,116,117*). The second method is reaction of phosphomolybdic acid with capsaicinoids to give the blue color of molybdenum blue (*95,103,118,119*). The third method is based on the reaction of diazobenzenesulfonic acid with capsaicinoids to give a red color (*120–122*). Because the first method developed by Fodor had shortcomings (the color development was unstable and not specific to capsaicinoids) this method could not be widely accepted in spite of proposals of modified procedures (*79,123,124*). Jonczyk evaluated the spectrophotometric method and colorimetric method as sufficiently accurate for the determination of capsaicin in *Capsicum* fruits (*125*). The color reaction of the second method also is not specific to capsaicinoids. In comparison with the former two methods, the third method using diazobenzenesulfonic acid is superior; however, it still has the defect of nonspecific color development.

The diazo colorimetric method and Gibbs' colorimetric method as the substituent of the diazo method were established as recommendable determination methods of capsaicinoids by The Joint Committee (*80*). In Gibbs' colorimetric method, 2,6-dichloro-*p*-benzoquinone-4-chloroimine in methyl alcohol (known as Gibbs' reagent) (*126*) is used in alkali to give a blue color with an absorption maximum at 595 nm and an $E^{1\%}_{1cm}$ value of 470 by reaction with phenolic compounds. The limit of detection by Gibbs' method is at the microgram level.

Even after the presentation of recommended procedures for capsaicinoid determination, modified methods of colorimetry using vanadium reagents or molybdate reagents were proposed (*127–130*). Marquis reagent using one drop of 40% formalin in 1.5 ml of conc sulfuric acid is used to detect capsaicinoids in fresh plant tissue at the microscopic level (*131*); however, it too is not capsaicinoid specific. Among the colorimetric methods, no method specific to capsaicinoids has been reported. Therefore, samples must be purified and freed from interfering substances prior to colorimetric determination. With the development of modern techniques for the capsaicinoid determination, colorimetric methods are becoming less preferable, however, a considerable numbers of papers using colorimetric methods for capsaicinoid determination are still being published (*85,106,129,130,132–135*). Visualization of capsaicinoids on a TLC plate followed by densitometric assaying has been increasing (*1,64,135–137*).

2. Spectrophotometric Determination

In the early days of capsaicinoid research, spectrometric assay methods were less valuable. However, in contrast to colorimetric assay methods, the importance of spectrophotometric methods in combination with modern instrumental techniques such as HPLC and TLC is increasing. Capsaicin and its analogs give characteristic absorptions in the ultraviolet (UV) region. The Joint Committee investigated the UV absorption characteristics of capsaicinoids (*78*). Capsaicin, which should now be regarded as a mixture of capsaicin and its analogs, in ethanolic solution shows absorption maxima at 229 nm with $E^{1\%}_{1cm}$ values of ~245 to 258 and at 280 nm with an $E^{1\%}_{1cm}$ value of 102. In a methanolic solution, the maximum at 279 nm with an $E^{1\%}_{1cm}$ value of 102 is characteristic. In 0.1 *N* sodium hydroxide solution, the absorption maxima of capsaicinoids are shifted to 248 nm ($E^{1\%}_{1cm}$ = 330) and 294 nm ($E^{1\%}_{1cm}$ = 136), respectively. Spectrophotometric characteristics of capsaicinoids in 0.02 *N* sodium hydroxide–80% methyl alcohol, and in 0.01 *N* hydrochloric acid–80% methyl alcohol were also determined by The Joint Committee. Kosuge and collaborators isolated capsaicin and its analogs and determined their spectrophotometric characteristics (*24,31,138*). It was shown that individual capsaicin analogs exhibit almost the same spectrophotometric characteristics as the mixture of capsaicin and congeners.

The Joint Committee proposed a direct spectrophotometric method and a spectrophotometric difference method as recommendable procedures (*80*). In the second paper of their recommendation, the $E_{1cm}^{1\%}$ values for the spectrophotometric difference method at 248 and 296 nm were revised to 313 and 127, respectively. To examine contamination of interfering material, The Joint Committee recommends checking the ratios of the optical density readings at 270 nm/280 nm and 290 nm/280 nm for capsaicin. If the ratios are 0.60 for 270 nm/280 nm and 0.53 for 290 nm/280 nm contamination with interfering substances is negligible (*80*). The Joint Committee concluded that the spectrophotometric difference method gave the most reliable results among all the colorimetric and spectrophotometric methods tested. Nevertheless, when contaminants with impurities of similar absorption characteristics exist, an accurate capsaicinoid determination by spectrophotometry is difficult. Therefore, preliminary purification by TLC or column chromatography is essential. Spectrophotometric determination is now playing an important detection role in various chromatographic techniques (*67,139–141*). The detection limit of capsaicinoids by spectrophotometric methods is at the microgram to subnanogram level. A spectrophotometric difference method for practical use was reported by DiCecco (*142*).

3. Other Spectrometric Determination Methods

a. Fluorometry. The detection limits of both colorimetric and spectrometric methods are at the microgram to subnanogram level; sensitivity at the nanogram level or even lower may be obtained by fluorometric detection. Some papers dealing with fluorometric detection have been reported in which fluorometry was used in combination with HPLC (*83,143,144*). Saria *et al.* (*83*) reported that as little as 3 ng of capsaicin could be measured by fluorometric detection using an excitation wavelength of 270 nm and an emission wavelength of 330 nm.

b. Infrared Spectrometry. Infrared (IR) spectrometry has been used for identification of capsaicin and analogs (*24,26,31,145*). Since natural capsaicin shows an absorption at 970 cm^{-1} due to a trans double bond present in the acyl moiety, some trials to determine the content of capsaicin by measuring IR absorption were made (*80,146,147*). Datta and Susi reported that the composition of a mixture of natural capsaicin and vanillyl-*n*-nonamide could be determined by measuring the ratio of the base line absorbance of the 970-cm^{-1} and 1040-cm^{-1} bands (*146*). Later, The Joint Committee approved Datta and Susi's method as a recommendable determination method of synthetic capsaicin analogs (*80*). Müller-Stock *et al.* also reported a quantitative determination of capsaicinoids with a trans double bond in the acyl moiety (*147*). However, the accuracy of quantitation by IR spectrometry is poor, and determination of individual analogs is impossible.

c. NMR Spectrometry. As well as IR spectrometry, NMR spectrometry has also been used as a powerful tool for the identification of capsaicin and analogs (*30,31,33,36,148*). On the other hand, NMR spectrometry was also applied for a quantitative determination of the olefinic and saturated components ratio in a capsaicinoid mixture by Müller-Stock *et al.* (*147*). The ratio of capsaicin analogs with saturated acyl moieties to those with olefinic moieties may be determined by their methods; however, the amount of the individual analog could not be established. Therefore, NMR spectrometry is now rarely applied for the determination of capsaicin analogs. Kirby's group, however, successfully used NMR spectrometry in studying the biosynthesis of capsaicin (*29,149*).

d. Mass Spectrometry. To obtain information on the molecular weight and chemical structure, mass spectrometry (MS) is the most informative tool. In fact, the chemical structure of capsaicin and its analogs have been confirmed by MS in cooperation with NMR, IR, and UV spectrometries and elementary analysis. Before introduction of GC–MS, unidentified capsaicinoids were subjected to mass spectrometry by using a direct inlet system (*29,31,98*). For MS analysis using the direct inlet system, the sample must first be purified to a single component. Jurenitsch *et al.* used preparative scale polyamide–cellulose TLC for purification (*33*). With improvement of MS, the identification and quantitation of capsaicin and its analogs by the direct inlet system using field desorption, chemical ionization, fast atom bombardment, or laser microprobe mass analysis will increase in future.

D. Chromatographic Analyses

Since crude capsaicinoid extracts often include pigments and materials that interfere with colorimetric determination and are hard to remove by solvent partition extraction or by recrystallization from nonpolar solvents, especially when the amount of sample is inadequate, data on capsaicinoid content obtained by colorimetric or spectrophotometric assay methods in the early days of capsaicinoid research were frequently unreliable. Therefore, effective methods for purification of the crude capsaicinoid extracts were developed. The progress in modern analytical chemistry brought with GLC, GC–MS, PC, TLC, and HPLC techniques has enabled capsaicinoid analyses to be more accurate, more sensitive, and more time efficient.

1. Column Chromatography

Column chromatography was mainly used to purify crude capsaicinoid extracts. Schenk (*150*) reported a simple method to isolate capsaicinoids by running them through a column of aluminum oxide-activated carbon (100 : 1, w/w) developed with ethyl alcohol. A similar method was used by Hollo *et al.* (*151*) for purification of capsaicinoids from paprika. Schulte and Krüger, on the other hand, used a column packed with separate layers of alumina and charcoal (3 : 1, w/w) to elute capsaicinoids with ethyl alcohol (*120*). Suzuki *et al.* (*77*) prepared a column packed with separate layers of acidic alumina, basic alumina, and charcoal mixed with an equal weight of Kieselguhr, and capsaicinoids were eluted with absolute methyl alcohol. The capsaicinoid amount in the eluates was determined by measuring light absorption at 280 nm.

Charcoal participates as an effective adsorbent to hold undesirable colored materials. However, fats and waxes present in the crude extracts cannot be removed with an alumina column by elution with alcohol. Brauer and Schoen (*152*) also used an aluminum oxide column to separate capsaicinoids from carotenoid pigments. Capsaicinoids eluted with isopropyl alcohol–water (94 : 6, v/v) were measured spectrophotometrically at 282 nm. The panel of The Joint Committee established a column chromatographic purification method (*78*) by modifying Schulte and Krüger's method (*153*). The recommended method uses a column in which aluminum oxide and a mixture of activated carbon and Kieselguhr are separately packed. Capsaicinoids were eluted with absolute methyl alcohol. Schulte and Krüger's method was found to be effective by Rangoonwala (*97*). An activated alumina column to purify the crude capsaicinoid extract from *Capsicum* species was used by Bennett and Kirby (*29*) and more recently by DiCecco (*142*) and Bajaj and Kaur (*85*).

Jentzsch *et al.* (*30*) used a polyamide column to separate capsaicin (**2**) and dihydrocapsaicin (**3**) on a preparative scale by eluting with 0.2 *N* sodium hydroxide. The polyamide column was also used by Sykulska *et al.* (*154*) to purify capsaicin (*2*).

Müller-Stock *et al.* (*98*) fractionated capsaicin and its analogs on a silicic acid column by eluting with various solvent systems. Nordihydrocapsaicin (**4**), nonylic acid vanillylamide (**11**), and capsaicin (**2**) were successfully separated from dihydrocapsaicin (**3**) and homodihydrocapsaicin (**5**).

2. Paper Chromatography

Paper chromatography (PC) is no longer used for capsaicinoid analysis as a major technique, however, the contribution of PC to capsaicinoid research should not be overlooked. Paper chromatography first enabled Kosuge and collaborators to recognize the heterogeneity of "capsaicin" prepared by conventional techniques (*23*). Fujita *et al.* also reported the purification and isolation of capsaicinoids by PC (*103*), and capsaicinoids located at R_f 0.13 were separated from carotenoids on a filter paper developed with petroleum benzin (bp 60–70°C) saturated with methyl alcohol. Capsaicinoids were visualized by spraying with 3% phosphomolybdic acid and 0.1 *N* sodium hydroxide. Two dimensional PC first developed with petroleum benzin saturated with methyl alcohol followed by methyl alcohol saturated with petroleum benzin showed a good separation of capsaicinoids from carotenoids, phenolic compounds, amino acids, organic acids, sugars, and other impurities. Fujita *et al.* (*103*) chose petroleum benzin saturated with methyl alcohol for purification of capsaicinoids by PC in which capsaicinoids were located at R_f 0.69. In their solvent system, however, neither mutual separation of capsaicin and analogs nor a separation of vanillin from capsaicinoids was achieved.

Kosuge and collaborators first used PC to examine the purity of isolated capsaicin from *Capsicum* fruits. They chromatographed the isolated "pure capsaicin" with a mp of 64.5°C on cellulose filter paper by developing with an equal volume mixture of 0.1 *M* sodium hydroxide and 0.05 *M* sodium carbonate and located two spots with R_f values of 0.68 ± 0.01 and 0.55 ± 0.01 with Folin–Looney spray reagent, thus showing that crystalline capsaicin prepared from *Capsicum* fruit by the conventional method was a mixture of two compounds. In a following paper, they obtained similar results confirming the heterogeneity of so-called "pure capsaicin" by developing with 0.4 *N* hydrochloric acid saturated with benzene (*24*).

For the separation of capsaicin analogs by PC, 0.2 *M* sodium hydroxide–0.1 *M* sodium carbonate–0.4 *M* sodium acetate (3 : 2 : 5, by vol.) (*155*) and 0.1 *M* sodium hydroxide–0.05 *M* sodium carbonate (3 : 2, v/v) (*119*) were found suitable. Interfering substances in ethyl ether or acetone extracts of *Capsicum* fruits were removed by PC developed with acetic acid–acetone–water (1 : 3 : 6, by vol.) for short distance, then impurities remaining at the origin are removed together with the filter paper by cutting with scissors. The capsaicinoids at the solvent front were further developed with 0.2 *M* sodium hydroxide–0.1 *M* sodium carbonate–0.4 *M* sodium acetate (3 : 2 : 5, by vol.) to obtain capsaicin and dihydrocapsaicin (*119*). By this method, any impurities except trace amounts of coloring materials were separated from capsaicinoids. For visualization of capsaicinoids Kosuge *et al.* found that the Folin-Ciocalteau reagent (*156*) and the ferric chloride–potassium ferricyanide reagent (*157*) were suitable reagents with a detectability of 2 μg or less (*119*). Thus, PC was a useful tool contributing to capsaicinoid analysis throughout the brilliant works of Kosuge and co-workers on the pungent principle of hot red peppers (*25,26,70,145*).

Kowaleski reported a separation method for capsaicinoids by PC (*158*), however, in his method tailing of the capsaicin spot was pointed out to be a shortcoming. Mutual separation of capsaicin analogs cannot be performed by his system either. Paper chromatographic separation has been used relatively recently by Govindarajan *et al.* (*65,67,159*), however, it has been replaced today with TLC or GLC.

3. Thin-Layer Chromatography

In comparison to PC, thin layer chromatography (TLC) is highly versatile, speedy, and sensitive, so PC has rapidly been replaced by TLC. Since the first report by Tiechert *et al.* (*160*) an enormous number of papers on capsaicinoid studies using TLC have been reported.

a. Silica Gel. Among TLC systems, the one using silica gel as the adsorbent has been most widely used. More than 20 papers using silica gel plates for capsaicinoid analysis have been reported since the 1960s (*1,66, 98,104,137,139,140,161–172*). In most cases TLC on silica gel plates is used for separation of capsaicinoids in order to get a single spot free from other impurities so that capsaicinoids can be recognized, quantitated, and/or isolated.

Various combinations of solvent systems have been examined for better separation of capsaicinoids. When the adsorbent, silica gel, is used without any treatment, TLC works in a mode of adsorption chromatography. Therefore, as in most adsorption chromatography, mixtures of organic solvents with different polarities are used as developing solvents. Tyihak *et al.* used an equal mixture of benzene and ethyl acetate for the chromatography on silica gel G and H layers (*1*). Spanyar and Blazovich used chloroform–ethyl alcohol (99 : 1, v/v) as a fast method for capsaicinoid analysis by TLC (*104*). In a new method for qualitative identification of capsaicin in drugs, Stahl and Kraus proposed a method (*168*) that rapidly identifies capsaicin in pharmaceutical preparations on silica gel GF_{254} plates developed with cyclohexane–chloroform–acetic acid (60 : 30 : 10, by vol.). Jentzsch *et al.* (*164*) used ethyl ether in preparation of capsaicinoids on silica gel HF_{254} (0.25 mm thick). In another paper, Jentzsch used a mixture of chloroform and ethyl ether (3 to 1) as the developing solvent for the quantitation of capsaicinoids (*169*). A mixture of chloroform–ethyl acetate–isopropyl alcohol (33 : 9 : 3, by vol.) on a silica gel G was used by Rios and Duden (*139*) for qualitative determination of capsaicinoids. Haguenoer *et al.* reported TLC of capsaicinoids extracted from pimentos and *C. fastigiatum* on silica gel with chloroform–methyl alcohol–acetic acid (95 : 1 : 5, by vol.) (*170*).

Müller-Stock *et al.* (*98*) tried to obtain better separation of capsaicinoids by applying a number of different combinations of solvent systems such as ethyl ether–ethyl alcohol–25% ammonia, benzene–ethyl alcohol–25% ammonia, toluene–ethyl alcohol–25% ammonia, chloroform–ethyl alcohol–25% ammonia, and chloroform–ethyl alcohol–trimethylamine, and found that chloroform–ethyl alcohol–25% ammonia (9 : 2 : 1, by vol.) and ethyl acetate–25% ammonia (9 : 1, v/v) are suitable solvent systems for satisfactory separation of capsaicin analogs. Polesello and Pizzocaro (*140*) separated capsaicin on silica gel G with chloroform–methyl alcohol (95 : 5, v/v) (*106*). Although Tyihak *et al.* (*1*) reported that capsaicinoid mixtures from *Capsicum* fruits could be separated into two components on silica gel G and H layers when developed with an equal mixture of benzene and ethyl acetate, capsaicin and its analogs could not be separated from each other by two-dimensional TLC on silica gel plates without any treatment (*98,141*). Pankar and Magar (*173*) prepared multiband thin layer plates coated with Kieselguhr G containing 8% activated charcoal between silica gel G in the horizontal direction and developed the compounds with absolute methyl alcohol–acetic acid (49 : 1, v/v). Pure capsaicinoids were isolated on a preparative scale with 92.5 to 96.2% recovery.

For visualization of capsaicinoids on silica gel plates, ferric chloride–potassium ferricyanide reagent (*1,104*), methanolic 2,6-dichloroquinone chlorimide solution, and subsequent exposure to ammonia vapor (*164,165*), modified Gibbs' reagent (*173,174*), Folin–Denis reagent (*137*), or UV measurement at 280–281 mμ (*139,140,169*) have been applied for both qualitative and quantitative use at μg level. For quantitative purposes, measuring light absorption at visible wavelengths was applied (*173,175,176*). Heusser reported the precision of photometric determination in TLC to be fairly high, with a standard deviation of $\pm$2.9% for the Folin–Denis reagent (*177*).

b. Kieselguhr. For the isolation of pure capsaicinoids on preparative scale, kieselguhr TLC has also been used. Karawya *et al.* used TLC on kieselguhr plates developed with petroleum ether–ethyl alcohol (99 : 1, v/v) for purification of capsaicinoids prior to spectrophotometric determination (*64*). In their method the recovery of capsaicinoids from thin-layer plates by solvent extraction was almost 100%. Debska and Okulicz-Kozarynowa, on the other hand, reported the separation of capsaicin on mixed plates of silica gel PF–kieselguhr G (1 : 2, w/w) developed with *n*-heptane–chloroform–butyl alcohol–water (60 : 34 : 6 : 0.2, by vol.) and subsequent determination by colorimetry (*176*).

c. Polyamide. For the separation of capsaicin and its analogs, polyamide thin-layer plates have been used. Rangoonwala preferred polyamide TLC for the analysis of capsaicin and analogs (*27,178,179*) in which the compounds were separated from each other with water–dioxane (2 : 1, v/v), or with 0.1 *N* sodium hydroxide–dimethylformamide (9 : 1, v/v), In a TLC system using cellulose powder DC MN 300–polyamide DC (1 : 9, w/w) developed with 0.1 *N* sodium hydroxide–dimethylformamide (*179*), a satisfactory separation of synthetic *cis*-capsaicin, capsaicin, *N*-vanillylperalgonic acid amide, and dihydrocapsaicin was achieved. Fast blue and potassium hydroxide (*27*), 0.5% diazobenzenesulfonic acid, and 1 *N* sodium hydroxide in methyl alcohol were used as color reagents for the visualization of capsaicinoids on the polyamide plates. Polyamide plates are superior to silica gel plates and Kieselguhr plates for separation of individual analogs, and running chromatography on a preparative scale is also possible (*33*). Jurenitsch *et al.* used polyamide TLC to isolate pure capsaicin analogs for their identification by MS (*33*).

d. Silver Ion Complexation TLC, Reversed-Phase TLC, and High Performance TLC. Todd *et al.*, during the course of seeking a suitable TLC method for the qualitative analysis of natural and synthetic capsaicin analogs, compared three methods and found that polyamide plates using silver ion in the developing solvent was the quickest of the three (*180*). Capsaicin and its analogs were separated from each other by either silver nitrate-impregnated silica gel TLC or by silicone-treated reversed-phase silica gel TLC. Bennett and Kirby first adopted the silver nitrate-impregnated silica gel GF_{254} for the separation of capsaicin and its analogs (*29*). According to their paper, thin-layer plates were prepared from a mixture of Silica Gel G (Merck), silver nitrate, boric acid, and water. A portion of the sample was subjected to chromatography on the plates and then eluted with a 1 : 1 mixture of chloroform and ethyl acetate. Capsaicin (R_f 0.32) and dihydrocapsaicin (R_f 0.48) were located as yellow spots on a brownish background by spraying with a solution of potassium permanganate in a 2% sodium carbonate solution (*29*). Kosuge and Furuta (*31*) fractionated the crystalline pungent principles from the Japanese *Capsicum* into capsaicin (R_f 0.50), dihydrocapsaicin (R_f 0.32), and two unidentified spots with R_f values of 0.42 and 0.21 by silver nitrate TLC on tetralin-coated silica gel developed with an equal mixture of methyl alcohol and 2% silver nitrate. A separation of capsaicin analogs by reversed-phase TLC was also reported by Todd (*180*) in which capsaicin moves together with nordihydrocapsaicin to give R_f 0.52 while dihydrocapsaicin and synthetic homocapsaicin (**6**) move together to be located at R_f 0.41, and homodihydrocapsaicin (**5**) shows R_f 0.31.

For analysis of capsaicin analogs at the nanogram level, reversed-phase high performance TLC (RP–HPTLC) can be a convenient and powerful tool (*181,182*). Iwai and colleagues used a Merck RP-8 reversed-phase silica gel G plate for the separation of capsaicin and its natural and synthetic analogs by developing with mixture of 0.05 *M* silver nitrate and 0.05 *M* boric acid in 85% methyl alcohol. Location and quantitation were done with Gibbs' reagent or by reading the absorption at 235 nm with a thin-layer chromatogram scanner. Homodihydrocapsaicin, dihydrocapsaicin, synthetic *cis*-capsaicin, synthetic *cis*-homocapsaicin, and natural capsaicin were satisfactorily separated. Better separation was obtained on the same plate by two dimensional TLC developed with 85% methyl alcohol in the first development, followed by with the mixture of 0.05 *M* aqueous silver nitrate and 0.05 *M* aqueous boric acid in 85% methyl alcohol (*181*). Jurenitsch also reported the identification in *Capsicum* fruits and pharmaceuticals of adulterations with nonylic acid vanillylamide, *N*-(4-hydroxy-3-methoxybenzyl)-nonylamide, by high performance silica gel RP-8 plates developed with 1.6% silver chlorate in acetonitrile–water (8 : 2, v/v) as the mobile phase, and obtained satisfactory results (*182*). Jurenitsch proposed RP–HPTLC developed with a solvent system containing silver as an official analytical method for capsaicinoids in *Capsicum* extracts and pharmaceuticals.

4. Gas Chromatography

Because of its high sensitivity, separability, and speed, gas chromatography (GLC) has also been a useful tool for capsaicinoid analysis along with TLC and HPLC. Capsaicinoids were first analyzed with GLC in form of fatty acid methy esters after splitting them into fatty acids and vanillylamine by acid hydrolysis (*31,99*). Analysis of fatty acid methyl esters continues to be used for the quantitation of nonlylic acid vanillylamide and other capsaicinoids in extracts from *Capsicum* fruits with glass capillary columns (*89*).

Gas chromatographic analysis of free capsaicinoids was also reported by Morrison (*183*) in which free capsaicin analogs were applied on a column coated with a nonpolar liquid phase, and chromatographed at a fairly high temperature. In general, when capsaicin and its analogs are injected to columns in the free form, tailing of peaks is frequently observed and reproducibility of peak areas is relatively poor. Grushka and

Kapral (*184*), however, reported that sufficient separation of three major components (capsaicin, dihydrocapsaicin and nordihydrocapsaicin) from oleoresin of *Capsicum* fruits was obtained, without any derivation of capsaicinoids, on either 10 or 15% Dexsil 300 on silanized Chromosorb W at 240–250°C.

In general, a weakly polar or nonpolar liquid phase is preferred for analysis of capsaicinoids, but a polar liquid phase is occasionally used (*185*). Frequently used liquid phases are silicone SE-30 (*102,115,148*), silicone SE-52 (*66,105*), JXR-silicone (*98,186*), Carbowax 20M (*187*), and Apiezone grease L (*188*), which are coated on a silicone inactivated support such as Chromosorb W.

Because trimethylsilyl (TMS) derivatives of capsaicinoids give better peaks than free capsaicinoids, even at lower column temperatures, GLC of capsaicinoids has mostly analyzed TMS derivatives on a moderate or nonpolar liquid phase column (*66,98,102,148,186–190*). Trimethylsilylation is generally accomplished by the use of mixture of trimethylchlorosilane and hexamethyldisilazane in pyridine, by *N,O*-bis(trimethylsilyl)acetamide in acetonitrile, or by *N,O*-bis(trimethylsilyl)trifluoroacetamide (*32,102,105,115,186*).

Capsaicinoid peaks may be detected and quantitated by a thermal conductivity detector (*99*), flame ion detector (*32,148,186*), or alkaline flame detector (*102*). Among these methods the flame ion detector has been most widely used because of its high sensitivity. The alkaline flame detector has been reported to show much higher sensitivity than the flame ion detector, however, it has not become so popular as the flame ion detector.

By using a column packed with 3% silicone SE-30 on Chromosorb GHP (100–120 mesh) in 2 m × 2 mm i.d., Todd *et al.* (*115*) obtained satisfactory separation of natural and synthetic capsaicinoid mixtures that were tested as TMS derivatives. Synthetic vanillyl octamide, vanillyl peralgonamide, and vanillyl capramide were separated from natural capsaicinoids. For natural capsaicinoids, nordihydrocapsaicin, homocapsaicin, and homodihydrocapsaicin were also separated satisfactorily. Capsaicin and dihydrocapsaicin were partially separated, enabling a quantitation of the individual analogs.

A comparatively good separation of capsaicin and dihydrocapsaicin was also obtainable with a 2 m × 3 mm i.d. glass U-tube packed with 3% SE-30 on 60–80 mesh, acid washed, silanized Chromosorb W by raising the temperature from 170 to 230°C with an increasing rate of 2°C/min (*32*). Complete separation of capsaicin and dihydrocapsaicin on a normal-sized packed column, however, is still difficult. For complete separation of individual analogs use of a capillary column coated with a high temperature-resistant liquid phase such as the silicone OV series would be useful.

The benefit of GLC is its high performance in simultaneous identification and quantitation of capsaicin and analogs in relatively short time. Gas chromatography of capsaicinoids does not necessarily require strict pretreatment to remove interfering substances as in the case of colorimetry and spectrophotometry, however, preliminary treatment to remove impurities before subjection to GLC should be favored.

5. Gas Chromatography–Mass Spectrometry

Gas chromatography is an excellent technique for rapid analysis of capsaicin and its analogs, and it is possible in part to obtain information on the structure of unidentified capsaicin analogs by comparing relative retention time on the gas chromatogram (*31*). However, the information obtained from the retention time is by no means conclusive as that from mass spectra and NMR spectra. When contaminant peaks are involved, prediction of the structure of a capsaicinoid is quite difficult and sometimes entirely impossible. Mass spectrometry (MS) is a powerful method for identification of unknown compounds, however, it requires a pure sample. As described in Section III,C,3,d, time-consuming and sophisticated pretreatment has to be done in order to obtain purified capsaicin and analog for MS. Gas chromatography–mass spectrometry (GC–MS) spares the analyst the sophisticated and tedious pretreatment by making it possible to perform simultaneous separation and identification of capsaicinoids.

The idea proposed by Bennett and Kirby (*29*) that the pungent principle of *C. annuum* was composed of at least five closely related compounds was directly confirmed by Masada *et al.* (*32*) using GC–MS. In mass spectra, the molecular ions of capsaicin, dihydrocapsaicin, nordihydrocapsaicin, homodihydrocapsaicin, and homocapsaicin were observed at *m/z* 305, 307, 293, 321, and 319, respectively (*32*). TMS derivatives of capsaicin, dihydrocapsaicin, nordihydrocapsaicin, homodihydrocapsaicin, and homocapsaicin showed molecular ions at *m/z* 377, 379, 365, 393, and 391, respectively, together with the most abundant common fragment peak at *m/z* 209 derived from 4-*O*-TMS-3-methoxybenzyl ion. The M−15 ions are also commonly observed in TMS derivatives of capsaicinoids. The occurrence of caprylic acid vanillylamide, or *N*-(4-hydroxy-3-methoxybenzyl)octanamide, along with capsaicin, dihydrocapsaicin, and nordihydrocapsaicin was confirmed by Jurenitsch *et al.* (*191*) using GC–MS.

The detection limit of GC–MS is generally around 0.1 μg for a peak, but, however, it can be improved to the nanogram to subpicogram level by using a selective ion monitoring technique (mass fragmentography) (*105*) or mass chromatography (*86*). By selective ion monitoring or mass chromatography, plural numbers of capsaicin analogs are identified and quantitated simultaneously even when two peaks are inseparable on the chromatogram. Moreover, even when the capsaicinoid peaks are contaminated with impurities, both selective ion monitoring and mass chromatography enable the determination of the individual analog provided that characteristic mass fragments for the individual analog are not overlapped and obscured by those from impurities.

6. High Performance Liquid Chromatography

High performance liquid chromatography (HPLC) is a powerful method that is far superior to conventional analytical techniques applied to capsaicinoids. The basic principle of HPLC is essentially the same as that of conventional column chromatography. However, owing to the introduction of carefully selected high performance column packings, high pressure loading, and highly sensitive detectors, the separation efficiency, analysis time, detection limit, and sample amount required for chromatography were marvelously improved in comparison to conventional liquid chromatography. Beside improvements in column and pumping systems, the introduction of highly sensitive detectors in HPLC has enabled it to detect nanogram quantities of capsaicinoids using fluorometric detection (*83,143,144*). Analysis of capsaicinoids by HPLC was first reported by Iwai and colleagues (*105*) in the isolation of pure capsaicin from commercial capsaicin on a reversed-phase column coated with octadecyltrimethoxysilane (Permaphase ODS) by developing a linear elution gradient by increasing the methyl alcohol concentration in water. In their method, separation of the analogs was accomplished in only 20 min. Eluted capsaicin and dihydrocapsaicin were detected by monitoring the UV absorption at 254 nm. The advantage of capsaicinoid analysis by HPLC is that it does not require modification of capsaicinoids prior to injection.

In Sticher *et al.*'s report (*192*), the separation and quantitative determination of capsaicin, dihydrocapsaicin, nordihydrocapsaicin, and homodihydrocapsaicin in *Capsicum* fruits was performed on a reversed-phase μBondapak C_{18} column developed with methyl alcohol–water (53 : 47,v/v) at a flow rate of 1.5 ml/min. Capsaicin and analogs are detected at the 100-ng level by UV monitoring at 279 nm. It takes ~40 min to finish a single chromatographic run. However, HPLC does not serve only for determination of trace amounts of capsaicinoids. Iwai *et al.* (*86*) used a high performance silica gel (Zorbax SIL) column for the isolation of microgram levels of capsaicinoid by developing with isopropyl alcohol–*n*-hexane–methyl alcohol (10 : 90 : 1, by vol.) at 50 kg/cm^2. Eluates were collected by monitoring the elution profile at 235 nm. With the solvent system mentioned above, capsaicin, dihydrocapsaicin, and nordihydrocapsaicin were eluted together. The collected eluates were thus provided for further determination of the individual analogs by GC–MS. A similar method was proposed by Heresch and Jurenitsch (*34*) in which a reversed-phase LiChrosorb RP-8 column eluted with dioxane–water (44 : 56,v/v) was used for the separation. Eluates were monitored at 280 nm. Confirmation of the structure of homodihydrocapsaicin II as 8-methyldecanoic acid vanillylamide, or *N*-(4-hydroxy-3-methoxybenzyl)-8-methyldecanamide, by MS and NMR was also preceded by preparative HPLC (*36*). Satisfactory separation of individual capsaicin analogs has also been accomplished by Jurenitsch *et al.* (*88*). Vanillylamide caprylate (**10**), nordihydrocapsaicin (**4**), vanillylamide nonylate (**11**), capsaicin (**2**), dihydrocapsaicin (**3**), vanillylamide decylate (**12**), homodihydrocapsaicin I (**5**), and homodihydrocapsaicin II (**8**) were separated on a reversed-phase LiChrosorb RP-8 column eluted with dioxane–water (44 : 56, v/v) or with methyl alcohol–water (57 : 43,v/v). They found a direct linear relation between capacity values (log K') and numbers of carbon atoms of the side chain within two homologous series of capsaicinoids.

Since capsaicin analogs may be separated on a TLC plate by complexation with silver, similar attempts were made with HPLC developed with silver-containing solvent systems (*144*). Five separate peaks that were presumed to be capsaicin (**2**), nordihydrocapsaicin (**4**), homocapsaicin (**6**), dihydrocapsaicin (**3**), and homodihydrocapsaicin (**5**) were separated on a reversed-phase MicroPak MCH-10 column with a gradient elution of 50–80% methyl alcohol containing 0.05 *M* silver nitrate. Jurenitsch and Kampelmühler (*193*) also reported successful separation of capsaicin and analogs on a LiChrosorb RP-8 column developed with 52% aqueous methyl alcohol containing silver ion. The critical pair capsaicin (**2**) and nonylic acid vanillylamide (**11**) can be separated by their method. As mentioned by Jurenitsch and Kampelmühler (*193*), the silver complexation HPLC can be a more precise, reproducible, and convenient method of determination than GLC for capsaicin and its analogs for routine quality control of *Capsicum* fruits and extracts in the near future, assuming that HPLC apparatus will become cheaper.

High performance liquid chromatographic methods have recently been used for routine analysis of capsaicin in tear gas (*194,195*), and in animal feed, human urine, and waste water (*84*). To determine and separate capsaicin analogs in animal tissues, Saria *et al.*'s method using reversed-phase HPLC also is recommendable (*83*).

E. Miscellaneous Methods

Capsaicin and its analogs are polarographically inactive. However, their reaction products with sodium nitrite in acid medium are reducible.

By taking advantage of this property, Spanyar *et al.* proposed a method involving polarometric titration with *p*-diazobenzenesulfonic acid, which is applicable to samples containing more than 30 mg of capsaicin per 100 g raw material and less than 20% moisture (*196*). Bozsa *et al.* (*197*) also established a capsaicin determination method by using polarography. An application of the polarographic determination to capsaicinoids in anti-rheumatic tapes was reported by Miedzinski and Stachowiak (*198*).

Earlier, a fluorometric titration method was proposed by Nogrady (*199*) in which capsaicin was determined by titrating with picric acid and fluorescence desorption. This method has been scarcely applied for the practical use due to its poor reliablity.

IV. Synthesis of Capsaicin and Analogs

Chemical synthesis of capsaicin and its analogs was started from two different motives: (A) to assign the proposed chemical structure of capsaicin isolated from *Capsicum* fruits, and (B) to reveal the relationship between pungency and chemical structure. Later, because of its usefulness to the spice industry, to pharmacy, and even as a chemical weapon, efforts in developing new chemical synthetic procedures for capsaicinoid syntheses have been made. The first paper on chemcial synthesis of capsaicinoids was reported by Nelson and examined the relationship between pungency and chemical structure (*200*). He synthesized undecylenoyl vanillylamine, or *N*-(4-hydroxy-3-methoxybenzyl)undecylamide, by reacting 2 mol free vanillylamine and 1 mol of undecylenoyl chloride in etheral solution. The product was recovered in crystalline form by recrystallization from petroleum ether.

Ott and Zimmerman (*201*) reacted vanillylamine hydrochloride and undecylenoyl chloride in the presence of sodium carbonate, but failed to obtain undecylenoyl vanillylamide in crystalline form. In an effort to reveal the relationship between pungency and chemical structure, Kobayashi (*202*) synthesized a number of capsaicin analogs and homologs in crystalline form and with high yield using either Nelson's procedure (*200*) or Ott and Zimmermann's procedure (*201*). Mitter and Ray also synthesized various capsaicin homologs (*203*). Chemical syntheses of capsaicinoids for similar purposes were reported by Nakajima (*204*) and Tsai *et al.* (*189*).

In 1920, Nelson resynthesized capsaicin from decenoic acid vanillylamine obtained from capsaicin (*19*), and afterwards succeeded in a synthesis of 8-methylnonoyl vanillylamide, or *N*-(4-hydroxy-3-methoxybenzyl)-nonamide, from 8-methylnonoyl chloride and vanillylamine (*20*).

A total chemical synthesis of capsaicin (**2**) was reported by Späth and Darling (*205*). According to their method, isobutyl zinc iodide was reacted

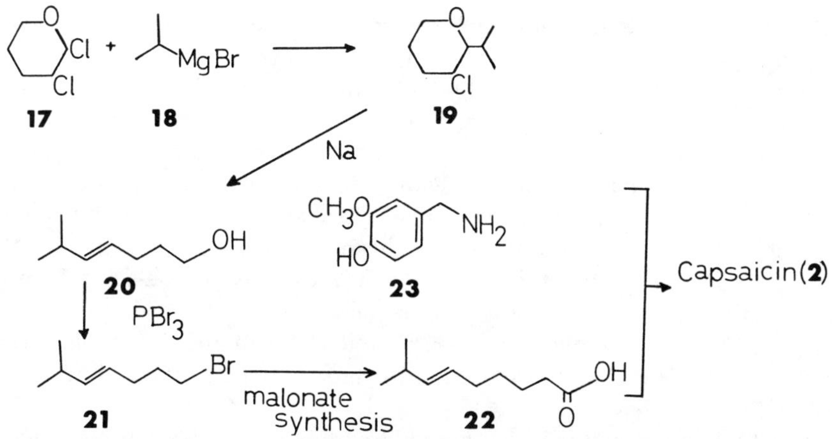

SCHEME 1. Chemical synthesis of capsaicin (**2**) (*22*).

with ethyladipyl chloride in cold toluene to yield the ester of 8-methylnonan-6-on-1-ic acid. The free acid was reduced to 8-methylnonan-6-ol acid and then brominated to 8-methylnonan-6-bromo acid. The methyl ester of 8-methylnonan-6-bromo methyl ester was dehydrobrominated to the methyl ester of 8-methyl-6-nonen-1-ic acid by distillation in quinoline at 200–220°C. After saponification and subsequent conversion to the acid chloride, the 8-methyl-6-nonenoyl chloride was reacted with vanillylamine in diethyl ether to give capsaicin (**2**). The capsaicin synthesized by Späth and Darling's procedure was a mixture of structural and geometrical isomers, and their method afforded less than 6% yield of capsaicin as the mixture of isomers.

The procedure proposed by Crombie *et al.* (*22*) does not require such sophisticated chemistry to obtain pure *trans*-capsaicin. The outline of chemical synthesis of *trans*-capsaicin is shown in Scheme 1. Isopropylmagnesium bromide (**18**) in diethyl ether is treated with 2,3-dichlorotetrahydropyran (**17**) to yield crude 3-chlorotetrahydro-2-isopropylpyran (**19**) as a mixture of stereoisomers. The crude ether is then added to powdered metallic sodium in diethyl ether, reacted with methyl alcohol, and quenched with water. The diethyl ether layer yielded 6-methylhept-*trans*-4-en-1-ol (**20**). Ozonization of **20** yielded isopropylaldehyde, which was characterized. Compound **20**, when treated with phosphotribromide in pyridine at 0°C for 0.5 hr, afforded 6-methylhept-*trans*-4-enyl bromide (**21**), which was used to alkylate ethylmalonate in ethyl alcohol in the presence of sodium. The crude alkenylmalonate was hydrolyzed with aqueous potassium hydroxide to the alkenylmalonic acid and the latter decarboxylated at 160–180°C for 2 hr to 8-methylnon-*trans*-6-enoic acid (**22**). Treatment of **22** with thionyl chloride afforded the acid chloride and, when reacted in diethyl ether with freshly prepared vanillylamine, gave crude capsaicin (**2**). Pure material was obtained by repetitive recrystallization from light petroleum (bp 40–60°C)–diethyl ether (9 : 1, v/v) in colorless leaflets. The total yield of capsaicin was 4.7%.

To synthesize dihydrocapsaicin (**3**), 8-methylnon-*trans*-6-enoic acid (**22**) was hydrogenated in dioxane with palladium oxide. To avoid complicated multistage reaction procedures, Takahashi *et al.* (*206*) modified Crombie *et al.*'s synthesis. In the Takahashi synthesis, the acyl moiety of **3** was obtained in a one-step reaction in 30% yield using an electrochemical coupling method (*207*) of isovaleric acid with pimelic acid monoethyl ester. 8-Methylnon-6-enoic acid (**22**), similarly prepared from 5-methylhex-*trans*-3-enoic acid and glutaric acid monoethyl ester, afforded only 6% of **2**.

Nordihydrocapsaicin (**3**) and homodihydrocapsaicin (**5**) were first synthesized by Kosuge and Furuta (*31*). A total yield of 16% and 21% for **3** and **5**, respectively, was obtained by using Crombie *et al.*'s procedure (*22*).

The synthesis of *cis*-capsaicin (**73**), which is not present in nature, was reported by Rangoonwala and Seitz who obtained a 70% yield by reaction of 2 mol of vanillylamine with 1 mol of the *cis* acid chloride (*208*). *cis*-8-Methylnon-6-enoic acid (**72**) was synthesized starting from dibromopentane (**69**) via monobromopentanoyl cyanide (**70**), triphenyl-phosphorpentanoyl cyanogen bromide, and 8-methylnon-6-enyl cyanide (**71**). By this method, the total yield was 16%. The outline of *cis*-capsaicin synthesis is shown in Scheme 2.

Vig *et al.* (*209*) proposed a new method for the synthesis of capsaicin (**2**): 4-methyl-1-pentene-3-ol, prepared by the reaction of vinylmagnesium bromide with isopropylaldehyde, was converted to 3-vinyloxy-4-methyl-1-pentene, which was then rearranged stereospecifically to *trans*-6-methyl-4-heptenal by a Claisen rearrangement and converted to *trans*-6-methyl-4-heptenol by reduction with $LiAlH_4$. Bromination of the *trans*-6-methyl-4-heptenol and alkylation of sodium malonate afforded diethyl-2-(*trans*-6′-methyl-4′-heptenyl)-malonate. Decarboxylation of the above compound by a mixture of aqueous dimethylsulfoxide and sodium chloride gave 8-methylnon-6-enoic acid ethyl ester. After saponification and reaction of 8-methylnon-6-enoic acid (**22**) with phosphorous trichloride, followed by reaction with vanillylamine, capsaicin (**2**) was obtained.

Nelson synthesized vanillylamine from vanillyloxime by reduction with metallic sodium in acetic acid (*18*). Takahashi *et al.* (*206*) further improved this procedure.

For the study of capsaicinoid biosynthesis, radioactively labeled precursors are often used to reveal biosynthetic pathways. Iwai and colleagues, in their paper studying enzymatic capsaicinoid synthesis, re-

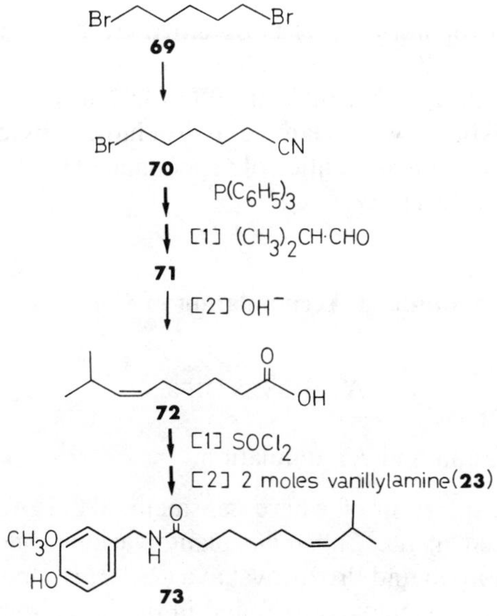

Br⁄⁄⁄Br
69

Br⁄⁄⁄CN
70

$P(C_6H_5)_3$

[1] $(CH_3)_2CH{\cdot}CHO$

71

[2] OH^-

OH

72

[1] $SOCl_2$

[2] 2 moles vanillylamine(**23**)

CH_3O HO N H

73

SCHEME 2. Chemical synthesis of *cis*-capsaicin (**73**) (*208*).

ported a method for chemical synthesis of 4-hydroxy[3-^{14}C]-methoxybenzylamine (*211*). In the chemical synthesis of the compound as outlined in Scheme 3, the principal step has been based on the method previously reported (*212*). In the first step, pyrocatechol (**25**) is methylated to ^{14}C-labeled veratole (**26**) with $^{14}CH_3I$. In the last step, ^{14}C-labeled

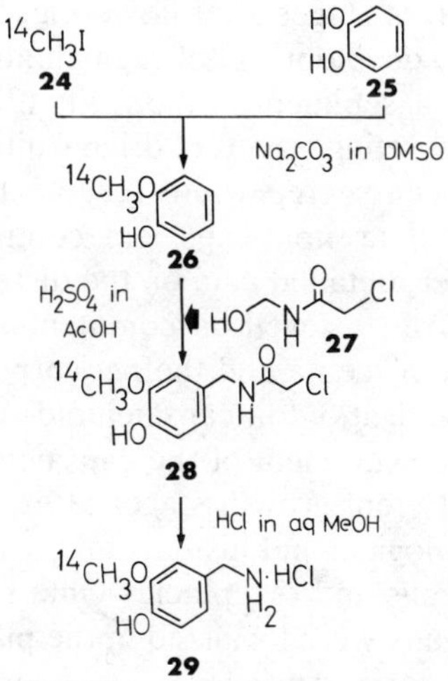

$^{14}CH_3I$
24

HO HO
25

Na_2CO_3 in DMSO

$^{14}CH_3O$ HO
26

H_2SO_4 in AcOH

HO N H Cl
27

$^{14}CH_3O$ HO N H Cl
28

HCl in aq MeOH

$^{14}CH_3O$ HO $N{\cdot}HCl$ H_2
29

SCHEME 3. Synthesis of [^{14}C]methoxyvanillylamine hydrochloride (*211*). (By permission of *The Agr. Chem. Soc. Japan*.)

vanillylamine hydrochloride (**29**) is obtained by heating 4-chloroacetyl-aminoethyl guaiacol (**28**) in methyl alcohol containing conc hydrochloric acid (1 : 0.2, v/v) in a sealed tube at 70°C for 2 days. The product was identified as ^{14}C-labeled vanillylamine hydrochloride by comparing its IR spectrum with that of an authentic cold specimen. The yield, however, by this method was only 11.4%.

V. Formation and Accumulation in *Capsicum* Plants

A. Biosynthesis

1. Sites of Formation and Accumulation

As regards the question of where capsaicinoids are formed, accumulated, and secreted in the *Capsicum* plant, various theories have been presented. Prokhorova and Prozorovskaya (*213,214*) reported that a maximum amount of capsaicinoid was found in the inner walls (placenta and dissepiment) of *Capsicum* fruits. They obtained results showing that pericarp and seeds contained practically no capsaicinoid.

It was Furuya and Hashimoto who first described the distribution of capsaicinoid secretory organs in the *Capsicum* plant (*215*) at the microscopic level. They examined the content and distribution of capsaicinoid in the *Capsicum* plant at various stages after flowering and reported that capsaicinoid secretory cells were fruit-specific at any stage throughout cultivation. Carefully examining the morphological changes in tissues of *Capsicum* fruit at different stages after flowering, Furuya and Hashimoto found the characteristic development of segmentation and proliferation of epidermal cells of the dissepiment in the fruits (*215*). Capsaicin crystals were observed in the secreting organs of dried fruits. They speculated that capsaicin must have been secreted into receptacles from the epidermal cells of the dissepiment as lipid-capsaicin complexes. In Furuya and Hashimoto's brief paper, detailed data on the distribution of capsaicinoid in organs other than fruit as well as comments on the distribution of secretory organs in the placenta and the pericarp were not presented.

Later, a similar investigation on capsaicinoid-secreting organs and receptacles along with measurement of the capsaicinoid content in various parts of the fruit at different growth stages after flowering was made by Ohta (*131*). From histological and histochemical observations of the capsaicinoid-secreting organs and receptacles Ohta concluded that the capsaicinoid-secreting organs were localized in the placenta and dissepiment of *Capsicum* fruits. By comparing capsaicinoid content in pericarp, seeds, and placentae with interlocular septae (dissepiments) at the top, middle, and base of the fruit, Ohta found that placentae and interlocular septae were the major portions that contain much more capsaicinoid than any other organs. The small amount of capsaicinoid detected in pericarp and seeds, however, were presumed to be an adulterant in the form of adherence of a small quantity of capsaicinoid from the receptacles due to disruption. Ohta's histological data on capsaicinoid-secreting organs and receptacles were essentially the same as those revealed by Furuya and Hashimoto (*215*). In addition, Ohta revealed that the development of a receptacle in a young fruit began as soon as 10 days after flowering (*131*).

Unfortunately, the reports presented both by Furuya and Hashimoto and by Ohta were written in Japanese, so their ideas do not seem to have widely circulated in the world. Even after their reports were published, there appeared papers insisting that the pericarp and seeds were the major organs of capsaicinoid secretion and accumulation (*216,217*); however, in these cases unsuitable experimental procedures seem to have led to the wrong conclusions. It should be kept in mind that the site of capsaicinoid-secreting organs and receptacles may be species specific, so we should not blindly conclude that the sites of capsaicinoid secretion and accumulation exist in the placenta and dissepiment in all *Capsicum* plants. In fact, although capsaicinoid occurs predominantly in the pericarp, the capsaicinoid-synthesizing activity is found in particulate subcellular fractions from sweet pepper (*C. annuum* var. *grossum*) fruit (*218*). Nevertheless, judging from the papers reported after Furuya and Hashimoto, it seems likely that the major site of capsaicinoid secretion and accumulation in most cases is the placenta and dissepiment (*219–223*).

Neumann was the first person who tried to locate the site of capsaicinoid formation in *Capsicum* species using grafting and tracer techniques with ^{14}C-labeled CO_2 (*224*). It has generally been recognized that the secondary metabolites of Solanaceae family are derived as a whole or in part from the roots; in other words, the precursors or end products of the secondary metabolites are formed in the roots, then transported to the specific organs. Neumann applied reciprocal grafting between *C. annuum* and *Lycopersicon esculentum* (tomatoes). No capsaicin was detected, however, in the grafted ripened tomato fruit (the root of which was *C. annuum*), but in the reciprocal graft, capsaicin was isolated only from the *Capsicum* fruit. Neumann therefore concluded that capsaicin should be synthesized in the fruit of *Capsicum* plants.

Iwai and colleagues afterward confirmed Neumann's speculation in their study using a radioisotopic tracer technique (*222*). They compared the incorporation of DL-[3-^{14}C]phenylalanine into capsaicinoid molecules during incubation with various isolated organs of *C. annuum* var. *annuum* cv. Karayatsubusa, namely, root, stem, leaf, pericarp of fruit, and placenta of fruit. After incubation for 40 hr at 30°C, the highest incorporation of DL-[3-^{14}C]phenylalanine into capsaicinoid (54,800 cpm/g of tissue) was found in the placenta of fruit. The radioactivity of capsaicinoid recovered from pericarp, on the other hand, was only 1,800 cpm/g of tissue; root showed 4,000 cpm/g of tissue, which might suggest some capsaicinoid formation in the root. They confirmed the accumulation site of capsaicinoid to be the placenta of fruit, however, their term "placenta" should be regarded as "placenta with dessepiment" because they did not strictly distinguish dissepiment from placenta but rather took for granted that the dissepiment is an organ fundamentally belonging to the placenta.

After confirming the placenta and dissepiment to be the major site of capsaicinoid formation, Iwai and colleagues further investigated the intracellular localization of capsaicinoid by microscopic techniques involving transmission electron microscopy (*225*). Prior to examination with electron microscopy, they investigated the morphological development of *Capsicum* fruit after flowering at the light-microscopic level, which confirmed the results reported by Furuya and Hashimoto (*215*) and Ohta (*131*) that some morphological changes occur, mainly in the epidermal tissue of the placenta during maturation. As shown in Fig. 1, they recognized elongation of the epidermal cells [Fig. 1(1)] and many osmiophilic granules in the epidermal cells of the placenta [Figs. 1(2) and 1(3)] that suggesting the formation and accumulation of capsaicinoid in the epidermal cells of the placenta. By ultraviolet microscopy the osmiophilic granules were shown to absorb at 280 nm, which fulfills the requirement for capsaicinoid. The electron-microscopic observation showed that the electron-dense granules stained with glutaraldehyde and osmium tetroxide, probably capsaicinoid, were observed both in small vesicles and in vacuoles of epidermal cells of placenta specifically [Fig. 1(3)].

In their subsequent advanced work to reveal the intracellular localization of capsaicinoid, Iwai and co-workers used protoplasts prepared from the placenta for subcellular fractionation by Percoll density gradient centrifugation (*226*). The distribution of capsaicinoid in the fractionated subcellular organs and the electron micrograph are shown in Fig. 2. In Fig. 2, protoplasts prepared from the placenta are shown in A. Capsaicinoid in the protoplasts was located primarily at the top of the gradient [Fig. 2(F)]; therefore, they collected the capsaicinoid-rich fraction for light- and transmission electron microscopy. It was found that the fraction with the highest amount of capsaicinoid consisted of vacuoles [Fig. 2(B) and 2(E)], electron microscopy of which also revealed vesicle and vacuole-like structures [Fig. 2(D)] similar to those observed in the intact epidermal cells of the placenta. No electron-dense granules with a structure similar to that of epidermal cells was observed in any other fraction than that

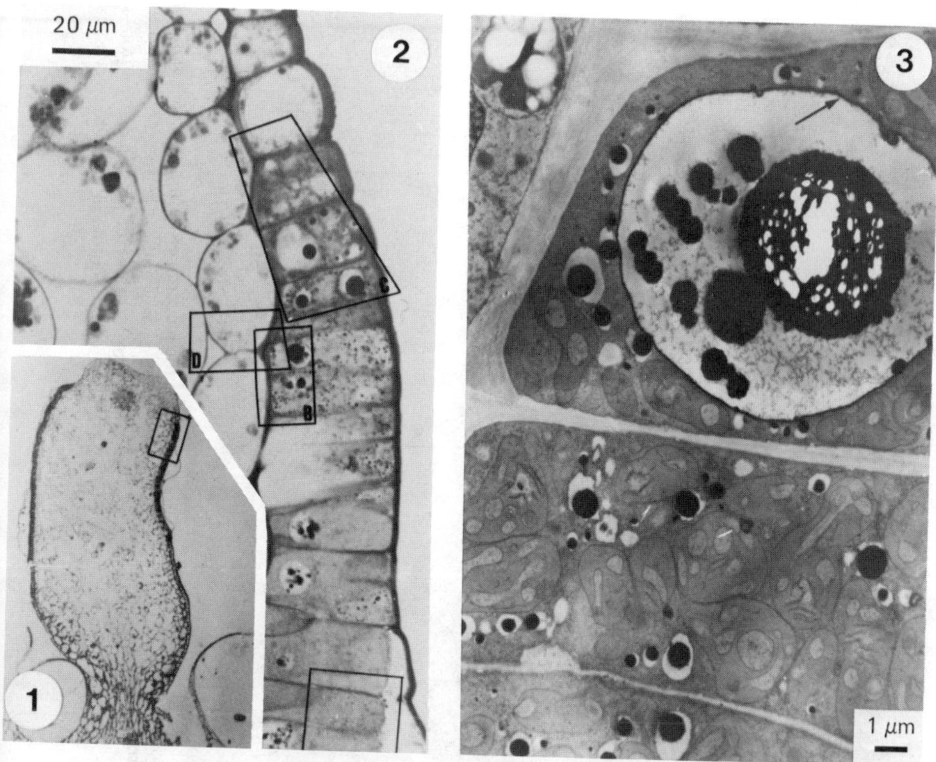

FIG. 1. Photo- and transmission electron micrographs of the placenta of *C. annuum* var. *annuum* cv. Karayatsubusa fruit 30 days after flowering. (1) Low magnification photomicrograph. (2) Higher magnification photomicrograph of the region enclosed by the rectangle in photo 1. (3) Transmission electron micrograph of the region in rectangle B. (Reproduced by permission of the Japanese Society of Plant Physiologists.)

mentioned above. Therefore, they reached the conclusion that capsaicinoid is located mostly in the vesicles or vacuole-like subcellular organs of the epidermal cells of placenta, isolated from other subcellular organs. Since capsaicinoid itself could also be considered toxic to plant mitochondria by inhibiting the oxidative phosphorylation, as reported in mammalian tissue (*227*), such localization of capsaicinoid in the plant tissue seems to be quite reasonable.

2. Fluctuation of Capsaicinoid Content in *Capsicum* Fruit at Different Growth Stages after Flowering

For a long time the highest concentration of capsaicinoid had been known from brownish green fruit (*228*) even though it was about 20 years

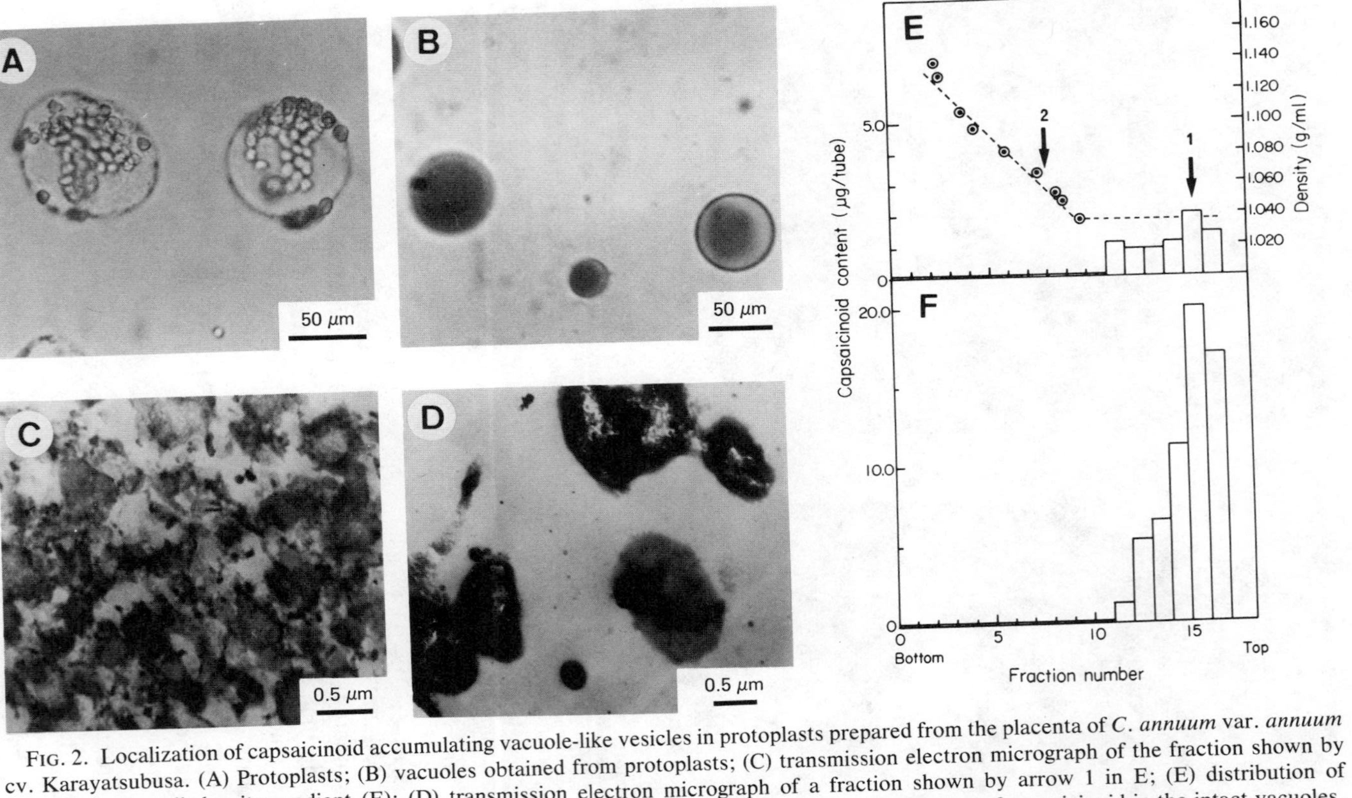

FIG. 2. Localization of capsaicinoid accumulating vacuole-like vesicles in protoplasts prepared from the placenta of *C. annuum* var. *annuum* cv. Karayatsubusa. (A) Protoplasts; (B) vacuoles obtained from protoplasts; (C) transmission electron micrograph of the fraction shown by arrow 2 in Percoll density gradient (E); (D) transmission electron micrograph of a fraction shown by arrow 1 in E; (E) distribution of capsaicinoid in the protoplasts (the dotted line represents the density gradient profile); (F) distribution of capsaicinoid in the intact vacuoles. (Reproduced by permission of the Japanese Society of Plant Physiologists.)

ago when fluctuations in the capsaicinoid content of *Capsicum* fruit were systematically surveyed (*70,229*). Ohta investigated the fluctuations of capsaicinoid content in the course of maturation of fruits of *C. annuum* L., *C. frutescens* L., *C. pendulum* Willd., *C. chacoens* Hunz., and *C. pubescens* Ruis et Pavion and revealed that the secretion of capsaicinoid started at remarkably early stages after flowering, i.e., 1 week after flowering in *C. chacoens,* and 2 weeks in the remaining species. Ohta further revealed that the maximum capsaicinoid content in dry matter appeared 2 to 4 weeks after flowering in *C. chacoens, C. annuum, C. pendulum,* and *C. frutescens,* while a marked delay was observed in *C. pubescens* (maximum of 10 weeks after flowering). The total capsaicinoid per fruit reached a plateau 4 weeks after flowering in all species except *C. pubescens.*

Schratz and Ragoonwala also investigated the variability of capsaicinoid content in different *Capsicum* species at different growth stages after flowering (*63*); they found that capsaicin content was low in the young green stage, reached its maximum in the mature green stage, but dropped somewhat in the ripening stage. Saga and Tamura also compared capsaicinoid content of three *Capsicum* varieties, a hot variety and two sweet varieties, by harvesting every 10 days after flowering. The result they obtained was that total capsaicinoid in the fruit of the hot variety increased up to 60 days after flowering, with a cessation of increase between the thirtieth and fortieth day after flowering. Their data were expressed as relative proportion to fruit, so the continuous increase in capsaicinoid content does not necessarily mean the increase of absolute amount of capsaicinoid. The increase in the percentage of capsaicinoid after the thirtieth day may be due to decreased moisture in the fruit or increased extraction efficiency because of disintegration of tissues due to drying. That accumulation of capsaicinoid continues after reaching maturation is very suspicious.

Kosuge and Inagaki, on the other hand, investigated the fluctuation in the ratio of capsaicin and dihydrocapsaicin with the growth stage of *Capsicum* fruit after flowering (*70*). They reported that capsaicinoid began to be accumulated around 10 to 20 days after flowering and that the ratio of capsaicin to dihydrocapsaicin remained constant around 1.0 : 0.4–0.5 throughout stages following flowering, which would indicate that both capsaicin and dihydrocapsaicin are formed from the beginning in a constant ratio. In their data the maximum capsaicinoid content was also recognized around 30 to 40 days after flowering both in *C. annuum* var. *annuum* cv. Takanotsume or *C. annuum* var. *annuum* cv. Yatsubusa with a slight decrease of capsaicinoid content after the fortieth day. Iwai *et al.* (*222*) also obtained results similar to those of Kosuge and Inagaki (*70*) by analyzing capsaicinoid content and its proportion at various growth stages after flowering in *C. annuum* var. *annuum* cv. Karayatsubusa. Decrease in capsaicinoid content could be explained by enzymatic degradation and/or chemical degradation, however, the question has remained unresolved.

In work feeding $^{14}CO_2$ to fruits of various ages, Neumann (*224*) reported that incorporation of $^{14}CO_2$ into the capsaicin molecule took place only in young fruits and that the total amount of capsaicin synthesized during a few days of $^{14}CO_2$ incorporation remained constant until maturity.

3. Biosynthetic Pathways and Enzymes Involved in Capsaicinoid Biosynthesis

Research on the biosynthetic pathways of capsaicin and its analogs were started contemporarily by British scientists Bennett and Kirby (*29*) and by American scientists Leete and Louden (*3*). Applying radioisotopic techniques to the fruit of *C. annuum,* Bennett and Kirby revealed that phenylalanine (**30**), *p*-coumaric acid (**32**), ferulic acid (**34**), caffeic acid (**33**), and vanillylamine (**23**), the aromatic rings of which were labeled with tritium, were incorporated biosynthetically into the benzyl residue (C-6 to C-1 unit) of capsaicin during a 7-day *in vivo* period of incubation of *Capsicum* fruit. In contrast to the efficient incorporation of phenylalanine into the capsaicin molecule, the incorporation of tyrosine was far less. Kirby and colleagues further revealed that hydroxylation of cinnamic acid (**31**) at the C-4 position involves migration of hydrogen to the neighboring carbon (*149*). Another interesting result was the fact that the minor saturated fraction contained 67% of the total activity, a specific activity as much as 5 times higher than that found in the major olefinic fraction (*29*). Bennett and Kirby speculated that the incorporation of phenylalanine into capsaicin appears to proceed by the usual pathway, known as the phenylpropanoid pathway (*231*), via various hydroxylated cinnamic acids.

On the other hand, shortly after Bennett and Kirby's report Leete and Louden also presented a paper in which *C. frutescens* was fed DL-[3-^{14}C]phenylalanine, DL-[3-^{14}C]tyrosine, and L-[methyl-^{14}C]methionine, separately *in vivo* and incubated 2 weeks to revealing the biosynthetic pathway of the vanillylamine residue of capsaicinoid (*3*). The results showed that only DL-[3-^{14}C]phenylalanine labeled the methylene group of vanillylamine, while DL-[3-^{14}C]tyrosine was poorly incorporated into the capsaicinoid moiety. The labeled methyl residue of L-methionine was found to be incorporated into the methoxy group of the vanillylamine residue. With respect to biosynthesis of acyl residues of capsaicinoid, they predicted that leucine and/or mevalonic acid would be incorporated into 8-methylnonenoic acid via 3-isopropylacrylic acid. The feeding study using DL-[2-^{14}C]mevalonic acid and DL-[1-^{14}C]leucine, however, gave no positive results. Meantime, when L-[U-^{14}C]valine was fed, radioactive capsaicin was obtained. They explained the incorporation of L-[U-^{14}C]valine into the acyl residue of capsaicin as occurring via isobutyryl-CoA by sequential chain elongation reactions with acetate units.

Rangoonwala (*232*) afterward obtained the same results as Bennett and Kirby's 1968 report with respect to the incorporation of [3-^{14}C]phenylalanine into the aromatic residue of capsaicin. However, data on the incorporation of [2-^{14}C]mevalonate into the acyl residue of capsaicin was not reconcilable with Leete and Louden's results (*3*); that is, in contrast to Leete and Louden's result that mevalonate was not incorporated into the acyl residue of capsaicin, Rangoonwala obtained the result indicating that [2-^{14}C]mevalonic acid was a more effective substrate for synthesis of the acyl residue than [2-^{14}C]acetate. Rangoonwala confirmed Leete and Louden's hypothesis of the origin of the methoxy group in the vanillylamine moiety. Furthermore, Rangoonwala, by using ^{15}N-labeled amino acids, revealed that the nitrogen in capsaicin is not derived from the amino group of glutamine or glutamic acid but from phenylalanine.

After the biosynthetic route of capsaicinoids was outlined by Bennett and Kirby (*29*), Leete and Louden (*3*) and by Rangoonwala (*232*) separately, no valuable work on the pathway of capsaicinoid biosynthesis had been reported until relatively recently. Studies on capsaicinoid biosynthesis have been separately resumed by Jurenitsch and collegues (*35,233,234*) and by Iwai and colleagues (*236–238*). During the course of studies on the formation and metabolism of the pungent principles of *Capsicum* fruits, Iwai *et al.* (*218*) found that the capsaicinoid-synthesizing enzyme which catalyzes the condensation reaction between vanillylamine and fatty acids distributes in particular fractions in sweet pepper fruits (*C. annuum* var. *grossum*). However, the enzyme activity of the capsaicinoid-synthesizing enzyme in the sweet pepper fruits was so low they could not reveal the detailed enzymatic properties. Therefore, they introduced an alternative *Capsicum* species with remarkably high potential for capsaicinoid formation, *C. annuum* var. *annuum* cv. Karayatsubusa, as the material.

By using the newly adopted species, Iwai and colleagues succeeded in revealing various unknown facts of capsaicinoid biosynthesis. The condensation reaction between various fatty acids and vanillylamine was examined at the cell-free level (*211*). The fatty acids examined were synthetic free or the thioester with coenzyme A of branched or straight chain fatty acids with chain lengths of 7–12 carbons. They thought that the reaction between free acid and vanillylamine to form capsaicinoid should be performed by two-step reactions *in vitro,* i.e., a fatty acid activating reaction and a subsequent condensation reaction. The iso type fatty acid activation reaction was found to require coenzyme A, ATP, and magnesium ion as cofactors. It was also revealed that 7-methyloctanoyl–CoA, the acyl residue for nordihydrocapsaicin (**4**), was used most effectively as the acyl donor for capsaicinoid formation *in vitro*. On the other hand, 5-methylhexanoyl–CoA, 6-methylheptanoyl–CoA, and 10-methylundecanoyl–CoA were not used as the acyl donor. When equivalent amounts of

7-methyloctanoyl–CoA, 8-methylnonanoyl–CoA, 8-methyl-6-nonenoyl–CoA, and 9-methyldecanoyl–CoA were added together as acyl donors, the major product was nordihydrocapsaicin. In nature, the proportion of nordihydrocapsaicin in the total capsaicinoid is only a small percentage, and a mechanism to form capsaicin and dihydrocapsaicin as the major natural components in most *Capsicum* fruits has not yet been found.

It had been fairly difficult to reveal the biosynthetic mechanism of capsaicinoid in the intact plant using subcellular organelles and/or cell-free extracts prepared by conventional techniques in enzymology because the subcellular organelles, especially vacuoles, the main site of capsaicinoid formation and accumulation, are mostly disrupted or inactivated during cell fractionation. Even if the cell-free system could be used efficiently, the data do not necessarily reflect the biochemical status *in vivo*. Iwai and colleagues tried to ease the difficulty by replacing the conventional materials with macerated cells such as spheroplasts or protoplasts (*236,237*). Since both spheroplasts and protoplasts are groups of intact cells, the biological systems inside the cells are expected to reflect intact status more than the cell-free system, and the efficiency of incorporation of substrate into cells is much better than in whole fruits, tissue homogenates, or slices (*239*).

Macerated cells are effectively prepared from young *Capsicum* fruits 10 to 20 days after flowering; however, the recovery rate of protoplasts from matured fruits is terribly low. As expected, radiolabeled intermediates of the vanillylamine moiety were easily recovered with HPLC from the incubation medium containing 10^5 protoplasts which were fed L-[U-^{14}C]phenylalanine and L-[3,4-^{3}H]valine. Judging from the distribution of radiolabeled intermediates *trans*-cinnamic acid (**31**), *trans-p*-coumaric acid (**32**), *trans*-caffeic acid (**33**), and *trans*-ferulic acid (**34**) and from previous information on the phenylpropanoid pathway, a proposed pathway of synthesis of the vanillylamine moiety (**23**) of capsaicinoid from L-phenylalanine (**30**) and the enzymes involved were presented by Iwai and colleagues (*237*). They further investigated the intracellular distribution of enzymes and intermediates involved in the biosynthesis of capsaicin and its analogs by subfractionation of the protoplasts from the placentae of *Capsicum* fruits. The activities of *trans*-cinnamic acid 4-monooxygenase, which catalyzes the reaction from **31** to **32**, and capsaicinoid-synthesizing enzyme (capsaicinoid synthetase), which catalyzes the condensation reaction between acyl-CoAs and vanillylamine (**23**), were found to exist in the vacuole fraction; the activity of phenylalanine ammonia-lyase, which catalyzes the conversion from **30** to **31** was found in the cytosol fraction. After feeding L-[U-^{14}C]phenylalanine (**30**) to the protoplasts, newly synthesized *trans-p*-coumaric acid (**32**) and capsaicinoids were found in the vacuole fraction while *trans*-cinnamic acid (**31**) was not.

With respect to biosynthesis of the branched chain fatty acid residues of capsaicinoids, Jurenitsch's group and Iwai's group almost simultaneously in separate experiments revealed that acyl residues with even-numbered and odd-numbered iso type chains and anteiso type fatty acids are synthesized from valine (**35**), leucine (**38**), and isoleucine (**41**), respectively (*35,233,236*). The acyl moieties of individual capsaicinoids in *Capsicum* fruits are synthesized by pathways similar to those proposed for animals and bacteria by Kaneda (*240*); that is, L-valine (**35**) is converted to α-ketoisovalerate (**36**), then to isobutyryl CoA (**37**) after decarboxylation, followed by conversion to even-numbered acyl moieties after chain elongation. L-Leucine (**38**) and L-isoleucine (**41**) are similarly converted to α-ketoisocaproic acid (**39**), to α-keto-β-methylvaleric acid (**42**), and then to isovaleryl CoA (**40**) and α-methylbutyryl CoA (**43**), leading to odd-numbered and anteiso odd-numbered acyl CoA, respectively. In contrast to Iwai's report, Jurenitsch's group found that valine was incorporated not only into even-numbered branched chain acyl residues but also into odd-numbered iso type branched chain acyl residues; however, this discrepancy could be due to too long incubation (14 days compared to 3 days in Iwai's experiment).

Iwai *et al.* (*222*) speculated that interconversions among capsaicinoids, such as from dihydrocapsaicin to capsaicin, were unlikely to occur because the proportion of capsaicinoid composition did not appreciably change with the growth stage. Their speculation was later confirmed by Kopp and Jurenitsch (*234*) who examined the interconversion of administered ^{14}C-labeled capsaicin to dihydrocapsaicin in the mesocarp of immature *Capsicum* fruits. The interconversion of saturated fatty acid to unsaturated fatty acid must, therefore, take place at the free fatty acid level. Concerning the position of the double bond in homocapsaicin, it is more feasible for the double bond to be introduced at C-6 than at C-7 by the general rule of unsaturation in fatty acids that the initial reaction between the substrate and enzyme (desaturase) is at the carboxyl end of the fatty acid chain (*241*) and the fact that even-numbered branched chain acyl residues and odd-numbered ones are biosynthesized independently from different amino acids.

The biosynthetic pathways for capsaicin and analogs are outlined in Scheme 4. The detailed pathway between ferulic acid (**34**) and vanillylamine (**23**) has not yet been worked out, and the enzymes, their properties, and the reaction mechanisms involved in individual steps have also remained unknown. The biosynthetic mechanism of capsaicin and its analogs has not yet been fully revealed in many respects. Regarding the intracellular formation of capsaicinoid, Iwai and colleagues hypothesize that capsaicinoid is synthesized from vanillylamine (**23**) and several iso type fatty acids, probably in the form of acyl CoAs, on the tonoplast and that it may then penetrate through the tonoplast into the vacuole and accumulate in the vacuole (*238,242*). Once the capsaicinoid in the vacuole

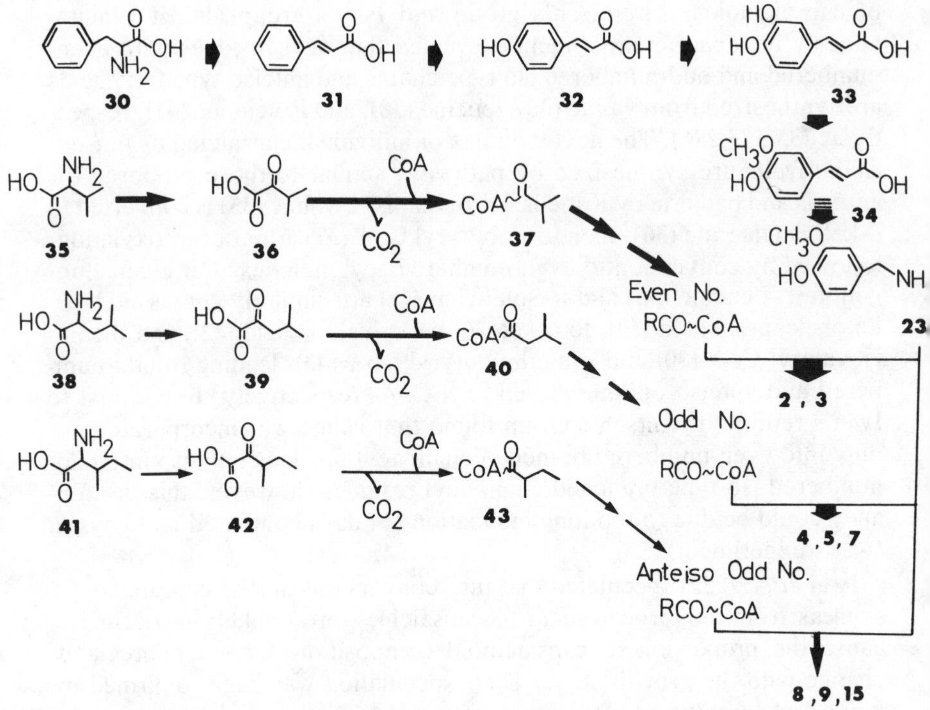

SCHEME 4. Biosynthetic pathways of capsaicin and analogs (*234,236–238*).

reaches a certain level, it must be secreted out of the cell and accumulates in the receptacle as a lipid–capsaicinoid mixture as previously suggested by Furuya and Hashimoto (*215*), Ohta (*131*), and by Suzuki *et al.* (*225*).

B. BIOLOGICAL SIGNIFICANCE

The simple question why the hot pepper forms and accumulates capsaicinoid inside the fruit is easy to ask but very difficult to answer. The pungent principle of *Capsicum* fruit has been used as a spice or pharmaceutical drug for a long time, however, nobody knows the biological significance of capsaicinoid in the plant. There have been no reports elucidating the biological significance of capsaicinoid in the plant by showing reasonable scientific evidence. Some say that capsaicinoid is nothing more than a waste product without any biological significance, others speculate that it would have been formed to keep enemies such as birds or insects away from *Capsicum* fruits to protect seeds, and still others insist that capsaicinoids may have been formed to act as a phytoalexin or as an antibiotic. Several reports that may hint at the biological significance of capsaicinoids in *Capsicum* fruits have been reported.

The antifungal effect of capsaicin has been reported against *Zygosaccharomyces* species (*243*) and *Mycoplasma agalactiae* (*244*). It has also been known that sporulation of *Bacillus anthracis* is induced by capsaicin (*245,246*). Gal (*247,248*) investigated the origin of the antibacterial activity of paprika spice by comparing two components, a steroid saponin, capsicidin (*249*), and capsaicin. In contrast to the weak antibacterial activity of capsicidin, capsaicin showed bacteriostatic activity against several bacterial species, remarkably against *B. cereus* and *B. subtilis,* at 1/10,000 dilution. However, no effect of capsaicin was recognized against *Staphylococcus aureus* or *Escherichia coli.*

In contrast to the results that seem to suggest a bacteriostatic effect of capsaicin, there have, on the other hand, appeared papers showing a negative or rather stimulatory effect of capsaicin on bacterial production (*250,251*). Kosuge and Takeuchi showed that capsaicin had no inhibitory effect on the growth of *S. cerevisiae, Zygosaccharomyces sojae, Lactobacillus casei, B. subtilis, Aspergillus niger, Penicillium chrysogenum,* and *Rhizopus usami* at 1000 μg/ml. *Aspergillus niger, P. chrysogenum,* and *R. usami,* on the contrary, decomposed capsaicin in a medium within 10 days' incubation. Mikhailova indicated that at low doses, capsaicin as well as piperine stimulated the reproductive capacity of *B. anthracis*. The inhibitory effect was shown only by a large dose far beyond the actual content in the *Capsicum* fruit.

Recently, Gutsu *et al.* (*252*) proposed a hypothesis that capsaicinoid, the secondary metabolite of red pepper, may be functioning as an immunity factor. They realized that resistance to diseases (such as *Verticillium* wilt and mosaic virus) in the *Capsicum* plant seems to be related to capsaicinoid content, i.e., that disease resistance increased in accordance with increase of capsaicinoid content in *Capsicum* fruit.

A sesquiterpenoid, capsidiol, is a known phytoalexin in sweet pepper (*253,254*); however, Gutsu *et al.* speculated that capsaicinoid might be a substitute for capsidiol because capsidiol is formed only in sweet pepper varieties containing no or little capsaicinoid. Nevertheless, it is seriously doubtful that the capsaicinoid, which is found only in the placenta and dissepiment, is able to prevent disease in other organs. Anyway, there is no distinctive theory on the biological significance of capsaicinoid in *Capsicum* species.

C. Factors Affecting Formation and Accumulation

It is well known that capsaicinoid content is not species specific but varies with environmental factors such as temperature, light exposure, fertilization, or other chemical and/or physical treatments. Several reports have been published from the viewpoint of plant genetics and horticulture. Basic research from the plant physiological point of view has been done, too.

1. Inheritance Factor

The absolute capacity of capsaicinoid formation and accumulation for the individual species is determined by the hereditary factor, genotype, under the rule of heredity. Heredity of pungency of *Capsicum* species has long been a subject for debate among plant geneticists. It has been believed that the pungent gene is dominant over the nonpungent gene because the phenotype of heterozygotes were pungent in the first filial generation. In other words, pungent character in *Capsicum* fruit was passed to the next generation by transmission of the pungent characteristics as the dominant gene over the nonpungent gene, in the manner of Mendel's law, when the pungent species and nonpungent species were crossed (*255–259*). While Odland reported mosaicism in a phenotype of a hybrid from a pungent species and nonpungent species (*260*), Ohta (*261*) investigated quantitative inheritance of pungency using several different *Capsicum* varieties. Crossing pungent and nonpungent parents gave an F_1 hybrid containing approximately the same amount of capsaicinoid as the parent within a narrow range of variation. However, in the second generation various degrees of pungency were segregated from pungent to nonpungent according to Mendel's first law. No metaxenia was observed in fruits obtained from the heterozygote of nonpungent (female) × pungent (male).

During the course of study on the capsaicinoid content of the first filial generation of 13 *C. annuum* varieties, Michina obtained results showing considerable heterosis as measured by capsaicinoid content (*262*). Hybrids obtained by crossing low-capsaicin and high-capsaicin species showed a remarkable heterosis effect. The heterosis effect was not recognized in crossing low-capsaicin varieties.

With respect to any correlation of fruit size and pungency, the capsaicin content in *Capsicum* fruit had once been considered to be inversely proportional to the size of the fruit (*263*). However, it is now believed that the size of fruit has nothing to do with the pungent potency (*57,73,262*). *C. frutescens* seems to be the most promising species for breeding more pungent varieties (*57*).

2. Temperature

The fact that the capsaicinoid content of the same species is variable under different environmental conditions has been widely recognized (*16,257*). Ohta investigated the relationship between pungency and environmental factors, especially temperature (*264*). Capsaicinoid formation and accumulation were influenced by temperature, and it was highly likely that higher nighttime temperatures brought higher capsaicinoid content. Cultivation below 20°C severely affected blooming and capsaicinoid accumulation. Reduction in photosynthesis caused repression of plant growth and blossom number; however, capsaicinoid content remained unchanged (*264*).

3. Light

Whether insufficient light exposure depresses capsaicinoid formation and accumulation or not remains controversial. In contrast to Ohta's results, several papers have suggested that light exposure may be an important factor in capsaicinoid formation and accumulation. Iwai and colleagues in the course of study on the formation of capsaicinoid in postharvest ripening sweet pepper fruit obtained data indicating light exposure to be essential for capsaicinoid formation in sweet pepper fruit (*265,266*). Capsaicinoid was produced in the placenta of *C. annuum* var. *grossum* (sweet pepper) during postharvest ripening under continuous light for 4 to 7 days, while no capsaicinoid was detected in postharvest ripened fruits in the dark (*265*). Similar results were obtained from other experiments such as incubating premature sweet pepper fruits with vanillylamine and isocapric acid in the light or dark at 30°C (*266*); no capsaicinoid was formed in the sweet pepper fruits incubated in the dark. Since all these results were obtained from postharvest ripened fruits isolated from intact whole plants, they should not be compared directly with Ohta's data obtained from the *in vivo* experiment.

Kim *et al.*, on the other hand, investigated the effect of light quality on capsaicinoid formation in *C. annuum* grown in a polyethylene film greenhouse (*267*), and showed that capsaicinoid as well as carotenoid content was higher in fruits grown in the portion of the polyethylene house covered with red-colored film than those grown under transparent film. However, the same author reported somewhat contradictory results in a subsequent paper, namely that the capsaicinoid content was slightly increased by blue light and that red light had no effect on capsaicinoid formation (*268*).

The exact effect of gamma-ray irradiation on capsaicinoid formation has also been controversial. Zderkiewicz (*269*) reported a stimulatory effect of gamma rays at 5,000 to 10,000 roentgens (R) on the capsaicinoid content of fruits. On the contrary, in the report of Izvorska, although irradiation of *Capsicum* seeds at 500 to 5,000 R increased the rate of growth, total blossom number, and pepper yield, the capsaicinoid content decreased (*270*).

4. Fertilization

Nutrient requirements to obtain the highest fruit yield and pungency of *Capsicum annuum* have been investigated. Galcz *et al.* (*271*) reported that doses of three major nutrients in a certain ratio brought the highest fruit yield and capsaicinoid content. Mineral fertilizers did not affect the proportion of capsaicinoid in the fruit. Nowak, on the other hand, showed that yields of fruits and of capsaicinoid per *C. annuum* plant grown in hydroponic culture were considerably affected by varying the proportion of microelements such as copper, boron, or molybdenum (*272,273*). Bubicz *et al.* (*274*) showed that the highest capsaicinoid content of *C. annuum* fruits was obtained by applying 130 kg/ha of potassium on a background of 50 kg/ha of nitrogen and 22 kg/ha of phosphorous.

Beside fertilizer and micronutrients, the effect of chemicals has also been studied. *Capsicum* plants grown from seeds that had been treated with 0.3, 0.6, or 0.125% ethyleneimine, which is now regarded as a carcinogen, brought fruits of greater than normal size, weight, and capsaicinoid content (*269*). Ethephon (compound No. in Chemical Abstracts: 16672-87-0) at 500 ppm also increased the capsaicinoid content of fruits (*275*). Phenylalanine at 200 ppm did not affect ripening of the fruit but increased capsaicinoid content when added together with ethephon (*275*).

The level of valine and leucine in the placenta of *Capsicum* fruit seems to be closely related to the content and composition of capsaicinoid (*276*). Lipid and fatty acid compositions of *Capsicum* fruits have been analyzed to obtain information on regulation of capsaicinoid composition; however, no valuable information has yet been obtained heretofore excepting one paper by Haymon and Aurand on the volatile constituents of tabasco peppers (*277*).

VI. Nutritional, Metabolic, and Pharmacological Studies

A. Influence on Nutrient Absorption

The effect of capsaicin, either chronic or acute, on nutrient absorption is in most cases inhibitory (*278*). However, the effect of red hot pepper fruit has been controversial (*278,279*). The effect of long-term feeding of red hot pepper fruits on growth seems to be dose dependent.

The effect of red pepper fruit or capsaicin on intestinal absorption of saccharides has recently been investigated by Monsereenusorn and others. The effects of capsaicin and crude extracts of *Capsicum* on intestinal glucose absorption by isolated intestinal preparations *in vitro* were shown to be stimulatory at 7.0 mg/100 ml on glucose absorption in the jejunum of rat and hamster, but inhibitory at 14 or 21 mg/100 ml after a 30-min incubation period (*280*). Since inhibition was relieved by addition of ATP to the mucosal side of incubated preparations, the inhibitory effect of capsaicin on glucose absorption was suggested as due to a reduction in ATP content in the intestinal mucosa by inhibition of mitochondrial ATP production (*280*).

During further investigation of intestinal glucose utilization and lactic acid production caused by capsaicin in everted sacs of jejunum from rat and hamster (*281*), Monsereenusorn showed that glucose transport was significantly inhibited in both preparations when 14 mg% capsaicin was added and incubated at pH 7.4 for 60 min: capsaicin brought about a significantly greater glucose utilization and lactic acid formation than the control, which could partly explain the depression mechanism of intestinal glucose transport. Monsereenusorn and Glinsukon further suggested that capsaicin might inhibit the ATPase-dependent sodium pump of the basolateral membrane, which led to the inhibition of intestinal glucose absorption (*235,282*). It has been confirmed that capsaicin inhibits sodium exit through the serosal pole of the epithelium of the jejunum (*283*). A similar inhibitory effect of capsaicin on glucose absorption has been reported *in vivo* (*284*): the increase of plasma glucose level after oral glucose ingestion was inhibited by capsaicin in the diet. Jonietz has recently reported that the activity of either disaccharidase in jejunum (*285*) or duodenal alkaline phosphatase and Mg^{2+}- and Ca^{2+}-activated ATPases in rats were not adversely affected by capsaicin equivalent to the average per capita daily consumption by Thai people (*286*).

The effect of capsaicin on fat absorption or serum cholesterol levels has also been studied (*278,287–290*). After feeding 1 mg capsaicin/kg/day, which is equivalent to the daily consumption of red hot pepper fruit by the average Thai person, to rats on 28- or 56-day diets, Nopanitaya observed a decrease in the number of fat droplets in epithelial absorption cells, indicating that capsaicin reduced fat absorption (*287*). When capsaicin was fed to rats on a low-protein diet, fat absorption and growth rate was slower than in rats either on a high-protein diet or on a low-protein diet without capsaicin.

Sambaiah *et al.* (*288,290*), on the other hand, reported that neither whole red peppers nor capsaicin adversely affected fat absorption in the rat. Feeding 5% red pepper fruit, equivalent to 15 mg% capsaicin, together with a high-fat and low-protein diet showed a lipotropic effect similar to that induced by choline with simultaneous elevation of serum total lipid. Lowering of blood cholesterol levels by capsaicin administration has been shown (*278,289*). When high cholesterol diets were fed to female albino rats together with red pepper or an equivalent amount of capsaicin, the rise in liver cholesterol levels was significantly depressed and accompanied by an enhancement of fecal excretion of both free cholesterol and bile acids. The preventive effect of capsaicin on rise in serum cholesterol, on the other hand, was not significant as in the case of liver cholesterol (*289*).

Although conflicting data had once been reported (*291*), the stimulatory effect of capsaicin on gastric juice secretion has generally been accepted (*292,293*). Limlomwongse *et al.* suggested that capsaicin caused the gastric acid secretion and mucosal blood flow to release the endogenous gastric secretagogs, which should increase both tissue perfusion and the secretory activity (*294*).

The effect of capsaicinoids on nutrient absorption and variation in the cholesterol level are more or less directly or indirectly, related to physiological changes in the tissues. Studies on the mechanism of action of capsaicin on the gastrointestinal tract and blood cholesterol will be described from a pharmacological point of view in Section VI,C.

B. Metabolism

In comparison to study on nutrition and pharmacology, study of the absorption and metabolism of capsaicinoids in mammalian tissues had been neglected until Monsereenusorn and others began the study. Since hamster is more sensitive to capsaicin given orally than rat, Monsereenusorn (*295*) compared the *in vitro* intestinal absorption of capsaicin in rats and hamsters to elucidate the different susceptibility to the enterotoxin capsaicin. Hamsters were shown to have greater absorbability of capsaicin than rats. The absorption of capsaicin in intestinal mucosa was suggested to be conducted by passive diffusion, which has recently been demonstrated by Iwai and co-workers (*296,297*). Iwai and co-workers have further shown that capsaicinoid is absorbed rapidly in the stomach and small intestine *in vivo*. About 85% of the dose was absorbed in the gastrointestinal tract within 3 hr. It has also been found that within 60 min of oral administration of capsaicinoid into stomach, jejunum, and ileum, about 50, 80, and 70% of the dose disappeared from the lumen, respectively. The main route of capsaicinoid absorption has been shown to be via the portal pathway, in which some portions of capsaicinoids were found as metabolite(s) combining to albumin in the portal blood. No significant reduction of [^{3}H]dihydrocapsaicin uptake was observed in jejunum when an inhibitor, 2,4-dinitrophenol or NaCN was added, which supports passive diffusion of capsaicinoid in the tract. Iwai and coworkers concluded that capsaicinoid must be transported readily through the gastrointestinal tract by passive diffusion to the portal vein and partially metabolized during absorption and/or transportation.

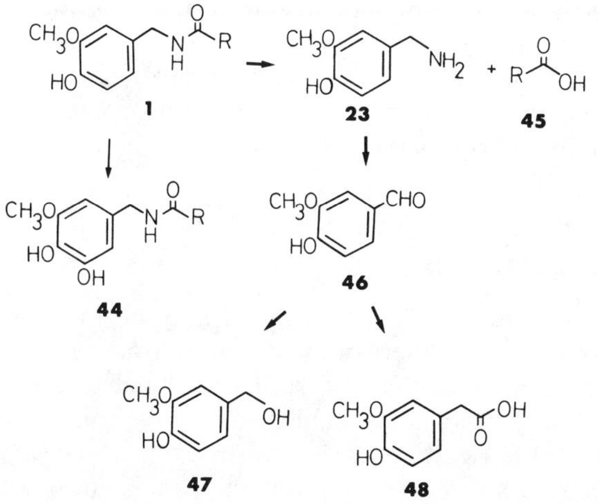

SCHEME 5. Metabolism of capsaicinoids in mammalian tissues (*296,302*).

Kim and Park (*298*) showed orally administered capsaicin reached its maximum concentration in the plasma in 40 min. During the course of study on the cytotoxicity of capsaicinoid, it was suggested that gastrointestinal absorption of capsaicin is considerably poorer than absorption when administered by other routes, such as intravenously, intraperitoneally, and subcutaneously (*298,299*).

Lembeck and co-workers (*300*) investigated the distribution of capsaicin in various organs of rats after systemic administration. They found a considerable amount of capsaicin in the spinal cord and brain at 3 and 10 min after intravenous administration of 2 mg capsaicin/kg to rats. The low levels of capsaicinoid in the liver were explained by detoxification to form conjugates with glucuronic acid or sulfuric acid. Their speculation has recently been confirmed by Iwai and Kawada (*296*). In the same paper the latter investigators chased after orally administered dihydrocapsaicin and its metabolites in urine and feces with HPLC, and proposed metabolic pathway for capsaicinoid in mammalian tissue: capsaicinoid (**1**) is hydrolyzed, mainly in the liver, to fatty acids (**45**) and vanillylamine (**23**), which is reduced to vanilin (**46**) then further metabolized to vanillic acid (**48**) and/or vanillyl alcohol (**47**) (see Scheme 5). In addition, a considerable amount of each compound is thought to be conjugated with glucuronic acid and/or sulfuric acid.

Involvement of a mixed-function monooxidase system in drug-treated liver has also been reported in capsaicinoid metabolism (*301,302*). It has been reported that capsaicinoids were converted to *N*-(4,5,dihydroxy-3-methoxybenzyl)acylamides (**44**) in rat liver by a mixed-function monooxidase system induced by hexobarbital injection. Capsaicinoid itself has been reported to be a compound capable of inducing the monooxidase system in rat liver (*301*). In Lee's papers, *N*-(4,5-dihydroxy-3-methoxybenzyl)acylamide (**44**) seems to be the only metabolite identified in the incubation medium containing rat liver homogenate and capsaicin (*303,304*). Details on capsaicinoid metabolism have mostly remained unknown.

C. Pharmacological Studies

Pharmacological studies on capsaicinoid may be the most attractive topic for the majority of readers of this chapter. More than 200 out of 500 papers concerning capsaicinoids have been related to pharmacological studies, most of which have been published over the past 15 years. Unfortunately we do not have enough space to review all the disciplines of the study, so the authors will briefly outline the competent pharmacological studies on capsaicinoid. Those who wish to know more about the pharmacological studies on capsaicinoid should read reviews (*4,8,305–310*) or the original papers cited in the references.

1. Toxicity

Capsaicinoid not only causes strong irritation and inflammation of the skin, mucous membrane, and eyes, but also shows acute fatal toxicity when a large dose of capsaicinoid is given at one time (*310a*). A large amount of orally administered capsaicinoid causes damage to the gastrointestinal tract (*311,312*). The most pronounced ultrastructural alterations have been observed as matrix swelling and disorganization of cristae in mitochondria and deformation of nuclei, which developed in both short-term and long-term feeding of capsaicin with a low-protein diet. Reported LD_{50} values for orally administered capsaicin to mice range from ~122 to 294 mg/kg, while those for intravenous injection are 0.36 to 0.87 mg/kg (*299*). To elucidate the toxicity of capsaicinoids, various studies have been done (*227,313,314*). Chudapongse and Janthasoot studied the toxic effect of capsaicin on rat liver mitochondria (*227,314*), which led to the conclusion that capsaicin inhibits oxidation of exogenous NADH by retarding electron flow from NADH to coenzyme Q.

Beside fatal toxicity, capsaicinoid induces desensitization to various stimuli especially when neonatally administered (*316,316a*). The mode of action of capsaicinoid as a neurotoxin has been vigorously studied by many neurophysiologists. Fine structural changes in mitochondria of sensory nerve cells by subcutaneously injected capsaicin (*317–319*), changes in vascular permeability (*320*), and selective degeneration of type-B sensory ganglion cells and the neonatal dorsal horn by capsaicin treatment in laboratory animals have been revealed by various neurochemical techniques and microscopies (*321–326*). Degeneration and glial engulfment of buttons and unmyelinated axons in the dorsal horn were observed in neonatal rats after subcutaneous capsaicin injection (*327,328*). Degeneration of numerous nerve processes has been reported in the nerve elements of the small intestine in neonatal cats after capsaicin treatment (*329*). Dubois traced the neurotoxic effect of capsaicin to a reversible blockage of one component of the potassium ion current in the node of Ranvier (*330*). Hoyes *et al.* (*331*) examined *in vitro* exposure of rat trachea to capsaicin. Significant changes were observed in the intraepithelial axons, too.

2. Gastrointestinal Tract

The long-term and acute effects of capsaicin on the gastrointestinal tract have already been mentioned in Sections VI,A and VI,C,1. As mentioned above, the most profound effect of capsaisin on the gastrointestinal tract is stimulation of gastric juice secretion, which may occasionally cause ulcers. Szolcsanyi and Bartho (*332*) investigated the impaired defense mechanism to peptic ulcers in capsaicin-desensitized rats, and revealed that capsaicin in low concentrations enhances the defense mechanism of the stomach.

It has also long been known that capsaicin shows a laxative effect on the colon (*333,334*). The effect of capsaicin on nerve systems of the small intestine has been vigorously studied by Molnar and co-workers (*335–338*), and capsaicin has been shown to have no effect on movement of the small intestine in the cat. However, contraction was evoked in an isolated guinea pig ileum preparation.

Investigations on the site of action of capsaicin in guinea pig small intestine have been done by Szolcsanyi and Bartho (*339,340*). Holzer and Lembeck investigated the effect of capsaicin on the contraction of isolated guinea pig ileum induced by rapid cooling (*341*). In subsequent papers, they indicated a possible relationship between substance P and capsaicin (*342,343*). It has been proved that the contractile response of guinea pig ileum preparations to capsaicin and mesenteric nerve stimulation is mediated by a release of substance P (*343–345*).

3. Cardiovascular and Respiratory Systems

The effect of capsaicinoids on the cardiovascular and respiratory systems has been studied for a long time. The effects of capsaicin on blood pressure and respiration were reported as early as 1928 (*346*) and 1935 (*347*). Extreme stimulation of respiration as well as vasoconstriction caused by capsaicin were subsequently demonstrated separately by Porszasz *et al.* (*348*) and Toh *et al.* (*291*). The pulse increase in blood pressure caused by injection of capsaicin to the perfusion fluid was accounted for as a reflex increase of blood pressure similar to that produced by acetylcholine (*350–352*). Significant increases in mean aortic pressure, heart rate, cardiac output, and respiratory minimum volume are caused by capsaicin injections into a neurally intact donor-perfused hindlimb (*353*).

On the other hand, it has also been known that capsaicin administration causes hypotension and bradycardia shortly after hypertension. The distribution of capsaicin-sensitive receptors in the pulmonary artery and extrapulmonary parts of its main branches, which produce systemic hypotension and bradycardia, was investigated (*354,355*). It was shown by Molnar and György (*356*) that capsaicin causes a rapid onset and high intensity of pulmonary arterial pressure which is not achieved even with high doses of histamine, bradykinin, serotonin, adrenaline, or noradrenaline. Brender and Webb-Peploe showed that capsaicin injected into the right or left side of circulation causes cutaneous vasodilatation in dogs, which could be elucidated by stimulation of baroreceptors in the low-pressure and high-pressure circulations as peripheral receptors probably situated in skeletal muscle (*357*). They further reported that the vascular reflexes caused by capsaicin injection were similar to those caused by electric stimulation of afferent nerve fibers from hindlimb muscles (*358*). It has been generally accepted that receptors in the pulmonary vascular bed, as well as arterial basoreceptors, are involved in regulating the cardiovascular and respiratory systems (*359*).

Ueno demonstrated transient apnea in anesthetized dogs to which capsaicin had been administered intravenously (*360*). Porszasz *et al.* (*361*) suggested that apnea induction as well as bradycardia, and hypotension caused by capsaicin in dogs were due to excitation of pressoreceptors rather than chemoreceptors. It is now well recognized that intravenous administration or direct injection of capsaicin into the bifurcation of the pulmonary artery of anesthetized cats and dogs causes a Bezold–Jarish reflex; direct injection of capsaicin into the carotid artery of the cat does not cause a Bezold–Jarish reflex but produces hypertension and tachypnea instead. The strong pressor and respiration stimulations caused by intracisternal capsaicin administration in the cat are inhibited by ganglionic blocking agents such as tetraethylammonium chloride, which suggests direct action of capsaicin on the central nervous system. Since vagotomy and sectioning of the sinus nerve abolish the apnea and hypotension produced by capsaicin, participation of vagal reflex mechanisms or sensitization or excitation of the carotid sinus baroreceptors are probably involved.

Russell and Lai-Fook (*362*) confirmed that reflex bronchoconstriction is produced by pulmonary receptors sensitive to capsaicin that are accessible via pulmonary circulation. Molnar *et al.* also discussed the bronchoconstrictive action of capsaicin on the guinea pig (*363*). They showed that capsaicin treatment abolishes the hypotensive effect of subsequent intravenous capsaicin injection or the depressive effect of histamine (*364*). Atrium contractions caused by capsaicin was reported by Fukuda and Fujiwara in guinea pig (*365*).

The prophylactic action of capsaicin against chemically induced acute skin inflammation has been discussed in connection to substance P released from local sensory nerve terminals (*366–376*). Irritation and subsequent desensitization of the respiratory tract caused by capsaicin were reported by Alarie and Keller (*377*) and Hoyes *et al.* (*331*). Andoh *et al.* examined the effects of intraarterially administered capsaicin and its various analogs on vocalization in guinea pigs (*378*).

Interesting data was presented by Ki *et al.* (*379*): capsaicin at 2.3 mg/kg decreased levels of cholesterol in aorta myocardium tissues and blood serum of turkeys in a 9-day experiment. A slight decrease in high-density lipoprotein was observed at 32 hr after cholesterol feeding, but no difference was seen at 96 hr.

4. Thermoregulation

A drop in body temperature induced by capsaicin administration in dogs was reportedly observed as early as in 1878 by Hogyes. The steep drop in body temperature of mice, rats, and guinea pigs caused by intraperitoneal or subcutaneous capsaicin administration was afterward demonstrated by Jancso-Gabor and others (*380–383*). The phenomenon itself seemed similar to that caused by histamine, however, calorimetric measurements suggested that the capsaicin-induced hypothermic effect was different from that of a histamine-induced one (*381*), i.e., capsaicin did not produce any detectable change in the metabolic rate.

The work by Jancso-Gabor, Szolcsanyi, and Jancso on capsaicin-induced irreversible impairment of thermoregulation brought out a number of new facts (*315*). Subcutaneous or intraperitoneal injection of capsaicin to rats and guinea pigs produces remarkable hypothermia associated with skin vasodilatation. Large doses of capsaicin, on the other hand, cause desensitization to the hypothermic action of the drug. Although capsaicin-densensitized animals are still able to respond to cold, they are no longer able to protect themselves against overheating. Capsaicin was then thought first to stimulate the hypothalamic warmth detectors which was followed by desensitization (*380*). Positive data supporting their idea were presented afterward (*384–387*). In further investigations, they provided electron-microscopic evidence to show a characteristic swelling of mitochondria in the perikaryon of a single type of nerve cell in the preoptic area of the hypothalamus that was induced by subcutaneous administration of capsaicin to rats (*318*).

Other facts such as the strong effect of narcotics on capsaicin-desensitized animals (*388*) suggest the impairment of thermoregulatory systems in the preoptic area of hypothalamus. Rats chronically pretreated with peripheral or intracerebral capsaicin injections did not exhibit the hypothermic effect. Szolcsanyi and Jancso-Gabor (*391*) proposed a similar idea for the chronic effect of capsaicin administration as previously mentioned. The latest hypothesis proposed by Szolcsanyi (*392*) is that desensitization of the peripheral and extrahypothalamic warmth sensors rather than the hypothalamic sensors may be involved in an inhibition of the heat escape reaction in rats treated with capsaicin subcutaneously. Cabanac *et al.* (*393*) earlier tried to explain impairment of the thermoregulatory capability under high temperatures by damage to the salivary secretion system.

Frens (*394*) revealed that the effect of capsaicin on the heat sensor pathway in thermoregulation is mediated by release of serotonin. Subcutaneous injection of capsaicin permanently reduces the capacity of rats to withstand a hot environment (*393*). Mechanical nociceptive thresholds were increased in neonatally capsaicin-treated rats, however, thermal nociceptive thresholds remained unchanged (*395*). Jancso and Wollerman further speculated that the pharmacological effect of capsaicin in the preoptic area would be mediated through activation of hypothalamic adenylate cyclase linked to the thermo-regulation system (*396,397*).

Thermal analgesia caused by capsaicin has recently been studied in connection to substance P or other neurotransmitters (*398–400*). Other studies on the effect of capsaicin on the thermoregulation system may be found in references *389, 390,* and *401–406*.

5. Nociception and Anesthesia

The fact that the pungent principle of *Capsicum* fruit is extremely irritating to the skin, mucous membrane, and tongue is known by everybody. However, it was not so long ago that studies on the nociceptive effect of capsaicinoid began. A number of papers on the nociceptive effect of capsaicinoid have been reported over the past 10 years (*319,349,407–410*).

The mechanism of nociception caused by capsaicinoids was first discussed in connection with ACTH secretion (*411,412*). In a comparison of the nociception caused by capsaicin with that caused by formalin, the action site of capsaicin was speculated to be only partially local because the drug raised the plasma corticosteroid level even in denervated areas.

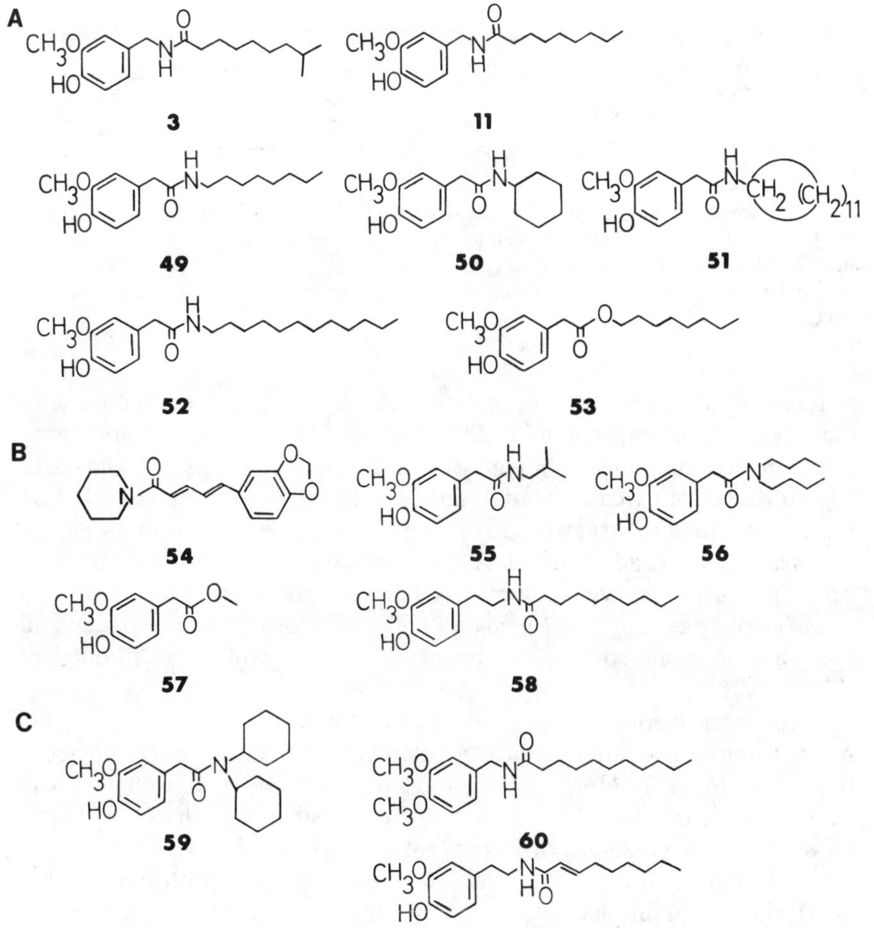

FIG. 3. Relationship between chemical structure and irritability. (A) Capsaicin congeners causing strong irritation to rats' eyes. (B) Capsaicin congeners causing weak irritation. (C) Capsaicin congeners causing almost no irritation. From Szolcsanyi and Jancso-Gabor (*427*).

Initial stimulation of primary and secondary muscle spindle endings by intraarterial injections of capsaicin was observed in cats (*413*). The nociceptive effect of capsaicinoids has been investigated in order to locate the selectively sensitive pain receptor (*414–420*). Capsaicin was shown to act on nerve terminals of certain primary afferent fibers to cause depolarization and calcium-dependent release of substance P (*421–423*). Jancso *et al.* have demonstrated that local capsaicin treatment of peripheral nerves selectively damages the chemosensitive nerve fibers by exhausting their substance P. Stimulation of prostaglandin E biosynthesis by capsaicin has also been reported (*424–426*).

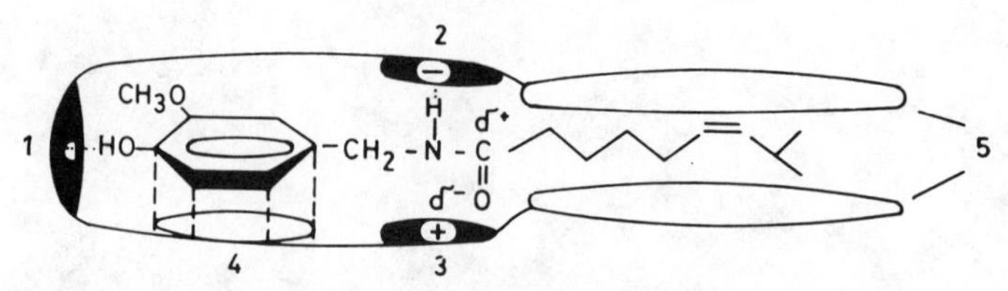

FIG. 4. Schematic representation of the hypothetical capsaicin receptor. (1) H-Bonding site for the OH group; (2) electronegative site for the H of the NH group and for the C^+ atom; (3) electropositive site for the oxygen of the carbonyl group; (4, 5) apolar areas bound by van der Waals forces. By permission of Editio Cantor and Drs. Szolcsanyi and Jancso-Gabor (*427*).

It has been shown that pain-inducing activity is dependent on the chemical structure of capsaicinoid (*427*) (Fig. 3). Those which exert strong irritation are compounds shown by chemical structures of **3**, **11**, and **49**, to **53** and those with weak irritation are **54** to **58**, while compounds **59** to **61** have no pain-inducing activity. The hydroxy group on the aromatic ring is necessary to exert nociception. A hypothetical pain receptor for capsaicin on pain sensory nerve endings has been proposed by Szolcsanyi and Jancso-Gabor as shown in Fig. 4 (*427*). The relationship between pain and pungency in connection with chemical structure will be mentioned in Section VII,A.

Chronic and neonatal capsaicin treatment induce a long-lasting or permanent densensitization against almost every type of chemical nociception (*316,316a,428–431*). Preliminary capsaicin administration has also been reported to affect the duration time of anesthetics such as hexobarbital or morphine (*432,433*) or to prevent neurogenic edema (*377,434,435*). Since desensitization by capsaicin has been observed with human skin (*436*), the compound has been used as a medicine to relieve muscle pain (*437*). In contrast to the short duration of analgesia caused by morphine, that induced by capsaicin remains for 7 to 10 days (*437*).

Lembeck and co-workers explained desensitization and elevation of the nociceptive threshold induced by neonatal capsaicin pretreatment by the degeneration of afferent nerve fibers (*438*). Morphological and immunohistological investigations have shown the irreversible degeneration of chemosensitive primary sensory neurons (*324,326,327,439,440*).

The antinociceptive effect induced by neonatal capsaicin treatment has lately been discussed in connection with a decrease of substance P in primary sensory neurons of the central nervous system (*370,371,400,405,441–456*). In connection with substance P, capsaicin has been considered to induce depletion of substance P and/or impairment of substance P-immunoreactive primary sensory neurons, which consequently lead to long-lasting desensitization. However, the antinociceptive effect of capsaicin cannot necessarily always be explained by depletion of substance P namely because antinociception without depletion of substance P has recently been observed (*457–459*). Tervo suggests that capsaicin is not a specific agent of substance P depletion but rather a neurotoxin which may block the release of substance P in the nerve (*457*). Jancso *et al.* have shown similar results that also strongly suggest that the effect of capsaicin is not confined to substance P (*460,461*). Involvement of other neurotransmitters have also been investigated (*230,462–464*).

Since capsaicin acts on various nervous systems to exert a number of interesting pharmacological activities, it has been utilized as a useful tool to elucidate information processing in nerve cells (*210,465–472*).

VII. Capsaicin and Analogs as Food Constituents

A. Pungency and Chemical Structure

The relationship between pungency and chemical structure of capsaicin analogues has been studied since 1919 (*20,21,155,189a,201–204,427,473–475*). Reviews on this aspect have been published by Newman (*476–478*) and others (*8,56*).

Kobayashi (*202*) synthesized a number of capsaicin analogs and homologs to reveal the relationship between pungency and chemical structure. As shown in Fig. 5, the vanillylamide moiety and acyl residues with appropriate chain lengths seem to be required for exerting pungency. Before Kobayashi, Nelson (*200*), comparing vanillylamide derivatives, disclosed that pungency depends on chain length as shown in Table I. The alkyl residue with C_8H_{17}(**11**) showed the strongest pungency, and pungency decreased both with longer and shorter alkyl chain lengths

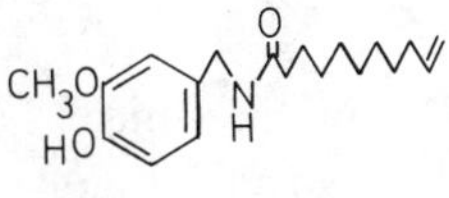

62 (1,000) 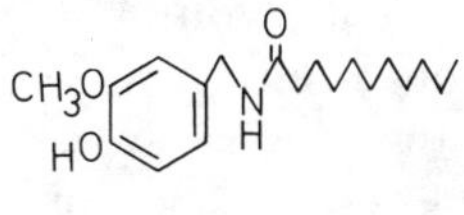 **14** (1,000) 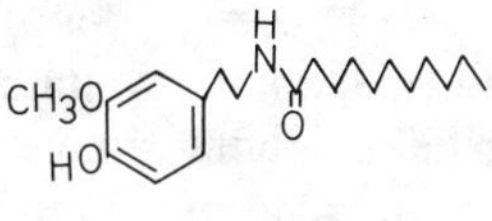 **63** (0)

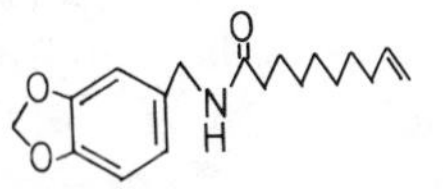

64 (0) 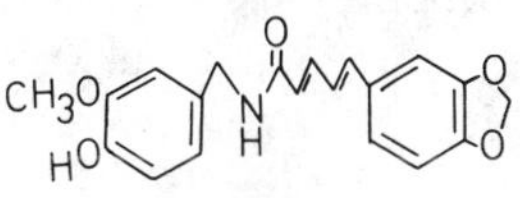 **65** (100) 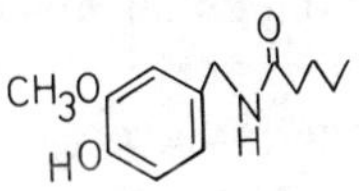 **66** (20)

FIG. 5. Relationship between structure and pungency. Numbers in parentheses are pungency relative to undecylenic vanillylamide (**62**) as 1000. From Kobayashi (*202*).

TABLE I
RELATIONSHIP BETWEEN CHAIN LENGTH OF THE ALKYL MOIETY OF CAPSAICINOIDS AND PUNGENCY[a]

67	$R = C_5H_{11}$	(5)
11	$R = C_8H_{17}$	(100)
14	$R = C_{11}H_{23}$	(25)
68	$R = C_6H_{13}$	(25)
12	$R = C_9H_{19}$	(50)
10	$R = C_7H_{15}$	(75)
13	$R = C_{10}H_{21}$	(25)

[a] From Nelson (*200*). R, Alkyl residue of capsaicinoid (**1**). Numerals in parentheses are pungency relative to the oleoresin of *Capsicum* fruit as 100.

(*200,479*). The double bond in the alkyl chain was found to have nothing to do with pungency (*20,155,202*). Pungency seems to depend not only on the presence of the methoxy group but also on the position of the hydroxy group (*201*).

Szolcsanyi and Jancso-Gabor have recently studied the relationship between chemical structure of capsaicin analogs and pungency or desensitization by using as many as 47 different analogs (*427,475*). They revealed that (i) neither pungency nor nociceptive activity to rats' eyes was affected by inversion of the acylamide linkage; (ii) by insertion of a methylene group between the aromatic ring and acylamide, the degree of pain production was not affected although pungency was considerably reduced; (iii) both pungency and pain-producing activity seem to respond in a parallel manner to capsaicin analogs with alkyl chains of different lengths. They concluded that participation of the 3-methoxy-4-hydroxybenzyl residue in the molecule should be very important for pungency and that the hydroxy group at C-4 of the aromatic ring must be essential for both pungency and pain production.

Unlike other primary tastes such as sweet, salt, sour, and bitter, the pungent taste has been regarded as a neuroresponse caused by nonspecific stimulation of afferent sensory nerves (*480*). However, even if this is true, mechanisms of pain production and pungency production seem to be somewhat different at the molecular level. Electrophysiological studies on the pungency of capsaicin have mostly remained undone as yet.

B. ANTIOXIDATIVE EFFECT

Spices in general have been recognized to have antioxidative activity (*481,482*). Hot red pepper fruits also show some antioxidative activity

(*483*). However, in comparison to the antioxidant activity of rosemary and sage, the activity of *Capsicum* fruit is low. The antioxidative effect of capsaicin and analogs has been studied by Sethi and Aggarwal (*484*) and Kaneda and co-workers (*479*). An antioxidative effect of capsaicin on peanut oil was reported by Sethi and Aggarwal (*484*).

During the course of study to develop new antioxidants, Kaneda and co-workers have shown that the antioxidant activities of capsaicin analogs were not affected by varying saturated acyl chain lengths from 9 to 12 carbons. The addition of 0.02 mol/kg of capsaicin analogs that have no pungency showed an antioxidative effect equivalent to that of butyrated hydroxytoluene at the 0.02% level (*479*). The antioxidative activity of capsaicin analogs is considered to come from the hydroxy group on the aromatic ring.

C. Deterioration during Preservation and Processing

Capsaicinoids have generally been thought to be fairly stable against microbial, thermal, light, and oxidative degradation. However, they are not so stable as we expect. Capsaicinoids are decomposed by fungi, light irradiation, heat, and oxidants under appropriate conditions.

The decomposition of capsaicinoids by *Aspergillus* species has been reported separately by Japanese (*485,486*) and Korean researchers (*303,304,487–490*). Despite use of the same species, the degradation products were different. The product obtained by Onozaki *et al.* from *A. niger* and *A. oryzae* were vanillylamine and fatty acids (*486*). In contrast, transformation products of the mixture of capsaicin and dihydrocapsaicin after a 48-hr incubation with *A. niger* as reported by Lee and You were *trans*-8-hydroxymethyl-*N*-vanillyl-6-nonenamide and 8-hydroxymethyl-*N*-vanillylnonamide, which would be formed by ω-hydroxylation (*491*). Reasonable explanation of these different results has not yet been given.

In addition to microorganisms, capsaicinoid is also deteriorated by heat drying, ultraviolet light, and oxidants. Several studies have been reported on the effect of drying conditions on capsaicinoids and other constituents of *Capsicum* fruit (*58,491–498*). Hot-air drying of harvested *Capsicum* fruit with minimum exposure to sunlight or in the dark has given the highest capsaicin yield (*492,496,499*). Exposure to a widely used fumigant, ethylene oxide gas, for 40 days at 40°C brought about an extreme decrease in the capsaicin content of *Capsicum* fruit (K. Iwai, unpublished data, 1982). Capsaicin in red pepper paste seems to be fairly labile because as much as 30% of the original content was decomposed during 15 min of cooking (*495*). Since capsaicinoid is selectively adsorbed to clay, clay bleaching has been proposed to remove capsaicinoid from oleoresin (*498*).

According to Charazka *et al.* (*500,501*), hot pepper extracts are well preserved for 9 months at room temperature, but preservation below 0°C is required for longer preservation.

D. Mutagenicity and Carcinogenicity

Mutagenicity of capsaicin in *Leptospira* (*502*) and occurrence of small amounts of volatile nitrosamines in meat products cured with cayenne pepper (*503*) had once been reported. It was concluded, however, that capsaicin and analogs have no mutagenic effect in the *Salmonella* test system over a wide range of concentrations (*504,505*). Data on the carcinogenicity of capsaicin and analogs have been insufficient.

Acknowledgments

The authors are grateful to Prof. Dr. M. Tanaka, Director of Plant Germplasm Institute, Faculty of Agriculture, Kyoto University, for his kind suggestion on the description of the taxonomy of *Capsicum* species. Special thanks are due to Prof. S. Oka and Dr. K. Nakamura, Institute for Chemical Research, Kyoto University, for their kind help in the arrangement of Chapter IV. Thanks are also due to Dr. H. Fujiwake and Mr. T. Kawada for their assistance in collecting information on a number of capsaicinoid studies.

References

1. E. Tyihak, A. Gulyas, and K. Juhasz, *Herba Hung.* **5,** 225 (1966).
2. E. Tyihak, K. Juhasz, A. Gulyas, and J. Szoeke, *Abh. Dtsch. Akad. Wiss. Berlin, Kl. Chem., Geol., Biol.* 623 (1966).
3. E. Leete and C. L. Louden, *J. Am. Chem. Soc.* **90,** 6837 (1968).
4. A. G. Mathew, Y. S. Lewis, N. Krishnamurthy, and E. S. Nambudiri, *Flavour Ind.* **2,** 693 (1971).
5. B. Wojciechowska and E. Dombrowicz, *Diss. Pharm. Pharmacol.* **18,** 61 (1966).
6. B. Broda, B. Wojciechowska, and E. Dombrowicz, *Diss. Pharm. Pharmacol.* **18,** 501 (1966).
7. P. G. Jones and G. R. Fenwick, *J. Sci. Food Agric.* **32,** 419 (1981).
8. J. Molnar, *Arzneim.-Forsch.* **15,** 718 (1965).
9. S. Kosuge, *Kagaku* **24,** 60 (1969).
10. W. Gavern, *Manuf. Chem. Aerosol News* **39,** 35, 42 (1968).
11. Trench (1876), cited in L. F. Tice, *Pharm. J.* **7,** 21 (1933).
12. J. C. Thresh, *Pharm. J. Trans.* **7,** 21, 259, 473 (1876).
13. J. C. Thresh, *Pharm. J. Trans.* **8,** 187 (1877).
14. F. A. Flückiger and D. Hanburg, "Pharmacologia Verlag." Macmillan, London, cited in Ref. *8*. (1874).
15. K. Micko, *Z. Unters. Nahr,- Genussm. Gebrauchsgegenstaende* **1,** 818 (1898).
16. K. Micko, *Z. Unters. Nahr.- Genussm. Gebrauchsgegenstaende* **2,** 411 (1899).
17. E. K. Nelson, *J. Ind. Eng. Chem.* **2,** 419 (1910).
18. E. K. Nelson, *J. Am. Chem. Soc.* **41,** 1115 (1919).

19. E. K. Nelson, *J. Am. Chem. Soc.* **42,** 597 (1920).
20. E. K. Nelson and L. E. Dawson, *J. Am. Chem. Soc.* **45,** 2179 (1923).
21. A. Lapworth and F. A. Royle, *J. Chem. Soc.* **115,** 1109 (1919).
22. L. Crombie, S. H. Dandegaonker, and K. S. Simpson, *J. Chem. Soc.* 1025 (1955).
23. S. Kosuge, Y. Inagaki, and S. Uehara, *Nippon Nogei Kagaku Kaishi* **32,** 578 (1958).
24. S. Kosuge, Y. Inagaki, and S. Niwa, *Nippon Nogei Kagaku Kaishi* **32,** 720 (1958).
25. S. Kosuge, Y. Inagaki, and K. Kimura, *Nippon Nogei Kagaku Kaishi* **35,** 596 (1961).
26. S. Kosuge, Y. Inagaki, and H. Okumura, *Nippon Nogei Kagaku Kaishi* **35,** 923 (1961).
27. H. Friedrich and R. Rangoonwala, *Naturwissenschaften* **52,** 514 (1965).
28. K. Juhasz and E. Tyihak, *Acta Agron. Acad. Sci. Hung.* **18,** 113 (1969).
29. D. J. Bennett and G. W. Kirby, *J. Chem. Soc. C* 442 (1968).
30. K. Jentsch, H. Pock, W. Kubelka, and O. Saiko, *Monatsh. Chem.* **99,** 661 (1968).
31. S. Kosuge and M. Furuta, *Agric. Biol. Chem.* **34,** 248 (1970).
32. Y. Masada, K. Hashimoto, T. Inoue, and M. Suzuki, *J. Food Sci.* **36,** 858 (1971).
33. J. Jurenitsch, M. David, F. Heresch, and W. Kubelka, *Planta Med.* **36,** 61 (1979).
34. F. Heresch and J. Jurenitsch, *Chromatographia* **12,** 647 (1979).
35. B. Kopp and J. Jurenitsch, *Planta Med.* **43,** 272 (1979).
36. J. Jurenitsch and R. Woginger, *Sci. Pharm.* **50,** 111 (1982).
37. M. Windholz, ed., "The Merck Index," 9th ed., p. 224. Merck & Co., Rahway, New Jersey, 1976.
38. O. Weissbach, "Beilstein's Handbuch der Organisdnen Chemie," 13, p. 2192. Springer-Verlag, Berlin and New York, 1973.
39. A. A. Newman, *Chem. Prod.* **16,** 413 (1953).
39a. C. D. Sauer, *in* "Handbook of South American Indians" (J. H. Steward ed.), Vol. 6, p. 521. Cooper Square Publisher, Inc., New York, 1963.
40. C. B. Heither, Jr., cited in "The Book of Spices" by F. Rosengarten, Jr., p. 138. Pyramid Books, New York, 1973.
41. B. Pickersgill, *Evolution* **25,** 683 (1971).
42. T. Tanaka, *in* "The Origin of Cultivated Plants," NHK Books 245, p. 201. Nippon Hososhuppan Kyokai, Tokyo, 1975.
43. P. G. Smith and C. B. Heither, Jr., *Am. J. Bot.* **38,** 362 (1951).
44. P. G. Smith and C. B. Heither, Jr., *Bull. Torrey Bot. Club* **84,** 413 (1957).
45. P. G. Smith, C. M. Rick, and C. B. Heither, Jr., *Bull. Torrey Bot. Club* **57,** 339 (1951).
46. C. B. Heither, Jr. and P. G. Smith, *Proc. Am. Soc. Hortic. Sci.* **52,** 331 (1948).
47. J. Y. Wilson, *Nature* (*London*) **183,** 1142 (1959).
48. C. B. Heither, Jr. and P. G. Smith, *Brittania* **10,** 194 (1958).
49. P. G. Smith and C. B. Heither, Jr., *Proc. Am. Soc. Hortic. Sci.* **70,** 286 (1957).
50. F. F. Campos and D. T. Morgan, Jr., *J. Hered.* **49,** 134 (1958).
51. P. A. Peterson, *Am. Nat.* **92,** 111 (1958).
52. J. W. Parry, "Spices—Their Morphology, Histology and Chemistry," 1st ed. Chem. Publ. Co., New York, 1962.
53. J. W. Parry, "The Story of Spices." Chem. Publ. Co., New York, 1953.
54. F. Rosengarten, Jr., ed., "The Book of Spices," p. 138. Pyramid Books, New York, 1973.
55. J. Jurenitsch, W. Kubelka, and K. Jentsch, *Planta Med.* **35,** 174 (1979).
55a. J. Jurenitsch, *Sci. Pharm.* **49,** 321 (1981).
56. V. S. Govindarajan, *ACS Symp. Ser.* **115,** 53 (1979).
57. Y. Ohta, *Seiken Jiho* **11,** 63 (1960).
58. T. S. Anantha Samy, V. N. Kamat, and H. G. Pandya, *Curr. Sci.* **29,** 271 (1960).
59. K. Csedo, J. Fuzi, Z. Kisgyorgy, and M. P. Horvath, *Rev. Med.* (*Tirgu-Mures, Rom.*) **8,** 485 (1962).

60. A. R. Deb, S. Ramanujam, G. S. R. Krishina Marti, and D. K. Thirumalachar, *Indian J. Technol.* **1,** 59 (1963).
61. D. K. Thirumalachar, *Curr. Sci.* **36,** 269 (1967).
62. K. Csedo and E. Kopp, *Pharmazie* **19,** 541 (1964).
63. E. Schratz and R. Rangoonwala, *Sci. Pharm., Proc. Congr., 25th, 1965* Vol. 1, p. 365 (1966).
64. M. S. Karawya, S. I. Balbaa, A. N. Girgis, and N. Z. Youssef, *Analyst* **92,** 581 (1967).
65. V. S. Govindarajan and S. M. Ananthakrishnar, *J. Food Sci. Technol.* **7,** 212 (1970).
66. M. P. Quaglio, M. Romagnolo, and G. Sandri Cavicchi, *Farmaco, Ed. Prat.* **26,** 349 (1971).
67. A. Trejo-Gonzalez and C. Wold-Altamirano, *J. Food Sci.* **38,** 342 (1973).
68. D. S. Pankar and N. G. Magar, *J. Maharashtra Agric. Univ.* **3,** 116 (1978); *CA* **91,** 73329c (1979).
69. F. Yuste, V. Castro, and F. Walls, *Rev. Soc. Quim. Mex.* **24,** 166 (1980); *CA* **94,** 190489c (1981).
70. S. Kosuge and Y. Inagaki, *Nippon Nogei Kagaku Kaishi* **36,** 251 (1962).
71. L. Quagliotti and E. Ottaviano, *Genet. Agrar.* **23,** 50 (1969).
72. M. B. Kvachdze, *Soobshch. Akad. Nauk Gruz. SSR* **82,** 169 (1976); *CA* **85,** 59767 (1976).
73. P. S. Arya and S. S. Saini, *Indian J. Agric. Res.* **10,** 223 (1976).
74. B. S. Sooch, *Indian Food Packer* **31,** 9 (1977).
75. P. Nowaczyk, *Herba Pol.* **27,** 25–30 (1981); *CA* **95,** 147322b (1981).
76. C. B. Jordan, E. W. Rebol, and H. O. Thompson, *Bull. Natl. Formul. Comm.* **10,** 49 (1942).
77. J. I. Suzuki, F. Tausig, and R. E. Morse, *Food Technol.* **11,** 100 (1957).
78. The Joint Committee of the Pharmaceutical Society and the Society for Analytical Chemistry on the Methods of Assay of Crude Drugs, *Analyst* **84,** 603 (1959).
79. K. Csedo, M. P. Horvath, and S. Z. Nagy, *Farmacia (Bucharest)* **8,** 435 (1960).
80. The Joint Committee of the Pharmaceutical Society and the Society for Analytical Chemistry on the Methods of Assay of Crude Drugs, *Analyst* **89,** 377 (1964).
81. J. S. Pruthi, *Adv. Food Res. Suppl.* **4,** 123 (1980).
82. K. Csedo, M. P. Horvath, and S. Nagy, *Orv. Sz.* **6,** 235 (1960); *CA* **54,** 23185 (1960).
83. A. Saria, F. Lembeck, and G. Skoftisch, *J. Chromatogr.* **208,** 41 (1981).
84. E. K. Johnson, H. C. Thompson, Jr., and M. C. Bowman, *J. Agric. Food Chem.* **30,** 324 (1982).
85. K. L. Bajaj and G. Kaur, *Mikrochim. Acta* **1,** 81 (1979).
86. K. Iwai, T. Suzuki, H. Fujiwake, and S. Oka, *J. Chromatogr.* **172,** 303 (1979).
87. K. Rajaraman, C. S. Narayanan, M. A. Sumathy Kutty, B. Sankarikutty, and A. G. Mathew, *J. Food Sci. Technol.* **18,** 101 (1981).
88. J. Jurenitsch, E. Bingler, H. Becker, and W. Kubelka, *Planta Med.* **36,** 54 (1979).
89. J. Jurenitsch and R. Leinmüller, *J. Chromatogr.* **189,** 389 (1980).
90. N. L. Sass, M. Rounsavill, and H. Combs, *J. Agric. Food Chem.* **25,** 1419 (1977).
91. R. Adamski and A. Socha, *Farm. Pol.* **23,** 435 (1967); *CA* **68,** 16101y (1968).
92. Ya. S. Meerov, A. N. Katyuzhanskayz, and N. F. Dyuban'kova, *Khim. Prir. Soedin.* 481 (1974); *CA* **82,** 14004j (1975).
93. J. C. Thresh, *Pharm. J. Trans.* **6,** 941 (1876).
94. J. C. Tice, *Am. J. Pharm.* **105,** 320 (1876).
95. H. North, *Anal. Chem.* **21,** 934 (1949).
96. G. Schenk, *Sci. Pharm.* **23,** 241 (1955).
97. R. Rangoonwala, *Planta Med.* **13,** 490 (1965).
98. A. Müller-Stock, R. K. Joshi, and J. Büchi, *J. Chromatogr.* **79,** 229 (1973).

99. P. H. Todd, Jr. and C. Perun, *Food Technol.* **15,** 270 (1961).
100. G. Gombtoko, *Elelmez. Ip.* **9,** 313 (1955); *CA* **52,** 8457 (1958).
101. B. Borkowski, Z. Kowaleski, and B. Pasichowa, *Biul. Inst. Rosl. Lecz.* **3,** 216 (1957).
102. K. T. Hartman, *J. Food Sci.* **35,** 543 (1970).
103. M. Fujita, T. Furuya, and A. Kawana, *Yakugaku Zasshi* **74,** 766 (1954).
104. P. Spanyar and M. Blazovich, *Analyst* **94,** 1084 (1969).
105. K. R. Lee, T. Suzuki, M. Kobashi, K. Hasegawa, and K. Iwai, *J. Chromatogr.* **123,** 119 (1976).
106. T. Tomova, M. Simova, and M. Kirkova, *Pharmazie* **34,** 448 (1979).
107. W. L. Scoville, *J. Am. Pharm. Assoc.* **1,** 453 (1912).
108. E. H. Wirth and E. N. Gathercoal, *J. Am. Pharm. Assoc.* **13,** 217 (1924).
109. J. C. Munch, *J. Assoc. Off. Anal. Chem.* **13,** 383 (1930).
110. American Spice Trade Association, "Official Analytical Methods for Spices," 2nd ed. ASTA, New York, 1968.
111. Indian Standards Institution, "Indian Standards," IS: 8104. ISI, New Delhi, 1976.
112. V. S. Govindarajan, S. Narasimhan, and S. Dhanaraj, *J. Food Sci. Technol.* **14,** 28 (1977).
113. P. H. Todd, Jr., *Food Technol.* **12,** 468 (1958).
114. N. C. Rajpoot, V. S. Govindarajan, and S. Venkatesa, *J. Assoc. Off. Anal. Chem.* **64,** 311 (1981).
115. P. H. Todd, Jr., M. G. Bensinger, and T. Biftu, *J. Food Sci.* **42,** 550, 680 (1977).
116. K. V. Fodor, *Kiserletugyi Kozl.* **33,** 155 (1930).
117. K. V. Fodor, *Z. Unters. Lebensm.* **61,** 94 (1931).
118. J. Büchi and F. Hippenmeier, *Pharm. Acta Helv.* **23,** 327, 353 (1948).
119. S. Kosuge, Y. Inagaki, and M. Nishimura, *Nippon Nogei Kagaku Kaishi* **33,** 915 (1959).
120. K. E. Schulte and H. M. Krüger, *Z. Anal. Chem.* **147,** 266 (1957).
121. P. Spanyar, E. Kevei, and M. Kiszel, *Elelmez. Ip.* **10,** 52 (1956).
122. P. Spanyar, E. Kevei, and M. Kiszel, *Acta Chim. Acad. Sci. Hung.* **11,** 137 (1957).
123. N. T. Prokhorova and L. L. Prozorovskaya, (1939). *Dokl. Vses. Akad. S'kh. Nauk im. V.I. Lenina* No. 16, 41; *CA* **36,** 3573 (1942).
124. G. Schenk, *Farmacognosia* **17,** 3 (1957).
125. J. Jonczyk, *Herba Pol.* **13,** 120 (1967); *CA* **69,** 12908 (1968).
126. H. D. Gibbs, *J. Biol. Chem.* **72,** 649 (1927).
127. R. Adamski and A. Socha, *Farm. Pol.* **23,** 603 (1967); *CA* **68,** 98677 (1968).
128. A. S. L. Tirimanna, *Analyst* **97,** 372 (1972).
129. R. Palacio and J. Jorge, *J. Assoc. Off. Anal. Chem.* **60,** 970 (1977).
130. R. Palacio and J. Jorge, *J. Assoc. Off. Anal. Chem.* **62,** 1168 (1979).
131. Y. Ohta, *Jpn. J. Breed.* **12,** 43 (1962).
132. M. Miedzinski and Z. Stachowiak, *Chem. Inz. Chem.* **13,** 251 (1979).
133. T. Altinkurt, *Eczacilik Bul.* **22,** 22 (1980).
134. K. L. Bajaj, *J. Assoc. Off. Anal. Chem.* **63,** 1314 (1980).
135. M. R. Srinivasan, M. N. Satyanarayana, and M. V. L. Rao, *Res. Ind.* **26,** 180 (1981).
136. M. S. Karawya, S. I. Balbaa, and A. N. Girgis, *J. Pharm. Sci. U.A.R.* **9,** 27 (1968).
137. A. G. Mathew, E. S. Nambudiri, S. M. Anathakrishna, N. Krishnamurthy, and Y. S. Lewis, *Lab. Pract.* **20,** 856 (1971).
138. S. Kosuge and Y. Inagaki, *Nippon Nogei Kagaku Kaishi* **33,** 918 (1960).
139. V. M. Rios and R. Duden, *Lebensm.-Wiss. Technol.* **4,** 97 (1971).
140. A. Polesello and F. Pizzocaro, *Riv. Sci. Tecnol. Alimenti Nutr. Um.* **6,** 305 (1976); *CA* **87,** 83266m (1977).

141. J. G. Mendez and J. G. Gonzalez, *Rev. Cubana Farm.* **12,** 335 (1978); **91,** 120382b (1979).
142. J. J. DiCecco, *J. Assoc. Off. Anal. Chem.* **62,** 998 (1979).
143. J. E. Woodbury, *J. Assoc. Off. Anal. Chem.* **63,** 556 (1980).
144. E. L. Johnson, R. E. Majors, L. Werum, and P. Reiche, *in* "Liquid Chromatographic Analysis of Food and Beverages" (G. Charalambous, ed.), Vol. 1, p.17. Academic Press, New York, 1979.
145. S. Kosuge, Y. Inagaki, and T. Ino, *Nippon Nogei Kagaku Kaishi* **34,** 811 (1960).
146. P. R. Datta and H. Susi, *Anal. Chem.* **33,** 148 (1961).
147. A. Müller-Stock, R. K. Joshi, and J. Büchi, *Helv. Chim. Acta* **56,** 799 (1973).
148. A. Müller-Stock, R. K. Joshi, and J. Büchi, *Pharm. Acta Helv.* **47,** 7 (1972).
149. W. R. Bowman, I. T. Bruce, and G. W. Kirby, *Chem. Commun.* 1075 (1969).
150. G. Schenk, *Sci. Pharm.* **23,** 241 (1955).
151. J. Hollo, I. Gal, and J. Suto, *Fette, Seifen, Anstrichm.* **59,** 1048 (1957).
152. O. Brauer and W. J. Schoen, *Angew. Bot.* **36,** 25 (1962).
153. K. E. Schulte and H. M. Krüger, *Mitt. Dtsch. Pharm. Ges.* **27,** 202 (1957).
154. Z. Sykulska, M. Galcqynska, Z. Kosiewicz, and M. Kwaiatleowska, *Acta Pol. Pharm.* **32,** 213 (1975); *CA* **83,** 19786h (1975).
155. S. Kosuge and Y. Inagaki, *Nippon Nogei Kagaku Kaishi* **33,** 470 (1959).
156. O. Folin and V. Ciocalteau, *J. Biol. Chem.* **73,** 627 (1922).
157. G. M. Barton, R. S. Evans, and G. A. F. Gardner, *Nature (London)* **170,** 249 (1952).
158. Z. Kowaleski, *Acta Pol. Pharm.* **16,** 309 (1959).
159. V. S. Govindarajan and S. M. Ananthakrishna, *Flavour Ind.* **5,** 176 (1974).
160. K. Tiechert, E. Mutschler, and H. Rochelmeyer, *Z. Anal. Chem.* **181,** 325 (1961).
161. D. Heusser, *Planta Med.* **12,** 237 (1964).
162. D. Heusser, *Pharm. Tijdschr. Belg.* **42,** 263 (1965).
163. P. Spanyar and M. Blazovich, *Konzerv- Paprikaip.* 11 (1969).
164. K. Jentsch, W. Kubelka, and H. Pock, *Sci. Pharm.* **37,** 153 (1969).
165. L. Kraus and E. Stahl, *Cesk. Farm.* **18,** 535 (1969).
166. M. Blazovich and P. Spanyar, *Elelmiszervizsgalati Kozl.* **15,** 358 (1969).
167. M. Blazovich and P. Spanyar, *Elelmiszertudomany* **3,** 127 (1969).
168. E. Stahl and L. Kraus, *Arzneim.-Forsch.* **19,** 684 (1969).
169. W. Kubelka, H. Pock, and K. Jentzsch, *Sci. Pharm.* **40,** 198 (1972).
170. J. M. Haguenoer, A. Senellart, and E. Couvreur, *Bull. Soc. Pharm. Lille* 37 (1973).
171. A. Laszlo and M. Laszlo, *Acta Aliment. Acad. Sci. Hung.* **4,** 113 (1975).
172. E. Kozma-Kovacs, E. Kevei-Pichler, and I. Lendvai, *Acta Aliment. Acad. Sci. Hung.* **6,** 1 (1977).
173. D. S. Pankar and N. G. Magar, *J. Chromatogr.* **144,** 149 (1977).
174. Bhagya, *J. Food Sci. Technol.* **14,** 176 (1977).
175. M. Miedzinski and Z. Stachowiak, *Chem. Inz. Chem.* **13,** 265 (1979).
176. W. Debska and B. Okulicz-Kozarynowa, *Chem. Anal. (Warsaw)* **18,** 291 (1973).
177. D. Heusser, *J. Chromatogr.* **33,** 400 (1968).
178. R. Rangoonwala, *Dtsch. Apoth.-Ztg.* **109,** 273 (1969).
179. R. Rangoonwala, *J. Chromatogr.* **41,** 265 (1969).
180. P. H. Todd, Jr., M. Besinger, and T. Biftu, *J. Chromatogr. Sci.* **13,** 577 (1975).
181. T. Suzuki, T. Kawada, and K. Iwai, *J. Chromatogr.* **198,** 217 (1980).
182. J. Jurenitsch, *Sci. Pharm.* **50,** 64 (1982).
183. J. I. Morrison, *Chem. Ind. (London)* 1785 (1967).
184. E. Grushka and P. Kapral, *Sep. Sci.* **12,** 415 (1977).
185. J. Hollo, E. Kurcz, and J. Bodor, *Lebensm.-Wiss. Technol.* **2,** 19 (1969).

186. A. Müller-Stock, R. K. Joshi, and J. Büchi, *J. Chromatogr.* **63,** 281 (1971).
187. J. J. DiCecco, *J. Assoc. Off. Anal. Chem.* **59,** 1 (1976).
188. K. Sagara, S. Kakizawa, K. Kasuya, T. Misaki, and H. Yoshizawa, *Chem. Pharm. Bull.* **28,** 2796 (1980).
189. P. L. Tsai, H. Ueda, and C. Tatsumi, *Nippon Shokuhin Kogyo Gakkaishi* **16,** 430 (1969).
189a. P. L. Tsai, H. Ueda, and C. Tatsumi, *Nippon Shokuhin Kogyo Gakkaishi* **16,** 346 (1969).
190. J. Jurenitsch, *Sci. Pharm.* **47,** 31 (1979).
191. J. Jurenitsch, W. Kubelka, and K. Jentzsch, *Sci. Pharm.* **46,** 307 (1978).
192. O. Sticher, F. Soldati, and R. K. Joshi, *J. Chromatogr.* **166,** 221 (1978).
193. J. Jurenitsch and I. Kampelmühler, *J. Chromatogr.* **193,** 101 (1980).
194. J. Krebs, R. J. Prime, and K. Leung, *J. Can. Soc. Forensic Sci.* **15,** 29 (1982).
195. T. Fung, W. Jeffery, and A. D. Beveridge, *J. Forensic Sci.* **27,** 812 (1982).
196. P. Spanyar, J. Kevei, and J. Kiszel, *Elelmiszervizsgalati Kozl.* **11,** 257 (1956).
197. I. Bozsa, M. Mosonyi, and G. Vastagh, *Gyogyszereszet* **17,** 134 (1973).
198. M. Miedzinski and Z. Stachowiak, *Chem. Inz. Chem.* **13,** 259 (1979).
199. G. Nogrady, *Kiserletugyi Kozl.* **46,** 160 (1943).
200. E. K. Nelson, *J. Am. Chem. Soc.* **41,** 2121 (1919).
201. E. Ott and K. Zimmermann, *Justus Liebig's Ann. Chem.* **425,** 314 (1921).
202. S. Kobayashi, *Rikagaku Kenkyusho Hokoku* **4,** 527 (1925).
203. P. C. Mitter and S. C. Ray, *J. Indian Chem. Soc.* **14,** 421 (1937).
204. M. Nakajima, *J. Pharm. Soc. Jpn.* **66,** 13 (1946).
205. E. Späth and S. F. Darling, *Ber. Dtsch. Chem. Ges.* **63,** 737 (1930).
206. M. Takahashi, K. Osawa, T. Ueda, and K. Okada, *Yakugaku Zasshi* **96,** 137 (1976).
207. W. S. Greaves, R. P. Linstead, B. R. Shephard, S. L. S. Thomas, and B. C. L. Weedon, *J. Chem. Soc.* 3326 (1950).
208. R. Rangoonwala and G. Seitz, *Dtsch. Apoth.-Ztg.* **110,** 1946 (1970).
209. O. Vig, R. C. Aggarwal, and S. D. Sharma, *Indian J. Chem., Sect. B* **17B,** 558 (1979).
210. H. S. Ahn and M. H. Makman, *Brain Res.* **153,** 636 (1978).
211. H. Fujiwake, T. Suzuki, S. Oka, and K. Iwai, *Agric. Biol. Chem.* **44,** 2907 (1980).
212. H. E. Zaugg and W. B. Martin, *Org. React.* **14,** 1 (1965).
213. N. T. Prokhorova and L. L. Prozorovskaya, *Dokl. Vses. Akad. S'kh. Nauk im. V.I. Lenina* 41 (1939).
214. N. T. Prokhorova and L. L. Prozorovskaya, *Khim. Ref. Zh.* 65 (1940).
215. T. Furuya and K. Hashimoto, *Yakugaku Zasshi* **74,** 771 (1954).
216. G. L. Tandon, S. V. Dravid, and G. S. Siddappa, *J. Food Sci.* **29,** 1 (1964).
217. L. Velarde-Cruz, *An. Fac. Farm. Bioquim., Univ. Nac. Mayor San Marcos* **8,** 125 (1957); *CA* **53,** 22735 (1959).
218. K. Iwai, K. R. Lee, M. Kobashi, T. Suzuki, and S. Oka, *Agric. Biol. Chem.* **42,** 201 (1978).
219. V. L. Huffman, E. R. Schadle, B. Villalon, and E. E. Burns, *J. Food Sci.* **43,** 1809 (1978).
220. B. J. Nagle, *Diss. Abstr. Int. B.* **41,** 2545 (1981).
221. Z. Kisgyorgy and K. Csedo, *Rev. Med.* (*Tirgu-Mures, Rom.*) **27,** 175 (1981); *CA* **97,** 69308b (1982).
222. K. Iwai, T. Suzuki, and H. Fujiwake, *Agric. Biol. Chem.* **43,** 2493 (1979).
223. K. Saga and T. Tamura, *Hokkaido Daigaku Nogakubu Hobun Kiyo* **7,** 294 (1970).
224. D. Neumann, *Naturwissenschaften* **53,** 131 (1966).
225. T. Suzuki, H. Fujiwake, and K. Iwai, *Plant Cell Physiol.* **21,** 839 (1980).

226. H. Fujiwake, T. Suzuki, and K. Iwai, *Plant Cell Physiol.* **21,** 1023 (1980).
227. P. Chudapongse and W. Janthasoot, *Toxicol. Appl. Pharmacol.* **37,** 263 (1976).
228. B. Borkowski, H. Gertig, and M. Olszak, *Acta Pol. Pharm.* **15,** 283 (1957).
229. Y. Ohta, *Jpn. J. Genet.* **37,** 86 (1962).
230. D. Dawbarn, A. J. Harmar, and C. J. Pycock, *Neuropathology* **20,** 341 (1981).
231. D. R. McCalla and A. C. Neish, *Can. J. Biochem. Physiol.* **37,** 537 (1959).
232. R. Rangoonwala, *Pharmazie* **24,** 177 (1969).
233. B. Kopp, J. Jurenitsch, and W. Kubelka, *Planta Med.* **39,** 289 (1980).
234. B. Kopp and J. Jurenitsch, *Sci. Pharm.* **50,** 150 (1982).
235. Y. Monsereenusorn and T. Glinsukon, *Toxicol. Lett.* **4,** 393 (1979).
236. T. Suzuki, T. Kawada, and K. Iwai, *Plant Cell Physiol.* **22,** 23 (1981).
237. H. Fujiwake, T. Suzuki, and K. Iwai, *Agric. Biol. Chem.* **46,** 2591 (1982).
238. H. Fujiwake, T. Suzuki, and K. Iwai, *Agric. Biol. Chem.* **46,** 2685 (1982).
239. T. Suzuki, H. Fujiwake, and K. Iwai, *Abstr. Pap., Annu. Meet. Agric. Chem. Soc. Jpn.* p. 85 (1982).
240. T. Kaneda, *Bacteriol. Rev.* **41,** 391 (1977).
241. M. I. Gurr and A. T. James, *in* "Lipid Biochemistry," p. 43. Chapman & Hall, London, 1975.
242. H. Fujiwake, Doctor's Thesis, Kyoto University, Japan, 1982.
243. K. S. Sim, *Yakhak Hoechi* **8,** 69 (1964).
244. I. Kuyumdzhiev, *C. R. Acad. Bulg. Sci.* **18,** 79 (1965).
245. L. Mikhailova, *C. R. Acad. Bulg. Sci.* **20,** 577 (1967).
246. L. Mikhailova, *Izv. Mikrobiol. Inst., Bulg. Akad. Nauk* **21,** 291 (1970).
247. I. E. Gal, *Z. Lebensm.-Unters. -Forsch.* **138,** 86 (1968).
248. I. E. Gal, *Elelmiszervizsgalati Kozl.* **15,** 80 (1969).
249. I. E. Gal, *Experientia* **21,** 383 (1965).
250. S. Kosuge and T. Takeuchi, *Nippon Shokuhin Kogyo Gakkaishi* **9,** 69 (1962).
251. L. Mikhailova, *Izv. Mikrobiol. Inst., Bulg. Akad. Nauk.* **21,** 277 (1970).
252. E. V. Gutsu, N. N. Baleshova, G. V. Lazur'evskii, and O. O. Timina, *Izv. Akad. Nauk Mold. SSR, Ser. Biol. Khim. Nauk* 24 (1982).
253. E. W. B. Ward, C. H. Unwin, and A. Stoessl, *Can. J. Bot.* **52,** 2481 (1974).
254. P. M. Molot, P. Mas, M. Conus, H. Ferriere, and P. Ricci, *Physiol. Plant Pathol.* **18,** 379 (1981).
255. H. J. Webber, *Am. Breed. Assoc., Annu. Rep.* **7,** 188 (1912).
256. K. Ramiah and M. R. Pillai, *Curr. Sci.* **4,** 236 (1935).
257. J. C. Miller and Z. M. Fineman, *Proc. Am. Soc. Hortic. Sci.* **35,** 544 (1937).
258. R. B. Deshpande, *Indian J. Agric Sci.* **5,** 513 (1935).
259. P. G. Smith, *J. Hered.* **41,** 138 (1950).
260. M. L. Odland, *Stn. Bull.—Minn., Agric. Exp. Stn.* **179,** 1 (1948).
261. Y. Ohta, *Jpn. J. Genet.* **37,** 169 (1962).
262. M. Michina, *Hodowla Rosl. Aklim. Nasienn.* **12,** 313 (1968); *CA* **70,** 112456e (1969).
263. R. Tenov and S. Khristov, *Nauchni Tr.—Nauchnoizsled. Inst. Konservna Prom-st., Plovdiv* **4,** 101 (1966); **67,** 31731 (1967).
264. Y. Ohta, *Seiken Jiho* **11,** 73 (1960).
265. K. Iwai, K. R. Lee, M. Kobashi, and T. Suzuki, *Agric. Biol. Chem.* **41,** 1873 (1977).
266. K. Iwai, T. Suzuki, K. R. Lee, M. Kobashi, and S. Oka, *Agric. Biol. Chem.* **41,** 1877 (1977).
267. K. S. Kim, S. D. Kim, J. R. Park, S. M. Roh, and T. H. Yoon, *Hanguk Sikp'um Kwahakh. Chi* **10,** 8 (1978).

268. K. S. Kim, S. M. Roh, and J. R. Park, *Hanguk Sikp'um Kwahakh. Chi* **11,** 162 (1979); *CA* **92,** 109314f (1980).
269. T. Zderkiewicz, *Acta Agrobot.* **24,** 343 (1971).
270. N. Izvorska, *Izv. Inst. Fiziol. Rast., Bulg. Akad. Nauk.* **18,** 79 (1973); *CA* **81,** 74047 (1974).
271. L. Galcz, S. Kordana, and R. Zalecki, *Herba Pol.* **16,** 107 (1970); *CA* **74,** 98987 (1971).
272. T. J. Nowak, *Acta Agrobot.* **33,** 59 (1980); *CA* **94,** 138414k (1981).
273. T. J. Nowak, *Acta Agrobot.* **33,** 73 (1980).
274. M. Bubicz, A. Korzen, and I. Perucka, *Rocz. Nauk Roln., Ser. A* **104,** 43 (1981); *CA* **96,** 103068 (1982).
275. S. W. Lee, K. S. Kim, and S. D. Kim, *Hanguk Sikp'um Kwahakh. Chi* **7,** 194 (1975); *CA* **84,** 131400 (1976).
276. T. Suzuki, T. Kawada, and K. Iwai, *Agric. Biol. Chem.* **45,** 535 (1981).
277. L. W. Haymon and L. W. Aurand, *J. Agric. Food Chem.* **19,** 1131 (1971).
278. M. R. Srinivasan, K. Sambaiah, M. N. Satyanarayana, and M. V. L. Rao, *Nutr. Rep. Int.* **21,** 455 (1980).
279. C. K. Hahn, *New Med. J.* **4,** 1305, 1345 (1961).
280. Y. Monsereenusorn and T. Glinsukon, *Food Cosmet. Toxicol.* **16,** 469 (1978).
281. Y. Monsereenusorn, *Toxicol. Lett.* **3,** 279 (1979).
282. Y. Monsereenusorn and T. Glinsukon, *Toxicol. Lett.* **4,** 399 (1979).
283. Y. Monsereenusorn, *J. Pharmacobio-Dyn.* **3,** 631 (1980).
284. Y. Monsereenusorn and T. Glinsukon, *Varasarn Paesachasarthara* **7,** 9 (1980).
285. P. Jonietz, *J. Sci. Soc. Thailand* **8,** 53 (1982).
286. P. Jonietz, *J. Natl. Res. Counc. Thailand* **15,** 15 (1983).
287. W. Nopanitaya, *Growth* **37,** 269 (1973).
288. K. Sambaiah, M. N. Satyanarayana, and M. V. L. Rao, *Nutr. Rep. Int.* **18,** 521 (1978).
289. K. Sambaiah and M. N. Satyanarayana, *Indian J. Exp. Biol.* **18,** 898 (1980).
290. K. Sambaiah and M. N. Satyanarayana, *J. Food Sci. Technol.* **19,** 30 (1982).
291. C. C. Toh, T. S. Lee, and A. K. Kiang, *Br. J. Pharmacol.* **10,** 175 (1955).
292. L. C. Gabor-Makara, R. Frankl, Z. Sanfai, and K. Szepeshazi, *Acta Med. Acad. Sci. Hung.* **21,** 213 (1965).
293. A. Csontos and C. Csedo, *Viata Med.* **17,** 399 (1970).
294. L. Limlomwongse, C. Chaitanchawong, and S. Tongyai, *J. Nutr.* **109,** 773 (1979).
295. Y. Monsereenusorn, *Toxicol. Appl. Pharmacol.* **53,** 134 (1980).
296. K. Iwai and T. Kawada, *Abstr. Pap., Southeast Asian West. Pac. Reg. Meet. Pharmacol., 3rd, 1982* p. 111 (1982).
297. T. Kawada, T. Suzuki, M. Takahashi, and K. Iwai, *Toxicol. Appl. Pharmacol.* **72,** 449 (1984).
298. N. D. Kim and C. Y. Park, *Yakhak Hoe Chi* **25,** 101 (1981); *CA* **96,** 135245 (1982).
299. T. Glinsukon, V. Stitmunnaithum, C. Toskulkao, T. Burannuti, and V. Tangkrisanavinont, *Toxicon* **18,** 215 (1980).
300. A. Saria, G. Skofitsch, and F. Lembeck, *J. Pharm. Pharmacol.* **34,** 273 (1982).
301. M. H. Kim, N. D. Kim, and S. S. Lee, *Yakhak Hoe Chi* **23,** 111 (1979).
302. S. S. Lee and S. Kumar, *Microsomes, Drug Oxid., Chem. Carcinog.* [*Int. Symp. Microsomes Drug Oxid.*], *4th, 1979* p. 1009 (1980).
303. S. S. Lee, *Recent Adv. Nat. Prod. Res., Proc. Int. Symp., 1979* p. 15 (1980).
304. S. S. Lee, *Saengyak Hakhoc Chi* (*Hanguk Saengyak Hakhoe*) **11,** 123 (1980).
305. R. M. Virus and G. F. Gebhart, *Life Sci.* **25,** 1273 (1979).
306. T. J. Haley, *Dangerous Prop. Ind. Mater. Rep.* **1,** 4 (1981).

307. J. I. Nagy, *Trends Neurosci.* (*Pers. Ed.*) **5,** 362 (1982).
308. Y. Monsereenusorn and P. D. Pezalla, *CRC Crit. Rev. Toxicol.* **10,** 321 (1982).
309. Y. Monsereenusorn, *in* "Adverse Effects of Foods" (E. F. P. Jelliffe and D. B. Jelliffe eds.), p. 195. Plenum, New York, 1982.
310. J. Szolcsanyi, *J. Congr. Hung. Pharmacol. Soc.* **2,** 167 (1974).
310a. C. L. Winek, D. C. Markie, and S. P. Shanor, *Drug Chem. Toxicol.* **5,** 89 (1982).
311. W. Nopanitaya, *Am. J. Dig. Dis.* **19,** 439 (1974).
312. W. Nopanitaya and S. W. Nye, *Toxicol. Appl. Pharmacol.* **30,** 149 (1974).
313. J. G. Smith, Jr., R. G. Crounse, and D. Spence, *J. Invest. Dermatol.* **54,** 170 (1970).
314. P. Chudapongse and W. Janthasoot, *Biochem. Pharmacol.* **30,** 735 (1981).
315. J. Szolcsanyi and A. Jancso-Gabor, *in* "Proceedings of the Satellite Symposium on Pharmacology of Thermoregulation" (E. K. Schonbaum ed.), p. 395. Basel, Switzerland, 1972.
316. A. Jancso-Gabor, *Proc. Congr. Hung. Pharmacol. Soc.* **2,** 161 (1974).
316a. G. B. Makara, *Acta Physiol. Acad. Sci. Hung.* **38,** 393 (1971).
317. F. Joo, J. Szolcsanyi, and A. Jancso-Gabor, *Life Sci.* **8,** 621 (1969).
318. J. Szolcsanyi, F. Joo, and A. Jancso-Gabor, *Nature* (*London*) **229,** 116 (1971).
319. J. Szolcsanyi, A. Jancso-Gabor, and F. Joo, *Naunyn-Schmiedeberg's Arch. Pharmacol.* **287,** 157 (1975).
320. A. Jancso-Gabor and J. Szolcsanyi, *in* "Inflammation Biochemistry and Drug Interaction" (A. Bertelli, ed.), p. 21. Excerpta Med., Amsterdam, 1968.
321. A. Jancso-Gabor and E. Kiraly, *Brain Res.* **210,** 83 (1981).
322. G. Jancso, *Neurosci. Lett.* **27,** 41 (1981).
323. J. I. Nagy, S. R. Vincent, W. A. Staines, H. C. Fibiger, T. D. Reisine, and H. I. Yamamura, *Brain Res.* **186,** 435 (1980).
324. J. I. Nagy, S. P. Hunt, L. L. Iversen, and P. C. Emson, *Neuroscience* **6,** 1923 (1981).
325. N. N. Palermo, H. K. Brown, and D. L. Smith, *Brain Res.* **208,** 506 (1981).
326. A. D. Hoyes and P. Barber, *Neurosci. Lett.* **25,** 19 (1981).
327. P. Holzer, A. Bucsics, and F. Lembeck, *Neurosci. Lett.* **31,** 253 (1982).
328. M. Fitzgerald, *Brain Res.* **248,** 97 (1982).
329. E. Fehr and J. Vadja, *Acta Morphol. Acad. Sci. Hung.* **30,** 57 (1982).
330. J. M. Dubois, *Brain Res.* **245,** 372 (1982).
331. A. D. Hoyes, P. Barber, and H. Jagessar, *Neurosci. Lett.* **26,** 329 (1981).
332. J. Szolcsanyi and L. Bartho, *in* "Advancement of Physiological Science" (G. Mozsik *et al.,* eds.), p. 39. Akadémiai Kiadó, Budapest, 1980.
333. A. G. Nast, *J. Am. Inst. Homeopathy* **16,** 30 (1923).
334. A. Anuras, J. Christensen, and D. Templeman, *Gut* **18,** 666 (1977).
335. J. Molnar, L. A. Baraz, and V. M. Khayutin, *Tr. Inst. Norm. Patol. Fiziol. Akad. Med. Nauk SSSR* **10,** 22 (1967); *CA* **70,** 113781 (1969).
336. L. A. Faraz, V. M. Khayutin, and J. Molnar, *Acta Physiol. Acad. Sci. Hung.* **33,** 225 (1968).
337. A. Baraz, V. M. Khayutin, and J. Molnar, *Acta Physiol. Acad. Sci. Hung.* **33,** 327 (1968).
338. J. Molnar, L. Gyorgyi, G. Unyi, and J. Kenyeres, *Acta Physiol. Acad. Sci. Hung.* **35,** 369 (1969).
339. L. Bartho and J. Szolcsanyi, *Naunyn-Schmiedeberg's Arch. Pharmacol.* **305,** 75 (1978).
340. J. Szolcsanyi and L. Bartho, *Naunyn-Schmiedeberg's Arch. Pharmacol.* **305,** 83 (1978).

341. P. Holzer and F. Lembeck, *Naunyn-Schmiedeberg's Arch. Pharmacol.* **310,** 169 (1979).
342. P. Holzer, R. Gamse, and F. Lembeck, *Eur. J. Pharmacol.* **61,** 303 (1980).
343. L. Bartho, P. Holzer, F. Lembeck, and J. Szolcsanyi, *J. Physiol. (London)* **332,** 157 (1982).
344. L. A. Chahl, *Naunyn-Schmiedeberg's Arch. Pharmacol.* **319,** 212 (1982).
345. K. Tsou, G. Louie, and E. L. Way, *Eur. J. Pharmacol.* **81,** 377 (1982).
346. L. J. Boyd, *J. Am. Inst. Homeopathy* **21,** 7 (1928).
347. J. Lille and E. Ramiretz, *An. Inst. Biol., Univ. Nac. Auton. Mex.* **6,** 23 (1935).
348. J. Porszasz, L. György, and K. Gibiszer-Porszasz, *Acta Physiol. Acad. Sci. Hung.* **8,** 60 (1955).
349. M. L. Kirby, T. F. Gale, and T. G. Mattio, *Exp. Neurol.* **76,** 298 (1982).
350. E. Gores, *Naunyn-Schmiedebergs Arch. Exp. Pathol. Pharmakol.* **236,** 145 (1959).
351. E. Gores and F. Jung, *Acta Biol. Med. Ger.* **3,** 41 (1959).
352. N. Toda, H. Usui, N. Nishino, and M. Fujiwara, *J. Pharmacol. Exp. Ther.* **181,** 512 (1972).
353. S. C. Crayton, J. H. Mitchell, and F. C. Payene, *Am. J. Physiol.* **240,** H315 (1981).
354. J. A. Bevan, *Circ. Res.* **10,** 792 (1962).
355. J. H. Mitchell, D. N. Gupta, and S. E. Barnett, *Circ. Res., Suppl.* **1,** 192 (1967).
356. J. Molnar and L. György, *Eur. J. Pharmacol.* **1,** 86 (1967).
357. D. Brender and M. M. Webb-Peploe, *Am. J. Physiol.* **217,** 1837 (1969).
358. M. M. Webb-Peploe, D. Brender, and J. T. Shephard, *Am. J. Physiol.* **222,** 189 (1972).
359. H. M. Coleridge and J. C. G. Coleridge, *J. Physiol. (London)* **179,** 248 (1966).
360. A. Ueno, *Nippon Yakurigaku Zasshi* **67,** 572 (1971).
361. J. Porszasz, G. Such, and K. Porszasz-Gibiszer, *Acta Physiol. Acad. Sci. Hung.* **12,** 189 (1957).
362. J. A. Russell and S. J. Lai-Fook, *J. Appl. Physiol.: Respir., Environ. Excercise Physiol.* **47,** 961 (1979).
363. J. Molnar, G. B. Makara, L. György, and G. Unyi, *Acta Physiol. Acad. Sci. Hung.* **36,** 413 (1969).
364. G. B. Makara, L. György, and J. Molnar, *Arch. Int. Pharmacodyn. Ther.* **170,** 39 (1967).
365. N. Fukuda and M. Fujiwara, *J. Pharm. Pharmacol.* **21,** 622 (1969).
366. J. A. Kiernan, *J. Exp. Physiol. Cogn. Med. Sci.* **62,** 151 (1977).
367. S. E. Carpenter and B. Lynn, *J. Physiol. (London)* **310,** 69P (1981).
368. J. E. Bernstein, R. M. Swift, K. Soltani, and A. L. Lorincz, *J. Invest. Dermatol.* **76,** 394 (1981).
369. S. E. Carpenter and B. Lynn, *Br. J. Pharmacol.* **73,** 755 (1981).
370. G. Jancso, E. Kiraly, and A. Jancso-Gabor, *Int. J. Tissue React.* **2,** 57 (1980).
371. R. Gamse, P. Holzer, and F. Lembeck, *Br. J. Pharmacol.* **68,** 207 (1980).
372. F. Lembeck and J. Donnerer, *Naunyn-Schmiedeberg's Arch. Pharmacol.* **316,** 240 (1981).
373. R. E. Papka, J. B. Furness, N. G. Della, and M. Costa, *Neurosci. Lett.* **27,** 47 (1981).
374. R. M. Virus and M. M. Knuepfer, *Eur. J. Pharmacol.* **72,** 209 (1981).
375. J. Donnerer and F. Lembeck, *Naunyn-Schmiedeberg's Arch. Pharmacol.* **320,** 54 (1982).
376. G. A. Ordway and J. C. Longhurst, *Circ. Res.* **52,** 26 (1983).
377. Y. Alarie and L. W. Keller, *Environ. Physiol. Biochem.* **3,** 169 (1973).
378. R. Andoh, S. Sakurada, K. Kisara, M. Takahashi, and K. Ohsawa, *Nippon Yakurigaku Zasshi* **79,** 275 (1982).

379. P. Ki, J. A. Neguleseo, and M. Murnane, *IRCS Med. Sci.: Libr. Compend.* **10,** 446 (1982).
380. A. Jancso-Gabor, (1947), cited in A. Jancso-Gabor, J. Szolcsanyi, and N. Jancso, *J. Physiol.* (*London*) **206,** 495 (1970).
381. B. Issekutz, I. Lichtneckert, and H. Nagy, *Arch. Int. Pharmacodyn. Ther.* **81,** 35 (1950).
382. B. Issekutz, I. Lichtneckert, and M. Winter, *Arch. Int. Pharmacodyn. Ther.* **83,** 319 (1950).
383. H. V. Czetsch-Lindenwald, *Fette, Seinfen, Anstrichm.* **55,** 185 (1953).
384. A. Jancso-Gabor, J. Szolcsanyi, and N. Jancso, *J. Physiol.* (*London*) **208,** 449 (1970).
385. T. Hori, *in* "Pharmacology and Thermoregulation" (B. Cox *et al.*, eds.), p. 214. Basel, Switzerland, 1979.
386. L. S. Rabe, S. H. Buck, L. Moreno, T. F. Burks, and N. Datny, *Brain Res. Bull.* **5,** 755 (1980).
387. T. Hori, *Proc. Int. Congr. Adv. Physiol. Sci., 28th, 1980* p. 53 (1981).
388. G. A. Balint, *Kiserl. Orvostud.* **24,** 101 (1972).
389. J. Frens, *IRCS Med. Sci.: Libr. Compend.* **4,** 176 (1976).
390. M. Szekely and J. Szolcsanyi, *Acta Physiol. Acad. Sci. Hung.* **53,** 469 (1979).
391. J. Szolcsanyi and A. Jancso-Gabor, *Proc. Symp. Temp. Regul. Drug Action, 1974* p. 331 (1975).
392. J. Szolcsanyi, *Proc.* 28th *Int. Congr. Adv. Physiol. Sci., 28th, 1980* p. 61 (1981).
393. M. Cabanac, M. Cormareche-Leyder, and L. J. Poirier, *Pfluegers Arch.* **366,** 217 (1976).
394. J. Frens, *Drugs, Biog. Amines Body Temp., Proc. Symp. Pharmacol. Thermoregul., 3rd, 1976* p. 20 (1977).
395. F. Cervero and H. A. McRitchie, *Brain Res.* **215,** 414 (1981).
396. G. Jancso and M. Wollermann, *Brain Res.* **123,** 323 (1977).
397. A. G. Horvath and M. B. Tyres, *Brain Res.* **179,** 401 (1980).
398. A. G. Hayes and M. B. Tyres, *Brain Res.* **189,** 561 (1980).
399. S. H. Buck, P. P. Deshnukh, H. I. Yamamura, and T. F. Burks, *Neuroscience* **6,** 2217 (1981).
400. T. L. Yaksh, D. H. Farb, S. E. Leeman, and T. M. Jessell, *Science* **206,** 481 (1979).
401. T. Nakayama, M. Suzuki, Y. Ishikawa, and A. Nishino, *Neurosci. Lett.* **7,** 151 (1979).
402. F. Obal, Jr., M. Hajos, G. Benedek, F. Obal, and A. Jancso-Gabor, *Physiol. Behav.* **27,** 977 (1981).
403. B. Dib, *Pharmacol., Biochem. Behav.* **16,** 23 (1982).
404. M. Szikszay, F. Obal, Jr., and F. Obal, *Naunyn-Schmiedeberg's Arch. Pharmacol.* **320,** 97 (1982).
405. T. E. Salt, C. S. Crozier, and R. G. Hill, *Neuroscience* **7,** 1141 (1982).
406. J. I. Nagy, P. C. Emson, and L. L. Iversen, *Brain Res.* **211,** 497 (1981).
406a. M. Cabanac, M. Cormareche-Leydier, and L. J. Poirier, *Drugs, Biog. Amines Body Temp., Thermoregul., 3rd, 1976* p. 99 (1977).
407. A. Jancso-Gabor and J. Szolcsanyi, *J. Dent. Res.* **51,** Suppl. 2, 264 (1972).
408. R. Andoh, K. Shima, T. Miyazawa, S. Sakurada, K. Kisara, K. Osawa, and M. Takahashi, *Jpn. J. Pharmacol.* **30,** 599 (1980).
409. M. Fitzgerald and C. J. Woolf, *Neuroscience* **7,** 2051 (1982).
410. J. T. Williams, and W. Zieglgaensberger, *Brain Res.* **253,** 125 (1982).
411. G. B. Makara, E. Stark, and K. Mihaly, *Magy. Tud. Akad., 5* (*Otodik*) *Orv. Tud. Oszt. Kosl.* **18,** 119 (1967).
412. G. B. Makara, E. Stark, and K. Mihaly, *Can. J. Physiol. Pharmacol.* **45,** 669 (1967).

413. K. S. K. Murthy and S. S. Deshpande, *Brain Res.* **79,** 89 (1974).
414. G. Jancso and E. Knyihar, *Neurobiology* **5,** 42 (1975).
415. R. W. Foster and A. G. Ramage, *Br. J. Pharmacol.* **57,** 436P (1976).
416. J. Szolcsanyi, *J. Physiol.* (*London*) **73,** 251 (1977).
417. R. W. Foster and A. G. Ramage, *Neuropharmacology* **20,** 191 (1981).
418. P. D. Wall and M. Fitzgerald, *Pain* **11,** 363 (1981).
419. P. D. Wall, M. Fitzgerald, and C. S. Woolf, *Exp. Neurol.* **78,** 425 (1982).
420. M. P. Kaufman, G. A. Iwamoto, J. C. Longhurst, and J. H. Mitchell, *Circ. Res.* **50,** 133 (1982).
421. M. Yanagisawa, S. Konishi, T. Suzue, and M. Otsuka, *Int. Brain Res. Organ. Monogr. Ser.* **7,** 43 (1980).
422. B. Ault, *Br. J. Pharmacol.* **70,** 95P (1980).
423. T. E. Salt and R. G. Hill, *Neurosci. Lett.* **20,** 329 (1980).
424. H. Juan, F. Lembeck, S. Seewan, and U. Hack, *Naunyn-Schmiedeberg's Arch. Pharmacol.* **312,** 139 (1980).
425. H. O. J. Collier, W. J. McDonald-Gibson, and S. A. Saeed, *Lancet* **1,** 702 (1975).
426. K. E. Choi and S. S. Lee, *Seoul Taehakkyo Yakhak Nonmunjip* **4,** 68 (1979).
427. J. Szolcsanyi and A. Jancso-Gabor, *Arzneim.-Forsch.* **25,** 1877 (1975).
428. O. Benesova, *Naunyn-Schmiedebergs Arch. Exp. Pathol. Pharmakol.* **236,** 131 (1959).
429. E. Jung and I. Kruger, *Acta Biol. Med. Ger.* **5,** 128 (1960).
430. A. G. Hayes, M. Skingle, and M. B. Tyres, *Neuropharmacology* **20,** 505 (1981).
431. A. G. Hayes, M. Skingle, and M. B. Tyres, *Br. J. Pharmacol.* **70,** 96P (1980).
432. A. Jancso-Gabor, *Acta Physiol. Acad. Sci. Hung.* **55,** 57 (1980).
433. G. Jancso and A. Jancso-Gabor, *Naunyn-Schmiedeberg's Arch. Pharmacol.* **311,** 285 (1980).
434. R. T. Arvier, L. A. Chahl, and R. J. Ladd, *Br. J. Pharmacol.* **59,** 61 (1977).
435. C. R. Morton and L. A. Chahl, *Naunyn-Schmiedeberg's Arch. Pharmacol.* **314,** 271 (1980).
436. M. S. Dash and S. S. Deshpande, *Adv. Pain Res. Ther.* **1,** 47 (1975).
437. T. R. LaHann, U.S. Patent 4,313,958 (1982); U.S. Patent Appl. 200,102 (1980).
438. P. Holzer, I. Jurna, R. Gamse, and F. Lembeck, *Eur. J. Pharmacol.* **58,** 511 (1979).
439. G. Jancso, E. Kiraly, and A. Jancso-Gabor, *Nature* (*London*) **270,** 741 (1977).
440. G. Jancso, E. Kiraly, and A. Jancso-Gabor, *Naunyn-Schmiedeberg's Arch. Pharmacol.* **313,** 91 (1980).
441. H. Akagi, M. Otsuka, and M. Yanagisawa, *Neurosci. Lett.* **20,** 259 (1980).
442. N. Mayer, R. Gamse, and F. Lembeck, *J. Neurochem.* **35,** 1238 (1980).
443. C. J. Helke, J. A. DiMicco, D. M. Jacobowitz, and I. J. Kopin, *Brain Res.* **222,** 428 (1981).
444. A. C. Cuello, R. Gamse, P. Holzer, and F. Lembeck, *Naunyn-Schmiedeberg's Arch. Pharmacol.* **315,** 185 (1981).
445. A. G. Hayes, J. W. Scadding, M. Skingle, and M. B. Tyres, *J. Pharm. Pharmacol.* **33,** 183 (1981).
446. M. Schültzberg, G. J. Dockray, and R. G. Williams, *Brain Res.* **235,** 198 (1982).
447. J. V. Priestley, S. Bramwell, L. L. Butcher, and A. C. Cuello, *Neurochem. Int.* **4,** 57 (1982).
448. E. Theriault, M. Otsuka, and T. Jessell, *Brain Res.* **170,** 209 (1979).
449. R. Gamse, A. Molnar, and F. Lembeck *Life Sci.* **25,** 629 (1979).
450. R. G. Hill, M. L. Hoddinott, and P. Keen, *J. Physiol.* (*London*) **303,** 27P (1980).
451. A. Bucsics and F. Lembeck, *Eur. J. Pharmacol.* **71,** 71 (1981).
452. C. J. Helke, D. M. Jacobowitz, and B. T. Nguyen, *Life Sci.* **29,** 1779 (1981).

453. T. F. Burks, S. H. Buck, M. S. Miller, P. P. Deshmukh, and H. I. Yamamura, *Proc. West. Pharmacol. Soc.* **24,** 353 (1981).
454. D. Lackner, G. Gamse, and S. E. Leeman, *Naunyn-Schmiedeberg's Arch. Pharmacol.* **316,** 38 (1981).
455. M. S. Miller, S. H. Buck, I. G. Sipes, H. I. Yamamura, and T. F. Burks, *Brain Res.* **250,** 193 (1982).
456. R. G. Hill, M. L. Hoddinott, and P. M. Keen, *Int. Brain Res. Organ. Monogr. Ser.* **7,** 31 (1980).
457. K. Tervo, *Acta Ophthalmol.* **59,** 737 (1981).
458. M. S. Miller, S. H. Buck, I. G. Sipes, and T. F. Burks, *Brain Res.* **244,** 193 (1982).
459. R. J. Bodnan, A. Kirchgessner, G. Nilaver, J. Mulhern, and E. A. Zimmerman, *Neuroscience* **7,** 631 (1982).
460. G. Jancso, T. Hokfelt, J. M. Lundberg, E. Kiraly, N. Halasz, G. Nilsson, L. Terenius, J. Rehfeld, H. Steinbush, *J. Neurocytol.* **10,** 963 (1981).
461. R. Gamse, U. Petsche, F. Lembeck, and G. Jancso, *Brain Res.* **239,** 447 (1982).
462. P. Holzer, A. Saria, G. Skoftisch, and F. Lembeck, *Life Sci.* **29,** 1099 (1981).
463. E. A. Singer, G. Sperk, and R. Schmid, *J. Neurochem.* **38,** 1383 (1982).
464. R. Gamse, S. E. Leeman, P. Holzer, and F. Lembeck, *Naunyn-Schmiedeberg's Arch. Pharmacol.* **317,** 140 (1981).
465. J. A. Kessler and I. B. Black, *Proc. Natl. Acad. Sci. U.S.A.* **78,** 4644 (1981).
466. C. B. Camras and L. Z. Bito, *Invest. Ophthalmol. Visual Sci.* **19,** 423 (1980).
467. J. T. Williams and W. Zieglgaensberger, *Adv. Biochem. Psychopharmacol.* **33,** 423 (1982).
468. G. P. Mueller, *Life Sci.* **29,** 1669 (1981).
469. T. M. Jessell and G. D. Fischbach, *Proc. Int. Congr. Adv. Pharmacol. Ther., 8th, 1981* Vol. 2, p. 155 (1982).
470. R. Andoh, S. Sakurada, T. Sato, N. Takahashi, and K. Kisara, *Jpn. J. Pharmacol.* **32,** 81 (1982).
471. A. Mandahl and A. Bill, *Acta Physiol. Scand.* **112,** 331 (1981).
472. R. W. Foster, A. H. Weston, and K. M. Weston, *Br. J. Pharmacol.* **70,** 98P (1980).
473. H. Staudinger and H. Schneider, *Ber. Dtsch. Chem. Ges.* **56,** 699 (1923).
474. P. Hegyes and S. Foldeak, *Acta Phys. Chem.* **20,** 115 (1974).
475. J. Szolcsanyi and A. Jancso-Gabor, *Arzneim.-Forsch.* **26,** 33 (1976).
476. A. A. Newman, *Chem. Prod. Chem. News* **16,** 467 (1953).
477. A. A. Newman, *Chem. Prod. Chem. News* **17,** 14 (1954).
478. A. A. Newman, *Chem. Prod. Chem. News* **17,** 102 (1954).
479. K. Fujimoto, Y. Kanno, and T. Kaneda, *Abstr. Pap., Int. Congr. Food Sci. Technol.* p. 294 (1978).
480. M. Schneider, *in* "Einfuhrung in die Physiologie des Menschen," p. 676. Springer-Verlag, Berlin, 1964.
481. S. J. Bishov, Y. Masuoka, and J. G. Kaspalis, *J. Food Process. Preserv.* **1,** 153 (1977).
482. U. Bracco, J. Loliger, and J. L. Viret, *J. Am. Oil Chem. Soc.* **58,** 686 (1981).
483. J. W. Parry, "Spices," Vol. 1, pp. 9, 43. Chem. Publ. Co., New York, 1969.
484. S. C. Sethi and J. S. Aggarwal, *J. Sci. Ind. Res. Sect. B.* **15B,** 34 (1956).
485. H. Onozaki and K. Minami, *Eiyo to Shokuryo* **25,** 454 (1972); *CA* **78,** 1920 (1973).
486. H. Onozaki, K. Sasaoka, and H. Ezaki, *Hakko Kogaku Zasshi* **54,** 297 (1976); *CA* **85,** 30476 (1976).
487. S. S. Lee, *J. Pharm. Soc. Korea* **3,** 111 (1957).
488. K. D. Hahn and S. S. Lee, *J. Pharm. Soc. Korea* **4,** 60 (1959).
489. S. S. Lee and I. S. You, *Hanguk Saenghwa Hakhoe Chi* **10,** 135 (1977).

490. S. S. Lee, Y. H. Baek, S. Y. Ko, and S. Kumar, *Proc. Int. Conf. Chem. Biotechnol. Biol. Act. Nat. Prod., 1st, 1981* p. 189 (*1981*).
491. S. S. Lee and I. S. You, *Hanguk Saenghwa Hakhoe Chi* **10,** 135 (1977).
492. B. Borkowski, H. Gertig, and M. Olszak, *Acta Pol. Pharm.* **15,** 289 (1957).
493. P. Spanyar, J. Kevei, J. Siszel, and F. Simek, *Elelmez. Ip.* **10,** 193 (1956).
494. J. G. Lease and E. J. Lease, *Food Technol.* **16,** 104 (1962).
495. C. A. Yum, *Hanguk Yongyang Hakhoe Chi* **2,** 99 (1969).
496. K. H. Kim and J. K. Chun, *Hanguk Sikp'um Kwahakhoe Chi* **7,** 69 (1975).
497. C. R. Park, *Hanguk Yongyang Hakhoe Chi* **8,** 167 (1975).
498. J. C. Kim and J. S. Rhee, *Hanguk Sikp'um Kwahakhoe Chi* **12,** 126 (1980).
499. K. Iwai, unpublished data (1982).
500. Z. Charazka, *Zesz. Probl. Postepow Nauk Roln.* **243,** 205 (1980); *CA* **94,** 138080 (1981).
501. Z. Charazka, K. Karowska, and E. Kostrezewa, *Nahrung* **25,** 711 (1981).
502. I. Kuyumdzhier, *Izv. Mikrobiol. Inst., Bulgar. Akad. Nauk.* **13,** 5 (1961); *CA* **57,** 9006 (1962).
503. T. A. Gough and K. Goodhead, *J. Sci. Food Agric.* **26,** 1473 (1975).
504. J. Farkas, E. Andrassy, and K. Incze, *Abstr. Pap., Int. Congr. Food Sci. Technol.* p. 91 (1978).
505. R. L. Buchanan, S. Goldstein, and J. D. Budroe, *J. Food Sci.* **47,** 330, 333 (1981).

CHAPTER 5

AZAFLUORANTHENE AND TROPOLOISOQUINOLINE ALKALOIDS

KEITH T. BUCK

Fries & Fries Division, Mallinckrodt, Inc.
Cincinnati, Ohio

I. Introduction

The azafluoranthene and tropoloisoquinoline alkaloids contain highly condensed aromatic heterocyclic nuclei. They are considered together because of their original isolation from the same plant sources, two members of the genus *Abuta* (Menispermaceae) (*1,2*), which suggests a biosynthetic relationship. The chapter covers the literature on these two classes of alkaloids to May, 1984, and also includes some unpublished data, in part to show the development of the subjects, but also to reveal the often frustrating "human" side of research.

THE ALKALOIDS, VOL. XXIII

ISBN 0-12-469523-X

The azafluoranthene alkaloids are derivatives of indeno[1,2,3-*i,j*]isoquinoline (**1**), hereinafter referred to simply as "azafluoranthene." The

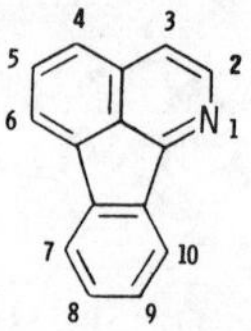

1

parent base is a component of coal tar (*3*) and cigarette smoke (*4*) and an identified air pollutant (*5*). The chemistry of **1** has received very little attention. The first alkaloids incorporating this nucleus were reported in 1972 (*6*), and five are known.

The tropoloisoquinoline alkaloids so far known are derivatives of the tautomers 10-hydroxy-9*H*-azuleno[1,2,3-*i,j*]isoquinolin-9-one (**2**) and 9-

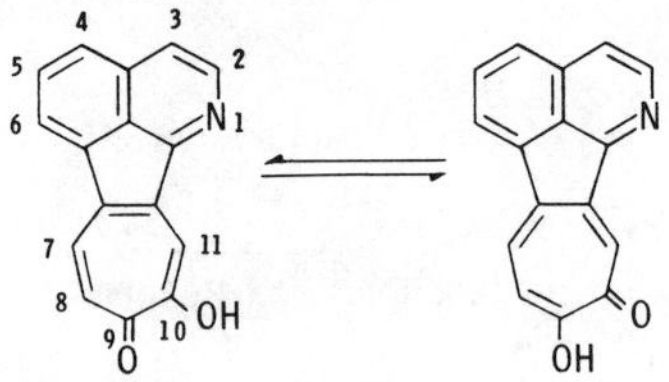

2 **3**

hydroxy-10*H*-azuleno[1,2,3-*i,j*]isoquinolin-10-one (**3**). The first of these alkaloids was identified in 1977 (*2*). Other than the two alkaloids and one artifact described in this chapter, no compounds containing the tropoloisoquinoline nucleus are known.

II. Occurrence and Structure of Azafluoranthene Alkaloids

A. Imeluteine

Imeluteine (**4**) was the first alkaloid of its class to be isolated and identified. (*1,6*). The basic extracts of the Brazilian climbing vines *Abuta imene* and *A. rufescens* (Menispermaceae) both afforded the alkaloid.

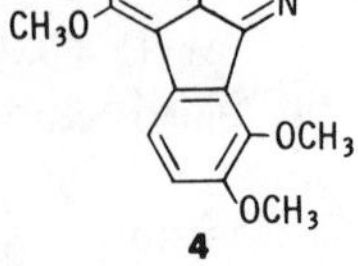

4

Imeluteine showed the following properties: yellow prisms, mp 146–147°C (isopropyl ether–methanol); IR 6.35 μm; UV λ_{max} 233 nm (log ε 4.48), 253 (4.49), 288 (4.43), 317 (3.75), 365 sh (3.72), 380 (3.85), 400 sh (3.72); NMR δ3.94, 4.02, 4.08, 4.10, 4.17 (all 3*H*, s), 6.91, 7.60 (both 1*H*, d, J = 8 Hz), 7.57, 8.65 (both 1*H*, d, J = 6 Hz); high resolution MS calculated for $C_{20}H_{19}NO_5$ 353.1262, found 353.1235; and MS (*7*) 353 (64%), 352 (41%), 338 (30%), 325 (20%), 324 (base), 322 (35%), 308 (50%), 307 (38%), 294 (45%), 292 (23%), 291 (15%), 280 (20%), 277 (28%), 264 (16%), 263 (27%), 237 (29%), 169 (27%), 162 (15%), 154 (16%).

Prior to the isolation of **4**, the major alkaloids of the two *Abuta* species had already been identified as the oxoaporhines imenine (**5a**) (*8*) and *O*-methylmoschatoline (**5b**) (*1*). Therefore an aporphinoid skeleton was ini-

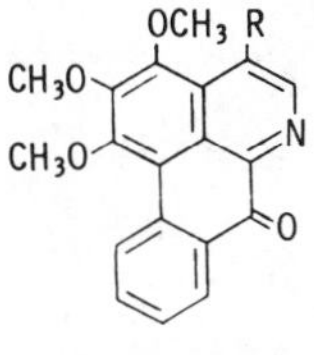

5a R = OCH_3
5b R = H

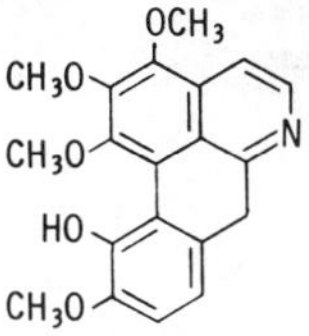

6

tially considered most likely for imeluteine. Although the MS of **4** indicates the correct parent ion, the base peak corresponds to loss of HCO. Since this behavior is typical of phenols (*9*), the first structure considered for imeluteine in our laboratories was **6**, an unprecedented B-ring aromatic aporphine (*7*). However, the questionable fit with the NMR data, the absence of behavior (such as UV base shift or extractability with NaOH) expected for even a hindered phenol, and the soon well documented ready aerial oxidation of B-ring aromatic aporphines to 7-oxoaporphines (*1,8,10–12*) all argued against such a structure.

It was therefore eventually concluded that all nonaromatic protons in imeluteine had to be present as methoxy groups, and hence that the ring structure was more condensed than initially thought. The indeno[1,2,3-*i,j*]isoquinoline nucleus was considered most likely. Following some unsuccessful model experiments in the 5,6-dimethoxyazafluoranthene series (see Section IV,E), **7** was chosen as the target structure for imeluteine

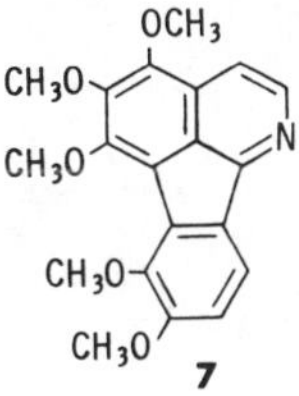

7

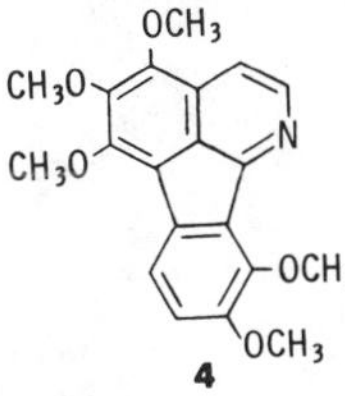

4

because of the similarity of its bottom ring substitution pattern to that of a number of D-ring dioxygenated aporphine alkaloids.

The synthesis of **7** was soon completed (*7*) (see Section IV,F). The amorphous yellow product had IR and NMR spectra (see Appendix) similar to those of imeluteine, suggesting that the alkaloid was the isomer **4**. This assignment was confirmed by total synthesis. It now appears that the troublesome loss of HCO in the MS of imeluteine can be explained by a mechanism such as that shown in Scheme 1.

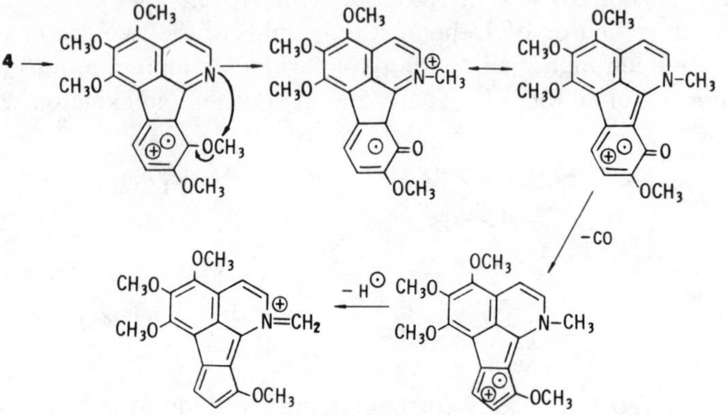

SCHEME 1

B. RUFESCINE

The second azafluoranthene alkaloid identified was rufescine (**8**). It was initially isolated from *Abuta rufescens,* and was soon shown to be present in *A. imene* (*1,6*).

Rufescine formed bright yellow, needle-shaped prisms, mp 88–90°C (hexane–ether): IR 6.15, 6.31 μm; UV λ_{max} 247 nm (log ε 4.52), 285 sh (4.31), 295 (4.34), 304 (4.29), 315 sh (3.84), 356 (3.65), 373 (3.78), 400 sh (3.32); NMR δ 3.94, 4.05, 4.11, 4.13 (all 3*H*, s), 7.63, 8.59 (both 1*H*, d, J = 6 Hz), 6.96 (1*H*, pair of doublets, J = 8 and 2 Hz), 7.68 (1*H*, d, J = 2 Hz), 7.82 (1*H*, d, J = 8 Hz); high resolution MS calculated for $C_{19}H_{17}NO_4$ 323.1157, found 323.1135; and MS (*7*) 324 (19%, M + 1), 323 (base), 308 (77%), 293 (22%), 278 (16%), 265 (38%), 250 (32%), 222 (33%), 194 (28%), 162 (21%), 147 (17%).

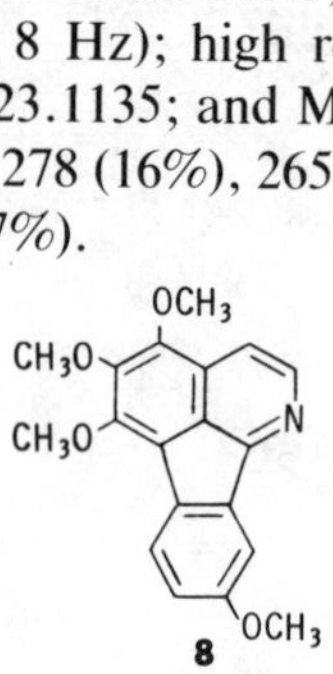

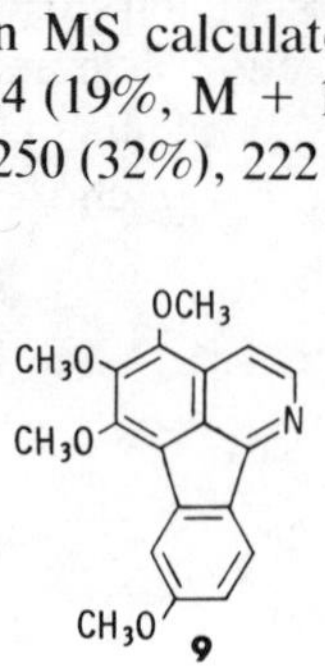

The spectral properties of rufescine and imeluteine (**4**) showed strong similarities, indicating that rufescine also contained the azafluoranthene nucleus but with one fewer methoxy substituents. The aromatic splitting pattern suggested one of the alternative structures **8** or **9**. From the available data it was not possible a priori to assign the correct structure, so synthetic work was undertaken. Compound **8** was chosen as the initial target by lot, and proved to be correct. The MS of rufescine shows the expected fragmentation of *O*-methyls, but no significant peak corresponding to loss of the elements of HCO because no rearrangement and ring contraction (as for **4**) is possible (*7*).

C. NORRUFESCINE

Norrufescine (**10**), an extremely polar alkaloid, was first isolated during preparative TLC from the chloroform-insoluble residues of *Abuta imene* and *A. rufescens* bases (*1*). It has also been obtained from *Telitoxicum peruvianum* (Menispermaceae) (*13*).

Norrufescine formed orange to brown plates, mp 235–238°C dec. (MeOH): IR 3–4 (broad), 6.21, 6.31 μm; UV λ_{max} 225 nm sh (log ε 3.56), 248 (3.83), 303 (3.68), 315 sh (3.36), 374 (2.87), after addition of NaOH 230 sh (3.64), 245 (3.77), 317 (3.85), 382 (2.60), 495 (2.30); NMR [$CDCl_3$ plus a drop of $(CD_3)_2SO$] δ 4.04, 4.08, 4.10 (all 3*H*, s), 7.59, 8.52 (both 1*H*, d, J = 6 Hz), 6.90 (1*H*, pair of doublets, J = 8 and 2 Hz), 7.53 (1*H*, d, J = 2 Hz), 7.71 (1*H*, d, J = 8 Hz); high resolution MS calculated for $C_{18}H_{15}NO_4$ 309.1001, found 309.0991; and MS (*14*) 310 (22%, M + 1), 309 (base), 294 (76%), 279 (20%), 251 (53%), 236 (20%), 208 (29%), 180 (26%).

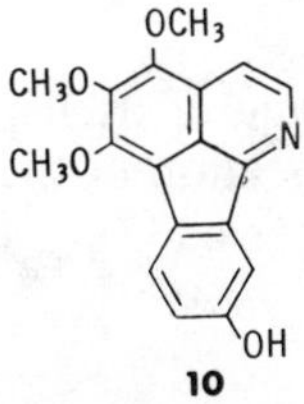

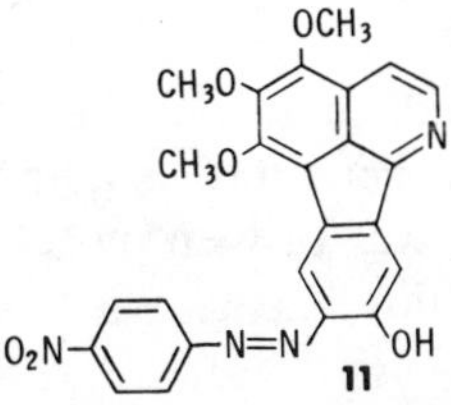

The oxygen substitution pattern of norrufescine was established by conversion of the alkaloid to rufescine (**8**) with diazomethane. A choice between the four possible isomeric monophenolic bases was fortuitously made since norrufescine formed a colored azo coupling product (probably **11**) with *p*-nitrobenzenediazonium chloride, thereby showing the hydroxy to possess a free *ortho* or *para* position. The indicated structure of norrufescine was confirmed by an X-ray crystallographic study (*15*). The data showed that the azafluoranthene nucleus is highly planar, with bond angles and distances not significantly different from those of fluoranthene (**12**). A total synthesis of the alkaloid was completed (*16*).

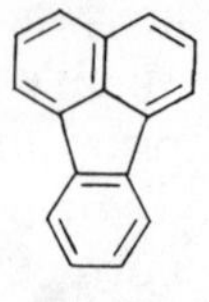

12

D. TRICLISINE

Triclisia gillettii (Dewild) Staner (synonomous with *Triclisia dictyophylla* Diels in England) (Menispermaceae) afforded the intensely yellow-green fluorescent alkaloid **13**. Originally unnamed (*17*), it has been designated as triclisine (*13, 17a*).

Triclisine formed yellow crystals, mp 155°C (benzene–petroleum ether): IR 6.14, 6.15, 6.77, 7.79, 8.33, 8.82, 9.78, 11.77, 13.30 μm: UV λ_{max} 245 nm (log ε 4.63), 263 (4.72), 282 sh (4.42), 296 (4.25), 308 (3.93), 317 (4.06), 352 (4.05), 363 (4.12) (values interpolated from graph); NMR δ 4.05, 3.92 (both 3*H*, s), 6.84 (1*H*, s), 8.52, 7.29 (both 1*H*, d, J = 5.8 Hz), 7.98, 7.38 (both multiplets, total of 4 aromatic H); high resolution MS calculated for $C_{17}H_{13}NO_2$ 263.0946, found 263.0952; and MS 263 (base), 248 (13%), 220 (33%), 205 (8%), 190 (11%), 177 (18%), 151 (11%).

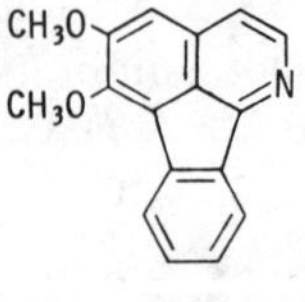

13

NMR study indicated that the bottom ring of triclisine was unsubstituted but did not definitely establish the location of the methoxys. That triclisine was **13** was proven by total synthesis (*17*).

A ^{13}C-NMR study has also been done (*17a*).

E. TELITOXINE

Telitoxine (**14**) was isolated from *Telitoxicum peruvianum* (Menispermaceae). It was among the first alkaloids identified in this genus (*13*). Telitoxine formed yellow crystals, mp 273–275°C (MeOH–CH_2Cl_2–$CHCl_3$): UV λ_{max} (EtOH) 233 nm (log ε 4.29), 243 (4.30), 277 (4.14), 288 (4.14), 298 (4.10), 307 sh (3.70), 322 (3.55), 350 (3.41), 367 (3.55), after addition of NaOH 233 (4.26), 245 (4.20), 255 (4.21), 279 sh (3.89), 288 sh (3.94), 307 sh (4.17), 314 (4.20), 345 sh (3.20), 355 (3.20), 373 (3.20), 448

(3.00); NMR (360 MHz, acetone-d_6) δ 4.06, 4.07 (both 3*H*, s), 6.92 (1*H*, dd, $J = 9$ and 2 Hz), 7.18 (1*H*, s), 7.57 (1*H*, d, $J = 2$ Hz), 7.82 (1*H*, d, $J = 9$ Hz), 7.53, 8.50 (each 1*H*, d, $J = 6$ Hz); high resolution MS calculated for $C_{17}H_{13}NO_3$ 279.0891, found 279.0890; and MS (*14*) 280 (27%, M + 1), 279 (base), 264 (16%), 236 (56%), 234 (22%), 206 (20%), 193 (17%).

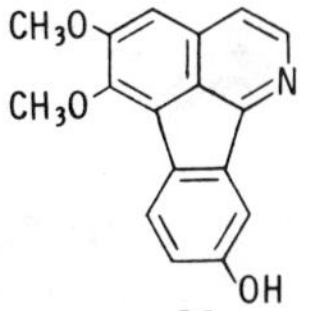

14

The alkaloid, soluble in aqueous base, showed the bathochromic shift on addition of NaOH typical of a phenolic alkaloid. The resemblance to norrufescine (**10**), also isolated from *T. peruvianum,* suggested structure **14** as most likely for telitoxine, but this assignment is not yet unequivocal.

III. Occurrence and Structure of Tropoloisoquinoline Alkaloids

A. Imerubrine

Imerubrine (**15**) was obtained from *Abuta imene* and *A. rufescens* (*1*). It was the first tropoloisoquinoline alkaloid to be isolated and identified.

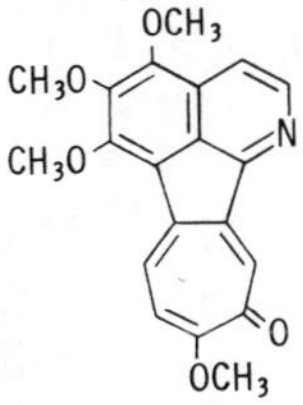

15

Imerubrine formed orange-red needles, mp 183–185°C (MeOH–ether): IR 6.35 μm; UV λ_{max} 250 nm (log ε 4.48), 267 (4.52), 295 (4.40), 350 (4.35), 372 sh (4.20), 394 (4.11), 450 (3.93); NMR δ 4.02, 4.07, 4.14, 4.16 (all 3*H*, s), 6.87, 8.06 (both 1*H*, d, $J = 9.4$ Hz), 8.29 (1*H*, s), 7.75, 8.68 (both 1*H*, d, $J = 5.8$ Hz) (*14,18*); high resolution MS calculated for $C_{20}H_{17}NO_5$ 351.1105, found 351.1098; and MS (*14*) 352 (M + 1, 27%), 351 (base), 323 (22%), 322 (47%), 308 (44%), 293 (17%), 292 (17%), 278 (20%), 265 (23%), 250 (26%), 222 (21%).

Imerubrine's striking color, the long-wavelength carbonyl band in the IR, and the NMR data indicated that the alkaloid was unusual, and originally structure **16** or **17** was proposed. Atypically substituted oxo-

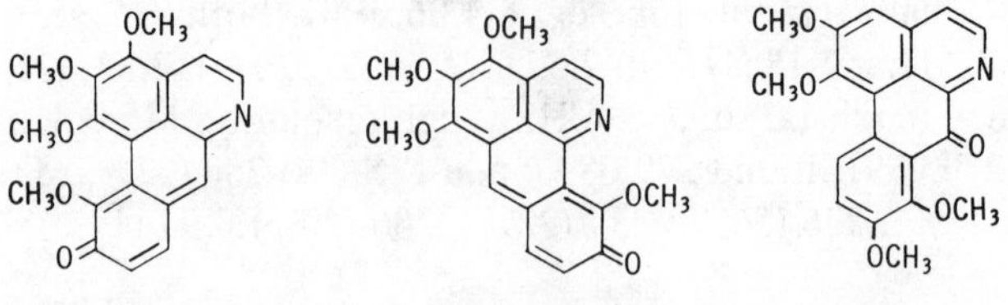

16 **17** **18**

aporphines such as **18** were also considered (*7*) although they are inconsistent with the IR and the large coupling constant (9.4 Hz) of the lower ring vicinal protons. That imerubrine was a tropoloisoquinoline was surmised by Cava soon thereafter, and the correct structure was assigned by X-ray crystallography (*2*).

B. Grandirubrine

Grandirubrine (**19**), isolated from *Abuta grandifolia* (Martius) Sandwith (*19*), was the first tropoloisoquinoline alkaloid shown to contain a free tropolone moiety. It achieved unexpected notoriety by being the five-millionth compound indexed by Chemical Abstracts Service Chemical Registry System (*20*). Grandirubrine formed reddish-brown needles, mp 201–203°C (chloroform–ethanol): IR 6.3 μm: UV λ_{max} 232 nm (log ε 4.96), 254 (4.79), 296 (4.58), 363 (4.72), 384 (4.41), 400 (4.19), 480 (3.90), 274 sh (4.66), 312 sh (4.46), 343 sh (4.51): NMR δ 4.06, 4.16, 4.20, (all 3*H* singlets), 7.42, 8.33 (both 1*H*, d, J = 10 Hz), 7.80, 8.73 (both 1*H*, d, J = 5 Hz), 8.41 (1*H*, s); high resolution MS calculated for $C_{19}H_{15}NO_5$ 337.0950, found 337.0949; and MS (*14*) (80 eV) 338 (22%), 337 (base), 322 (30%), 309 (12%), 294 (40%), 279 (20%), 251 (34%), 236 (21%), 208 (18%).

The structure of **19** was assigned by conversion of the alkaloid with diazomethane to a mixture of **20** and imerubrine (**15**) and NMR study of

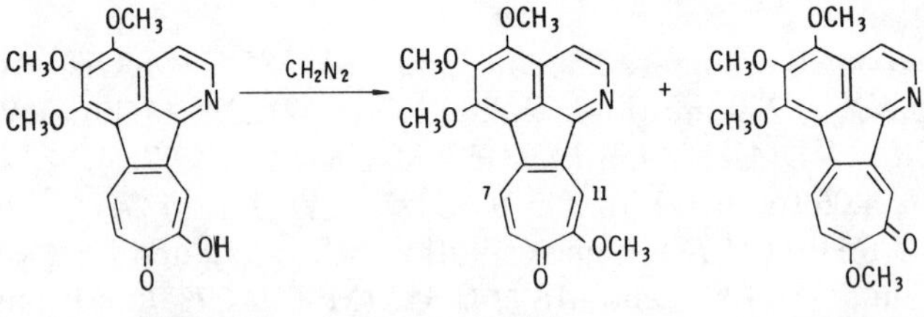

CH_2N_2

19 **20** **15**

the products. In the NMR, **19** showed a much closer resemblance to the major product **20** than to **15** (see Appendix); thus **19** is the major tautomer of grandirubrine. Analogy exists to the extensively documented NMR behavior of derivatives of colchicine (**21**) and isocolchicine (**22**) (*21*), but the C-11 and (especially) C-7 protons of the tropoloisoquinolines are more deshielded than the C-8 and C-12 protons of the colchicine and isocolchicine series.

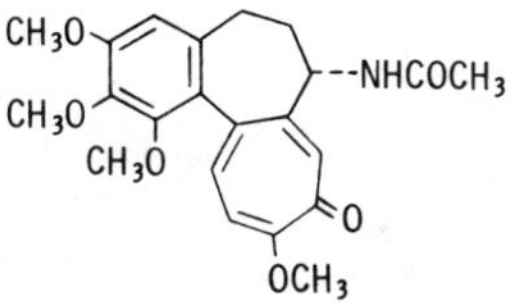

CH_3O CH_3O CH_3O $NHCOCH_3$ O OCH_3

21

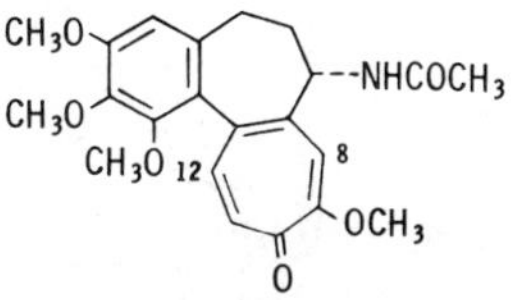

CH_3O CH_3O CH_3O 12 8 $NHCOCH_3$ OCH_3 O

22

IV. Synthesis of Azafluoranthene Alkaloids and Related Compounds

Retrosynthetic dissection of the azafluoranthene nucleus suggests four most likely choices for the final bond formation required to build the heterocyclic system (Fig. 1). The following examples demonstrate the use of all of these approaches to prepare azafluoranthene derivatives. All of the alkaloids so far synthesized were first prepared by closures of type a, apparently the most versatile of the four. An approach involving final construction of ring D was also successful.

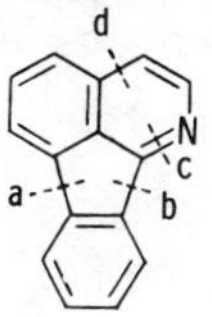

d N c a b

FIG. 1

A. Parent Base

At the time the first azafluoranthene alkaloids were isolated, only one synthesis of the azafluoranthene ring system, that of the parent base (**1**), had been reported (*22*). The starting material, **23**, was a known degradation product of fluoranthene (**12**), derived from coal tar distillate (*23*) (Scheme 2). This approach clearly lacked the generality necessary for straightforward synthesis of polyoxygenated azafluoranthene alkaloids.

B. Imeluteine

Imeluteine (**4**) was the first azafluoranthene alkaloid to be isolated and synthesized; a brief account of this work has been published (*6*). The route employed is shown in Scheme 3.

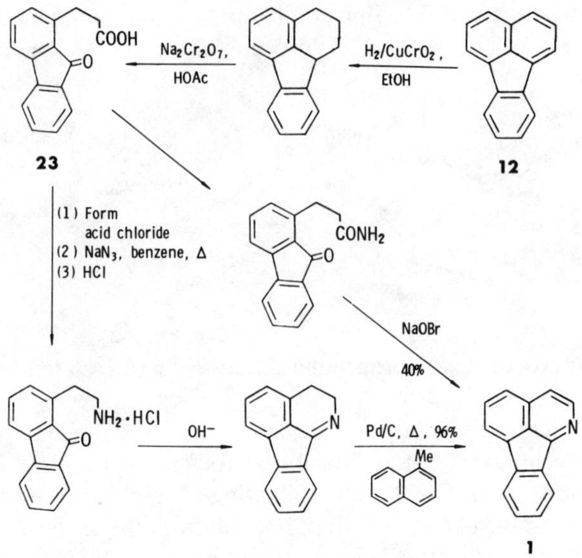

COOH
O
23
$Na_2Cr_2O_7$,
HOAc
$H_2/CuCrO_2$,
EtOH
12
(1) Form
acid chloride
(2) NaN_3, benzene, Δ
(3) HCl
$CONH_2$
O
NaOBr
40%
$NH_2 \cdot HCl$
O
OH^-
N
Pd/C, Δ, 96%
Me
N
1

SCHEME 2

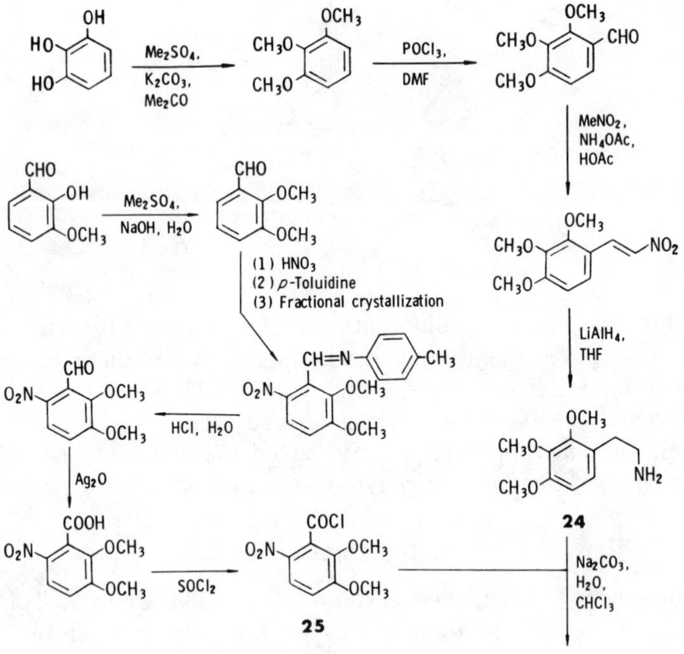

OH
HO
HO
Me_2SO_4,
K_2CO_3,
Me_2CO
OCH_3
CH_3O
CH_3O
$POCl_3$,
DMF
OCH_3
CH_3O
CHO
CH_3O
$MeNO_2$,
NH_4OAc,
HOAc
CHO
OH
OCH_3
Me_2SO_4,
NaOH, H_2O
CHO
OCH_3
OCH_3
(1) HNO_3
(2) p-Toluidine
(3) Fractional crystallization
OCH_3
CH_3O
NO_2
CH_3O
LiAlH$_4$,
THF
CHO
O_2N
OCH_3
OCH_3
HCl, H_2O
CH=N–C_6H_4–CH_3
O_2N
OCH_3
OCH_3
OCH_3
CH_3O
NH_2
CH_3O
24
Ag_2O
COOH
O_2N
OCH_3
OCH_3
$SOCl_2$
COCl
O_2N
OCH_3
OCH_3
25
Na_2CO_3,
H_2O,
$CHCl_3$

SCHEME 3

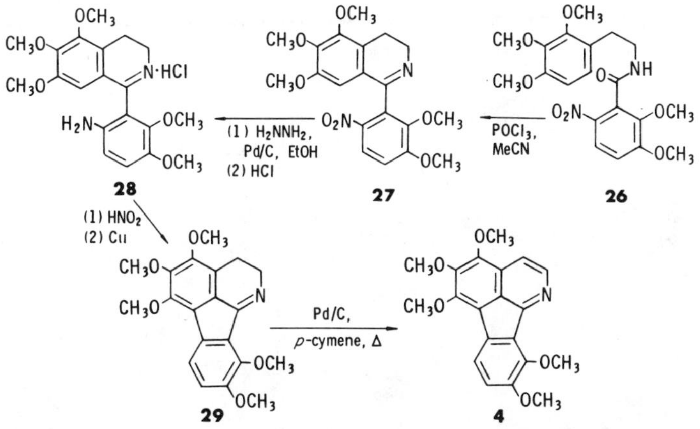

SCHEME 3 (*Continued*)

Amine **24** and the acid chloride **25**, prepared by modifications of known methods, were condensed to the amorphous amide **26**. Bischler–Napieralski cyclization of **26** with phosphorus oxychloride in acetonitrile gave **27** in 55% yield based on **24**. The $POCl_3$–CH_3CN reagent, first applied to isoquinoline synthesis in 1968 (*24*), appears to be the system of choice, especially in the electronically unfavorable cyclizations of 2,3,4-trialkoxy phenethylamides (see remarks on the synthesis of **7**, Section IV,F). Hydrazine reduction of **27** gave the bis amine, isolated in 81% yield as the hydrochloride **28**.

The key step of the synthesis was the Pschorr ring closure to form the five-membered ring. The Pschorr reaction has been widely employed to close six-membered rings as in the classic synthesis of aporphines but had not previously been used to prepare azafluoranthenes. However, the well-documented use of the Pschorr closure outside of alkaloid synthesis to form the five-membered rings of fluoranthenes, fluorenones, carbazoles, and dibenzofurans (*25*) suggested that such a reaction was feasible. The cyclization of **28** not only succeeded but went in 67% yield. As expected from the reported synthesis of **1**, **29** was easily dehydrogenated with Pd–C in *p*-cymene, affording 48% imeluteine (**4**).

An alternative synthesis of **4** was completed in 1984 (*25a*).

C. RUFESCINE

The synthesis of rufescine (**8**) paralleled that of imeluteine (**4**), and was carried out concurrently (*6*) (Scheme 4). Since this work was published, a more selective procedure giving an improved yield of 2-nitro-5-hydroxybenzaldehyde has appeared (*11*) (Scheme 5). Cyclization of **30**, as for **26**,

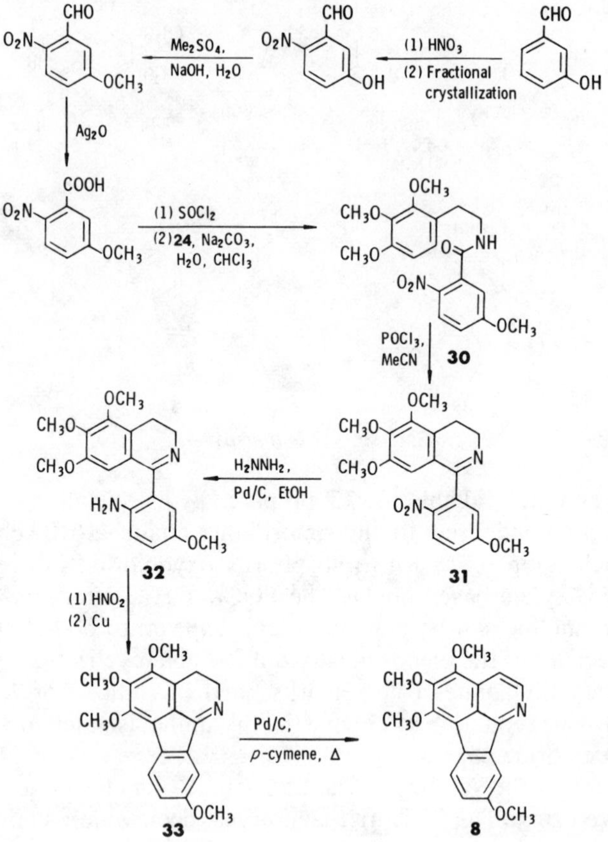

SCHEME 4

gave 51% **31**. Hydrazine reduction of **31** afforded 68% free base **32**. Pschorr cyclization of **32** gave, along with some deaminated material, a crude 56% yield of **33**, isolated by PTLC, and this was dehydrogenated in 49% crude yield to **8**. Both **33** and **8** were much more soluble than the corresponding compounds in the imeluteine series, leading to lower yields after purification (*7*).

A novel synthesis of **8** was reported in 1984 (*25a*).

D. NORRUFESCINE

The synthesis of norrufescine (**10**) was initially modeled after that of rufescine (**8**) but offered some unexpected difficulties, leading to the development of an alternative technique for closure of the five-membered ring (*16*) (Scheme 5). Thus aldehyde **34**, prepared as described (*11*), was

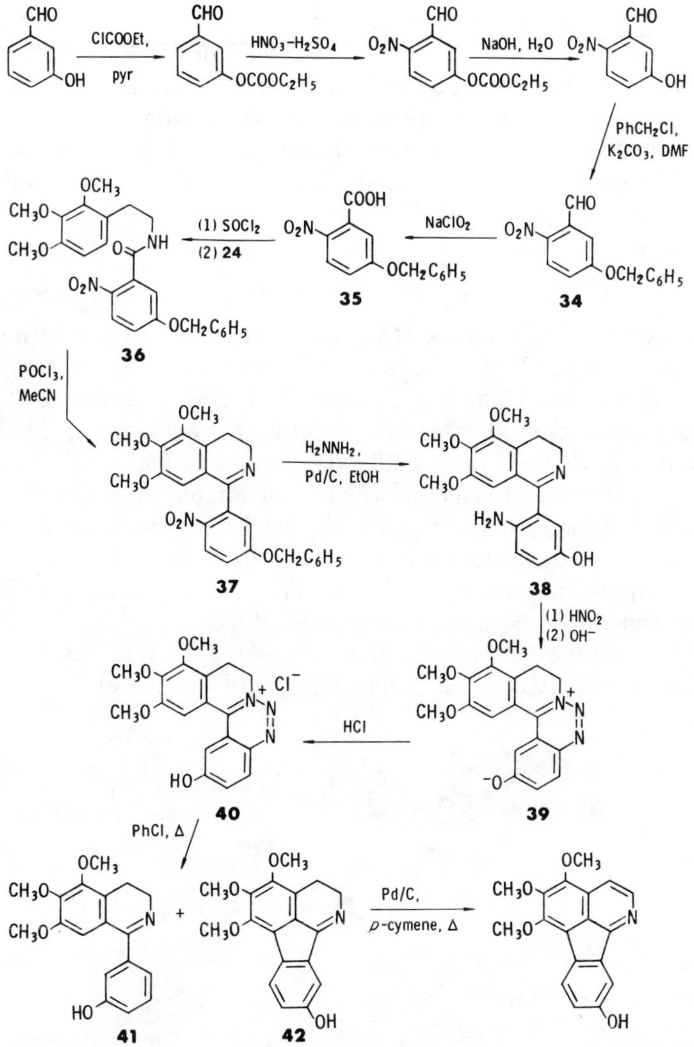

SCHEME 5

converted in 73% yield to the acid **35** with sodium chlorite. This oxidation (*26*) was more convenient, cheaper, and gave a more reproducible yield than the silver oxide method employed in the syntheses of imeluteine, rufescine, and **7**. (Schemes 3, 4, and 10). Preparation of the amide **36** in 90% yield was straightforward, and its cyclization gave a high yield (89%) (*27*) of **37**. Hydrazine reduction of the nitro group of **37** concomitantly removed the benzyl group, but the resulting amorphous aminophenol **38** was successfully employed in the synthesis. Thus direct diazotization of **38** gave a 90% yield of the stable triazinium dipole **39**, which was converted to the hydrochloride **40** with HCl. On heating in chlorobenzene, **40** gave a 24% yield of cyclization product **42**, accompanied by 17% of the deaminated material **41**. Dehydrogenation of **42** straightforwardly afforded 46% norrufescine (**10**). A small amount of **10** was produced directly from **40** during cyclization (*14*).

E. Triclisine

The synthesis of triclisine (**13**) was patterned after those of imeluteine (**4**) and rufescine (**8**) (*17*) (Scheme 6). Intermediates **43**, **44**, and **45** were known compounds. The ring closure of **43** to **44** occurred in good yield under well-established Bischler–Napieralski conditions (e.g., $POCl_3$–toluene, P_2O_5–xylene) (*28,29*), unlike those cases having a 2,3,4-trioxygenated phenethylamino residue in which cyclization must occur *meta* to two electron-donating groups (e.g., **26**, **30**, and **36**). Hydrazine reduction of **44** to **45** was also straightforward. Pschorr ring closure of **45** was reported to result in spontaneous aromatization of the products, giving **13** in 9% yield, accompanied by 52% deamination product **46**.

In fact, we used compound **45** as a model system for azafluoranthene alkaloid synthesis prior to publication (*6*) of our initial results. At that

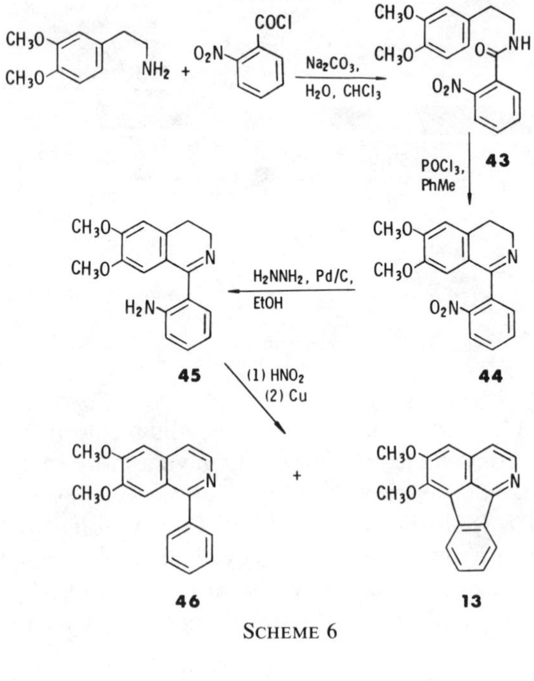

Scheme 6

time, attempts to effect the Pschorr ring closure of **45**, the corresponding tetrahydroisoquinoline, or the *N*-benzenesulfonyl derivative of the latter gave as the only identified relatively nonpolar products the corresponding deaminated materials. No cyclization or change in oxidation of the heterocyclic rings was observed (*7,30* [ref. *7*]). Part of this discrepancy in results seems to be due to the difference in the work-up employed. Thus, we used only weakly basic conditions (NH_4OH), whereas the Huls group resorted to aqueous NaOH, a system shown by us (*7*) to cause, in presence of air, significant aromatization of 1-phenyldihydroisoquinolines. The dihydro and tetrahydro cyclization products are apparently not easily distinguished from the predominant deaminated materials during TLC, and hence were overlooked by us. When work-up of the products of **45** was later carried out in our laboratories according to the Belgian procedure, the isolation was easily monitored under UV and visible light, and **13** and **46** were separated by chromatography as described (*7*).

In any case, deamination rather than cyclization is the predominant pathway in the 6,7-dimethoxyisoquinoline series. Thus the ease of formation of this system by Bischler–Napieralski reaction is more than offset by deactivation during the Pschorr closure step.

F. Other Derivatives of the Azafluoranthene Nucleus

Since so little work has appeared on the chemistry of azafluoranthene and its derivatives, we are including all available information on their synthesis. These syntheses, while not leading to known azafluoranthene alkaloids, nevertheless point out potentially useful techniques.

As part of a drug design program, **47** was prepared from fluorenone as indicated in Scheme 7 (*31*). The initial imine formation went in 91% crude

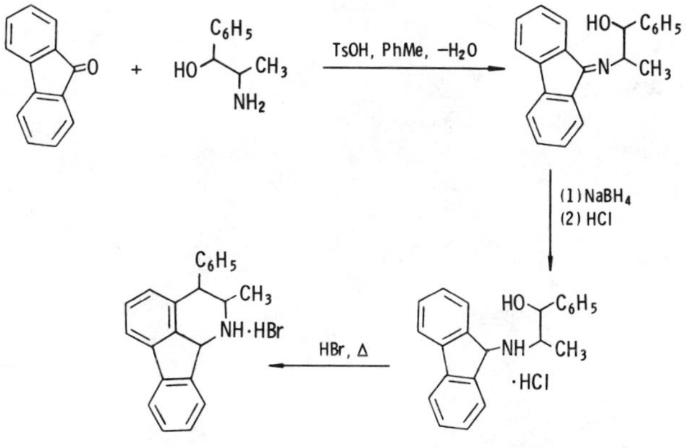

C_6H_5 HO CH₃ NH₂ — TsOH, PhMe, $-H_2O$ → HO C_6H_5 =N CH_3

(1) $NaBH_4$ (2) HCl

HO C_6H_5 NH CH_3 ·HCl — HBr, Δ → C_6H_5 CH_3 NH·HBr

47

Scheme 7

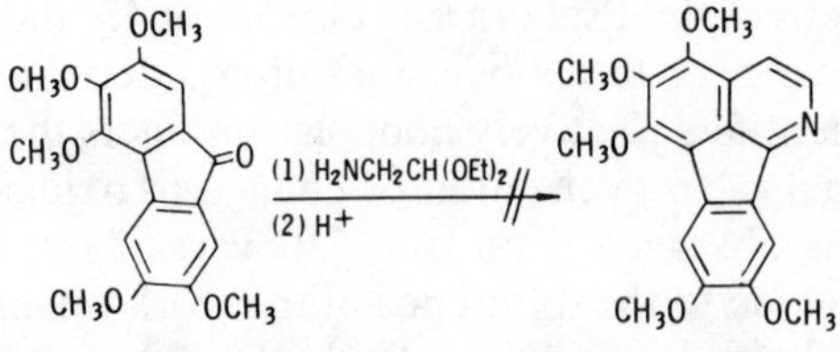

48 → (1) $H_2NCH_2CH(OEt)_2$ (2) H^+ ⇸ 49

SCHEME 8

yield, but the yield of the ring closure step was only 14%. In our laboratories, an attempt to carry out the related transformation of **48** (*32*) to **49** (Scheme 8) was unsuccessful, since the first step failed under a variety of conditions (*7*). In any case, the work of Patel and MacLean (*33*), in which the failure of **50** to cyclize with various acidic reagents was noted, suggests that this approach would not have worked anyway.

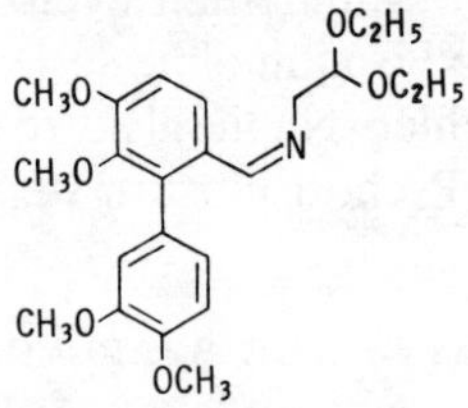

50

The Pschorr reaction of **51** gave, in addition to the oxidatively demethylated amine **52**, an orange crystalline compound tentatively identified as **53** (Scheme 9), although the isomeric structures **54** and **55** are also possible. Interestingly, the related amine **56** gave no azafluoranthene products

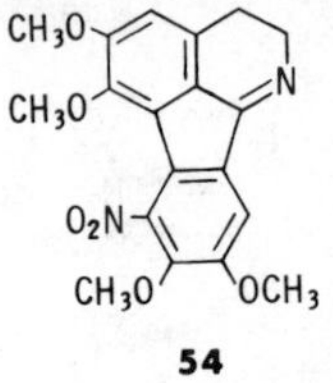

54 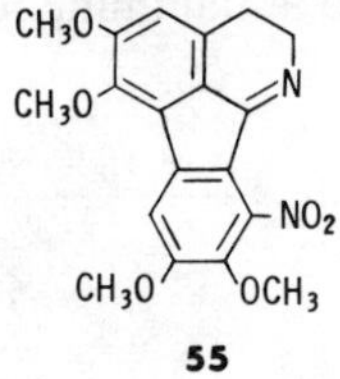 **55** 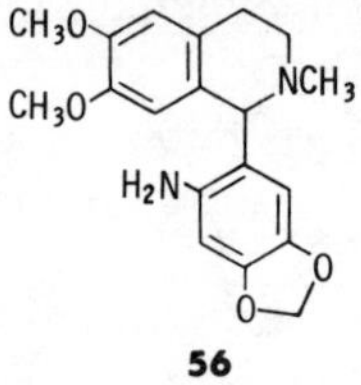 **56**

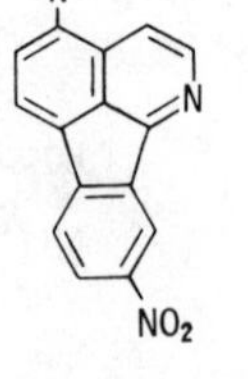

57a R = H
57b R = NO_2

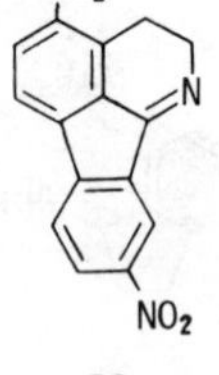

58

under the same conditions (*30*). The only other nitroazafluoranthenes known (**57a**, **57b**, and **58**) were prepared by nitration of azafluoranthene and 2,3-dihydroazafluoranthene (*22*).

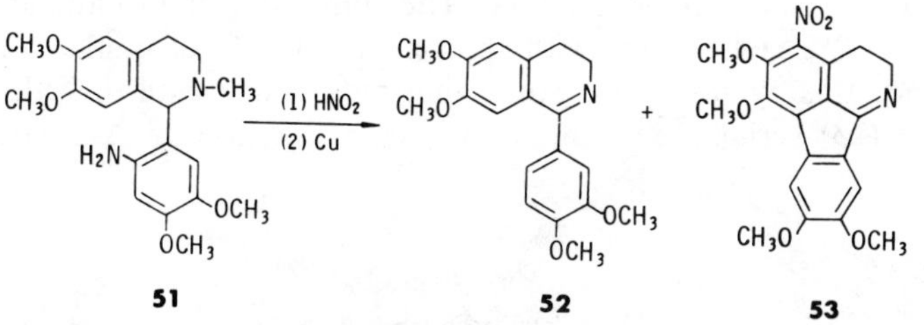

51 → (1) HNO_2 (2) Cu → **52** + **53**

SCHEME 9

The amorphous base **7**, isomeric with imeluteine (**4**) and the first oxygenated azafluoranthene to be synthesized, was prepared as shown in Scheme 10. The combined yield for the last two steps was 24%. This synthesis also demonstrated for the first time the clear superiority of the $POCl_3$–CH_3CN system for the Bischler–Napieralski reaction of *N*-(2,3,4-trimethoxyphenethyl)benzamides. Thus these reagents afforded a 53%

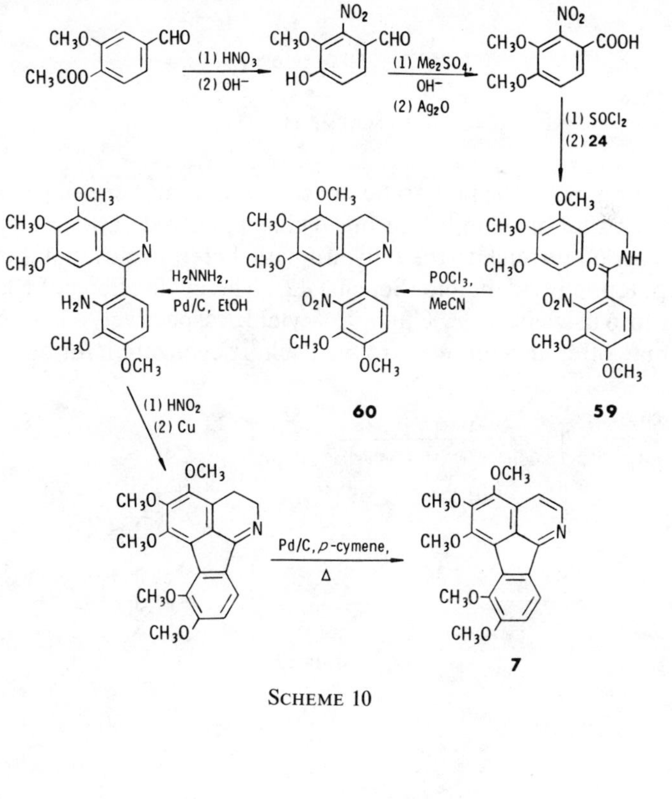

(1) HNO_3 (2) OH^-; (1) Me_2SO_4, OH^- (2) Ag_2O; (1) $SOCl_2$ (2) **24** → **59**; $POCl_3$, MeCN → **60**; H_2NNH_2, Pd/C, EtOH; (1) HNO_2 (2) Cu; Pd/C, *p*-cymene, Δ → **7**

SCHEME 10

yield of **60** from **59**, whereas only a 2.5% yield was achieved with the more traditional PCl_5 in refluxing chloroform (*7*).

As part of a drug development program (see Section VII), **62** was synthesized as shown in Scheme 11. The dihydroazafluoranthene **61** was formed in 32% yield, along with approximately 40% deaminated material **52**. Hence this cyclization is somewhat more successful than that in the triclisine (**13**) series (see Scheme 6). Aromatization of **61** gave 80% **62** (*34*).

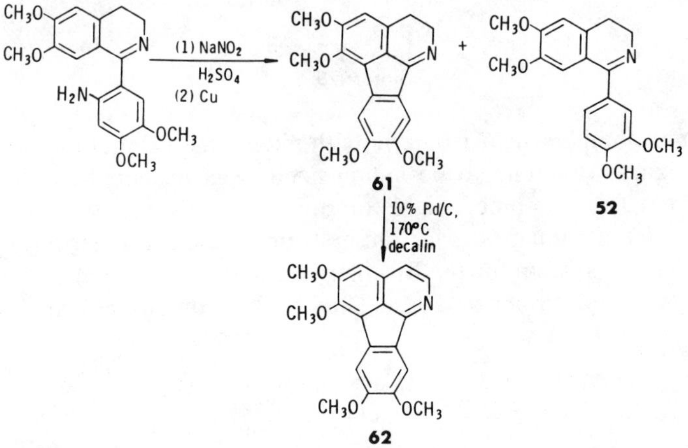
CH₃O
CH₃O
N
H₂N
OCH₃
OCH₃
(1) NaNO₂
H₂SO₄
(2) Cu
61
+
52
10% Pd/C,
170°C
decalin
62

SCHEME 11

The first report of the preparation of the azafluoranthene ring system by final ring closure of bond b (see Fig. 1) has appeared (*33*). The synthesis was not carried through to the fully aromatic heterocycles, but it is nevertheless potentially adaptable (Scheme 12). The cyclizations of **63a** to **64a** and **63b** to **64b** went in 96% and 92% yield, respectively. Thus this approach may offer an alternative to the Pschorr cyclization for the prepara-

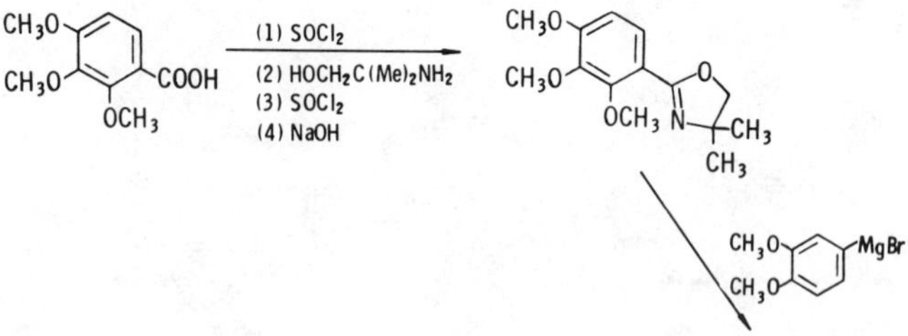
CH₃O
CH₃O
COOH
OCH₃
(1) SOCl₂
(2) HOCH₂C(Me)₂NH₂
(3) SOCl₂
(4) NaOH
O
N
CH₃
CH₃
MgBr

SCHEME 12

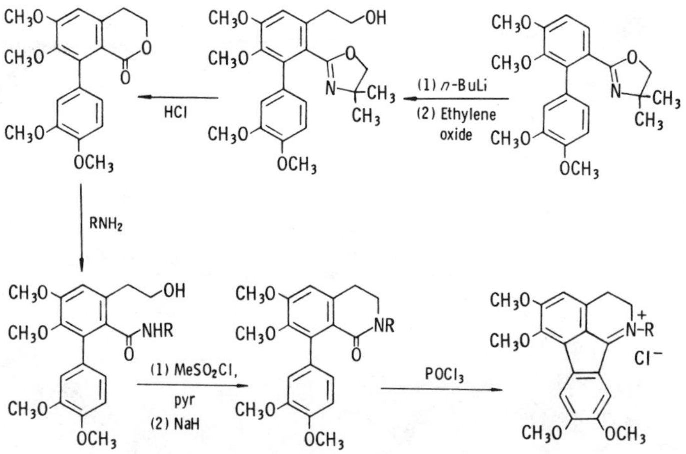

HCl
(1) n-BuLi
(2) Ethylene oxide
RNH_2
(1) $MeSO_2Cl$, pyr
(2) NaH
$POCl_3$

63a
63b

64a $R = CH_3$
64b $R = CH_2C_6H_5$

SCHEME 12 (*Continued*)

tion of alkaloids that lack an activating substituent at position 5 of the isoquinoline ring but allows only limited variation of the bottom ring.

Another synthesis of a simple azafluoranthene, **65**, that established the final ring closure at the nitrogen atom (Fig. 1, bond c) has also appeared (*35*). (Scheme 13), but the yield was mediocre and this method lacks

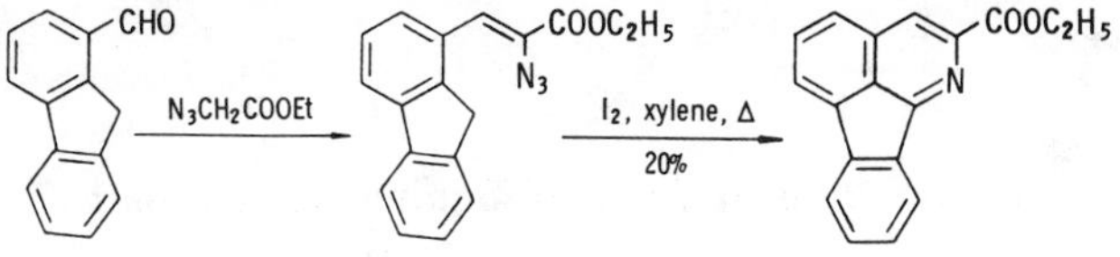

N_3CH_2COOEt
I_2, xylene, Δ
20%

65

SCHEME 13

versatility. The synthesis of the related alkaloid eupolauridine (1,6-diaza-fluoranthene, **66**) (*36*) may be mentioned (Scheme 14). An analogous approach to azafluoranthenes has not yet been attempted.

Other ring closures of type a (Fig. 1) beside the Pschorr may be feasible. One approach via a hydroxylamine intermediate (*37*) has been attempted with disappointing results. Thus **67** gave a 54% yield, after purification, of the reduction product **68** rather than cyclized material (*38*). Other related methods remain to be investigated, and we hope that a route that is both versatile and efficient will soon be discovered.

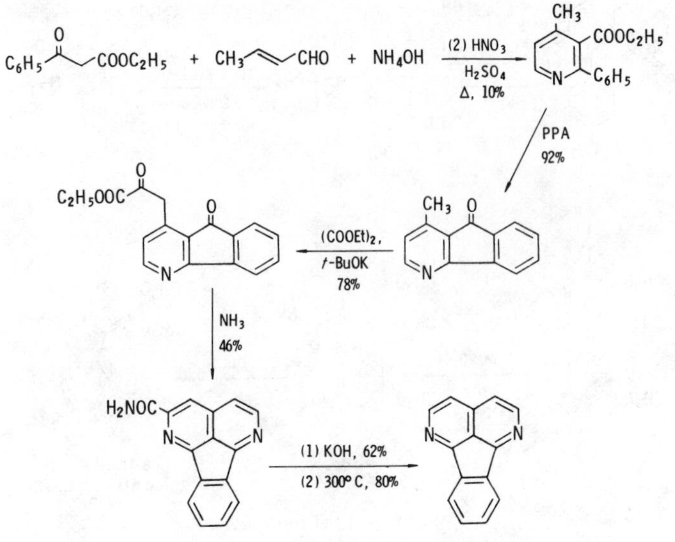

SCHEME 14

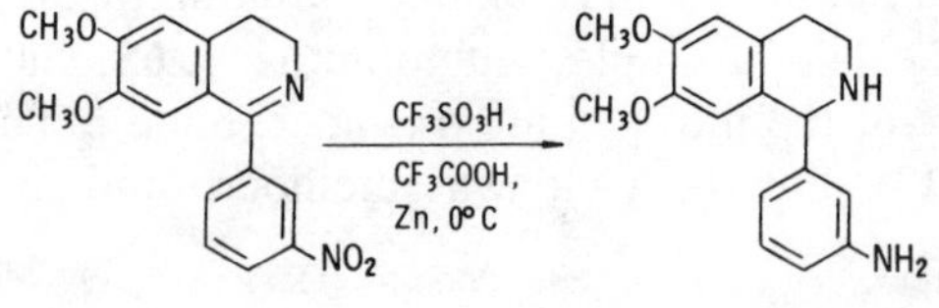

V. Synthesis of Tropoloisoquinoline Alkaloids

No work has yet appeared on the total synthesis of these alkaloids. The partial synthesis of imerubrine (**15**) and its isomer (**20**) by methylation of grandirubrine (**19**) with diazomethane (*19*) has been mentioned in Section III,B. Several approaches to **15** involving final construction of ring b were unsuccessful (*38a*).

VI. Biosynthetic Considerations

No work has yet been done on the biogenesis of azafluoranthene or tropoloisoquinoline alkaloids. As these two classes have so far been isolated from only one family (the Menispermaceae), and are known to cooccur in the genus *Abuta*, a convergent biosynthesis is conceivable. One possible route has already been proposed (Scheme 15) (*2*).

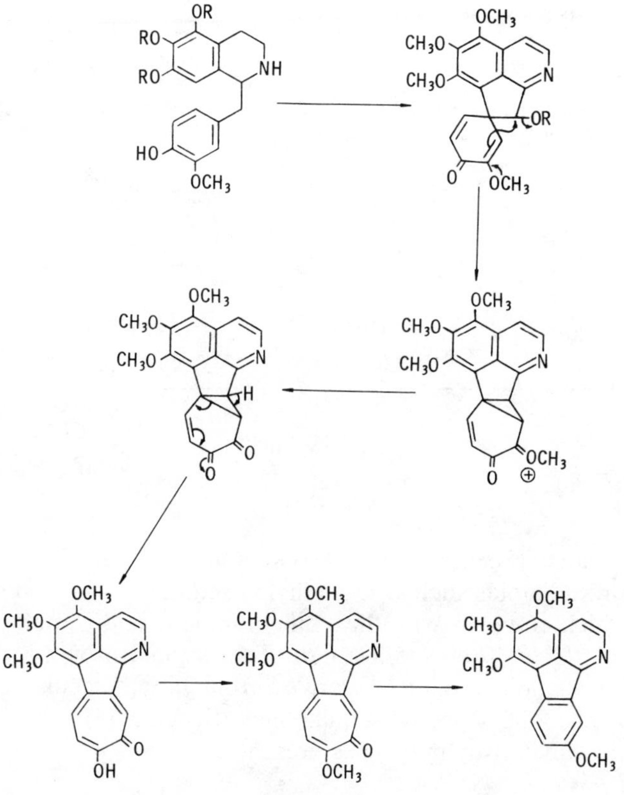

SCHEME 15

An alternative mechanism for the ring expansion has also been suggested (Scheme 16) (*39*). This scheme may necessitate methyl migration in order to achieve the final substitution pattern of imerubrine (**15**). Precedent for the conversion of imerubrine or grandirubrine to norrufescine (**10**) and rufescine (**8**), or of a more highly oxidized tropoloisoquinoline to imeluteine, already exists in the chemistry of the benzylisoquinoline-derived (*2*) colchicine family of alkaloids. For example, colchicine (**21**) readily undergoes contraction of the tropolone ring to produce *N*-acetylcolchicinol (**69**) (*40*).

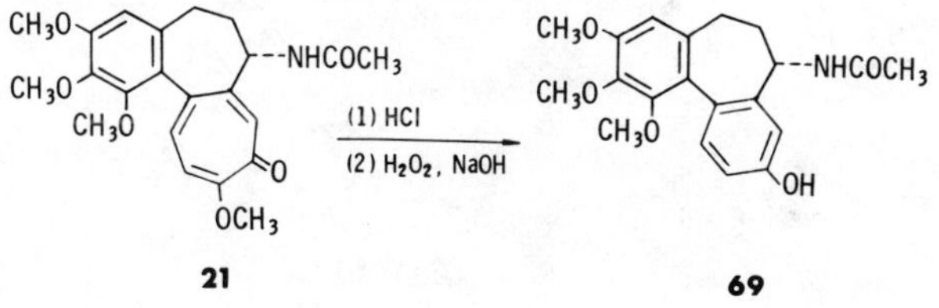

(1) HCl
(2) H_2O_2, NaOH

21 **69**

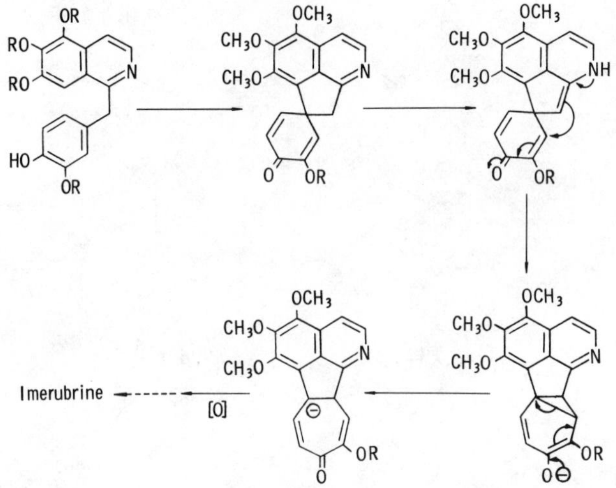

Imerubrine [O]

SCHEME 16

Our postulated precursor benzylisoquinoline **70** can also afford the oxoaporphine alkaloids such as *O*-methylmoschatoline (**5b**) and subsessiline (**71**), which co-occur with the azafluoranthene and tropoloisoquinoline alkaloids (*1,13*) (Scheme 17) (*38*). The original proposal that azafluoranthene alkaloids could be derived from intramolecular oxidative coupling of phenolic 1-phenylisoquinolines (Scheme 18) (*6*) fails to account for the other structural types present.

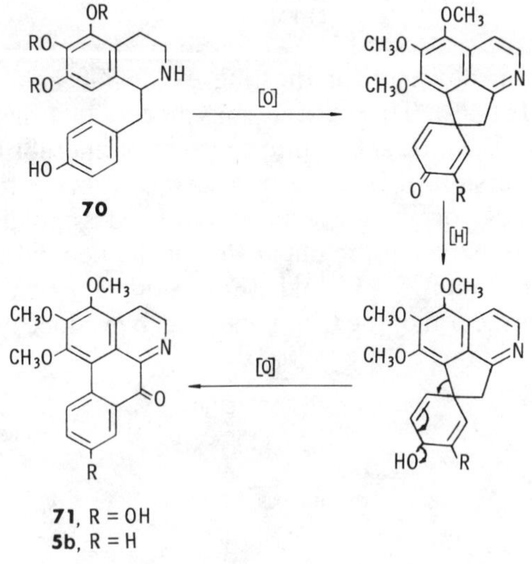

70

71, R = OH
5b, R = H

SCHEME 17

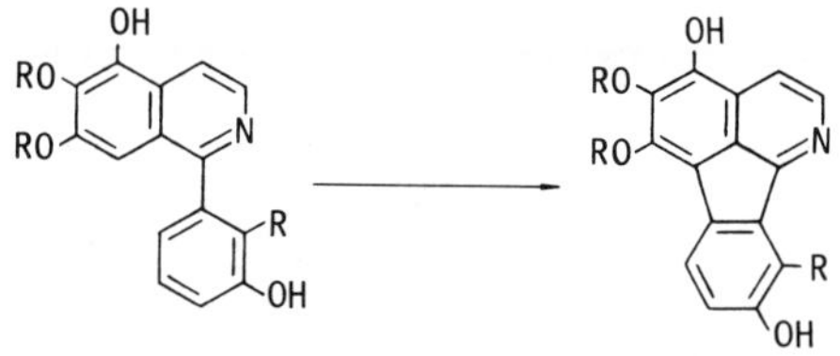

SCHEME 18

VII. Pharmacology

No pharmacological studies have been done directly on azafluoranthene or tropoloisoquinoline alkaloids. However, crude extracts of *Abuta* species containing these alkaloids have been reported as ingredients in arrow poisons used by the Juris of the Rio Japuras basin, Amazonas, Brazil (*41*). Azafluoranthene itself occurs widely in combustion products at low levels (*4,5*), but its health effects are unknown. The dihydroazafluoranthene **61** at concentrations of 10^{-4} *M* inhibits synthesis of DNA, but not RNA, by mouse thymus cells *in vitro,* and the fully aromatic compound **62** shows weaker but similar activity; thus these drugs resemble hydroxyurea more closely than papaverine in activity (*34a*). When administered intraperitoneally to mice, **47** functions as an antidepressant, inhibiting drug-induced ptosis (*31*).

Acknowledgments

The author is grateful to Prof. M. P. Cava, who directed much of the research reported here and offered suggestions during preparation of the manuscript. The author is grateful to Prof. M. D. Menachery, Dr. H. Sheppard, Dr. C. C. Wei, and Prof. D. B. MacLean for sharing details of their work. Thanks are due to Dr. M. V. Lakshmikantham for locating data and reference samples and Ms. M. Leyn Davis for clerical assistance. The author also thanks Mr. K. J. Strassburger for determining some of the mass spectra, Mr. S. Down for a copy of the MS of grandirubrine, and Dr. A. Brossi, Dr. H. C. Charles, Prof. M. Shamma, and Dr. M. J. Mitchell for helpful discussions. Thanks are also due Prof. D. L. Boger and Prof. D. A. Evans for information added in proof.

Appendix: Proton NMR Data

The following tables list ^1H-NMR data for azafluoranthene and tropoloisoquinoline alkaloids. Unless otherwise noted, δ values are in $CDCl_3$.

Azafluoranthene	H-2	H-3	H-4	H-7	H-8	H-9	H-10	Coupling constants
Imeluteine (**4**)	8.65	7.57		7.60	6.91			$J_{2,3} = 6$, $J_{7,8} = 8$ Hz
Rufescine (**8**)	8.59	7.63		7.82	6.96		7.68	$J_{2,3} = 6$, $J_{7,8} = 8$, $J_{8\text{-}10} = 2$
Norrufescine (**10**) (in $CDCl_3$ + DMSO-d_6)	8.52	7.59		7.71	6.90		7.53	$J_{2,3} = 6$, $J_{7,8} = 8$, $J_{8\text{-}10} = 2$
Triclisine (**13**)	8.52	7.29	6.84	7.38 + 7.98 (multiplets)				$J_{2,3} = 5.8$
Telitoxine (**14**) (in acetone-d_6)	8.50	7.53	7.18	7.82	6.92		7.57	$J_{2,3} = 6$, $J_{7,8} = 9$, $J_{8\text{-}10} = 2$
Isoimeluteine (**7**)	8.55	7.60				6.95	7.87	$J_{2,3} = 6$, $J_{9,10} = 8$
Compound **62**	8.57	7.32	6.90	7.67			7.56	$J_{2,3} = 6$

Tropoloisoquinoline	H-2	H-3	H-7	H-8	H-11	Coupling constants
Imerubrine (**15**)	8.68	7.75	8.06	6.87	8.29	$J_{2,3} = 5.8$, $J_{7,8} = 9.4$
Grandirubrine (**19**)	8.73	7.80	8.33	7.42	8.41	$J_{2,3} = 5$, $J_{7,8} = 10$
Isoimerubrine (**20**)	8.85	7.79	8.38	7.52	8.02	$J_{2,3} = 5$, $J_{7,8} = 11.5$

References

1. M. P. Cava, K. T. Buck, I. Noguchi, M. Srinivasan, and M. G. Rao, *Tetrahedron* **31,** 1667 (1975).
2. J. V. Silverton, C. Kabuto, K. T. Buck, and M. P. Cava, *J. Am. Chem. Soc.* **99,** 6708 (1977).
3. J. M. Schmitter, I. Ignatiadis, and G. Guiochon, *J. Chromatogr.* **248,** 203 (1982).
4. M. Dong, I. Schmeltz, E. Jacobs, and D. Hoffmann, *J. Anal. Toxicol.* **2,** 21 (1978); M. E. Snook, P. J. Fortson, and O. T. Chortyk, *Beitr. Tabakforsch. Int.* **11,** 67 (1981).
5. M. W. Dong, D. C. Locke, and D. Hoffmann, *Environ. Sci. Technol.* **11,** 612 (1977); M. Dong, I. Schmeltz, E. Lavoie, and D. Hoffmann, *Carcinog.—Compr. Surv.* **3,** 97 (1978).
6. M. P. Cava, K. T. Buck, and A. I. daRocha, *J. Am. Chem. Soc.* **94,** 5931 (1972).
7. M. P. Cava and K. T. Buck, unpublished observations.
8. M. P. Cava and I. Noguchi, *J. Org. Chem.* **38,** 60 (1973).
9. F. W. McLafferty, "Interpretation of Mass Spectra," 2nd ed., p. 118. Benjamin, New York, 1973.
10. J. W. Skiles, J. M. Saa, and M. P. Cava, *Can. J. Chem.* **57,** 1642 (1979).
11. J. W. Skiles and M. P. Cava, *J. Org. Chem.* **44,** 409 (1979).
12. L. Trifonov and A. Orakhovats, *Izv. Khim.* **11,** 297 (1978), *CA* **92,** 164129; photolysis of 6′-iodopapaverine in HCl in absence of oxygen gave the hydrochloride of the B-ring aromatic aporphine, which showed methylene resonance at δ 4.9 and rapidly oxidized in air (7).
13. M. D. Menachery and M. P. Cava, *J. Nat. Prod.* **44,** 320 (1981).
14. M. P. Cava and M. D. Menachery, private communication.

15. M. D. Klein, K. T. Buck, M. P. Cava, and D. Voet, *J. Am. Chem. Soc.* **100,** 662 (1978).
16. M. D. Menachery, M. P. Cava, K. T. Buck, and W. J. Prinz, *Heterocycles* **19,** 2255 (1982).
17. R. Huls, J. Gaspers, and R. Warin, *Bull. Soc. R. Sci. Liege* **45,** 40 (1976).
17a. H. Guinaudeau, M. LeBoeuf, and A. Cavé, *J. Nat. Prod.* **46,** 761 (1983). Because of prior use of the term "triclisine" (*CA* **30**, 5359), these authors advocate that the systematic name be adopted.
18. These NMR values are more accurate than those originally reported in ref. *1*.
19. M. D. Menachery and M. P. Cava, *Heterocycles* **14,** 943 (1980).
20. *Chem. Eng. News* **58**(40), 35 (1980); *CAS Rep.* No. 9, p. 9 (1980); *Heterocycles* **16,** No. 4 (1981), unpaginated frontispiece; cover photograph, CAS ONLINE brochure, Chemical Abstracts Service, (1981); cover photograph, *Chem. Eng. News* **61**(17) (1983).
21. C. D. Hufford, H.-G. Capraro, and A. Brossi, *Helv. Chim. Acta* **63,** 50 (1980).
22. N. Campbell and K. F. Reid, *J. Chem. Soc.* 4743 (1958).
23. G. M. Badger, W. Carruthers, and J. W. Cook, *J. Chem. Soc.* 4996 (1952).
24. S. Teitel and A. Brossi, *J. Heterocycl. Chem.* **5,** 825 (1968).
25. D. F. DeTar, *Org. React.* **9,** 409 (1957).
25a. D. L. Boger and C. E. Brotherton, submitted for publication. The syntheses of **4** and **8** utilized Diels–Alder addition of 3-carbomethoxy-2-pyrones to 8-keto-4,5,6-trimethoxycyclopenteno[1,2,3-ij] isoquinoline.
26. B. O. Lindgren and T. Nilsson, *Acta Chem. Scand.* **27,** 888 (1973).
27. The lower yield of **37** reported in Ref. *16* is incorrect.
28. S. Rajagopalan, *Proc. Indian Acad. Sci., Sect. A* **14A,** 126 (1941).
29. V. M. Rodionov and E. V. Yavorskaya, *J. Gen. Chem. USSR* (*Engl. Transl.*) **13,** 491 (1943).
30. T. Kametani, S. Shibuva, R. Charubala, M. S. Premila, and B. R. Pai, *Heterocycles* **3,** 439 (1975).
31. T. J. Schwan, U.S. Pat. 3,971,788 (1976); *CA* **85,** 192589.
32. V. A. Pol, S. M. Wagh, V. P. Barve, and A. B. Kulkarni, *Indian J. Chem.* **7,** 557 (1969).
33. H. A. Patel and D. B. MacLean, *Can. J. Chem.* **61,** 7 (1983).
34. C. C. Wei, Hoffmann-LaRoche, Inc., Nutley, New Jersey, private communication; also see Ref. *34a*.
34a. H. Sheppard, S. Sass, and W.-H. Tsien, *J. Cell Biol.* **83,** 298a (1979); H. Sheppard, private communication; also see ref. (*34*).
35. D. M. B. Hickey, C. J. Moody, and C. W. Rees, *J. Chem. Soc., Chem. Commun.* 3 (1982).
36. B. F. Bowden, K. Picker, E. Ritchie, and W. C. Taylor, *Aust. J. Chem.* **28,** 2681 (1975).
37. T. Ohta, R. Machida, K. Takeda, Y. Endo, K. Shudo, and T. Okamoto, *J. Am. Chem. Soc.* **102,** 6385 (1980).
38. K. T. Buck, unpublished observations.
38a. P. M. Koelsch, *Diss. Abstr. Int. B* **42,** 1457 (1981).
39. M. P. Cava, private communication.
40. J. Čech and F. Šantavý, *Collect. Czech. Chem. Commun.* **14,** 532 (1949).
41. B. A. Krukoff and H. N. Moldenke, *Brittonia* **3,** 1 (1938).

—CHAPTER 6—

MUSCARINE ALKALOIDS

PEN-CHUNG WANG AND

MADELEINE M. JOULLIÉ

Chemistry Department
University of Pennsylvania
Philadelphia, Pennsylvania

". . . I obtained by analysis a considerable quantity of a substance known as *muscarine,* which is the poisonous principle of a fungus, *Amanita muscaria,* or the Fly Agaric."

Dorothy L. Sayers

THE ALKALOIDS, VOL. XXIII

ISBN 0-12-469523-X

I. Introduction

The alkaloid muscarine achieved literary fame as a clue to a murder in the novel by Dorothy L. Sayers, "The Documents in the Case" (*1*). The solution to the Sayers mystery was based entirely on the difference in optical activity between the natural and synthetic products, the latter being racemic. Interestingly enough, the murder mystery appeared well before pure natural muscarine was isolated (*2*), and also before its correct structure was established unequivocally by X-ray analysis (*3*). Hart has published an interesting account of some details relating to the novel and its author (*4*).

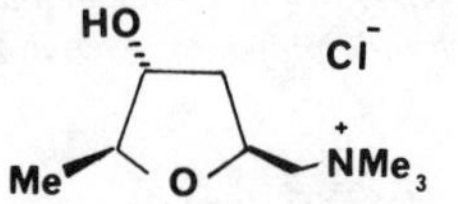

1 (2*S*,3*R*,5*S*) Muscarine

The alkaloid muscarine (**1**) is found in a variety of fungi along with choline, acetylcholine, and other basic physiologically active compounds (*5*). Included among these fungi are mushrooms of the very poisonous *Amanita* species: *A. phalloides* (death cap) and *A. muscaria* (fly agaric) (*6,7*). In mushrooms of the *Amanita* type, the base of the stem is enfolded in a volva and a ring is carried on the upper part of the stem. Their caps are decorated with scales. *Amanita muscaria* is one of the most attractive mushrooms that grow in autumnal woods. Its attractive appearance, however, is treacherous (*7*).

The widest distribution of fungal muscarine occurs in the genus *Inocybe* (*8,9*). The very poisonous *I. napipes* has a dark brown cap, sometimes with a silvery coating that tends to hide its color. It has a soapy feel and, fortunately, a characteristic unpleasant odor, which prevents it from being gathered by mistake. *Inocybe pudica* is bell-shaped and has a white cap flushed with salmon pink or red. It is found in conifer woods and, as with all *Inocybes*, should be avoided. The red staining helps in its identification.

Certain species of *Clitocybe*, such as *C. dealbata*, also contain muscarine (*10,11*). *Clitocybe dealbata* has a round, grayish white cap with a hollow center. Although not deadly, it causes unpleasant symptoms due to its muscarine content.

The effects of nutritional factors on muscarine production of *Clitocybe rivulosa* surface cultures were studied. Combinations of mannitol or sorbitol with beer wort afforded the best muscarine production (*12*). Screening of various isolates of basidiomycetes for their ability to produce muscarine in surface cultures showed that only an extract of *Clitocybe rivulosa* produced muscarinic responses in rats (*13,14*).

The chemical classes of compounds found in mushrooms that affect the nervous system are (1) muscarine, (2) acetylcholine, (3) ibotenic acid and muscimol, and (4) psilocybinpsilocin. The latter two structural types influence the central nervous system. Some are hallucinogens (*6*). Muscarine only has a peripheral action on the nervous system, and therefore produces distinctly different symptoms.

Muscarine is possibly the most elusive small molecule in the chemical literature. Many structural and synthetic studies have been directed at this compound, beginning with the first recorded attempts (1811) at the isolation of the active principles in mushrooms (*15*). These investigations are described in several excellent reviews (*5,15–19*). It took 150 years to determine the correct structure of muscarine (**1**) and to devise a synthetic route to this compound. The first synthesis of gram quantities of optically active muscarine was achieved in 1980 (*20*).

Muscarine, the best-known alkaloid of *Amanita muscaria,* has a very important place in pharmacology because it was the first drug known to have a selective action on organs innervated by the autonomic nervous system. Dixon suggested that a muscarine-like substance was liberated by the vagus nerve to act as a transmitter of its impulses. This hypothesis was supported by other investigators in classic studies on acetylcholine (*5*). The actions of acetylcholine and muscarine on smooth muscles and glands are so similar that the term "muscarinic action" was coined to describe and differentiate them from the effects of acetylcholine on ganglia and voluntary muscles, such action being referred to as "nicotinic action." The term "muscarinic activity" means direct peripheral action on cholinergic receptors in smooth muscles of the gastrointestinal tract, eye, exocrine glands, and heart.

The early work with muscarine was often controversial and inaccurate because it was based on either impure preparations or inadequate data (*16*). The isolation of pure muscarine chloride from *Amanita muscaria* by Eugster and Waser (*2*) began a new era of investigations that culminated in the elucidation of its structure, its synthesis, and reliable pharmacological data that could be extended to muscarine-like substances.

II. Natural Occurrence of Muscarine and Stereoisomers

The active principle of poisonous mushrooms has been a subject of great interest since the 1800s. Some investigators looked for the active substance in the red-colored material of the skin while others believed it must exist in the fatty contents of the flesh. Therefore, the presence of muscarine in poisonous mushrooms was suspected for many years (*18*).

Muscarine was first isolated from the fly agaric, *Amanita muscaria* (L. ex Fr.) Quél., a mushroom found in birch and pine woods from summer to late autumn. The mushroom has a brilliant orange-red cap, flecked with yellow-white spots, and is commonly found in Europe, northern Asia, North and South America, and South Africa. The fly agaric has been used by different cultures against a variety of unrelated diseases. It has also been used for religious rites and enjoyment, causing exaltation and hallucinations (e.g., the black variety of Mexican *A. muscaria*). Cases of poisoning with these mushrooms are relatively rare as they are easily recognized (*18*).

The content of muscarine in *Amanita muscaria* depends on location, climate, age, and season. The total amount is rather low, probably around 0.0003%. More muscarine is found in the red skin than in the flesh. *Amanita pantherina* D.C. ex Fr. is a smaller variety that has a yellow-brown cap and similar yellow-white spots. It contains minute amounts of muscarine (*21*). Other mushrooms, particularly those of the genus *Inocybe*, contain varying amounts of muscarine, some in amounts larger than in the fly agaric (*8,22–24*). *Inocybe lateraria* Ricken and *Inocybe patouillardi* Bres. have yellow-white or uneven buck-red caps with a dry shiny surface and a split brim rolled in. Their flesh is sweet and smells like a pear. These fungi are many times more poisonous than the fly agaric. Different varieties of *Inocybe* contain different amounts of muscarine as shown in Table I (*19*).

The symptoms of poisoning with muscarine-containing mushrooms are typical. Intoxication with *Inocybe* causes headaches, nausea, vomiting, salivation, lacrimation, diarrhea, bronchoconstriction, hypotension, and vasodilation. Death may occur unless the poisoning is diagnosed and promptly treated with atropine. Symptoms of poisoning with *Amanita muscaria* or one of its varieties depend upon the origin of the mushroom. Effects may include headache, nausea, hallucinations, delirium, convulsions, and spasms. The psychic symptoms, however, which are typical of intoxications with fly agaric, are not caused by poisoning with muscarine. Muscarine possesses only peripheral parasympathomimetic action, although its effects are increased by the presence of choline and acetylcholine, compounds also found in the mushrooms. Atropine antagonizes only the peripheral actions of muscarine.

Amanita muscaria has long had a reputation for insecticidal properties as the names muscaria, fly agaric, and fliegenpilz imply. The fungus had been used as a fly trap in Europe. Pure muscarine, however, is devoid of any action on houseflies. Isolation of fly-killing constituents in both Japanese fungi and *Amanita muscaria* were carried out by Takemoto *et al.* in

TABLE I
OCCURRENCE OF MUSCARINE[a]

Source	Yield (%)	(Reference)
Amanita muscaria (L. ex Fr.) Hooker	0.0002–0.0003[b,c]	(*25*)
Inocybe patouillardi Bres.	0.037[b,c]	(*26*)
I. fastigiata (Schff. ex Fr.) Quél.	0.01[b,c]	(*18*)
I. umbrina Bres.	0.003[b,c]	(*18*)
	0.27[d,f]	(*22*)
I. bongardi (Weinm.) Quél.	nil[b,c]	(*18*)
I. lilacina (Boudier) Kauffman	0.25–0.31[d,e], 0.38[d,f]	(*23,24*)
I. obscuroides Orton (*Stuntz* 3761)	0.32–0.52[d,e], 0.11–0.80[d,f]	(*23,24*)
I. sororia Kauffman	0.26–0.35[d,e] 0.13–0.28[d,f]	(*23,24*)
I. nigrescens Atkinson	nil[d,e], nil[d,f]	(*23,24*)
I. napipes Lange	0.23–3.15[d,e] 0.71[d,f]	(*23,24*)
I. picrosma Stuntz	0.005[d,e], nil[d,f]	(*23,24*)
I. kauffmanii A. H. Smith	0.486[d,e]	(*23,24*)
Stuntz 4292	0.144[d,e]	(*23,24*)
Stuntz 1790	0.255[d,e], nil[d,f]	(*23,24*)
Stuntz 1838	0.158[d,e], 0.03[d,f]	(*23,24*)
I. terrifera Kühner	0.269[d,e], 0.01[d,f]	(*23,24*)
I. geophylla (Fries) var. *geophylla* P. Karsten	0.259[d,e], 0.16[d,f]	(*23,24*)
I. pudica Kühner	0.117[d,e], 0.12–0.17[d,f]	(*23,24*)
I. agglutinata Peck	0.31–0.32[d,f]	(22)
I. stuntz 3691	0.08[d,f]	(22)
I. olympiana A. H. Smith	0.336[d,e]	(*23,24*)
I. subdestricta Kauffman	0.421[d,e] 0.22[d,f]	(*23,24*)
I. gausapata Kühner	0.438[d,e]	(*23,24*)
Stuntz 2147	0.476[d,e] 0.07–0.18[d,f]	(*23,24*)
Stuntz 3365	nil[d,f]	(22)
I. griseolilacina Lange	0.835[d,e], 0.17[d,f]	(*23,24*)
Stuntz 3399	0.116[d,e], 0.05[d,f]	(*23,24*)
Stuntz 1774	0.105[d,e], 0.06[d,f]	(*23,24*)
Stuntz 2907	0.28–0.30[d,f]	(22)
Stuntz 3983	0.12–0.13[d,f]	(22)
I. lacera (Fries) Quél.	0.846–1.00[d,e], 0.08[d,f]	(*23,24*)

(*Continued*)

TABLE I (*Continued*)

Source	Yield (%)	(Reference)
I. griseoscabrosa (Peck) Earle	nil[d,f]	(*22*)
I. pallidipes Ellis & Everhart sensu Kauff	0.16[d,f]	(*22*)
I. cinnamomea A. H. Smith	0.251[d,e],	
	0.03[d,f]	(*23,24*)
I. mixtilis (Britz.) Saccardo	1.33[d,e],	
	0.10[d,f]	(*23,24*)
I. xanthomelas Boursier & Kühner	0.09[d,e], nil[d,f]	(*23,24*)
I. praetervisa Quél.	0.107[d,e]	(*23,24*)
I. albodisca Peck	0.003[d,e], nil[d,f]	(*23,24*)
I. oblectabilis (Britz.) Saccardo	0.317[d,e]	(*23,24*)
fma. decemgibbosa Kühner		
Stuntz 3832	0.161[d,e]	(*23,24*)
I. decipientoides Peck	0.782[d,e]	(*23,24*)
Stuntz 1540	1.98[d,e],	
	0.23–0.24[d,f]	(*23,24*)
I. hirsuta var. *maxima* A. H. Smith	nil[d,f]	(*22*)
Stuntz 1187	0.10[d,f]	(*22*)
Stuntz 4291	0.11[d,f]	(*22*)
I. subexilis (Peck) Saccardo	nil[d,f]	(*22*)
Clitocybe dealbata (Sow. ex Fr.) Kümmer	0.15 ± 0.04[c,d,f]	(*10*)
C. rivulosa (Pers. ex Fr.) Kümmer	0.0–0.13[c,g]	(*11*)

[a] This table is taken from Ref. *19*. Other pertinent references are *8, 9, 21, 27, 28*.
[b] Relative to the fresh weight of the fruit body.
[c] Determined through isolation.
[d] Relative to the dry weight of the fruit body.
[e] Determined through biological tests.
[f] Determined through chromatographic spots comparison
[g] Based on the dry weight of the mycellium.

Japan (*29–32*) and Bowden and Drysdale in England (*33,34*). The English investigators observed the narcotic effect of the fly agaric. Flies were usually overcome within 10–15 min, and this effect lasted for 50 hr or longer. Both investigators isolated ibotenic acid (**2**) and muscimol (**3**) independently. Eugster and Müller had also isolated **2** and **3** during a

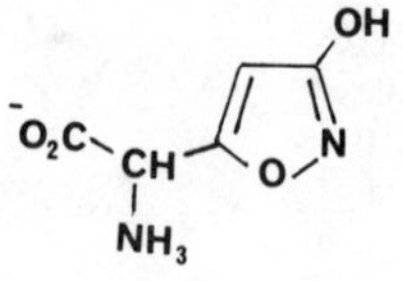

2 Ibotenic acid

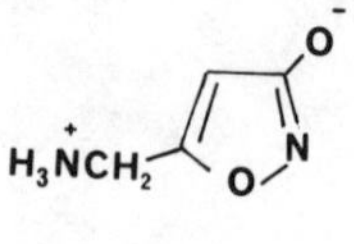

3 Muscimol

study of the narcotic potentiating nature of these compounds (*35*). The insecticidal effect of these substances was weak although greater on mosquitoes than on flies. The amounts required were too high for practical application. Both **2** and **3** were shown to act strongly on the central nervous system. The chemistry of all compounds isolated from fly agaric has been reviewed by Eugster (*19*).

All stereoisomeric muscarines occur in nature. (−)-Allomuscarine has been isolated from *Amanita muscaria* (*36*) and (+)-epimuscarine has been found in *Inocybe geophylla* (*27*). Investigations of the distribution of stereoisomeric muscarines in fruiting bodies of various species of agaricales disclosed pronounced variations in the relative amounts of the stereoisomers. Muscarine, epimuscarine, and allomuscarine were found in low concentrations, and epiallomuscarine was not detected in any of the species tested (*21*).

III. Isolation of Muscarine

Early attempts to isolate the active toxic principle of fly agaric and of other fungi were unsuccessful. The first preparations were heavily contaminated with choline and acetylcholine. Similarly, early syntheses produced a so-called "synthetic" muscarine that was obviously different from the natural product in its physiological action. Later attempts at the isolation of pure material met with varying degrees of success. Extensive pharmacological testing of both "synthetic" and natural muscarine was carried out. Eventually the synthetic material was found to be choline nitrous ester. The first fairly pure muscarine was prepared by King in 1922 (*37*). This investigator had examined the previous procedures carefully and was concerned with the presence of choline in the basic fractions. By modifying existing methods, King was able to separate muscarine from choline and thus obtain a pure product. King also determined the molecular weight for the base. In 1931, after a period of inactivity, Kögl and coworkers prepared what they considered pure material (*38*). These authors proposed possible structures for muscarine that were later proven wrong. Attempts to synthesize the postulated structures (*39*) did not afford compounds that resembled the natural alkaloid. Other synthetic approaches to molecules that might be identical to the natural product were carried out, but the compounds obtained were not as active as muscarine and exhibited a nicotine–curare action which was not shown by the alkaloid. The most interesting compound synthesized by such an approach was an acetal derivative prepared by Fourneau and colleagues (*40,41*). This compound is a very potent muscarinic substance and will be discussed further under analogs of muscarine.

The first pure crystalline muscarine chloride was obtained by Eugster and Waser (*2*). Muscarine and other bases were precipitated from the alcoholic extract of *Amanita muscaria* with Reinecke acid. The salts obtained were decomposed to the chlorides by the Kapfhammer method, and the chlorides were chromatographed on cellulose columns with different elution systems. The homogeneity of the fractions were tested by controlled toxicity tests and colorimetric methods. From 124 kg of mushrooms, 260 mg of pure muscarine chloride was isolated. Analytical data on the crystalline chloride, chloroaurate, and Reineckate salts of muscarine gave a new empirical formula, $C_9H_{20}O_2N^+X^-$. The optical rotation was larger than previously reported. At this time, several investigators realized the importance of chromatographic techniques in the isolation of pure muscarine, and shortly after the communication by Eugster and Waser (*2*), several publications appeared confirming their results. The identical observations obtained by several workers paved the way for determination of the structure of this deceptively simple molecule.

The investigations designed to establish the structure of muscarine were many and often overlapped. In this chapter we will describe only the most significant data that led to the correct structure.

IV. Structural and Conformational Studies of Muscarine

A. Structural Studies

Combustion analysis of muscarine chloride gave the empirical formula $C_9H_{20}O_2N^+Cl^-$, often with half a molecule of water. The melting point was reported to be 181–182°C and optical rotation in water $[\alpha]_D^{20.5}$ +6.7° (*16*); other investigators (*19*) report a melting point of 182–183°C, $[\alpha]_D^{25}$ +8.3° (ethanol); and $[\alpha]_D^{25}$ +8.1° (ethanol) (*42*). Muscarine iodide is much less hygroscopic than muscarine chloride and both salts are soluble in water. Muscarine is soluble in alcohol, insoluble in ether and is stable in aqueous alkaline solution but easily oxidized in acid solution. Muscarine salts may be precipitated from solution by phosphotungstic acid, Reinecke's acid, $KBiI_4$, $KHgI_4$, and sodium tetraphenyl boron.

Many chemical reactions were carried out on muscarine chloride in order to prove or disprove the present of certain functional groups. The presence of a trimethylammonium group was established by liberation of trimethylamine in molten alkali and by pyrolysis of muscarine chloride at 200°C *in vacuo*. The loss of methyl chloride afforded pharmacologically inactive normuscarine which could be quaternized to active muscarine chloride. Determination of the *N*-methyl groups showed that muscarine possesses three and normuscarine two methyl groups (*25*). Since muscarine was inert to periodate oxidation it was assumed by Kuehl and coworkers that there were no vicinal hydroxyl groups or adjacent hydroxyl and carbonyl groups (*42*). Acetylation of muscarine confirmed the presence of one hydroxyl group. No positive test results for an aldehyde group (Tollens reagent, Schiff reagent, Fehling solution, the Angel-Rimini test, and the malachite green test) were obtained, and no carbonyl group was detected by infrared spectroscopy (*43*). The Zeisel methoxyl test was negative, proving the absence of a methyl ether. The inert nature of the second oxygen suggested an ether linkage. By a series of degradation reactions, supported by evidence from infrared spectra, Eugster concluded that the second oxygen atom was located in a tetrahydrofuran ring (*25*). Oxidation of muscarine chloride by platinum dioxide in oxygen gave muscarone, whose infrared spectrum indicated the presence of a five-membered cyclic ketone (1754 cm^{-1}). This observation supported the position of the hydroxyl group in the tetrahydrofuran ring rather than on a side chain (*43*). This catalytic oxidation proved extremely valuable later in determination of the synthetic stereoisomers of muscarine chloride.

Elucidation of the carbon frame proved difficult. Degradation of muscarine chloride by chromic acid afforded only acetic acid. Treatment with phosphorus tribromide followed by reduction with zinc or lithium aluminum hydride gave propionic acid as the major product (*17*). However, degradation with hydriodic acid and phosphorus resulted in the formation of *n*-hexyltrimethylammonium iodide, a product containing the whole carbon chain (*44*). The main chemical reactions carried out on muscarine chloride are summarized in Fig. 1.

Finally, the X-ray diffraction analysis of muscarine iodide by Jellinek established beyond doubt the correct structure of the molecule (*3*). Bond angles and distances were determined. All atoms in the tetrahydrofuran ring were found to be in one plane except for C-4. The sequence C-4, C-5, C-6, N, and one of the methyl groups were virtually in a single plane. This crystallographic data also showed the cis arrangement of the 2-methyl and 5-quaternary ammonium groups and their trans relationship to the hydroxyl group (Fig. 2).

Muscarine possesses three asymmetric centers, and consequently has four pairs of enantiomers. All have been synthesized. The four racemates have been named and assigned the structures shown by Corrodi, Hardegger, and Kögl (*45*). The asymmetric centers are identified according to the Cahn, Ingold, and Prelog system (*46*) (Fig. 3).

The literature numbering of the tetrahydrofuran ring of muscarine is not always consistent. The numbering recommended by Chemical Abstracts will be used in this review. The numbering begins on oxygen and con-

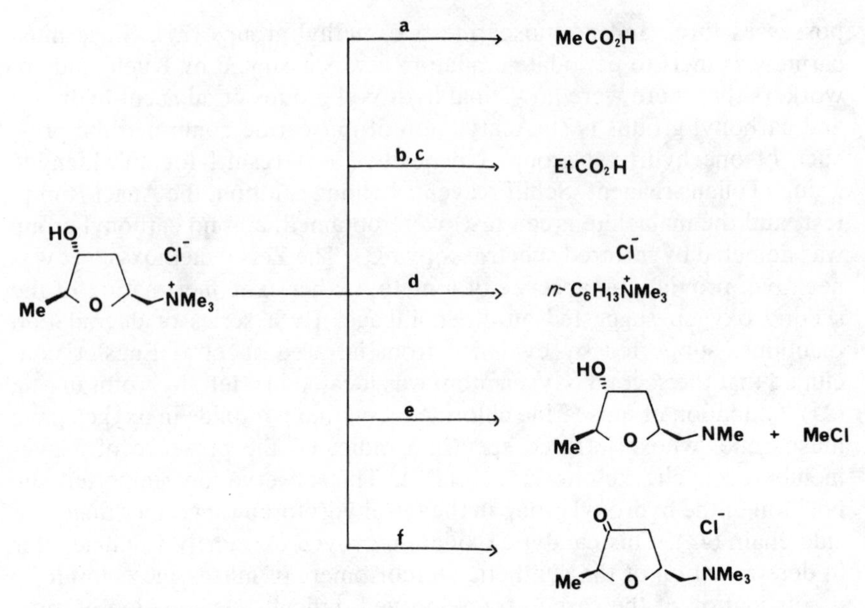

FIG. 1. Chemical reactions carried out on optically active muscarines. (a) CrO_3; (b) PBr_3; (c) Zn or $LiAlH_4$; (d) HI–P; (e) 200°C; (f) Pt–O_2.

tinues in a clockwise manner around the ring as seen in Fig. 3. The Chemical Abstracts registry number for muscarine is 300-54-9.

B. Conformational Studies

It is well documented that the conformation of naturally occurring substances has an important bearing on their biological activity. Therefore, the solution conformation of muscarinic agents has been the subject of several investigations.

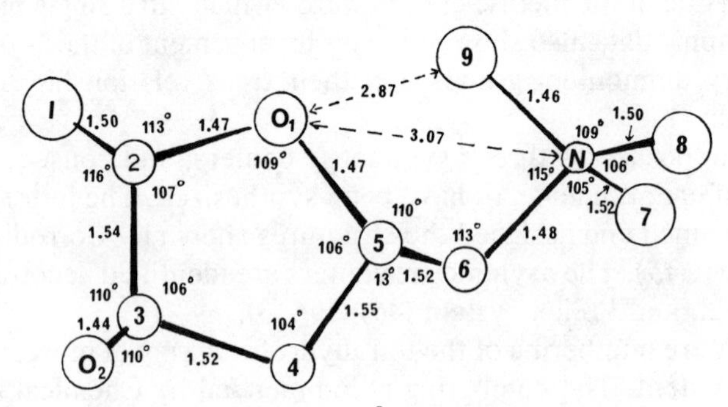

FIG. 2. Bond distances (in Å) and angles in muscarine.

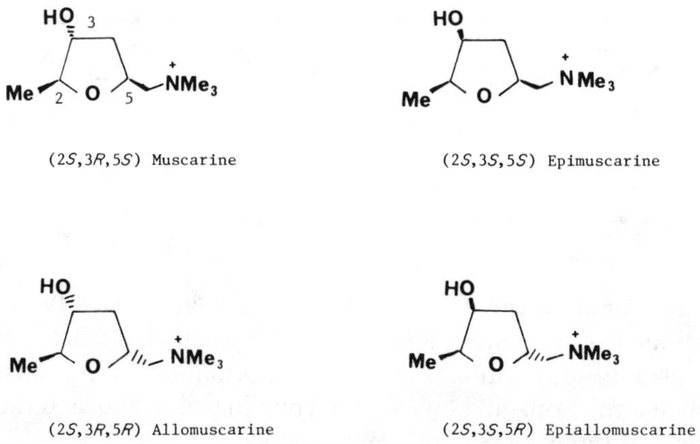

(2S,3R,5S) Muscarine

(2S,3S,5S) Epimuscarine

(2S,3R,5R) Allomuscarine

(2S,3S,5R) Epiallomuscarine

FIG. 3. Diastereomeric muscarines.

The conformation and electronic properties of muscarine were examined by the quantum mechanical method of perturbative configuration interaction using localized orbitals (*47*). These calculations indicated that the positive charge on the quaternary nitrogen atom was distributed over the surrounding *N*-methyl groups. Waser (*18*) proposed that the quaternary nitrogen was held over the tetrahydrofuran ring by electrostatic interaction with the ether oxygen. The X-ray diffraction analysis of muscarine iodide (*3,48*) and muscarone (*49*) afforded the interatomic distances and angles of these molecules in the solid state, and it showed that the cationic center was directed away from the ring. The approximate position of the cationic center in solution was examined using homonuclear intramolecular nuclear Overhauser effect techniques (*50*). The results indicated that the quaternary side chain of muscarine, in aqueous solution, was located in an extended conformation away from the ring, **a**, rather than in the conformation **b** proposed by Waser (*18*).

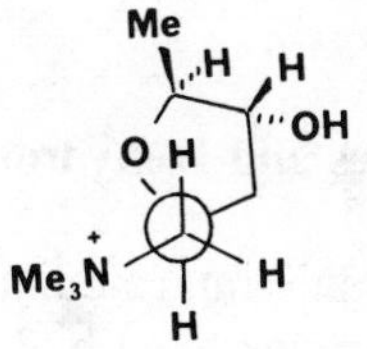

a 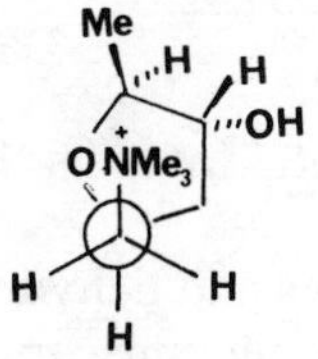 **b**

The common feature of all muscarines is a five-membered tetrahydrofuran ring that exists in dynamic equilibrium between two favored puckered conformations, a type-N conformer and a type-S conformer (*51*).

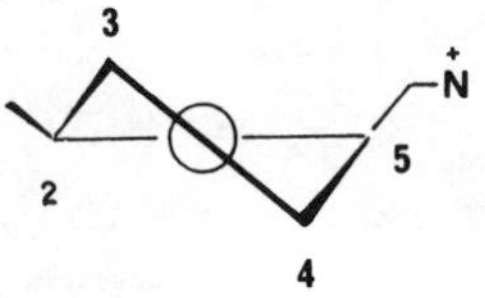

3 2 5 4 $\overset{+}{N}$ 4 2 3 5 $\overset{+}{N}$

C-3 endo C-4 exo

Type N

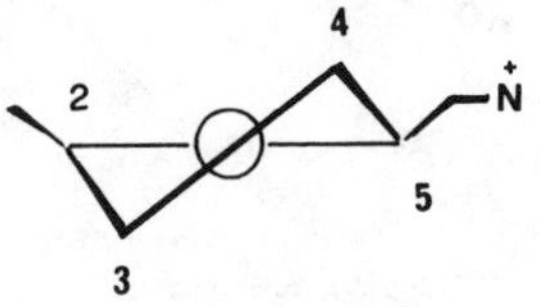

C-4 endo C-3 exo

Type S

Altona and Sundralingan introduced a pseudorotational analysis whereby five-membered ring conformations could be calculated from vicinal coupling constants (*52*). Studies on the conformation of furanose sugars established a relationship between the coupling constants and the relative percentage of the N or S conformations (*52*).

Huynh-Dinh and Pochet used the technique of pseudorotational parameters to determine the conformational population of L-(+)-muscarine and D-(−)-allomuscarine (*53*). For L-(+)-muscarine iodide, they observed a predominance of the S conformation (76%). For D-(−)-allomuscarine iodide, they observed equal presence of the two conformations S and N (45 and 55%, respectively). Employing the same technique, Brown and Mubarak concluded that the muscarine ring existed in rapidly interconverting puckered forms, with the S conformation being favored (71%) (*54*). These results are in good agreement with the crystal structure of muscarine iodide. Pseudorotational analysis of D-epiallomuscarine favored the N conformation over the S conformation in an approximate ratio of 67 to 33 (*55*).

As the rotational barriers between the N and S conformations of muscarine are energetically low, interconversion between the two forms occurs very rapidly. Therefore, it is unlikely that the pharmacological properties of muscarine depend on its conformational population, although there might be some correlation between degree of puckering and biological activity (*55*).

V. Synthetic Investigations of Muscarones and Dehydromuscarones

Muscarones and dehydromuscarones are key intermediates in the synthesis of racemic muscarines. Some muscarones have higher muscarinic activity than muscarine. Therefore, because of their synthetic and pharmacological importance, the preparation of these compounds will be treated separately.

Eugster *et al.* (*56–58*) synthesized the important intermediate 5-dimethylaminomethyl-2,3-dihydro-2-methyl-3-oxofuran (**8**) which was converted to the corresponding saturated ketone (**9**), an extremely useful

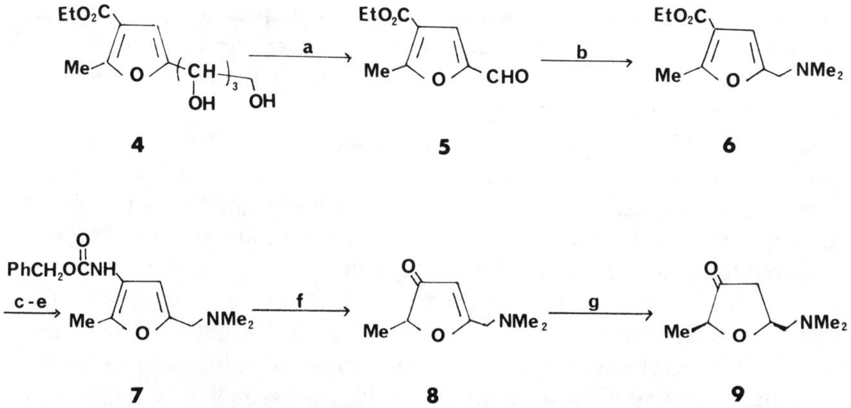

SCHEME 1. Reagents: (a) Pb_3O_4, 50% AcOH; (b) $HCONMe_2$, HCO_2H; (c) $N_2H_4 \cdot H_2O$; (d) HNO_2; (e) $PhCH_2OH$; (f) 2 *N* HCl; (g) Pd, H_2.

intermediate in the preparation of all stereoisomers of muscarine. This approach is shown in Scheme 1. The condensation of sugars with β-ketoesters afforded the furan derivative **4**. Cleavage of the hydroxyl side chain with lead tetraacetate gave the corresponding aldehyde (**5**) which was treated under Leuckart conditions to afford the dimethylaminomethyl derivative **6**. The ester group was then subjected to a Curtius degradation to give benzylurethane **7**. Hydrolysis with 2 *N* HCl at 100°C afforded 2,3-dihydrofuranone **8** which was catalytically hydrogenated to furanone **9**.

A more efficient synthesis of **8** was reported by Meister and Scharf (*59*). Elimination of methanol from 2-methyl-3,3,5-trimethoxytetrahydrofuran (**10**) afforded 2-methyl-3-methoxyfuran (**11**) which was converted to dimethylaminomethyl derivative **12** by a Mannich-type reaction. Hydrolysis of **12** gave **8** (Scheme 2).

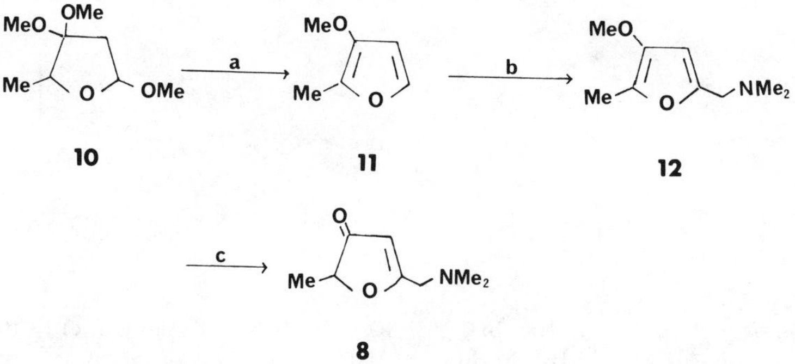

SCHEME 2. Reagents: (a) HCl, MeOH; (b) Me_2NH, HCHO, AcOH; (c) H_3O^+.

A review of the many methods available for preparing 3(2*H*)-dihydrofuranones has been published (*60*). Most of these, however, are more suitable for the preparation of structural isomers of muscarone than for the synthesis of muscarone itself.

The best known method for generating furanones is based on the Michael addition of anions derived from α-hydroxy esters to α,β-unsaturated substrates. Early use of this reaction by Hardegger *et al.* (*45*) led to the preparation of furanones (**14**), important precursors for the synthesis of allomuscarine, 2-methylmuscarine, and 2-methylmuscarone (**15**) (Scheme 3). Another approach to furanone **14a** was also devised by Hardegger *et al.* (*61*) utilizing the condensation of ethyl α-iodopropionate and diethyl α-hydroxybutanoate. The resulting substituted furanone (**13a**) was easily decarboxylated to a monocarboxylic acid which was reesterified with diazomethane. Conversion of furanone **14b** to 2-methylmuscarone iodide (**15**) is also shown in Scheme 3.

A cyclization that afforded a readily manipulable 3(2*H*)-dihydrofuranone (whose keto group was protected as a 1,3-dioxolane) was developed by Joullié and co-workers (*62*). Hydroxyenone **17** was obtained by

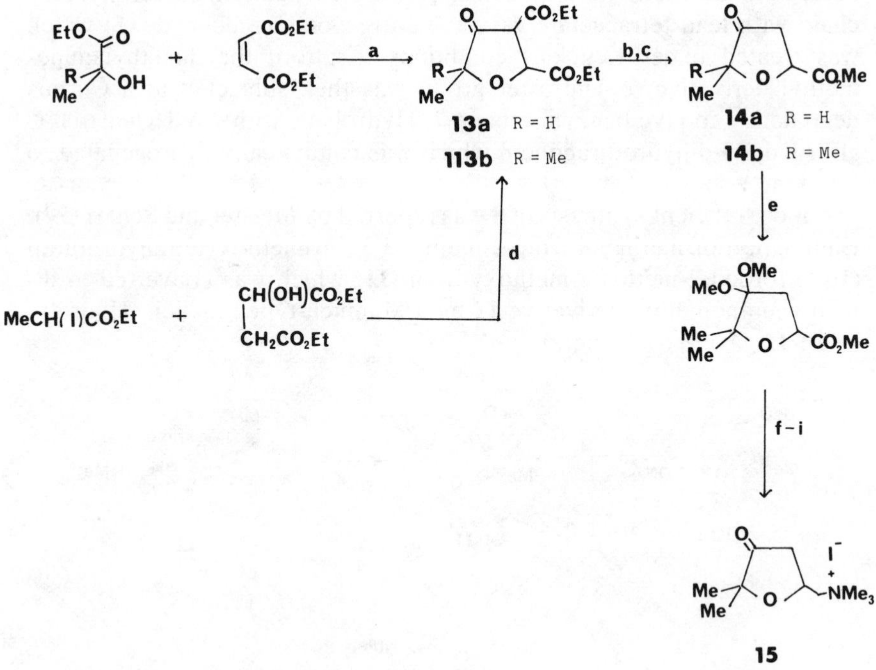

EtO, O, R, OH, Me + CO_2Et, CO_2Et —a→ O, CO_2Et, R, Me, O, CO_2Et —b,c→ O, R, Me, O, CO_2Me

13a R = H
113b R = Me

14a R = H
14b R = Me

$MeCH(I)CO_2Et$ + $CH(OH)CO_2Et$ / CH_2CO_2Et —d→ 13

e → OMe, MeO, Me, Me, O, CO_2Me

f–i → O, Me, Me, O, $\overset{+}{N}Me_3$ I^-

15

SCHEME 3. Reagents: (a) Na; (b) 2 *N* H_2SO_4; (c) CH_2N_2; (d) Ag_2CO_3; (e) $HC(OMe)_3$, H_3O^+; (f) Me_2NH; (g) $LiAlH_4$; (h) H_3O^+; (i) MeI.

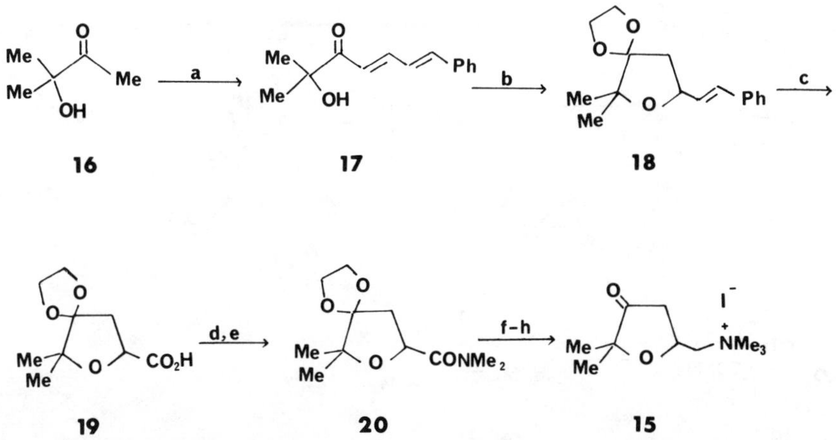

SCHEME 4. Reagents: (a) NaOEt, *trans*-cinnamaldehyde; (b) $(CH_2OH)_2$, *p*-TsOH; (c) $KMnO_4$, tricaprylmethyl ammonium chloride; (d) $(COCl)_2$; (e) Me_2NH; (f) $LiAlH_4$; (g) H_3O^+; (h) MeI.

aldol condensation of α-hydroxyketone **16** with *trans*-cinnamaldehyde. Cyclization of **17** with *p*-toluenesulfonic acid in the presence of ethylene glycol afforded the protected 3(2*H*)-dihydrofuranone (**18**). Phase-transfer catalyzed oxidation of **18** gave carboxylic acid **19** which was converted via its mixed anhydride into amide **20**. Reduction, deblocking of the 1,3-dioxolane group, and quaternization of the resulting amine afforded **15** (Scheme 4).

VI. Synthetic Investigations of Muscarines and Stereoisomers

A. TRANSFORMATIONS OF CARBOHYDRATE PRECURSORS

The design of synthetic routes to muscarine and its stereoisomers from carbohydrates must meet two requirements: (1) the method selected for C–C bond formation at the anomeric center must be stereocontrolled; (2) C-2 must be appropriately substituted or amenable to substitution, so as to permit construction of the quaternary ammonium salt function.

Hardegger and Lohse successfully completed a total synthesis of L-(+)-muscarine from L-glucosaminic acid in 1957 (*63*). The deamination of 2-amino-2-deoxy-L-gluconic acid (**21**) with nitrous acid produced L-chitaric acid (**22**) as shown in Scheme 5. After esterification with diazomethane, L-chitaric acid methyl ester was converted to dimethylamide **23** and then treated with *p*-toluenesulfonyl chloride to afford tritosylate **24**. Lith-

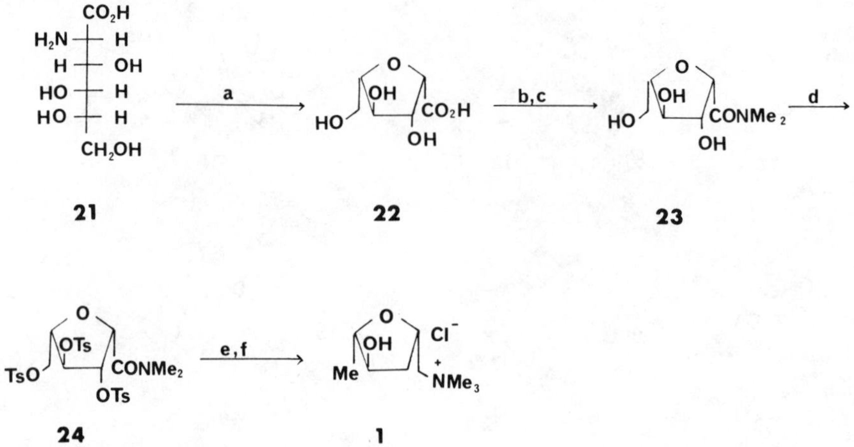

SCHEME 5. Reagents: (a) HNO_2; (b) CH_2N_2; (c) Me_2NH; (d) *p*-TsCl, C_5H_5N; (e) $LiAlH_4$; (f) MeCl.

ium aluminum hydride reduction of **24** gave a mixture of reduction products, including normuscarine. Purification via tetraphenylborate led to L-muscarine chloride, with the same configuration as natural muscarine chloride (Scheme 5). The unnatural D-muscarine was similarly synthesized from D-glucosaminic acid (*64*) and from 2-deoxy-D-ribose (*65*).

D-Epiallomuscarine iodide (**29**) was synthesized from α-D-glucose by Joullié and co-workers (*66,67*). 3,5-Anhydro-3,6-di-*O*-tosyl-L-idose dimethylacetal (**25**), obtained in 64% overall yield from α-D-glucose, was reduced by lithium aluminum hydride to afford a 3:2 mixture of alcohols (**26a** and **26b**). After separation of the alcohols, **26a** was converted to the corresponding acetate (**27**) with acetic anhydride. Hydrolysis of **27** produced a sensitive aldehyde which was immediately oxidized with Jones reagent to the corresponding carboxylic acid (**28**). Further elaboration of **28** afforded D-epiallomuscarine iodide (**29**) (Scheme 6).

A stereospecific preparation of L-(+)-muscarine chloride from D-mannitol was reported by Mubarak and Brown (*20*). Acid-catalyzed dehydration of D-mannitol gave 2,5-anhydro-D-glucitol which was converted to dibenzoate **30**. Reaction of **30** with excess tosyl chloride gave a ditosylate which when treated with sodium methoxide afforded epoxide **31** which was selectively reduced with sodium bis(2-methoxyethoxy)aluminum hydride to give a 12 : 1 mixture of diols **32a** and **32b**. Monotosylation of the diol mixture followed by chromatographic separation gave a major isomer (**33**) which on treatment with triethylamine and ion exchange gave L-(+)-muscarine chloride (Scheme 7).

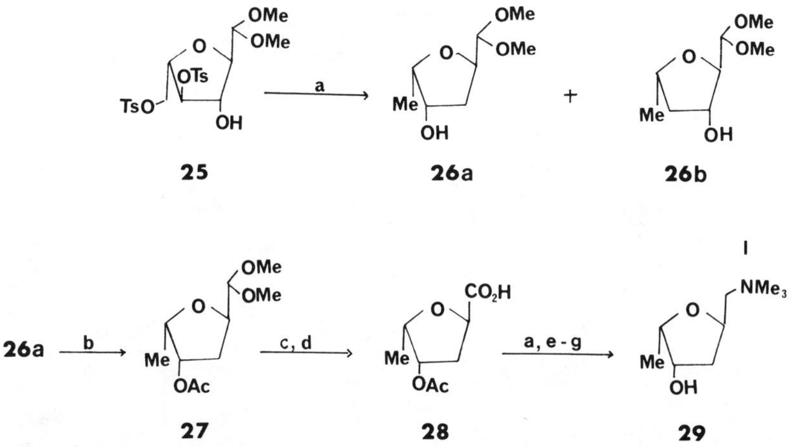

SCHEME 6. Reagents: (a) $LiAlH_4$; (b) Ac_2O, DMAP; (c) H_3O^+; (d) Jones reagent; (e) $(ClCO)_2$; (f) Me_2NH; (g) MeI.

Pochet and Huynh-Dinh synthesized L-(+)-muscarine and D-(−)allo-muscarine from 2-deoxy-L-ribose (*53*) (Scheme 8). 2-Deoxy-3,5-di-*O*-*p*-toluoyl-β- and α-L-erythropentofuranosyl 1-cyanides were prepared from 2-deoxy-L-ribose in three steps and separated by column chromatography. Treatment of the β cyanide with a methanolic hydrogen chloride solution gave a mixture of methyl esters **34a** and **34b**. Aminolysis of these esters afforded the corresponding amides (**35**). Selective tosylation of **35**, reduction of the tosylamide with lithium aluminum hydride, followed by quaternization with methyl iodide gave L-(+)-muscarine iodide. Treatment of

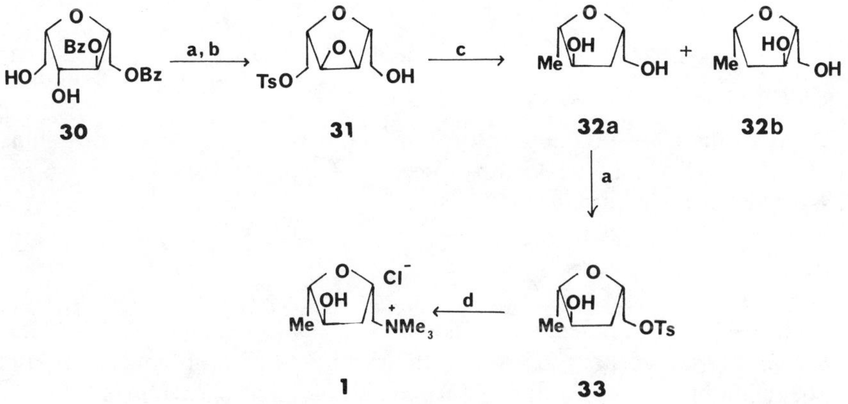

SCHEME 7. Reagents: (a) TsCl, C_5H_5N; (b) NaOMe, MeOH; (c) Red-al, THF; (d) Me_3N.

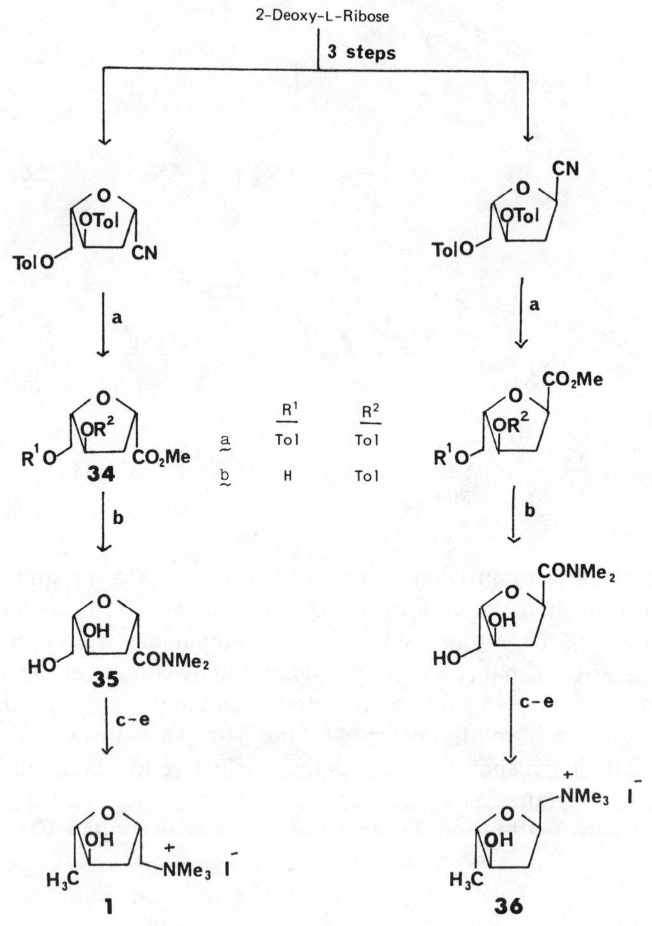

SCHEME 8. Reagents: (a) HCl, MeOH; (b) Me_2NH, MeOH; (c) TsCl; (d) $LiAlH_4$; (e) MeI.

the α cyanide by the same procedure gave D-(−)-allomuscarine iodide (*36*). D-(−)-**1** and L-(+)-**36** were prepared from D-2-deoxyribose via the same series of reactions (*53*).

B. CYCLIZATION METHODS

Use of an intramolecular Michael–Dieckmann condensation by Hardegger and co-workers (*68*) led to the syntheses of *dl*-allomuscarine and *dl*-epiallomuscarine. Furanone **14a** was prepared as shown in Scheme 3. Reduction of **14a** gave the corresponding alcohol (**37**). Fractional crystallization of the 3,5-dinitrobenzoates of **37** afforded four stereoisomers. The

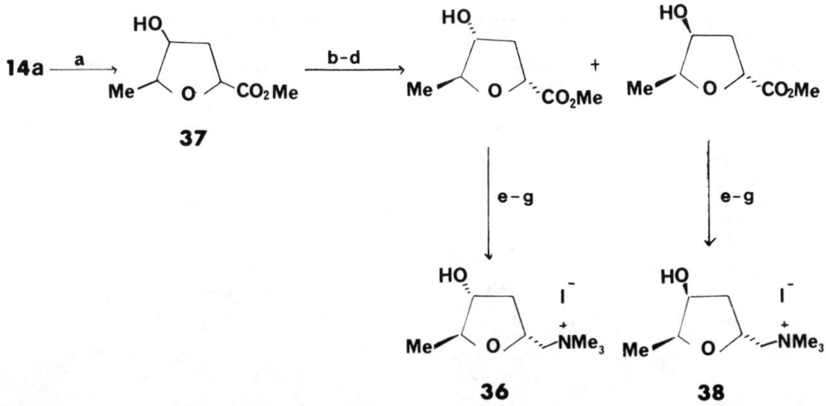

Reagents: (a) H_2, Raney Ni; (b) 3,5-dinitrobenzoyl chloride, C_5H_5N; (c) separation; (d) 0.1 *N* NaOH; (e) Me_2NH; (f) $LiAlH_4$; (g) MeI.

dl-allo and *dl*-epiallo configurations were confirmed by infrared and proton nuclear magnetic resonance spectroscopy and also by the syntheses of *dl*-allomuscarine iodide and *dl*-epiallomuscarine iodide. The results previously obtained by Corrodi *et al.* (*45*) do not agree with the results obtained for the salt of allomuscarine of established structure and configuration, and the presumed "allomuscarine" obtained by these authors probably does not possess the allo configuration.

Functionalized 1,4-diols are useful intermediates for the synthesis of tetrahydrofurans. Kögl, Cox, and Salemink (*69*) described a synthesis of *dl*-muscarine from δ-chloro-α-acetyl-γ-valerolactone (Scheme 9). The intramolecular dehydration of the functionalized 1,4-diol (**39**) afforded a

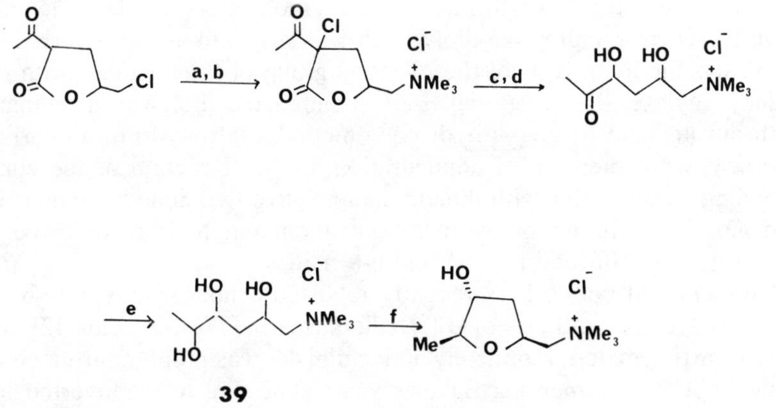

SCHEME 9. Reagents: (a) Me_3N; (b) SO_2Cl_2; (c) 20% HCl; (d) $CaCO_3$; (e) H_2, Raney Ni; (f) conc H_2SO_4.

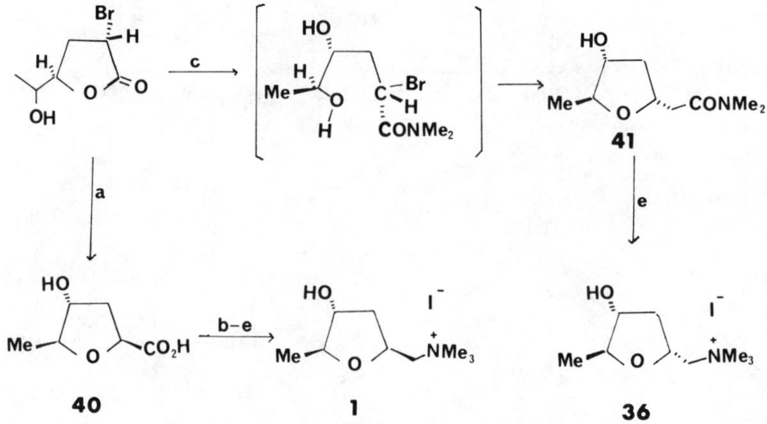

SCHEME 10. Reagents: (a) H_2O, 150–160°C; (b) CH_2N_2; (c) Me_2NH; (d) $LiAlH_4$; (e) MeI.

mixture of muscarine and its stereoisomers. The authors assumed, based on biological experiments, that the end product contained a considerable amount of *dl*-muscarine. Purification attempts revealed the presence of structurally different impurities. Neither muscarine nor any of its stereoisomers have been isolated in pure form from this reaction sequence (*17*).

In Matsumoto's synthesis of *dl*-muscarine and *dl*-allomuscarine (*70*), a bromolactone was stereospecifically converted to tetrahydrofuran **40** (CH_3/CO_2H cis) and dimethylamide **41** ($CH_3/CON(CH_3)_2$ trans). Compounds **40** and **41** were further transformed into D-**1** and D-**36** (Scheme 10).

A convenient synthesis of L-(+)-muscarine was reported by Belleau *et al.* (*71*) (Scheme 11). Performic acid oxidation of *N*-acetylcrotylglycine gave the corresponding 4,5-diol which was resolved at the α carbon by stereospecific hydrolysis of the *N*-acetyl group of the L-isomer with hog kidney acylase. The resulting α-L-(+)-aminoacid diol was deaminated with nitrous acid to give 4-hydroxy-5-methyl-2-tetrahydrofuran carboxylic acid with retention of configuration at C-2. Reaction of the corresponding methyl ester with dimethylamine gave two dimethylamides (**42** and **43**). Their reduction with lithium aluminum hydride followed by quaternization afforded L-(+)-**1** and L-(+)-**36**.

Grasselli and co-workers recently reported a noncarbohydrate-based enantioselective synthesis of D-(−)-allomuscarine (*72*) (Scheme 12). The chiral starting material, optically active diol **44**, was prepared from cinnamaldehyde and fermenting Baker's yeast. The diol was converted into diacetate **45** which was subjected to bromination with *N*-bromosuccinimide followed by an S_N2 displacement to afford azide **46**. The

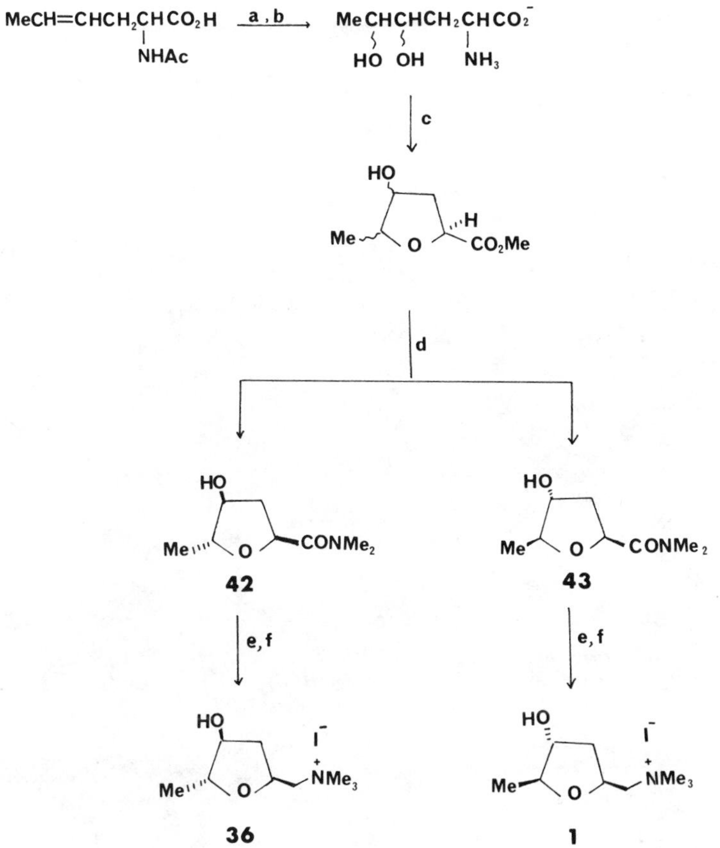

SCHEME 11. Reagents: (a) HCO_3H; (b) Hog kidney acylase, pH 8; (c) HNO_2–MeOH; (d) Me_2NH; (e) $LiAlH_4$; (f) MeI.

latter upon hydrogenation yielded a triacetyl derivative (**47**). The racemic mixture was ozonized in 90% formic acid. Oxidative work-up was followed by treatment with 4 *N* HCl at 100°C. Evaporation of the acid afforded the amino δ-lactone hydrochloride **48**. Deamination produced ester **49** which was converted to amide **50**. Since conversion of **50** to **36** had been previously accomplished, this synthesis was considered an enantioselective synthesis of **36** from **44**.

A chelation-controlled addition of Grignard reagents to a chiral α-alkoxy aldehyde to give *threo*-diol derivatives was used as the key step in Still's synthesis of *dl*-muscarine (*73*) (Scheme 13). Addition of methylmagnesium bromide to the corresponding cyclic hydrate diacetate of meso-bis(benzyloxymethyloxy)glutaraldehyde gave the *threo* triol deriva-

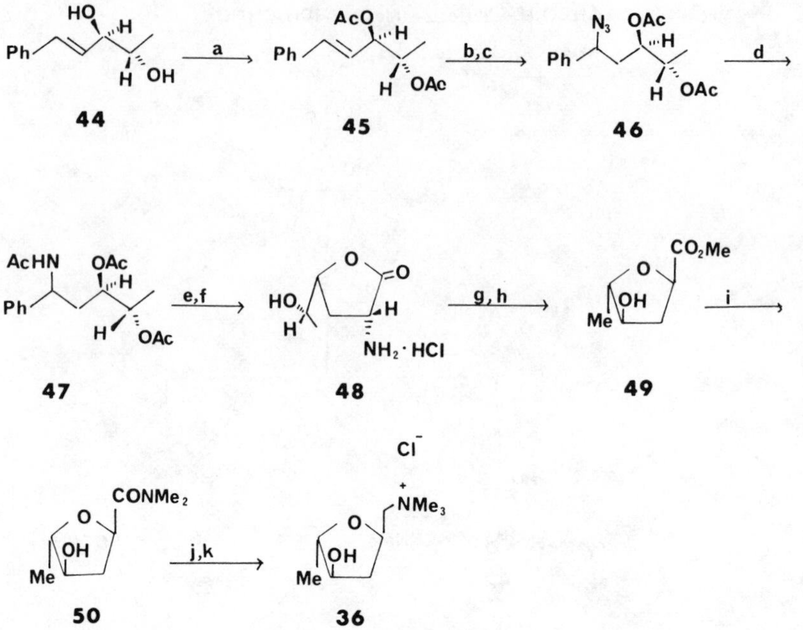

SCHEME 12. Reagents: (a) Ac_2O; (b) NBS, CCl_4; (c) NaN_3; (d) H_2/PtO_2; (e) O_3, 90% HCO_2H; (f) 4 *N* HCl; (g) HNO_2; (h) HCl, MeOH; (i) Me_2NH; (j) $LiAlH_4$; (k) MeI.

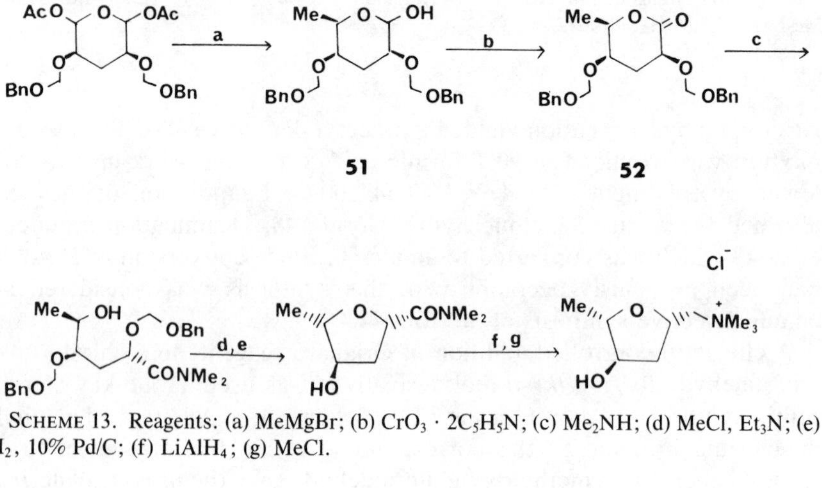

SCHEME 13. Reagents: (a) MeMgBr; (b) $CrO_3 \cdot 2C_5H_5N$; (c) Me_2NH; (d) MeCl, Et_3N; (e) H_2, 10% Pd/C; (f) $LiAlH_4$; (g) MeCl.

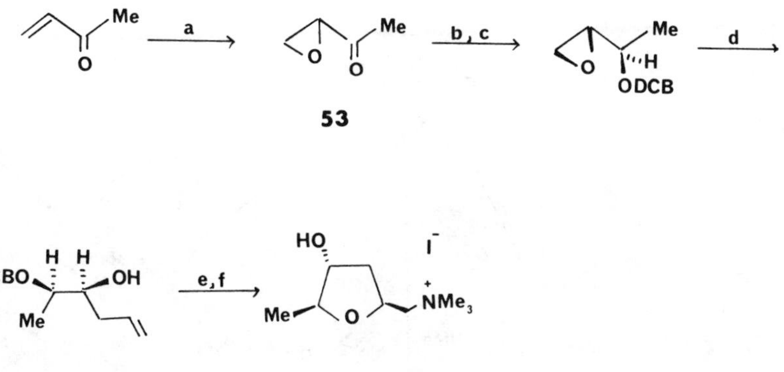

SCHEME 14. Reagents: (a) t-BuO_2H; (b) $Zn(BH_4)_2$; (c) NaH, 2,6-dichlorobenzyl bromide; (d) //\MgBr, 10% CuI; (e) I_2; (f) excess NMe_3.

tive (**51**). Oxidation of **51** to the corresponding lactone (**52**) was followed by conversion to *dl*-muscarine via amide formation, mesylation, deprotection, and lithium aluminum hydride reduction.

Using two highly stereoselective reactions, namely the zinc borohydride reduction of an α,β-epoxyketone (**53**) and the cyclization of an olefinic benzyl ether (**54**) with iodine, Chastrette and co-workers obtained *dl*-muscarine in six steps from methy vinyl ketone (*74*) (Scheme 14).

C. TRANSFORMATIONS OF FURAN DERIVATIVES

As mentioned in Section V, dehydromuscarones (**8**) obtained from furan derivatives were key intermediates in the preparation of muscarones that, in turn, were useful precursors in the synthesis of all four racemates of muscarines (*56–58,75*) (Fig. 3). Reduction of these ketones was studied extensively and the results of these investigations are summarized in Scheme 15. The various reductions led to mixed racemates.

A large number of experiments were designed to prepare pure quaternary salts from the reaction mixtures (*17*). Except for epinormuscarine, which could be obtained practically free of isomers by reduction of **8** with Raney nickel, most mixtures of quaternary salts could not be separated satisfactorily. Crystallization of muscarine iodides seldom led to pure muscarine isomers. Even partition chromatography over very large cellulose columns (1:1000 ratio of substance to cellulose) did not result in separation since the R_f values of the components were too similar even in different solvent systems.

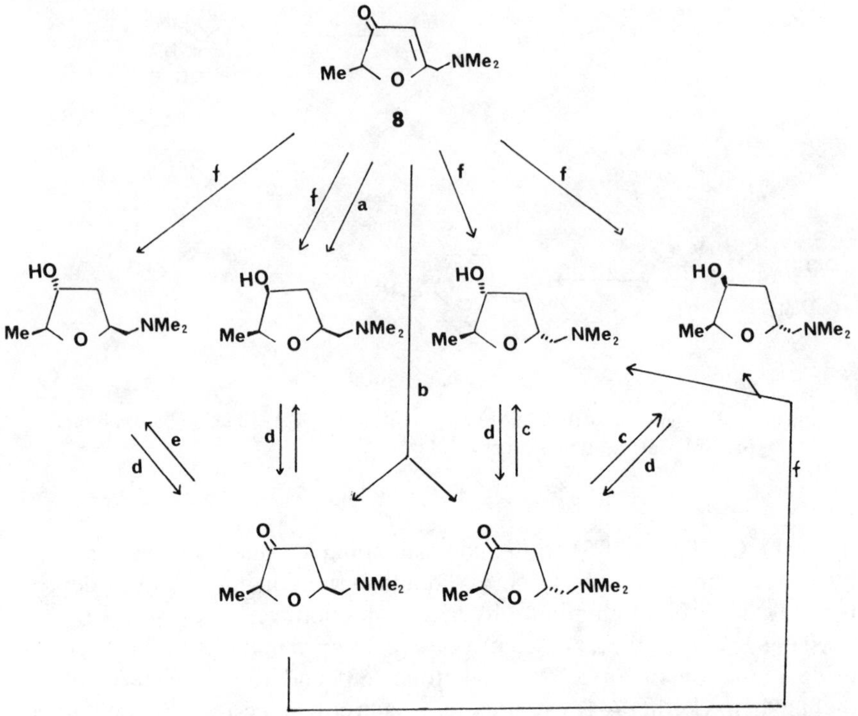

SCHEME 15. Reagents: (a) H_2, Raney Ni; (b) H_2, Pd; (c) KBH_4; (d) Pt, O_2; (e) $LiAlH_4$; (f) Al $(OCHMe_2)_3$.

Some separation was obtained by fractional crystallization of tetraphenylborates since the muscarine salt accumulated in the head fractions. Chromatography of these salts over neutral aluminum oxide in acetone solution, and of the chlorides in isopropanol over acidic aluminum oxide, also allowed some separation. These procedures were eventually abandoned when it was discovered that the tertiary norbases could be more easily separated than the salts either by fractional distillation or by adsorption chromatography using aluminum oxide as the adsorbent and benzene with increasing amounts of methanol or ether as eluent. Epinormuscarine was eluted first, followed by allonormuscarine. Epiallonormuscarine and normuscarine eluted much later. A similar order of elution (epi-, allo-, epiallo-, and normuscarine) was observed during gas chromatographic separations (*27*).

During studies designed to establish the chirality of all stereoisomeric muscarines, a number of racemic separations were accomplished (*9*). Enantiomeric di-*p*-toluenyltartaric acid was the reagent of choice. When

(±)-muscarine was treated with (−)-di-*p*-toluenyl-D-tartaric acid, the salt of (−)-muscarine was found in a fraction difficultly soluble in 2-propanol. Optical purity could usually be achieved after three to six recrystallizations of the diastereomeric salts from 2-propanol. It was noted, however, that these separations were still incomplete, regardless of the agreement of the rotations obtained with the reported values of the natural products. Purer preparations were obtained from reduction of optically active normuscarone. Experiments to separate the tertiary bases instead of the quaternary bases, using (−)-di-*p*-toluenyl-D-tartaric acid afforded (−)-allomuscarine in 100% optical purity after only two recrystallizations. Separation at the norbase stage could presumably be used to obtain other stereoisomers.

Infrared spectroscopy and proton nuclear magnetic resonance spectroscopy have been used extensively, in addition to optical rotations, to characterize the stereoisomeric muscarines. The infrared spectra of the norbases of stereoisomeric muscarines as well as the infrared spectra of the numerous salts of the quaternary compounds have been reported (*58,75*). Detailed NMR analyses of the quaternary compounds and their *O*-acetyl derivatives have been carried out (*77*). Couplings constants and chemical shifts for the ring protons of various muscarines and their derivatives have been reported (*19*), as well as other measurements (*68*). The yield and some physical constants reported for stereoisomeric muscarines are shown in Table II.

VII. Synthetic Investigations of Muscarine Analogs

Interest in the pharmacological properties of muscarines and muscarones stimulated the synthesis of many analogs since these compounds afforded some insight into the chemical nature of the active sites of cholinergic receptors. As many of these compounds are structurally very different from the stereoisomeric muscarines and muscarones, their synthesis will be treated separately.

A. Structural Isomers

In search of better cholinomimetic agents, a variety of muscarine structural isomers were synthesized. The simplest one was a tetrahydrofuran (**55**) in which the 3-hydroxyl group was absent and a 2-hydroxymethyl group replaced the 2-methyl group (*78*). As might be expected, this compound did not exhibit any muscarinic activity. The preparation of **55** is shown in Scheme 16.

TABLE II

YIELDS AND PHYSICAL CONSTANTS FOR STEREOISOMERIC MUSCARINES

Compound	Optical rotation (solvent)	mp (°C)	Overall yield (%)	(Reference)
(+)-(2*S*,3*R*,5*S*)-Muscarine chloride	+8.1°(EtOH)	178–179	—	(*76*)
	+7.4°(H_2O)	179–180	⩽1	(*63*)
(+)-(2*S*,3*R*,5*S*)-Muscarine tetraphenylborate	+6.5°(Acetone)	174	⩽1	(*63*)
(−)-(2*R*,3*S*,5*R*)-Muscarine chloride	−8.4°(EtOH)	179–180	—	(*76*)
	−7.6°(H_2O)	180	<1	(*64*)
	−7.0°(H_2O)	177	<1	(*65*)
(−)-(2*R*,3*S*,5*R*)-Muscarine tetraphenylborate	−8.2°(Acetone)	191	<1	(*64*)
(−)-(2*R*,3*S*,5*R*)-Muscarine iodide	−7.8°(H_2O)	144	<1	(*65*)
(−)-(2*R*,3*S*,5*R*)-Muscarine tetraphenylborate	−5.3°(Acetone)	173–174	<1	(*65*)
(+)-(2*R*,3*S*,5*R*)-Muscarine	+20.5°(EtOH)	—	<1	(*65*)
	+11.9°(EtOH)	—	—	(*76a*)
(−)-(2*R*,3*R*,5*R*)-Epimuscarine iodide	−38.9°(H_2O)	—	—	(*76a*)
(−)-(2*R*,3*R*,5*R*)-Epimuscarine	−53.5°(EtOH)	—	—	(*76a*)
(−)-(2*S*,3*R*,5*R*)-Allomuscarine iodide	−37.4°(H_2O)	—	—	(*76a*)
(−)-(2*S*,3*R*,5*R*)-Allomuscarine	−39°(EtOH)	—	—	(*76a*)
(+)-(2*S*,3*S*,5*R*)-Epiallomuscarine iodide	⩾0°(H_2O)	—	—	(*76a*)
(+)-(2*S*,3*S*,5*R*)-Epiallomuscarine	+16.7°(EtOH)	—	—	(*76a*)
(+)-(2*S*,3*R*,5*S*)-Muscarine chloride	+8.3°(EtOH)	182–183	5	(*25*)
(−)-(2*R*,3*S*,5*R*)-Muscarine iodide	−6.1°(H_2O)	145–146	4	(*53*)
(+)-(2*S*,3*R*,5*S*)-Muscarine iodide	+5.9°(H_2O)	145–147	3	(*53*)
(+)-(2*R*,3*S*,5*S*)-Allomuscarine iodide	+36.8°(H_2O)	127–128	5	(*53*)
(−)-(2*S*,3*R*,5*R*)-Allomuscarine iodide	−37.7°(H_2O)	125–126	<1	(*53*)
(+)-(2*S*,3*S*,5*R*)-Epiallomuscarine iodide	⩾0°(H_2O)	194–195	8	(*67*)
(+)-(2*S*,3*R*,5*S*)-Muscarine iodide	+3.1°(EtOH)	148–150	3	(*71*)
(+)-(2*R*,3*S*,5*S*)-Allomuscarine iodide	+2.9°(EtOH)	131–132	3	(*71*)
dl-Allomuscarine iodide	—	171	<1	(*69*)
dl-Epiallomuscarine iodide	—	186–187	<1	(*69*)
dl-Muscarine iodide	—	108–109	—	(*58*)
dl-Muscarine chloride	—	147–148	—	(*58*)
dl-Epimuscarine chloride	—	139.5–140.5	—	(*58*)

TABLE II (*Continued*)

Compound	Optical rotation (solvent)	mp (°C)	Overall yield (%)	(Reference)
dl-Epimuscarine iodide	—	130–131	—	(*58*)
dl-Allomuscarine iodide	—	131–132	—	(*57*)
dl-Allomuscarine chloride	—	153–154	—	(*57*)
dl-Epiallomuscarine iodide	—	159–160	—	(*75*)
dl-Muscarine iodide	—	116–118	<1	(*70*)
dl-Allomuscarine iodide	—	131–132	<1	(*70*)
dl-Muscarine iodide	—	109–110	14	(*74*)
dl-Muscarine chloride	—	146–148	26	(*73*)

The different stereoisomers of 3-(dimethylaminomethyl)-4-methyl-2-tetrahydrofuran-3-ol were prepared by Colovray and co-workers (*79*) from the Mannich reaction of 2-methyltetrahydrofuran-3-one or by amidification of 2-methyl-4-carbethoxytetrahydrofuran-3-ol (Scheme 17). Reduction of the resulting furanone (**56**) afforded a mixture of norbases (**57a–d**) which was separated using different chromatographic techniques.

In order to assess the importance of the 2-methyl group in determining muscarinic activity, Eugster and co-workers (*80*) synthesized desmethylnormuscarine by the route shown in Scheme 18. The methyl ester of 4-oxatetrahydro-2-furoic acid (**58**) was prepared by the cyclization method shown in Scheme 18. The keto group was protected by 1,3-dioxolane formation with ethylene glycol and the desmethylnormuscarone was obtained via a series of standard reactions. Reduction of desmethylnormuscarone (**59**) with potassium borohydride resulted in a mixture of *cis*- (**60a**) and *trans*-desmethylnormuscarines (**60b**) which were separated by fractional distillation and chromatography on alumina.

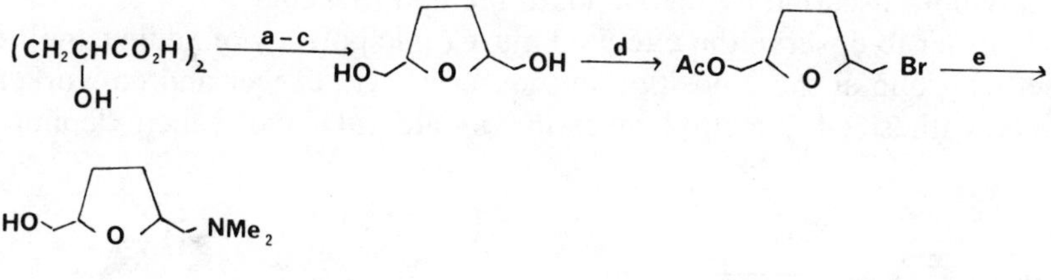

SCHEME 16. Reagents: (a) H_2O, 250°C; (b) EtOH, H_3O^+; (c) $LiAlH_4$; (d) AcOH, HBr; (e) Me_2NH.

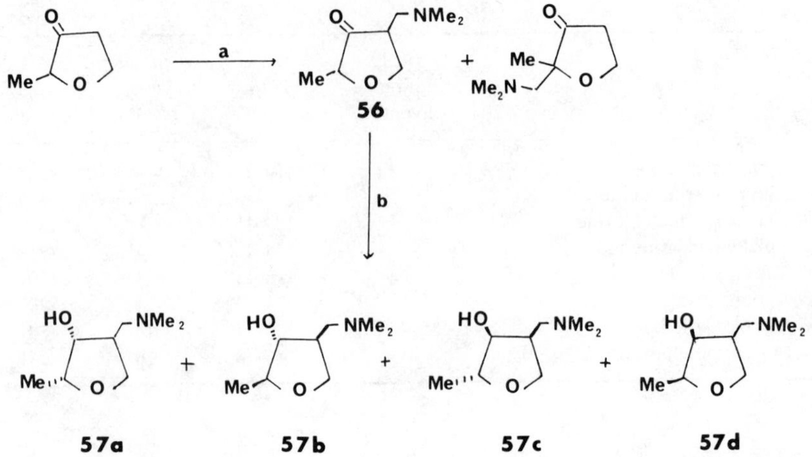

SCHEME 17. Reagents: (a) Me_2NH, HCHO, H_3O^+; (b) $LiAlH_4$.

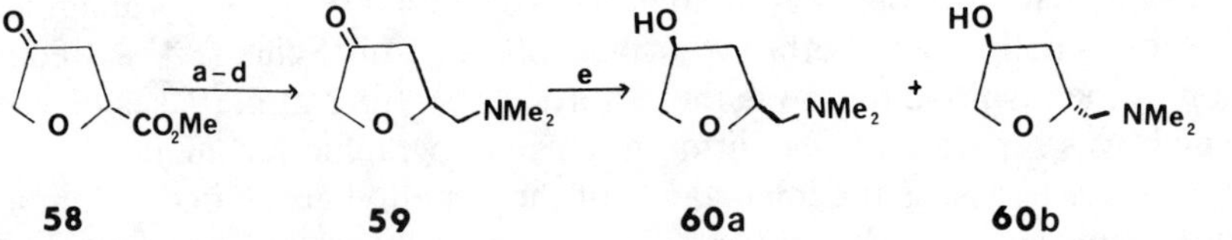

SCHEME 18. Reagents: (a) $(CH_2OH)_2$, *p*-TsOH; (b) Me_2NH; (c) $LiAlH_4$; (d) 1 *N* H_2SO_4; (e) KBH_4.

Matsumoto *et al.* (*81,82*) also prepared both *cis*- and *trans*-2-desmethylmuscarine by a modified procedure they had employed to prepare *dl*-muscarine and *dl*-allomuscarine. Ethyl α-allyl malonate was used instead of the ethyl α-crotyl malonate employed in the previous investigations. The resulting stereoisomeric 4-hydroxytetrahydrofuran-2-carboxylic acids were separated and converted to *cis*- and *trans*-desmethylnormuscarine by the standard method (Scheme 19).

In order to observe the effect on muscarinic potency of adding another methyl group at the 2 position of muscarine, Hardegger and co-workers (*61*) synthesized 2-methylmuscarine iodide (**61**) and 2-methylepimus-

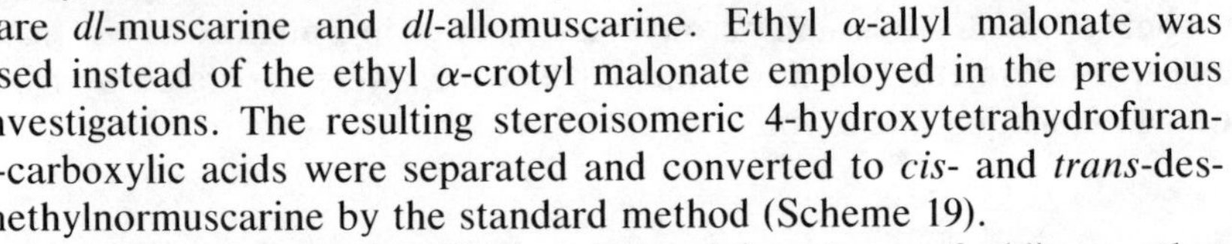

Reagents: (a) H_2, Raney Ni; (b) Me_2NH; (c) $LiAlH_4$; (d) MeI.

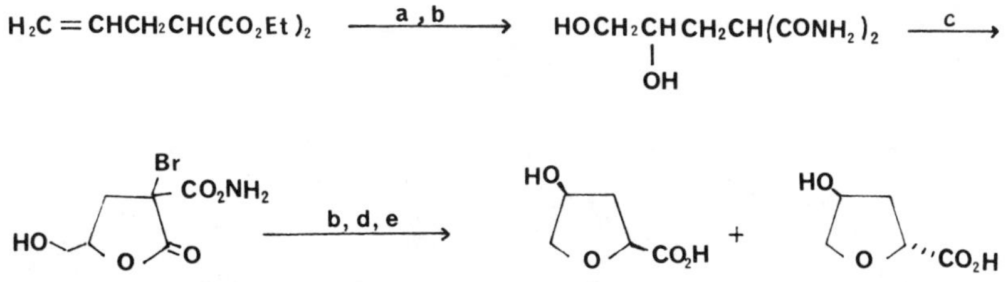

$H_2C{=}CHCH_2CH(CO_2Et)_2 \xrightarrow{a,b} HOCH_2CH(OH)CH_2CH(CONH_2)_2 \xrightarrow{c}$

$\xrightarrow{b,d,e}$

SCHEME 19. Reagents: (a) HCO_3H; (b) NH_3; (c) Br_2, AcOH; (d) 10% aq NaOH; (e) H_2O, Δ.

carine iodide (**62**). The key intermediate (**14b**) was prepared as shown in Scheme 3.

The synthesis of **61** and **62** was also carried out by Joullié and co-workers (*62*) (Scheme 20). The cyclization procedure described in Scheme 4 to prepare 2-methylmuscarone iodide was also used to synthesize 2-methylmuscarine iodide. The protected keto group (**18**) was deblocked, reduced, and acetylated to yield the corresponding acetate as a 1:1 mixture of cis and trans isomers (**63**). Ozonolysis of **63** followed by Jones oxidation of the aldehyde afforded the carboxylic acid which was converted to amide **64**. Reduction of the individual isomers followed by quaternization gave 2-methylmuscarine iodide and 2-methylepimuscarine iodide, respectively.

A general scheme for the synthesis of muscarine analogs having two alkyl groups at the 2 position or for the synthesis of novel spiromuscarine analogs was developed by Joullié and co-workers (*83*) (Scheme 21). The spiromuscarine analog **65** (R^1, R^2 = C_5H_{10}) was prepared from β,γ-unsaturated hydroxyketone **66** in ten steps via furanone **67**, which was obtained by cyclization of **66** with phenylselenenyl chloride.

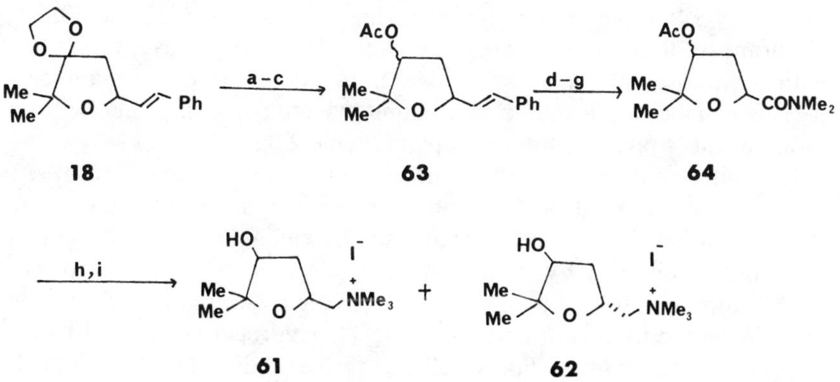

18 $\xrightarrow{a-c}$ **63** $\xrightarrow{d-g}$ **64** $\xrightarrow{h,i}$ **61** + **62**

SCHEME 20. Reagents: (a) H_3O^+; (b) $NaBH_4$; (c) Ac_2O, DMAP; (d) O_3; (e) Jones oxidation; (f) $(COCl)_2$; (g) Me_2NH; (h) $LiAlH_4$; (i) MeI.

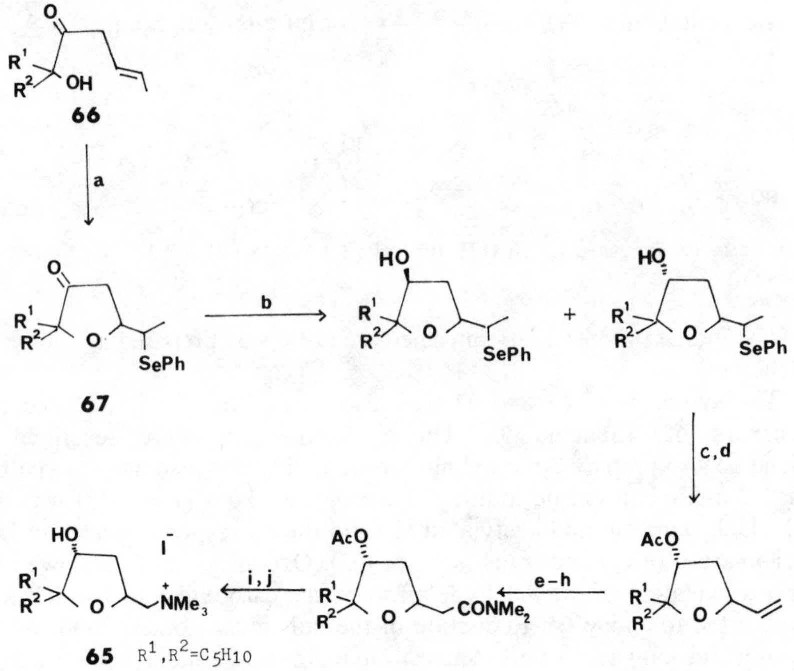

66

67

65 $R^1,R^2=C_5H_{10}$

SCHEME 21. Reagents: (a) PhSeCl, Et_3N; (b) $NaBH_4$; (c) Ac_2O, DMAP; (d) O_3, −78°C, Et_3N, CH_2Cl_2; (e) O_3, MeOH, −78°C, Me_2S; (f) Jones reagent; (g) $(ClCO)_2$; (h) Me_2NH; (i) $LiAlH_4$; (j) MeI.

As previous changes incorporated in the muscarine structure dealt mostly with variations at the 5-quaternary ammonium group or at the 2-methyl group, Wang and Joullié chose to introduce variations at C-3 and C-4 (*67*). They synthesized (2*S*,3*R*,5*S*)-isoepiallomuscarine (**68**) and (2*S*,3*S*,4*S*,5*S*)-3-hydroxyepiallomuscarine (**69**) (Schemes 22 and 23).

Furanose **70**, obtained in 64% overall yield from D-glucose, was treated with sodium selenophenolate to give diselenide **71** which was converted to D-isoepiallomuscarine, a compound which possesses the hydroxyl group at the 4 position, in six steps (Scheme 22).

Key intermediate **72** in the synthesis of **69** was prepared from α-D-glucose in 36% overall yield. Lithium aluminum hydride reduction opened the oxetane ring at C-6 to afford alcohol **73** which, after benzylation and hydrolysis, followed by oxidation, gave the corresponding carboxylic acid. The resulting carboxylic acid was treated with oxalyl chloride and dimethylamine to afford dimethylamide **74**. Hydrogenation of **74** yielded the corresponding *trans* diol which was further reduced with lithium aluminum hydride and quaternized with methyl iodide to afford D-3-hydroxy-epiallomuscarine (Scheme 23).

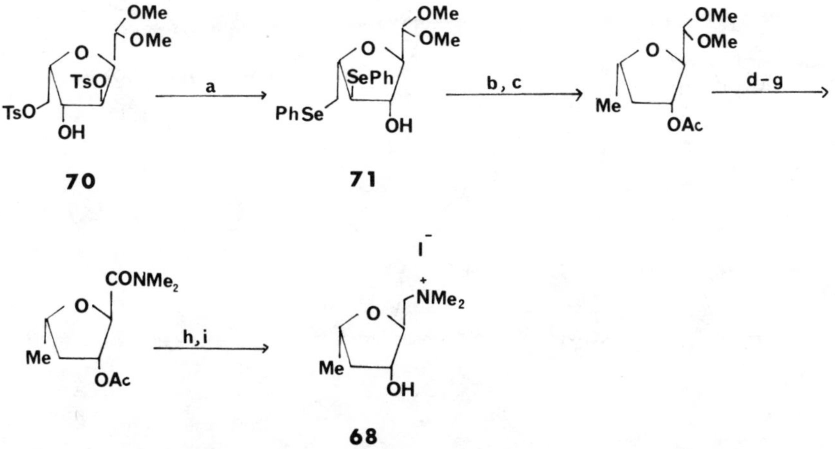

SCHEME 22. Reagents: (a) NaSePh; (b) Raney Ni; (c) Ac_2O, DMAP; (d) MeOH, 2% HCl; (e) Jones reagent; (f) $(ClCO)_2$; (g) Me_2NH; (h) $LiAlH_4$; (i) MeI.

cis-Bisquaternary ammonium salts (**75**) were prepared from D-chitaric acid (**76**) by the sequence shown in Scheme 24. The corresponding trans derivatives (**77**) were synthesized from 2,5-anhydro-D-mannosaccharic acid (**78**). The basic side chains were introduced by conventional methods (*84*).

From 2-methyl-4,5-dicarbethoxy-3-oxotetrahydrofuran, Hardegger and Halder prepared a bicyclic derivative of muscarine (**79**) which was found

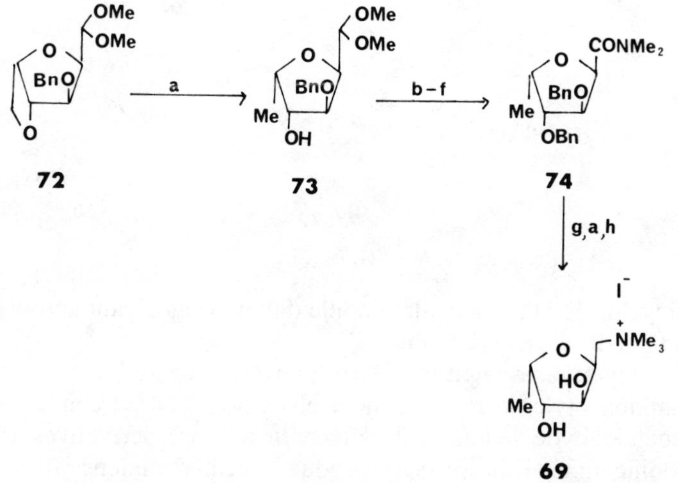

SCHEME 23. Reagents: (a) $LiAlH_4$; (b) NaH, BnCl; (c) H_3O^+; (d) Jones reagent; (e) $(ClCO)_2$; (f) Me_2NH; (g) H_2, Pd/C; (h) MeI.

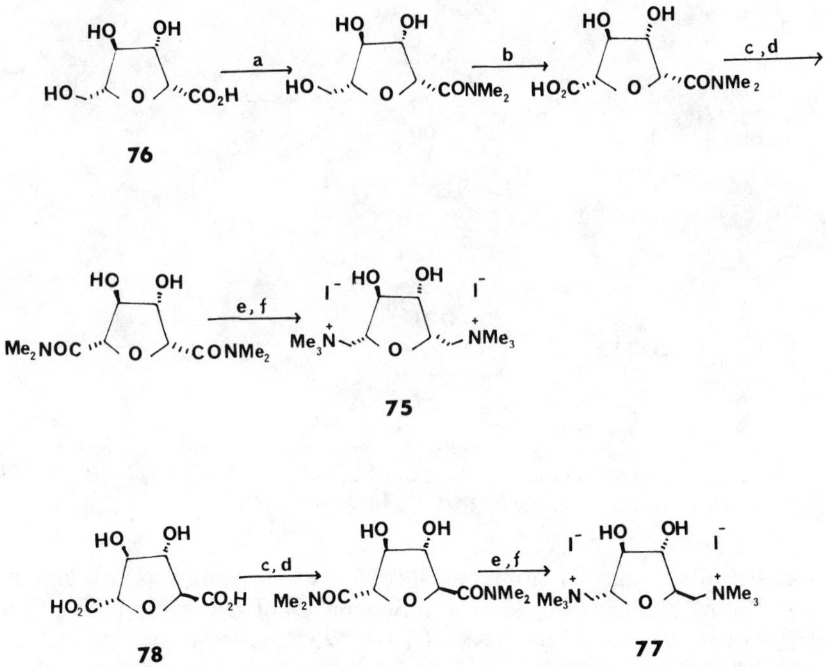

SCHEME 24. Reagent: (a) Me_2NH; (b) HNO_3; (c) CH_2N_2; (d) Me_2NH; (e) $LiAlH_4$; (f) MeI.

to possess muscarinic activity (*85*). Muscarine analogs in the 7-oxabicyclo[2.2.1]heptane series were prepared by Nelson and co-workers as rigid desoxymuscarine analogs. One of them, *endo*-2-dimethylaminomethyl-

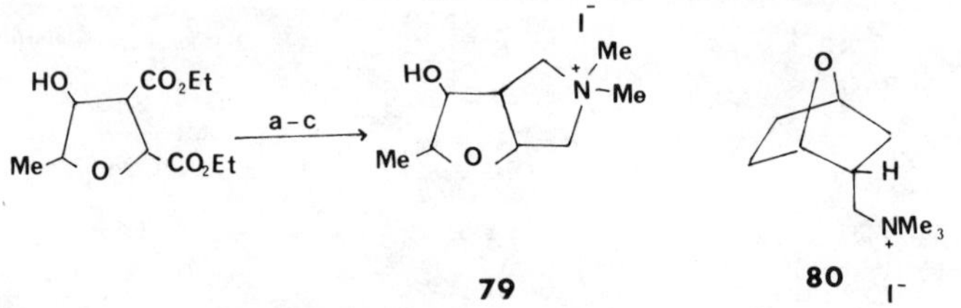

Reagents: (a) Me_2NH; (b) $LiAlH_4$; (c) MeI.

7-oxabicyclo[2.2.1]heptane methiodide (**80**) had significant activity, being 0.5% as potent as acetylcholine (*86*).

New compounds related to muscarine were prepared by cyclization of some aminoacetylenic glycols, themselves prepared by condensation of aminoacetaldehyde hemiacetal salts with alkynol derivatives (Scheme 25). Modification of the primary products of this condensation afforded muscarine analogs **81**,**82**,**83** (*87*).

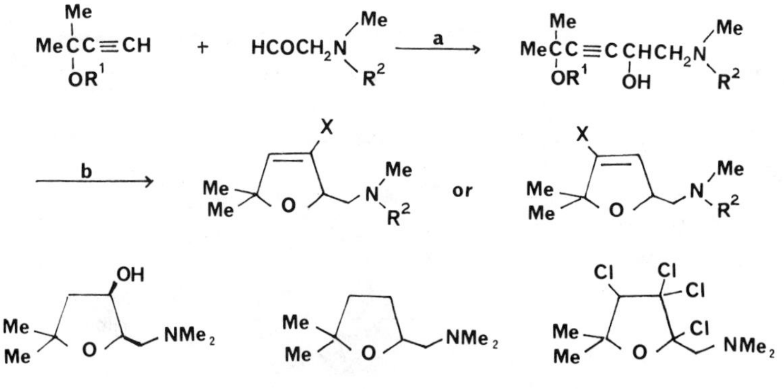

SCHEME 25. Reagents: (a) $NaNH_2$; (b) $HgSO_4 - H_2SO_4$ ($X = OH^-$); HCl (X = Cl); HBr (X = Br); HI (X = I).

B. TETRAHYDROTHIOPHENE ANALOGS

Substitution of oxygen by sulfur causes profound changes in bond angles, bond lengths, and electronic character. Such changes afford valuable data on the nature and topography of a receptor's active sites. Therefore, the synthesis of stereoisomeric thiomuscarines was undertaken by Eugster as early as 1962 (*88*). Although separation of the pure diastereomers was not achieved, the relationship of the C-3 and C-5 substituents were established. These synthetic studies, shown in Scheme 26, began with thiophanone dicarboxylate (**84**) that was prepared by the method of Fiesselmann and Schipprak (*89*). Hydrolysis and decarboxylation of the β-ester function afforded compound **85** which was protected as a dimethylacetal (**86**) during the conversion of the α-ester group to the dimethylaminomethyl group by conventional methods. Reduction of **87a** with lithium aluminum hydride afforded two epimeric norbases in a 3:7 ratio. Reduction of **87b** afforded two epimers in a 1:1 ratio. The relative configuration of these products was not established, but they were believed to have the more stable trans arrangement of the C-2 and C-5 substituents and were assigned the epiallo and allo configurations respectively (*88*).

In the muscarine series, retention of configuration at C-2 and C-5 of the norbases during hydride reduction of the corresponding 3-oxo compounds has been established. Enolization of C-2 does not occur under the reaction conditions (*90*). In the thiomuscarine series, the sulfur heteroatom probably increases the acidity of the hydrogen on C-2 so that facile epi-

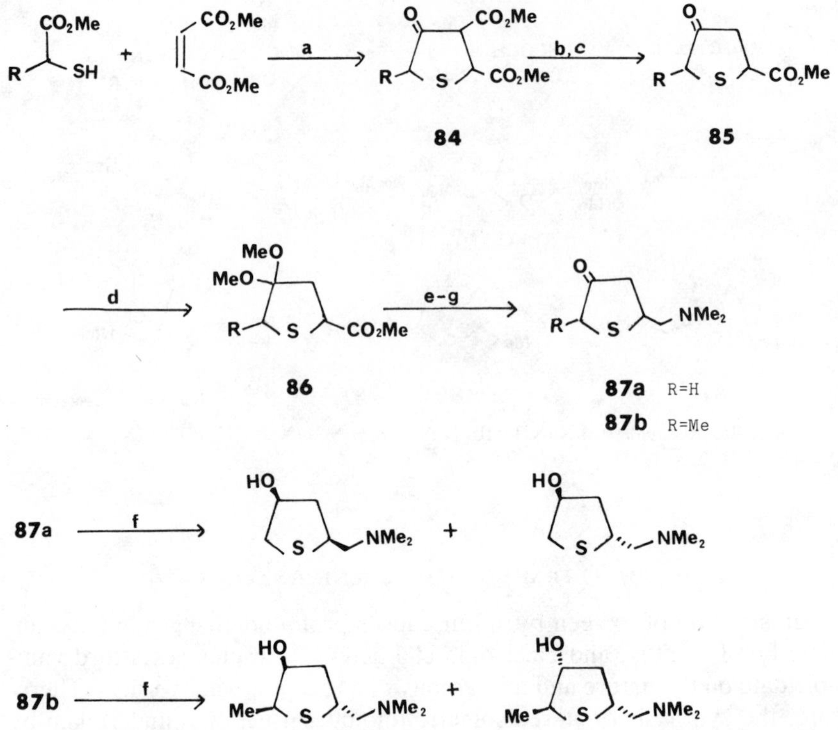

84 85

86 87a R=H
87b R=Me

SCHEME 26. Reagents: (a) $NaOCH_3$; (b) 10% H_2SO_4; (c) MeOH, H_3O^+; (d) $CH(OMe)_3$; (e) Me_2NH; (f) $LiAlH_4$; (g) H_3O^+.

merization of the oxo compounds (**88**) under these conditions is observed even at room temperature (*91*).

Because the mixture of 3-oxo compounds (**89**) after lithium aluminum hydride reduction yielded only three diastereomers (the epiallo series missing), the keto amides (**90**) were used as starting materials. The pure trans isomer (**90a**) was easily obtained by crystallization from the mixture while the cis isomer (**90b**) was isolated by chromatography with difficulty. Sodium borohydride reduction of the keto amides (**90**) afforded four diastereomeric hydroxy amides (**91a–d**) which were separated by chromatography. The pure compounds were isolated, reduced to the corresponding norbases (**92a–d**), and finally converted to the crystalline thiomuscarine iodides (**93a–d**) (Scheme 27) (*91*).

The assignment of the relative configurations was achieved with ^1H-NMR lanthanide induced shift experiments on the racemic norbases. The distribution of the reduction products (**91a–d**) suggested a favored attack of the hydride reagent trans to the amide group at low temperatures,

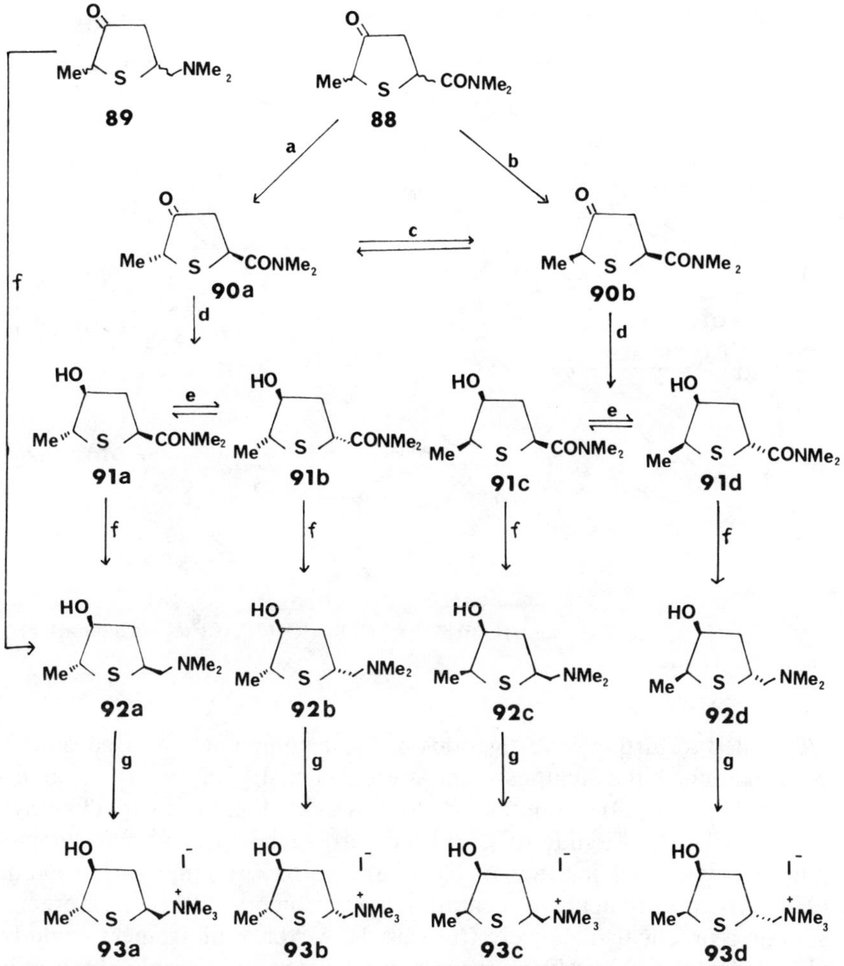

Scheme 27. Reagents: (a) crystallization from cyclohexane; (b) SiO_2 chromatography; (c) base, room temperature; (d) $NaBH_4$; (e) elevated temperature, base; (f) $LiAlH_4$; (g) MeI.

followed by epimerization at C-5 at higher temperatures. The epimerization provided a good route to preparation of isomers **91c** and **91d** from **91a** and **91b**.

C. 1,3-Dioxolane Analogs

The activity of the quaternary ammonium salts of 1,3-dioxolane derivatives of muscarine (oxamuscarines), discovered by Fourneau in 1944

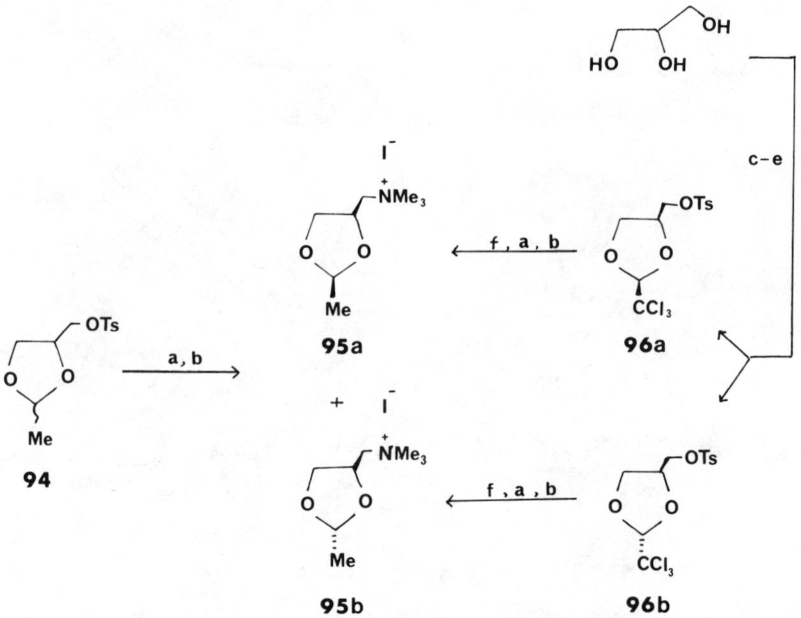

SCHEME 28. Reagents: (a) Me_2NH; (b) MeI; (c) CCl_3CHO; (d) TsCl; (e) separation; (f) H_2, Pd/C.

(*40,41*) led to further investigations of these compounds by Belleau (*92*) who examined the synthesis and stereochemistry of several oxamuscarines (Scheme 28). Monotosylate **94** was treated, in turn, with dimethylamine and methyl iodide to afford the corresponding quaternary ammonium salt as a 3:2 mixture of *cis*-(**95a**) and *trans*-oxamuscarine iodide (**95b**). Attempts to achieve separation of the isomers of **95** or their precursors on a practical scale were fruitless. However, both isomers could be obtained in pure forms by an indirect synthetic route. Fractional crystallization of the tosylate prepared from glycerol and chloral afforded pure cis (**96a**) and trans isomers (**96b**). Catalytic reduction converted the trichloromethyl group to a methyl group and the tosyl group was functionalized to a quaternary ammonium group. Pharmacological tests showed that the cis form had stronger activity than the trans.

The synthesis of optically active compounds was undertaken by both Harper *et al.* (*93*) and Belleau and Puranen (*94*). The first investigators used D-(+)- and L-(−)-glyceraldehyde as the precursors and converted these via reductive amination to enantiomeric dimethylaminopropanediols that were treated with acetaldehyde to afford a mixture of diastereomeric norbases (**97a–d**). After separation, these compounds were converted to the corresponding quaternary ammonium salts (**98a–d**). Belleau and Puranen (*94*) started from D-(−)-isopropylidene glycerol and converted it by previously described procedures to **98a**. Functionalization of the primary hydroxyl group in the optically active D-glycerol derivative afforded the L-(+)-enantiomer (**98d**) (Scheme 29).

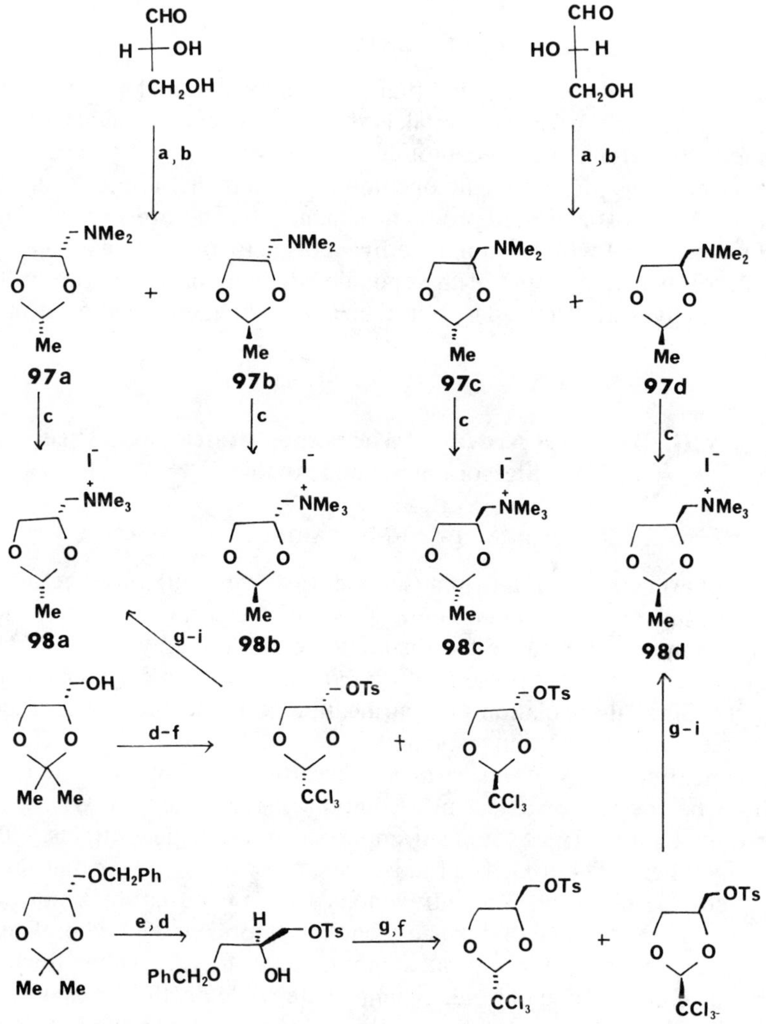

SCHEME 29. Reagents: (a) reductive amination; (b) MeCHO; (c) MeI; (d) TsCl; (e) H_3O^+; (f) CCl_3CHO; (g) H_2, Pd/C; (h) Me_2NH; (i) MeI.

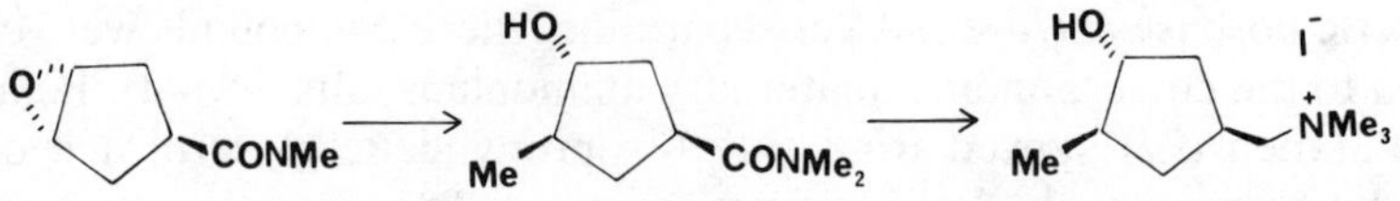

100 → 101 → 99

SCHEME 30. Reagents: (a) $LiCu(Me)_2$; (b) $LiAlH_4$; (c) MeI.

D. Cyclopentyl Analogs

Givens and co-workers found that the cyclopentyl analog (**99**) of muscarine (desether muscarine) parallels the activity and specificity of muscarine at the cholinergic receptor (*95*). The synthetic entry to desether muscarine is the stereospecific opening of a symmetrical intermediate to give a 1,2-trans-substituted product (Scheme 30). The *trans* epoxy amide (**100**) was treated with lithium dimethyl cuprate to give amido alcohol **101** via stereospecific opening of the epoxide. Reduction of **101** with lithium aluminum hydride followed by quaternization with methyl iodide afforded the desether muscarine (**99**).

VIII. Biological Activity of Muscarines, Muscarones, Their Stereoisomers, and Analogs

A. L-(+)-Muscarine

The pharmacology of L-(+)-muscarine was investigated well before the pure substance was available (*96*). This early experimental work, however, is unreliable because the results were due not only to the alkaloid but to other substances present in *Amanita muscaria,* in particular acetylcholine. Since the isolation and purification of **1**, the early observations were reevaluated and reinvestigated.

The pharmacology of muscarines, muscarones, and other related compounds has been the subject of a comprehensive review by Waser (*18*), and only a summary of the most important physiological effects will be described here. The toxicity of muscarine chloride is greater that that of acetylcholine chloride. The intravenous LD_{50} of muscarine chloride in mice is 0.23 mg/kg, and that of acetylcholine chloride is 33 mg/kg. Signs of poisoning are similar for both substances, but at the LD_{50} value, the lethal effect of acetylcholine chloride is immediate while death from muscarine takes 3–10 min. Although muscarine is stable to pepsin, **1** showed no oral activity in monkeys even at doses many times greater than those that cause poisoning in humans.

Muscarine chloride is approximately four times more potent than acetylcholine chloride in lowering blood pressure in rabbits, cats, and dogs. The heart rate is slowed considerably, and cardiac arrest will occur unless the action of muscarine is antagonized by atropine. In animals previously injected with atropine, higher doses of muscarine are needed for lowering blood pressure. This finding led to the conclusion that muscarine had little or no nicotinic action. The bronchoconstrictory effects of muscarine chloride are approximately four times more pronounced than those caused by acetylcholine chloride and are prevented by atropine.

Smooth muscles are highly sensitive to the action of the alkaloids but they are affected to different degrees depending on the organ, the bladder being the most sensitive. Pretreatment with muscarine changes the reactions of the muscles to other cholinomimetic drugs. In atropinized cats, muscarine in intravenous doses up to 500 g/kg never showed a neuromuscular blocking or enhancing action on the gastrocnemius muscle stimulated through the sciatic nerve, thereby showing no activity on skeletal muscles. This lack of activity is one of the major differences between muscarine and muscarine-like substances and also choline nitrite.

Muscarine chloride has no effect on the activity of cholinesterases from different sources and does not inhibit them in concentrations up to 10^{-4} *M*. These results led to the conclusion that muscarine chloride does not produce its characteristic effects via inhibition of cholinesterases (*18*).

To summarize, muscarine chloride is a potent parasympathomimetic drug that exerts the same effects as acetylcholine and other cholinomimetic drugs. However, its action is more potent and often of longer duration than that of acetylcholine. This behavior has been attributed to the fact that muscarine has no ester linkage and therefore is not hydrolyzed by cholinesterases. Although muscarine chloride acts on the same receptors as acetylcholine chloride, its action is restricted to peripheral effector organs innervated by the autonomic nervous system. Higher doses of muscarine chloride are needed to affect autonomic ganglionic synapses. No significant action on skeletal muscles has been observed. The physiological effects of the alkaloid are lowering of blood pressure, slowing of the heart rate, miosis, salivation, and bronchoconstriction. The cholinomimetic effects of muscarine are antagonized by atropine.

B. Muscarine Stereoisomers and Tetrahydrofuran Derivatives

Although the actions of muscarine, muscarone, their isomers, and a large number of muscarine analogs have been investigated, comparisons of activities are difficult because of the different animals and techniques that have been used. An excellent compilation of the most reliable data is found in the review by Waser (*18*). Both "muscarinic" and "nicotinic" actions have been examined, and all four pairs of enantiomers of muscarine have been investigated pharmacologically. The muscarinic effects of the stereoisomers of muscarine on the peripheral parasympathetic synapses are several hundred times weaker than that of the natural alkaloid.

The activity of the diastereomers varies with their configurations. Muscarinic activity appears to decrease whenever the 3-hydroxyl group is cis to either or both of the 2- and 5-substituents. Epimuscarine, in which all three substituents are on the same side of the ring, has the lowest cholinomimetic activity. Allomuscarine, in which the quaternary group is on the same side as the hydroxyl group, is somewhat more active; and epiallomuscarine, with only the relatively small methyl group on the same side as the hydroxyl group, is the most active of all the isomers. (+)-Muscarine, with its hydroxyl group trans to the other two substituents, is the most potent, and its enantiomer (−)-muscarine has only 0.1% of the activity of the natural alkaloid. The stereospecificity of the (+)-muscarine is remarkable. The correlation of activity to stereochemistry of the muscarine isomers is of great interest because their structures are rigid unlike other cholinomimetic agents, such as acetylcholine, which have flexible carbon chains and freely rotating groups. Therefore, the preferred conformations of all stereoisomers are of interest.

Differences in the anticholinesterase activities of the stereoisomers of muscarine are not as large as the differences in their muscarinic activity. It is believed that alkaloids of the natural series, muscarine and epimuscarine, have a closer approach to coplanarity of the hetero ring with both the 3- and 5-functional groups than those in the allomuscarine and epiallomuscarine series. Similar anticholinesterase activity was shown by 4,5-dehydromuscarine (**102**) and 4,5-dehydroepimuscarine (**103**) (*97*).

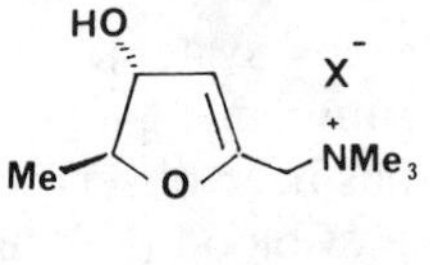

102 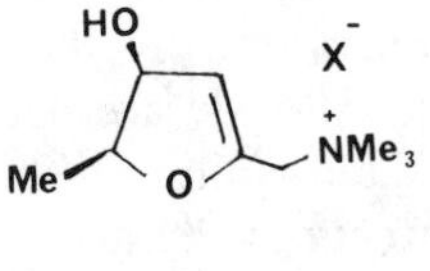 **103**

Muscarone (**104**) and allomuscarone (**105**) were highly active whereas 4,5-dehydromuscarone (**106**) showed considerably less potency. The most potent inhibitor of acetylcholinesterase was the chloride of *O*-acetylmuscarine (**107**), which is not hydrolyzed by the enzyme. These results indicate that the binding strengths of these compounds with acetylcholinesterase are the result of steric influences on the three substituted centers (C-2, C-3, and C-5) (*97,98*).

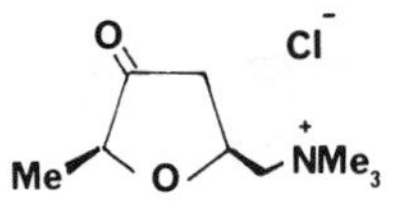

104 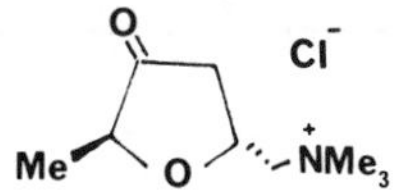 105 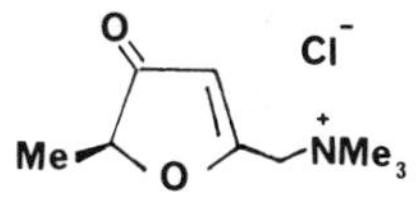 106 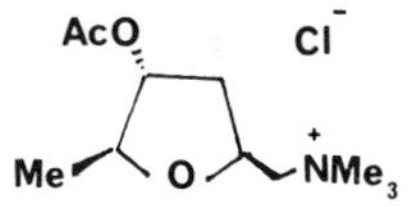 107

The atropine-like action of the stereoisomers of muscarine as their salts has also been investigated. The purely parasympathomimetic action of (±)-muscarine and (±)-epiallomuscarine was observed. (±)-Epiallomuscarine was parasympatholytic on the frog heart, and (±)-allomuscarine had a dual action, namely as a partial stimulant and a competitive antagonist with parasympatholytic properties.

Atropine-like action on the frog heart was observed with propyldesmethylmuscarine (**108**), 4,5-dehydroepimuscarine (**103**), thiomuscarine (**109**), epithiomuscarine (**110**), desmethylthiomuscarine (**111**), and desmethylepithiomuscarine (**112**) (*18,99*). The ability of these drugs to produce an effect on the more sensitive receptors of the rat intestines was the same. The affinities of the isomers to the receptors were in the order muscarine > epiallomuscarine > allomuscarine > epimuscarine.

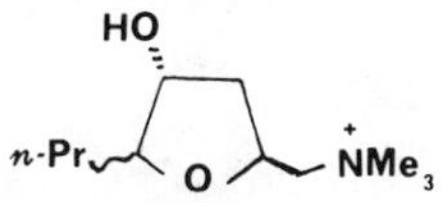

108 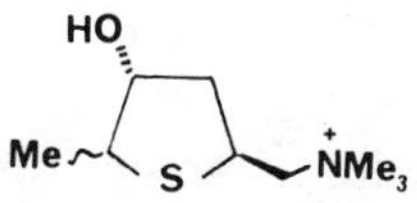 109 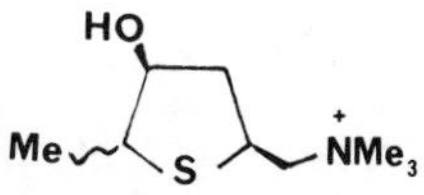 110

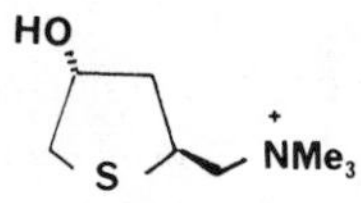

111 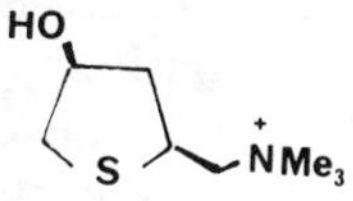 112

All muscarone derivatives resemble acetylcholine both in their structure and actions. They are more potent than **1** and its derivatives and exhibit strong stimulating and blocking effects on ganglionic synapses and neuromuscular junctions. In atropinized animals, higher doses block the sympathetic ganglia and the end plates in a curare-like fashion.

Stereospecificity is less important in the muscarone series. (+)-Muscarone and (±)-allomuscarone are only two or three times less active than (−)-muscarine and (±)-muscarone, respectively. It is noteworthy that the most active (−)-muscarone is related to the natural (+)-muscarine (*90*). All muscarine derivatives with the exception of 3-phenylmuscarine chlo-

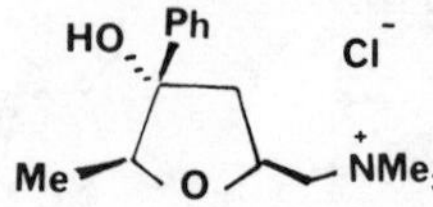

HO Ph Cl⁻ Me O NMe₃

113

ride (**113**) and the 4,5-dehydro derivatives (**102** and **103**) have muscarinic action and no nicotinic action, while all muscarone derivatives show both actions. *O*-Acetylmuscarine (**107**) and 3-phenylmuscarine (**114**) have muscarinic activity similar to that of acetylcholine on blood pressure.

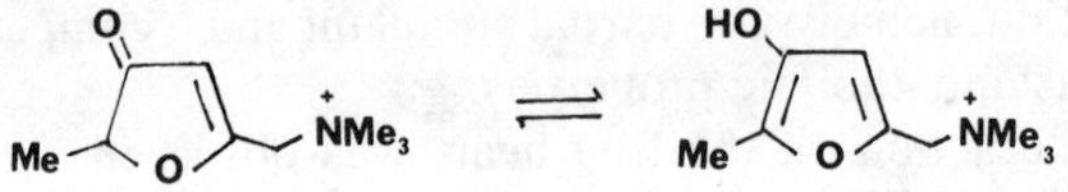

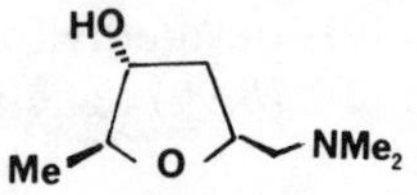

O Me O NMe₃ ⇌ HO Me O NMe₃

HO Me O NMe₂

114

Replacement of the 3-hydroxyl group by a carbonyl group as in the muscarones causes important changes in the properties of the molecule. The carbonyl group lies in the same plane as the ring and is therefore less subject to steric interferences by the substituents at the 2 and 5 positions. Enolization of muscarones under acid or basic conditions has been observed, and such a reaction is known to cause isomerization at C-2. The polarity and planarity of the carbonyl groups makes it amenable to attack by basic groups in the receptor. Such a mechanism has been proposed in the binding of acetylcholine to acetylcholinesterase.

Replacing the oxygen of the tetrahydrofuran ring by a sulfur atom resulted in a large reduction in muscarinic activity (1/1500 of the original potency). The reduction in activity is less noticeable in the muscarone compounds. The stereospecificity observed for muscarine is absent in the sulfur analog. The lesser electronegativity of sulfur may reduce the ability of the molecule to form hydrogen bonds with the receptors. The increase in ring size may also affect the fit of the molecule on the receptor. Acetylthiocholine is known to have 100 times less muscarinic activity than acetylcholine.

Removal of the 2-methyl group results in a decrease of cholinomimetic action in muscarine and to a lesser extent in muscarones. However, desmethylthiomuscarine (**111**) and desmethylepithiomuscarine (**112**) are three times more active than the corresponding methyl compounds. The presence of two methyl groups at the 2 position diminishes the activity of muscarine 3,000–10,000 times but that of muscarone only 10–50 times. Replacement of the 2-methyl group by longer alkyl chains such as propyl or isobutyl groups results in a marked decrease in activity.

The introduction of a double bond between C-4 and C-5 alters the shape of the furan ring and causes the C-5 substituent to be in a coplanar position. 4,5-Dehydromuscarine (**102**) and 4,5-dehydroepimuscarine (**103**) are almost as active at peripheral cholinoreceptive sites as muscarine. The potency of epimuscarine is increased 100-fold by the introduction of unsaturation at the C-4–C-5 position. 4,5-Dehydromuscarone is as potent as (±)-muscarone. All three dehydro compounds show a stronger nicotinic action on ganglionic and neuromuscular synapses than the corresponding saturated compounds.

Aromatization of the ring increases the electron availability in the ring but decreases that of the ring oxygen. The molecule is flattened by aromatization and all substituents become coplanar. Furan derivatives are reported to have higher muscarinic activity than the saturated analogs but equivalent nicotinic action. Enolization of 4,5-dehydromuscarone is capable of generating an aromatic ring, and keto–enol isomerism may afford the equilibrium mixture shown. While this process could cause total racemization, the enolization does not appear to have been well investigated.

The cationic part of the molecule appears to be very important for cholinergic action. Presumably the positive center of the molecule is bound to an anionic center on the receptor. Replacement of the quaternary ammonium group both in muscarine and muscarone resulted in almost total loss of activity. Normuscarine (**114**) and normuscarone (**9**) are practically devoid of cholinergic action.

Introduction of an additional quaternary ammonium group at the C-4 position of muscarone (**115**) did not increase activity. Replacement of one of the methyl groups by other groups (**116–118**) resulted in varying effects.

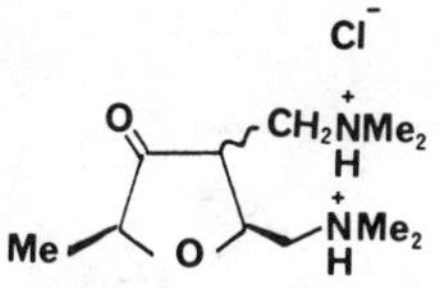

115

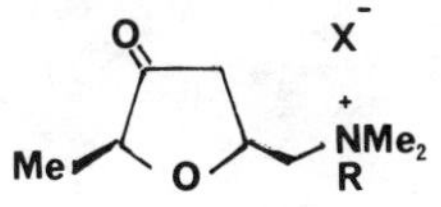

116 R = $PhCH_2CH_2$–

117 R = Allyl–

118 R = $PhCH_2$–

C. Heterocyclic Analogs

Several years before the structure of muscarine was known, the high cholinomimetic action of furans and tetrahydrofurans substituted with a methyl ammonium group at the α position of the ring was well known. When the ammonium group was substituted with three methyl groups, the activity of these compounds was comparable to that of acetylcholine, methacholine, and carbamylcholine. When larger alkyl groups were used as substituents, activity was strongly decreased. The reduced furans were somewhat less active but equally toxic. The fall of blood pressure, increased salivation, and increased tone of the intestine and bladder were overcome and prevented by atropine. The strong parasympathomimetic action of furfuryl trimethylammonium iodide (**119**) on man was investigated (*100*).

Of all the furan derivatives prepared, the most active were those in which the group at the 2 position was —H (**119**) or —CH_3 (**120**). 5-Methylfurtrethonium (**120**) was as active as acetylcholine. Moving the ammonium group from the α to the β position decreased the cholinomimetic activity. A bisquaternary derivative of tetrahydrofuran (**121**) had low muscarinic activity and only slight nicotinic activity. Other derivatives of tetrahydrofuran with two methoxyl groups at C-2 and C-5 or an ethyl group at C-2 were inactive.

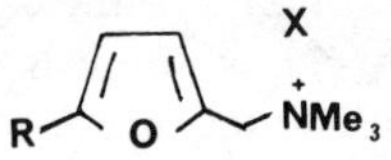

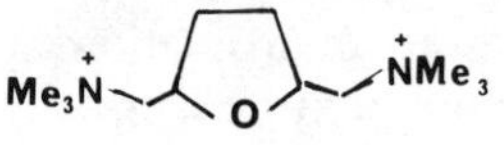

119 R = H-

120 R = Me-

121

In 1944, Fourneau described the strong muscarinic activity of acetals of propanediol with a trimethylammonium group (**97**, **98**, and **122–124**)

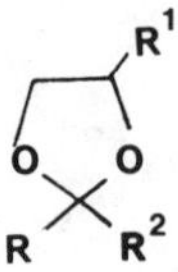

	R	R^1	R^2
122	H	H	$CH_2\overset{+}{N}Me_3$
98	H	Me	$CH_2\overset{+}{N}Me_3$
97	H	Me	CH_2NMe_2
123	CH_2NMe_3	H	H
124	CH_2NMe_3	H	Me

(*40,41*). The most active compound was the acetal of acetaldehyde (**98**), which lowered the blood pressure of dogs at smaller doses than acetylcholine and caused cardiac arrest at a dose of 1 g (*92,101*). The acetal of formaldehyde (**122**) was less potent, but less toxic, and was used clinically as a vasodilator in peripheral circulation disorders. This compound, however, also showed nicotinic actions as did all the propanediols derivatives shown (**125–127**). Lengthening the cationic side chain or introducing a phenyl group at the 2 position increased the nicotinic action, although this action was weaker than that of choline esters. The *N*-dimethyl derivative showed considerably less cholinomimetic activity than observed in the muscarine and muscarone series. Muscarinic activity was found in compounds **125–127**. Most propanediol derivatives also exhibit nicotinic action. Steric effects were found to decrease activity as was noted in other series.

The introduction of a quaternary ammonium group in five- or six-membered heterocycles (**128–137**) afforded some interesting results with re-

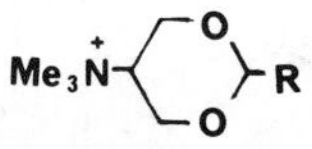

125 R = H
126 R = Me
127 R = Ph

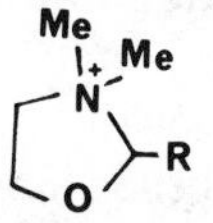

128 R = $CH_2CH(OH)Me$
129 R = CH_2CHMe_2
130 R = $CH_2CH_2CH_2Me$
131 R = $CH_2CH_2CH(OH)Me$

spect to cholinomimetic activity. Almost all quaternized oxazolidines (**128–131**) showed remarkably high, nonspecific activity (peripheral, ganglionic, or endplate). The position of the hydroxyl group in the side chain was influential on the activity (*18*).

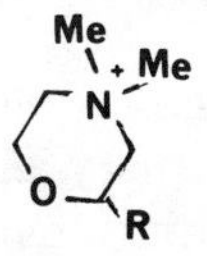

132 R = CH_2CH_2OH
133 R = CH_2CH_2Me
134 R = $CH_2CH(OH)Me$

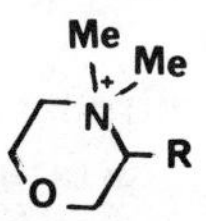

135 R = CH_2CH_2OH
136 R = $CH_2CH(OH)Me$

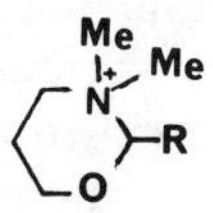

137 R = $CH_2CH(OH)Me$

A number of quaternized 1,3- and 1,4-oxazines (**132–137**) were also examined for muscarinic and nicotinic activity (*18*). Compounds with the side chain in the β position relative to nitrogen (C-2) were more active than those with the side chain in the α position (C-3). Almost all compounds showed both muscarinic and nicotinic activity, and some showed slight curare-like effects for short periods of time. Muscarinic activity was more variable than nicotinic activity. Peripheral cholinomimetic action was stronger for the five-membered heterocycles. α-Substituted 1,3-oxazines exhibited only weak muscarinic activity and some nicotinic and curare-like activities in cats.

IX. Structure–Activity Relationships in Muscarine, Its Stereoisomers, and Analogs

Before considering possible structure–activity relationships the highlights of the pharmacological findings will be summarized.

1. The enantiomers of muscarine differ greatly in their cholinomimetic properties, and the muscarinic activity is almost exclusively associated with the (+) isomer. The racemic forms of epi-, allo-, and epiallomuscarine possess only a fraction of the potency of (+)-muscarine. An enormous decrease in muscarinic activity results when the 3-hydroxyl group becomes cis to either one of the other two groups.
2. Muscarones, on the other hand, do not show the same stereospecificity as muscarines. (+)- and (−)-Muscarones and (±)-allomuscarone differ only slightly in their action on smooth muscle. In addition to their high muscarinic activity, muscarones exert effects at other synapses. They exhibit strong nicotinic activity and block ganglionic and neuromuscular transmissions. Muscarones, and in particular (−)-muscarone [which is related to (+)-muscarine], are more potent than (+)-muscarine and acetylcholine in stimulating postganglionic parasympathetic sites.
3. The introduction of a double bond at the 4,5 position has less effect on the pharmacological properties than changes in the stereochemistry of the muscarine substituents. Aromatization augments muscarinic activity but does not influence nicotinic action.
4. From the various modifications introduced in the muscarine nucleus, the following observations are noteworthy: (a) the quaternary ammonium group is essential to biological activity since the tertiary base normuscarine is totally inactive; (b) removal of the methyl group at C-2 results in a great decrease in activity; (c) the electronegative ring oxygen is essential for both muscarinic and nicotinic action since replacement of oxygen by sulfur reduces the diverse biological actions; (d) increasing the size of either the C-2 or C-5 side chain usually decreases activity; and (e) although modifications of the hydroxyl such as acetylation or introduction of a bulky group such as phenyl at the 3 position does not affect appreciably the activity of the resulting molecule, the trans arrangement of the

group at C-3 relative to the other ring substituents appears to be of utmost importance.

5. The inhibition of cholinesterase activity depends on features different from those required for muscarinic activity. Although muscarone and allomuscarone are strong cholinesterase inhibitors, the most potent inhibitor of acetylcholinesterase is (+)-*O*-acetylmuscarine. The differences in the anticholinesterase activities of the stereoisomers of muscarine are not as large as the differences in their muscarinic activity. Alkaloids of the natural series, muscarine and epimuscarine, where the 2 and 5 substituents are cis, appear to have a closer approach to coplanarity of the ring with both the 5-quaternary nitrogen atom and 3-hydroxyl group than those in the allo and epiallo series.

A comparison of the actions of different types of compounds with regard to their binding on a receptor area requires similar physiochemical properties. The analogs and stereoisomers of muscarine examined were all water soluble as expected of ammonium salts and they all had similar molecular weights. The shape of a molecule is of utmost importance in understanding the fit of a compound to a given receptor, and the large differences in activity of the various structural modifications investigated illustrates this point.

Early studies made use of the so-called 5-atom rule (*102,103*). This rule states that in any homologous series of parasympathomimetic drugs of general formula $RN(CH_3)_3$, the most active member should be the one that contains a five-atom chain in the R group (*104*). There are few exceptions to this rule, and it has held fairly well for most of the compounds examined, although no explanation has been advanced for this empirical finding. The exceptions noted might be due to steric effects, which are known to interfere with the formation of drug–receptor complexes.

Specific actions have been attributed to characteristic features. For instance, the carbonyl group is believed to be responsible for both muscarinic and nicotinic activity. Furan derivatives without the carbonyl group have low nicotinic potency. The charge distribution in the ring is believed to be of great importance, the saturated tetrahydrofuran ring only has *p*-electrons while the carbonyl and double bonds have π electrons. Charge distribution in the ring and dipolar forces are believed to be responsible for nicotinic action.

All compounds that contain trimethylammonium methyl side chains do not show a nicotinic effect on ganglia or neuromuscular synapses. On the other hand, compounds containing a carbonyl group (muscarones and most of their derivatives) exhibit even stronger muscarinic action and less stereospecificity. As the strongest effects of both muscarines and muscarones are muscarinic, their active structural features must act on the same cholinergic receptors and should therefore have similar interatomic distances. The conformations of muscarine and muscarone were examined in light of the position the active groups must have for fixation on muscarinic receptors.

Waser developed a theory that was based on the following observations (*18*). 3-Phenylmuscarine (**113**) had a high muscarinic effect but low nicotinic effect while 2-methylmuscarine and 2-methylepimuscarine had no peripheral cholinergic effect. These results were interpreted to mean that steric hindrance of the ring oxygen was more influential in decreasing activity than steric hindrance at the 3 position. In contrast, introduction of an additional methyl group in muscarone or removal of the 2-methyl group did not appear to be of great importance and was taken to mean that the 2 position in muscarone was of minor importance for bonding to the receptor. Although thiomuscarine (**109**) and desmethylthiomuscarine (**111**) have no muscarinic activity, thiomuscarone and desmethylthiomuscarone exhibit muscarinic activity 20–40 times less than that of muscarone and similar to that of desmethylmuscarone.

The above results led to the assumption that muscarinic action must occur via hydrogen bonding to the electronegative oxygen of the heterocycle. The 3-hydroxyl group and the 2-methyl group when in a trans arrangement are possibly held by hydrogen bonding to the receptor and thereby account for the high stereospecificity found in the muscarines. Steric effects that perturb hydrogen bonding to the ether oxygen decrease the muscarinic effect while steric hindrance at the 3-hydroxyl group does not much change the activity. Bonding to a nicotinic receptor does not seem possible. With muscarone, steric hindrance at the ether oxygen or replacement of oxygen by sulfur has little influence on the muscarinic effect because the hydrogen bonding now occurs preferentially on the oxygen of the carbonyl group.

Therefore, muscarinic action was believed to be due to ionic binding of the quaternary nitrogen group and to hydrogen bonding to the ether oxygen. Since (+)-muscarine and (−)-muscarone exhibit the greatest muscarinic effect, it was believed by Waser that their responsible functional groups must fit on the same cholinergic receptors and must therefore possess similar atomic distances. This author calculated that the maximum distance of the quaternary nitrogen to the ether oxygen of the muscarine was 3.8 Å, while in muscarone the minimum distance between the quaternary nitrogen and the carbonyl oxygen was 4.2 Å. Waser used this conformation to explain the higher activity of D-(−)-muscarone by assuming that the C-2 and C-5 substituents were in a similar steric position relative to the carbonyl group as they were to the ether oxygen in the highly active L-(+)-muscarine. However, such an argument is no longer valid since it was demonstrated by Eugster that (−)-muscarone has identical chirality at C-2 and C-5 as natural (+)-muscarine (*90*). The low stereospecificity of muscarone could not be explained by these conformational arguments.

Another picture of the cholinergic receptor, proposed by Beckett and co-workers (*105*), was based on a three-point attachment of muscarine with an interatomic distance of 3 Å between the quaternary nitrogen and ether oxygen. This picture does not account for the lack of stereospecificity of muscarones as compared to muscarines or for the nicotinic action of the muscarones. Waser proposed that maximal muscarinic action depended on the quaternary ammonium group as a cationic center and either an ether oxygen or a carbonyl group as a nucleophilic center. Nicotinic activity was believed to depend on the polarizability of the nucleophilic part of the molecule (carbonyl group) and another electron rich center (oxygen, double bond, or aromatic ring) in a symmetrical position relative to the quaternary nitrogen (*98*).

An X-ray structure analysis of 2-ethoxyethyltrimethyl ammonium chloride, a cholinergic agonist, showed that in the crystalline form, four conformations of the cation may occur. The anions were arranged stereospecifically with respect to the tetraalkylammonium group. This investigation of a cholinergic neurotransmitter showed that the anions could be used as a model for the anionic bonding site of the receptor. Furthermore, it was found that a triangle formed by the nitrogen of the quaternary ammonium group, the ether oxygen, and an anion occupying a specific tetrahedral face of a $Me_3\overset{+}{N}CH_2$ group, is characteristic of muscarinic activity. This arrangement, as shown in Fig. 4 for muscarine itself, is termed an "activity triangle" (*106*).

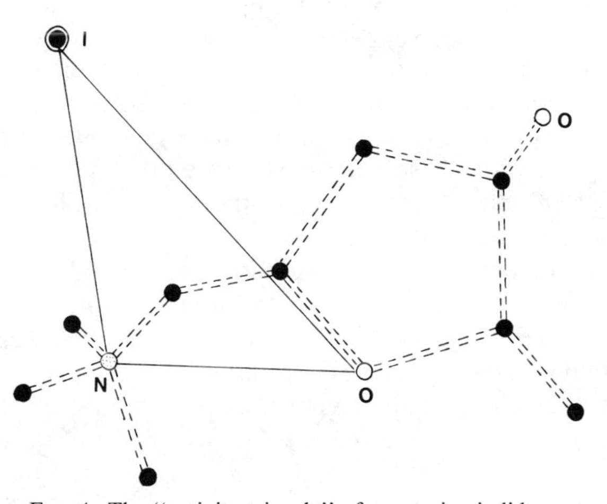

FIG. 4. The "activity triangle" of muscarine iodide.

X. Biosynthesis of Muscarine

Early investigations of the biogenesis of muscarine were based on the assumption that this compound was formed via cyclization and reduction of an hexose (*107*). Muscarine and the related muscaridine (**138**) incorporated the intact carbon skeleton of glucose.

138

However, the hexose precursor would have to be an L-hexose to afford L-(+)-muscarine, even though no sugars of the L series have been detected in fly agaric. Although *Amanita muscaria* does produce several ammonium compounds, conversion of an aldehyde group into a trimethylammonium function has not been observed often in other systems. Incorporation experiments with marked D-glucose on *Inocybe* species, which produce relatively larger amounts of muscarine, produced very low insertion rates. In addition, muscarine from either *Amanita* or *Inocybe* species is accompanied by varying amounts of stereoisomers. Therefore, the fungal organism would have to epimerize the hexose or different sugars should be involved for each stereoisomer. Up to date, all the stereoisomeric muscarines isolated from fungi have shown the same chirality at C-2 (2*S*), and the only muscarine yet to be isolated from natural sources is epiallomuscarine.

If the biogenesis of muscarine is independent of its source, it appears more probable that the biosynthetic precursors are smaller components than sugars. During investigations of the distribution and relative concentrations of muscarines in *Inocybe* species, Eugster and Catalfomo found that *I. cinnamomea* A. H. Smith, *I. geophylla* Karst., and *I. lacera* (Fr.) Quél. contained concentrations of epimuscarine equal to or higher than muscarine itself (*8*). These authors suspected that any precursor to muscarine must be able to accommodate the stereospecific configuration of the existing isomers.

Later, the incorporation of distribution of several ^{14}C-labeled simple compounds into muscarine with mycelial cultures of *Clitocybe rivulosa* was investigated (*108*). The carbon atoms of pyruvate were found to be incorporated into the muscarine at positions 2 and 3 of the ring and at the methyl at C-2. The carbons at positions 2, 3, and 4 of glutamate were incorporated into the muscarine at positions 4 and 5 of the ring and at the methylene at C-5; the carbons at positions 1 and 5 of the glutamate are lost during biosynthesis. Therefore, the same investigators concluded that the biosynthesis of muscarine did not take place through a hexose but rather

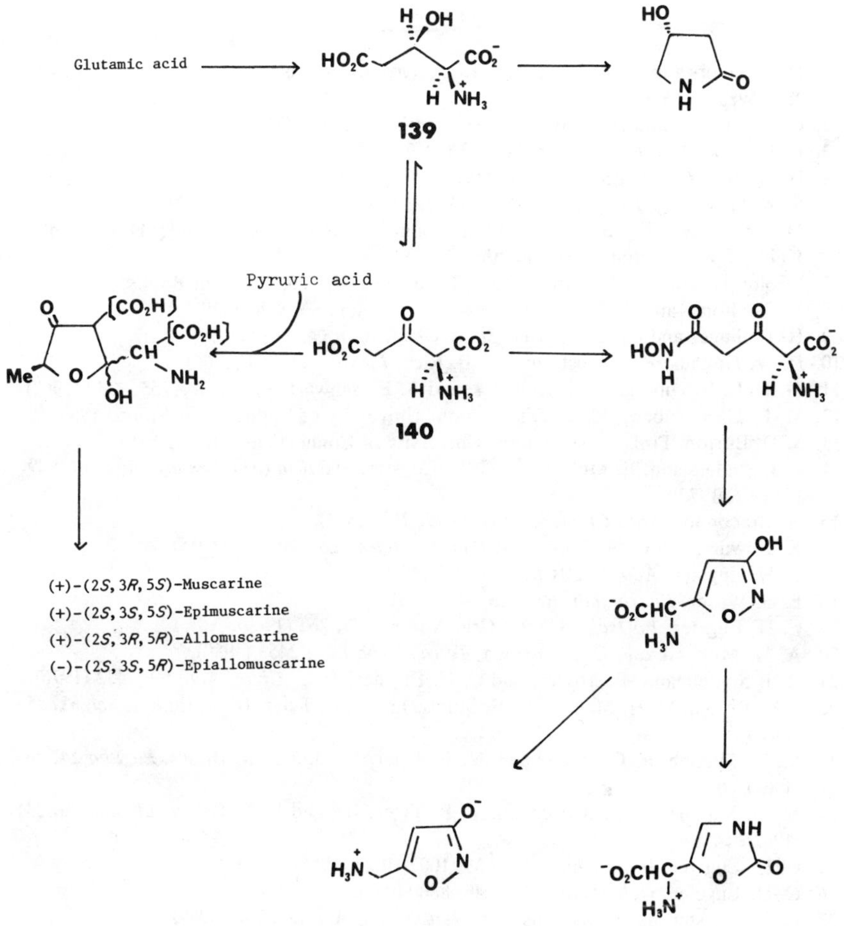

Glutamic acid → 139 → 4-hydroxy-2-pyrrolidone

139 ⇌ 140

Pyruvic acid + 140 →

(+)-(2*S*, 3*R*, 5*S*)-Muscarine
(+)-(2*S*, 3*S*, 5*S*)-Epimuscarine
(+)-(2*S*, 3*R*, 5*R*)-Allomuscarine
(−)-(2*S*, 3*S*, 5*R*)-Epiallomuscarine

SCHEME 31

through condensation of pyruvic acid and γ-ketoglutamic acid, with subsequent loss of the two carboxyl groups of the hypothetical product (Scheme 31). The proposed intermediates are *threo*-3-hydroxyglutamic acid (**139**) and 3-ketoglutamic acid (**140**). This biogenetic scheme also explains the formation of the other components of *Amanita muscaria:* ibotenic acid (**2**), muscazone (**141**), muscimol (**3**), and (−)-(*R*)-4-hydroxy-2-pyrrolidone (**142**), and is more plausible than the hexose hypothesis.

Acknowledgment

We are grateful to Professor C. H. Eugster for reviewing this manuscript and making several valuable suggestions.

References

1. D. L. Sayers and R. Eustace, "The Documents in the Case," pp. 139, 195, 214. Harper & Row, New York, 1930.
2. C. H. Eugster and P. G. Waser, *Experientia* **10,** 298 (1954).
3. F. Jellinek, *Acta Crystallogr.* **10,** 277 (1957).
4. H. Hart, *J. Chem. Educ.* **52,** 444 (1975).
5. S. Wilkinson, *Q. Rev., Chem. Soc.* **15,** 153 (1961).
6. M. McKenny, "The Savory Wild Mushroom" (revised and enlarged by D. E. Stuntz). Univ. of Washington Press, Seattle, 1976.
7. "Color Treasury of Mushrooms and Toadstools," p. 20. Crescent Books, 1973.
8. P. Catalfomo and C. H. Eugster, *Helv. Chim. Acta* **53,** 848 (1970).
9. H. Bollinger and C. H. Eugster, *Helv. Chim. Acta* **54,** 1332 (1971).
10. D. W. Hughes, K. Genest, and W. B. Rice, *Lloydia* **29,** 328 (1966).
11. M.-L. L. Swenberg, W. J. Kelleher, and A. E. Schwarting, *Science* **155,** 1259 (1967).
12. M.-L. L. Swenberg, Ph.D. Dissertation, University of Connecticut, Storrs, 1969.
13. S. D. Burton, Ph.D. Dissertation, University of Connecticut, Storrs, 1970.
14. R. J. Stadelmann, E. Müller, and C. H. Eugster, *Sydowia* (*Annales mycologici II*) **29,** 15 (1976–1977).
15. H. Braconnot, *Ann. Chim.* (*Paris*) [1] **79,** 265 (1811).
16. K. Bowden and G. A. Mogey, *J. Pharm. Pharmacol.* **10,** 145 (1958).
17. C. H. Eugster, *Adv. Org. Chem.* **2,** 427 (1960).
18. P. G. Waser, *Pharmacol. Rev.* **13,** 465 (1961).
19. C. H. Eugster, *Fortschr. Chem. Org. Naturst.* **27,** 262 (1969).
20. A. M. Mubarak and D. M. Brown, *Tetrahedron Lett.* 2453 (1980).
21. R. J. Stadelmann, E. Müller, and C. H. Eugster, *Helv. Chim. Acta* **59,** 2432 (1976).
22. J. K. Brown, M. H. Malone, D. E. Stuntz, and V. E. Tyler, Jr., *J. Pharm. Sci.* **51,** 853 (1962).
23. M. H. Malone, R. C. Robichaud, V. E. Tyler, Jr., and L. R. Brady, *Lloydia* **24,** 204 (1961).
24. M. H. Malone, R. C. Robichaud, V. E. Tyler, Jr., and L. R. Brady, *Lloydia* **25,** 231 (1962).
25. C. H. Eugster, *Helv. Chim. Acta* **39,** 1002, 1023 (1956).
26. C. H. Eugster, *Helv. Chim. Acta* **40,** 886 (1957).
27. C. H. Eugster and E. Schleusener, *Helv. Chim. Acta* **52,** 708 (1969).
28. P. Catalfomo and C. H. Eugster, *Bull. Narc.* **22,** 33 (1970).
29. T. Takemoto and T. Nakajima, *J. Pharm. Soc. Jpn.* **84,** 1183, 1230 (1964).
30. T. Takemoto, T. Yokobe, and T. Nakajima, *J. Pharm. Soc. Jpn.* **84,** 1186 (1964).
31. T. Takemoto, T. Nakajima, and T. Yokobe, *J. Pharm. Soc. Jpn.* **84,** 1232 (1964).
32. T. Takemoto, T. Nakajima, and R. Sakuma, *J. Pharm. Soc. Jpn.* **84,** 1233 (1964).
33. K. Bowden and A. C. Drysdale, *Tetrahedron Lett.* 727 (1965).
34. K. Bowden and A. C. Drysdale, *Nature* (*London*) **206,** 1359 (1965).
35. G. F. R. Müller and C. H. Eugster, *Helv. Chim. Acta* **48,** 910 (1965).
36. E. Schleusener and C. H. Eugster, *Helv. Chim. Acta* **53,** 130 (1970).
37. H. King, *J. Chem. Soc.* **121,** 1743 (1922).
38. F. Kögl, H. Duisberg, and H. Erxleben, *Justus Liebigs Ann. Chem.* **489,** 156 (1931).
39. F. Kögl and H. Veldstra, *Justus Liebigs Ann. Chem.* **552,** 1 (1942).
40. E. Fourneau, D. Bovet, F. Bovet, and G. Montezin, *Bull. Soc. Chim. Biol.* **26,** 516 (1944).
41. E. Fourneau, D. Bovet, G. Montezin, J. P. Fourneau, and S. Chantalou, *Ann. Pharm. Fr.* **3,** 114 (1945).

42. F. A. Kuehl, N. Lebel, and J. W. Richter, *J. Am. Chem. Soc.* **77,** 6663 (1952).
43. C. H. Eugster and P. G. Waser, *Helv. Chim. Acta* **40,** 888 (1957).
44. F. Kögl, C. A. Salemink, H. Schouten, and F. Jellinek, *Recl. Trav. Chim. Pays-Bas* **76,** 109 (1957).
45. H. Corrodi, E. Hardegger, and F. Kögl, *Helv. Chim. Acta* **40,** 2454 (1957).
46. R. S. Cahn, C. K. Ingold, and V. Prelog, *Experientia* **12,** 81 (1956).
47. B. Pullman, P. Courriere, and J. L. Coubeils, *Mol. Pharmacol.* **7,** 397 (1971).
48. F. G. Canepa, P. Pauling, and H. Sorum, *Nature (London)* **210,** 907 (1966).
49. P. Pauling and T. J. Pechter, *Nature (London)* **236,** 112 (1972).
50. D. L. deFontaine, B. Ternai, J. A. Zapan, R. S. Givens, and R. A. Wiley, *J. Med. Chem.* **21,** 715 (1978).
51. D. B. Davies and S. S. Danylak, *Biochemistry* **13,** 4417 (1974).
52. C. Altona and M. Sundaralingam, *J. Am. Chem. Soc.* **54,** 8205 (1972).
53. S. Pochet and T. Huynh-Dinh, *J. Org. Chem.* **47,** 193 (1982).
54. A. M. Mubarak and D. M. Brown, *Tetrahedron* **38,** 41 (1982).
55. A. Zemtsov, M.S. Thesis, University of Pennsylvania, Philadelphia, 1982.
56. C. H. Eugster, *Helv. Chim. Acta* **40,** 2462 (1957).
57. C. H. Eugster, F. Häfliger, R. Denss, and E. Girod, *Helv. Chim. Acta* **41,** 583 (1958).
58. C. H. Eugster, F. Häfliger, R. Denss, and E. Girod, *Helv. Chim. Acta* **41,** 205 (1958).
59. C. Meister and H.-D. Scharf, *Synthesis* 737 (1981).
60. J. E. Semple and M. M. Joullié, *Heterocycles* **14,** 1825 (1980).
61. H. Corrodi, K. Steiner, N. Halder, and E. Hardegger, *Helv. Chim. Acta* **44,** 1157 (1961).
62. J. E. Semple, A. E. Guthrie, and M. M. Joullié, *Tetrahedron Lett.* 4561 (1980).
63. E. Hardegger and F. Lohse, *Helv. Chim. Acta* **40,** 2383 (1957).
64. H. C. Cox, E. Hardegger, F. Kögl, P. Liechti, F. Lohse, and C. A. Salemink, *Helv. Chim. Acta* **41,** 229 (1958).
65. E. Hardegger, H. Furter, and J. Kiss, *Helv. Chim. Acta* **41,** 2401 (1958).
66. P. C. Wang, Z. Lysenko, and M. M. Joullié, *Tetrahedron Lett.* 165 (1978).
67. P. C. Wang and M. M. Joullié, *J. Org. Chem.* **45,** 5359 (1980).
68. E. Hardegger, N. Chariatte, and N. Halder, *Helv. Chim. Acta* **49,** 580 (1966).
69. F. Kögl, H. C. Cox, and C. A. Salemink, *Experientia* **13,** 137 (1957).
70. T. Matsumoto, A. Ichihara, and N. Ito, *Tetrahedron* **25,** 5889 (1969).
71. J. Whiting, Y. K. Au-Young, and B. Belleau, *Can. J. Chem.* **50,** 3322 (1972).
72. G. Fronza, G. Fuganti, and P. Grasselli, *Tetrahedron Lett.* 3941 (1978).
73. W. C. Still and J. A. Schneider, *J. Org. Chem.* **45,** 3375 (1980).
74. R. Amouroux, B. Gerin, and M. Chastrette, *Tetrahedron Lett.* **23,** 4341 (1982).
75. C. H. Eugster, F. Häfliger, R. Denss, and E. Girod, *Helv. Chim. Acta* **41,** 705 (1958).
76. C. H. Eugster and G. Müller, *Helv. Chim. Acta* **42,** 1189 (1959).
76a. H. Bollinger and C. H. Eugster, *Helv. Chim. Acta* **54,** 287 (1971).
77. G. Hansen, Dissertation, University of Zürich, 1965.
78. E. Gryszkiewicz-Trochimowski, O. Gryszkiewicz-Trochimowski, and R. S. Levy, *Bull. Soc. Chim. Fr.* 603 (1958).
79. G. Colovray, R. Durand, and G. Descotes, *Chim. Ther.* **3,** 116 (1968).
80. G. Zwicky, P. G. Waser, and C. H. Eugster, *Helv. Chim. Acta* **42,** 1177 (1959).
81. T. Matsumoto and A. Ichihara, *Bull. Chem. Soc. Jpn.* **33,** 1015 (1960).
82. A. Ichihara, T. Yamanaka, and T. Matsumoto, *Bull. Chem. Soc. Jpn.* **38,** 1165 (1965).
83. Z. Lysenko, F. Ricciardi, J. E. Semple, P. C. Wang, and M. M. Joullié, *Tetrahedron Lett.* 2679 (1978).
84. J. Kiss, H. Furter, F. Lohse, and E. Hardegger, *Helv. Chim. Acta* **44,** 141 (1961).

85. E. Hardegger and N. Halder, *Helv. Chim. Acta* **50,** 1275 (1967).
86. W. L. Nelson, D. R. Allen, and F. F. Vincenzi, *J. Med. Chem.* 698 (1971).
87. K. Bowden and B. H. Warrington, *J. Chem. Soc., Perkin Trans. I* 1493 (1978).
88. C. H. Eugster and K. Allner, *Helv. Chim. Acta* **45,** 1750 (1962).
89. H. Fiesselmann and P. Schipprak, *Chem. Ber.* **87,** 835 (1954).
90. H. Bollinger and C. H. Eugster, *Helv. Chim. Acta* **54,** 2704 (1971).
91. M. Giannella, M. Pigini, P. Ruedi, and C. H. Eugster, *Helv. Chim. Acta* **62,** 2329 (1979).
92. D. J. Triggle and B. Belleau, *Can. J. Chem.* **40,** 1201 (1962).
93. N. J. Harper, A. H. Beckelt, and R. J. Scott, *Chem. Ind.* (*London*) 1331 (1962).
94. B. Belleau and J. Puranen, *J. Med. Chem.* **6,** 325 (1963).
95. R. S. Givens, D. R. Rademacher, J. Kongs, and J. Dickerson, *Tetrahedron Lett.* 3211 (1974).
96. H. Fuhner, *Heffler's Handb. Exp. Pharmakol.* **1,** 640 (1923).
97. B. Witkop, R. C. Durant, and S. L. Friess, *Experientia* **15,** 300 (1959).
98. P. G. Waser, *Experientia* **14,** 356 (1958).
99. P. G. Waser and W. Hopff, *Adv. Pharmacol. Ther.* **3,** 261 (1979).
100. A. Myerson, M. Rinkel, J. Loman, and W. Dameshek, *J. Pharmacol.* **68,** 476 (1940).
101. B. Belleau and G. Lacasse, *J. Med. Chem.* **7,** 768 (1964).
102. G. A. Alles and P. K. Knoefel, *Univ. Calif., Berkeley, Publ. Pharmacol.* **1,** 187 (1939).
103. H. R. Ing, P. Kordik, and D. P. H. Tudor Williams, *Br. J. Pharmacol.* **7,** 103 (1952).
104. H. R. Ing, *Science* **109,** 264 (1949).
105. A. H. Beckett, N. J. Harper, J. W. Clitherow, and E. Lesser, *Nature* (*London*) **189,** 671 (1961).
106. A. Gieren and M. Kokkimidis, *Z. Naturforsch., C. Biosci.* **37C,** 977 (1982).
107. E. Leete, *in* "Biogenesis of Natural Products" (P. Bernfeld, ed.), p. 739. Macmillan, New York, 1963.
108. K. Nitta, R. J. Stadelmann, and C. H. Eugster, *Helv. Chim. Acta* **60,** 1747 (1977).

INDEX

D

E

I

J

K

L

M

U

V

W

Z